T0177637

Branching Space-Times

OXFORD STUDIES IN PHILOSOPHY OF SCIENCE

Branching Space-Times

Theory and Applications

NUEL BELNAP, THOMAS MÜLLER,
AND TOMASZ PLACEK

OXFORD
UNIVERSITY PRESS

OXFORD
UNIVERSITY PRESS

Oxford University Press is a department of the University of Oxford.
It furthers the University's objective of excellence in research, scholarship,
and education by publishing worldwide. Oxford is a registered trade mark of
Oxford University Press in the UK and in certain other countries.

Published in the United States of America by Oxford University Press
198 Madison Avenue, New York, NY 10016, United States of America.

Library of Congress Cataloging-in-Publication Data

Names: Belnap, Nuel D., 1930- author. | Müller, Thomas, 1969- author. |
Placek, Tomasz, 1960- author.
Title: Branching space-times / Nuel Belnap, Thomas Müller, and Tomasz
Placek.
Description: New York : Oxford University Press, [2020] | Includes
bibliographical references and indexes.
Identifiers: LCCN 2021033975 | ISBN 9780190884314 (hardback) | ISBN
9780190884338 (epub)
Subjects: LCSH: Determinism (Philosophy)–Mathematics. | Events
(Philosophy)–Mathematical models. | Mathematical physics. | Space and
time–Mathematical models. | Chaotic behavior in systems–Mathematical
models. | Probabilities.
Classification: LCC QC6.4.D46 B45 2020 | DDC 530.01–dc23/eng/20211001
LC record available at https://lccn.loc.gov/2021033975

1 3 5 7 9 8 6 4 2

Printed by Integrated Books International, United States of America

Contents

II. APPLICATIONS

III. APPENDIX

Preface

This book concerns indeterminism as a local and modal concept. At its heart lies the notion that our world contains events or processes with alternative, really possible outcomes. We assume that our world is indeterministic in this sense, and we set ourselves the task of clarifying what this assumption involves. We address this task in two ways. First, we provide a mathematically rigorous theory of local and modal indeterminism. Second, we support that theory by spelling out the philosophically relevant consequences of this formulation and by showing its fruitful applications in metaphysics. To this end, we offer a formal analysis of causation, which is applicable in indeterministic and non-local contexts as well. We also propose a rigorous theory of objective single-case probabilities, intended to represent degrees of possibility. Third, we link our theory to current physics, investigating how local and modal indeterminism relates to some aspects of physics, in particular quantum non-locality and spatio-temporal relativity. We also venture into the philosophy of time, showing how our resources can be used to explicate the dynamic concept of the past, present, and future based on local indeterminism.

The theory that we put forward and develop here was launched by Nuel Belnap in 1992, and is called *Branching Space-Times* (BST). As stated in the founding paper, the aim is to answer the following question: "How can we combine relativity and indeterminism in a rigorous theory?" The combination of non-relativistic time and indeterminism was already well-known from Arthur Prior's ideas of so-called Branching Time (BT) and related semantic models; the challenge was to accommodate space-like related events, as known from the special theory of relativity. One aspect of space-like relatedness, the absence of causal connections, was also needed in another research program that Nuel Belnap was involved in with his collaborators in the 1980s. This was the so-called *stit* program in the theory of agency and action, where "*stit*" is an acronym for the agentive notion of "seeing to it that" (see, e.g., Belnap et al., 2001). A careful analysis of causal dependencies and independencies is very much needed in the theory of collaborative actions. Some insights from this project then became useful for the formulation of BST. The central idea of BST, in agreement with BT, is to represent events as being partially ordered. A partial order leaves room for some events to be incomparable. BST shows that certain incomparable events can be understood to be space-like related, and these could, for example, represent the actions of independent agents. Thus, despite the present book's focus on applications to physics, it is worth keeping in mind that the initial inspiration for BST came from a problem in the theory of agency. Limitations of space means that we do not develop BST as a theory of agency in this work but there are, however, several promising ideas for combining the *stit* program with BST (see, e.g., Belnap, 2003a, 2005a, 2011).

Many of the formal results presented in this book, as well as the philosophical discussions that surround them, have been published in papers we have written (sometimes with collaborators) over the past twenty-five years. Although this book has been written entirely afresh, we acknowledge the particular papers that provided the basis for the content of several of our chapters.

Chapter 2 presents the results and discussion of the BST founding paper (Belnap, 1992), of which there is also a postprint providing additional information (Belnap, 2003b). A recent alternative formulation of BST is the topic of Chapter 3, which draws on the paper of Belnap et al. (2021). The BST theory of events and propositions was initially presented in Belnap (2002), sets of transitions were investigated in Müller (2010), topological issues in BST were studied in Placek and Belnap (2012), Müller (2013), and Placek et al. (2014)—these five papers are relevant for Chapter 4. Our investigation of non-local modal correlations in Chapter 5 builds on the earlier work reported in Belnap (2002, 2003c) and Müller et al. (2008). Chapter 6 extends Belnap's (2005b) theory of originating causes (*causae causantes*), making it applicable in non-local contexts as well. Chapter 7 gives a BST-based theory of objective single-case probabilities; our earlier publications relevant to this topic are Müller (2005), Weiner and Belnap (2006), and Belnap (2007). Chapter 8 then applies our accounts of non-local modal correlations and objective single-case probabilities to analyze non-local quantum correlations. Over the years, quite a number of papers have appeared on this topic, starting with Belnap and Szabó (1996) and including Placek (2000), Müller and Placek (2001), and Placek (2010). In Chapter 9, the sections on Minkowskian Branching Structures draw on Müller (2002), Wroński and Placek (2009), and Placek and Belnap (2012). Sections on the Hausdorff property and on bifurcating curves in General Relativity build on Luc and Placek (2020), whereas for our discussion of determinism and indeterminism, the relevant papers are Müller and Placek (2018) and Placek (2019). Finally, Chapter 10 integrates material from Müller (2006, 2019) and from Placek (2021).

This book and its authors owe a debt of gratitude to a number of people. Over a long period of time, we have benefited from discussions with our students and collaborators and from the feedback of the numerous audiences to whom we presented our ideas. We especially acknowledge the contributions of our co-authors on BST papers, including Kohei Kishida, Joanna Luc, László Szabó, Matthew Weiner, and Leszek Wroński. Most importantly, for this book, Joanna Luc read and checked all the formal proofs, finding errors and infelicities, providing us with suggestions for how to correct them, and supplying her valuable insights about philosophical issues as well. She also compiled the index. A full list of Joanna's contributions would be far too long, so we offer her our sincere thanks for her invaluable help. The formal proofs in earlier versions of Chapters 2 and 3 were also checked by David O'Connell. Of course, any remaining errors are our own. Aeddan Shaw helped with language correction, Sahra Styger double-checked the index, Louis Pfander helped with the galley proofs, and Marta Bielińska labored long and hard over the preparation of the diagrams and figures. Saul Kripke, Robert Stalnaker, and Craig Callender allowed Tomasz Placek to participate in their seminars and provided feedback on some of the issues discussed in the book. Agnieszka Barszcz, Antje Rumberg, Jacek Wawer, and Leszek Wroński also provided various important suggestions over the course of the creation of this book.

We gratefully acknowledge the generous financial support of the (Polish) National Science Centre research grant: Harmonia 8 number 2016/22/M/HS1/00160. We thank Peter Ohlin and his team at OUP for the smooth handling of the whole book production process. Last, but not least, we thank our families and especially our loved ones, Eva and Kasia, for bearing with us during the times of absences and stress that finishing a book implies.

Pittsburgh
Konstanz
Kraków

March 2020

PART I
THE CORE THEORY

1

Introduction

The subject of this book, to put it broadly, is the analysis of *real possibilities*, or *indeterministic possibilities for the future*. The theory of Branching Space-Times that we develop here describes with mathematical rigor how real possibilities can be anchored in a spatio-temporal world that is rudimentarily relativistic. The picture we attempt to paint represents indeterministic events as happening locally in our spatio-temporal world, and it portrays indeterminism modally, via *alternative possibilities for the future* that are open in particular circumstances. This idea can be traced back to Aristotle. In *De Interpretatione* 9 (19^a12–14) he says of a particular cloak, under particular circumstances, that the cloak might wear out, but that it could also be cut up first, that is, before wearing out. The currently dominant analysis of modalities in terms of possible worlds will represent the possibilities open to the cloak by multiple possible worlds, in some of which the cloak is cut up, whereas it wears out in the others. That analysis thus (1) postulates many possible worlds, (2) includes the notion of an actual world, and (3) identifies the actual world with one of the possible worlds. Branching Space-Times opposes this representation. Given that possible-worlds theories assume these three claims, Branching Space-Times is thus not a possible-worlds theory.

It is true that, in general, talk of possibilities only makes sense before a contrast between possibility and actuality. This idea can also be traced back to Aristotle, who observes in *Metaphysics* 9 (1048^b4–6) that, together, the actual and the potential form an antithesis. However, it is not generally true that actuality has to be structurally like the possibilities with which it is contrasted, that alternate possibilities have to be alternatives *to* a given actuality. In the case of real possibilities, actuality is given as the concrete situation here and now, while the possibilities are alternatives *for* the future of that concrete situation, with none of these possibilities being actual (yet). What is actual (i.e., the concrete given situation) is structurally different from the alternative possibilities for the future.

Branching Space-Times: Theory and Applications. Nuel Belnap, Thomas Müller, and Tomasz Placek, Oxford University Press. © Oxford University Press 2022. DOI: 10.1093/oso/9780190884314.003.0001

1.1 Real possibilities

Some things are really possible, while others are not. Real possibilities are special because they are future-directed. Our talk of possibilities does not always distinguish them from other types of possibilities, but the distinction is philosophically important and easy enough to make. Let us start with an example.

Consider Alice, who is sitting in a restaurant in Pittsburgh talking to the waiter who is taking her order for lunch. At the moment we are considering, the waiter has asked, "Do you want fries with that?", and Alice is thereby prompted to give the answer "yes" or "no". It is really possible for her to give either answer. At these particular circumstances (but not later), she has two alternative possibilities for the future, saying "yes" and saying "no". Many other things are, however, not really possible. For example, it is not really possible for her to turn her glass of water into a plate of fries. Nor is it really possible for her to have dinner in Kraków later on the same day, because there are no means of transportation available that could get her there in time (note that at noon in Pittsburgh it is already 6 pm in Kraków). If Alice was in Konstanz, the situation would, of course, be different: given a concrete situation of having lunch in Konstanz, it is really possible to have dinner in Kraków on the same day.[1]

Real possibilities as alternatives *for* rather than alternatives *to*. Put abstractly, given the situation at hand at some concrete moment, some things can really follow from those circumstances, while others cannot. Real possibilities are always future-directed and tied to a concrete situation constituting the "here and now". Real possibilities are, to make a point about English usage, alternatives *for* the future, for what can happen in the future. It makes no sense to say that they are alternatives *to* the future, because the future, being in the future, has not yet happened. None of the possibilities for the future is actual yet.[2]

---

[1] From Pittsburgh's Shadyside, it would take Alice at least an hour (more realistically, two to three) to be airborne, and the distance is such that current commercial aircraft take at least 9 hours to complete the journey to Kraków Airport. (More realistically, unless Alice has access to a private jet, she will have to stop over at least once, which would prolong the journey still further). From Konstanz, on the other hand, a direct flight from Zürich takes just 2 hours, and there is no difference in time zone.

[2] This point is made forcefully in Rumberg (2016b).

Real possibilities are special. We talk of other types of possibilities as well, and for many of those, alternative possibilities *are* alternatives *to* a given actuality. Logical, metaphysical and natural (physical) possibilities all provide examples in which actuality is structurally like the alternative possibilities. We provide an illustration in terms of epistemic possibilities, as these form a crucially important class.[3] Let us thus consider Bob, who is looking for his keys. For all he knows, they could be in his office, or they could be in his car. It is perfectly adequate to say that it is possible that the keys are in Bob's office, and it is possible that they are in his car. In actuality, they are either in one place or the other, and that is a settled matter. In fact, Bob operates under the explicit assumption that one of the options is settled to be true, and the other is settled to be false; he just does not know which is which. Bob is considering epistemic possibilities rather than real possibilities. What is epistemically possible for him is what is compatible with his evidence, and that will change as he learns things. Bob faces an actual situation and a number of alternatives to that actual situation, which he cannot distinguish based on his available evidence. He thus faces the task of finding out which of these alternatives is the actual one. In actuality, it is already settled that the keys are either in his office or in his car, and let us hope for Bob's sake that he will find out quickly which it is.

From an abstract point of view, Bob's predicament is one of self-location. Given his evidence, there is a space of possibilities that includes actuality.[4] His task is to rule out the non-actual alternatives so that he knows his location in that space of possibilities, which will enable him to retrieve his keys. The way he does this is by looking, by acquiring new evidence that will narrow down the space of alternative possibilities. A valid formal approach to modeling Bob's predicament is in terms of possible worlds (or perhaps,

---

[3] Linguistic data in fact seem to support the thesis that generally, when we say "it is possible that…", we are pointing to epistemic and not to real possibilities (Vetter, 2015). Real possibilities are more often expressed via the so-called root modality, as in "this dog could bite" or "Alice can choose to have fries". The root modality generally expresses either real possibilities or deontic possibilities, i.e., what one is allowed to do. It may be interesting to note that actuality need not be among the scenarios that are deontically possible (i.e., admissible) either.

[4] Generally speaking, epistemic (or rather, doxastic) possibilities might *exclude* actuality: Bob might have mistaken beliefs, which he would need to revise in order to properly self-locate. Bob's evidence might after all be false: the keys might in fact have been taken by a prankster in his office. We rule out these complications, noting that there is a large body of literature on belief revision that deals with cases in which agents have to revise their evidence (see, e.g., Spohn, 2012). For our purposes, the important point is that even if these issues are taken into account, it remains adequate to represent that relevant space of possibilities via possible worlds.

smaller entities such as possible situations). There is Bob's actual situation, and there are alternatives to it; actuality is just one of the possibilities.[5]

1.2 Representing possibilities via branching vs. possible worlds

Many kinds of possibilities can be represented via possible worlds. But for real possibilities, a representation in terms of possible worlds distorts the picture. Consider Alice in the restaurant again. Actually, Alice has yet not answered the waiter, so none of the alternatives for the future that she is facing has been realized yet. She can really give either answer. Her task is to decide what to do, whether to order fries or not. Her actual situation does not yet include a decision; her task is not to find out what is *true*, but to find out what is *good* for her, and then to make it happen. If one tries to formally represent Alice's possibilities for the future via possible worlds, there has to be an actual world among them, which means that one complete temporal course of events has to be actual. This, however, distorts the structure of temporal actuality, since none of the possibilities for the future is yet actual. A more adequate formal representation is not in terms of possible worlds, but in terms of a branching structure of possible courses of events. One such course of events represents Alice's answering "yes", and another one of those courses of events represents Alice's answering "no". At the branch point, which represents Alice's actual situation, both courses of events are really possible as continuations of the actual situation, and none of these possibilities has yet been actualized.

Branching Space-Times builds on the idea of multiple alternative possibilities open in concrete situations. It identifies *Our World* with the set of all real possibilities accessible from a given concrete event. The set is constructed in indexical fashion by starting with a concrete actual event and then including all of the really possible events that are accessible from it. For instance, taking Alice's situation in the restaurant in Pittsburgh as actual, we can say in indexical language that it is really possible that she will answer yes, or that she will take a walk after lunch, or that she will come back to the

---

[5] This verdict does not change if one explicitly includes time in the picture, as is common, for example, for tasks involving temporal self-location (see, e.g., Spohn, 2017). In that case, both actuality and the alternatives are properly represented via so-called centered possible worlds, which are possible worlds in which one point in time is singled out as the current one. The formal machinery is more or less the same.

same restaurant in a year's time. We can also say that it was possible that she would not go to the restaurant, but rather skip lunch, or that she could have left Pittsburgh for Kraków a week ago, or that at her first day at school many years back, she could have put on her blue socks instead of the green ones she actually put on then. Based on the plethora of real possibilities open at various really possible events accessible from Alice's concrete situation, we thus arrive at the concept of a complete possible course of events or, as we will say, a *history*. The farther back we go, the larger the set of possible courses of events that becomes accessible. Since courses of events are differentiated by alternative possibilities open in concrete situations, any two courses of events overlap, sharing a common past. A formal structure representing alternative real possibilities thus forms a unified whole, not a set of separate alternatives. It is possible to map out the whole structure of real possibilities by starting from one really possible point and going back and forth along the branching histories. The unity of the structure can be described in indexical terms starting from *any* one point.

It is illuminating to see what happens if one tries to force a formal representation of real possibilities in terms of possible worlds. Possible worlds are themselves complete, separate courses of events. It is, therefore, easy to extract a possible worlds representation from a branching representation; just put all the possible histories side by side without any overlap. Actuality then has to be represented by one of these worlds, perhaps together with an indication of a temporal now. How is Alice represented? Without entering the somewhat controversial topic of representing individuals,[6] the sensible majority option seems to be that Alice, our Alice that we can point to, is an inhabitant of precisely one of those possible worlds. Other worlds may contain Alice-lookalikes ("counterparts"), but not her. In this picture, Alice's decision of what to do is represented in exactly the same way in which we represent Bob's looking for his keys: there is an actual world, Alice's actuality, for which the past as well as the future are both fully fixed. Alice is just uncertain about her location in the total space of possibilities. Actuality is guaranteed to be one of them. Her actual past, which is (or at least could be) known to her, is compatible with many other possible worlds, while her future is such that she just has no information about it. In that possible worlds rendering of the situation, deciding what to do—which is tied, some way or other, to finding out what is *good*—amounts to gathering more evidence—that is,

---

[6] See Lewis (1986a, p. 199) and Kripke (1980, p. 45).

finding out what is *true*. Being tied to one possible world, Alice cannot, as it were, jump worlds to arrive at a better one. There is no agency, no room for practical rationality; there is just theoretical rationality tied to finding out what is true. This picture does not capture what is special about agency and therefore appears inadequate. A branching framework, on the other hand, can picture agents choosing between different possibilities for their future and therefore leaves room for them to select the good over the bad.

What we have just claimed might sound contentious, but it suffices, we think, to at least motivate the quest for a detailed representation of real possibilities in a branching framework.

1.3 Some thoughts on our modally rich world

We turn next to the philosophical vision underlying our construction of branching structures.

Our World. *Our World* is very big. For one, many things have already happened in *Our World*, and there is more to that particular past of *Our World* than these happenings. Even if we had a full specification of all of them, that would not provide us with complete information about the past. Things that once were really possible but did not happen belong to a full specification of *Our World* as well. Such real but never actualized possibilities are not necessarily derivable from what has happened. Similarly, all things that can really happen later on also contribute to a full characterization of *Our World*. *Our World* therefore has modal aspects that are not reducible to its non-modal features. Branching Space-Times decides to describe our modally rich world in terms of really possible events. Such events find their place in a coherent formal structure representing alternative real possibilities. That structure forms a unified whole, containing all really possible courses of events. And any two such courses of events overlap, sharing a common past.

Before we describe the formal structure in detail in subsequent chapters, we present a vision that underwrites the concept of a modally rich world, and we reflect on its coherence in particular. That vision is not strictly necessary to understand and to follow our formal constructions and their applications in the chapters to come. We nevertheless lay it down here to specify a useful metaphysics for the theory of Branching Space-Times.

To repeat, to arrive at the totality of all really possible events, one needs to start with some actual happening, like your actual eye-blinking (assuming you just blinked). There are events that can really happen later (i.e., after your eye blinking); we say that they belong to the future of possibilities of this actual event. There are also events that have already occurred, and events that could really have occurred. Your actual eye-blinking is linked to other events that are, from its perspective, really possible, via a pre-causal relation that can be expressed by a phrase like "something can really happen after something else". This relation is used to capture the totality of all really possible events from the perspective of our actual event. What we have just said might suggest that, in order to account for the totality of all really possible events from the perspective of a given actual event, it is enough to shift the perspective back in time and then forward along one of the open possibilities. It will transpire, however, that in a theory of branching histories that represents both space and time, we will need not just a V-shaped indexical reference back and forth, but an M-shaped zig-zagging chain of indexical references—see Fact 2.4. We therefore take *Our World* $\mathscr{W}$ to be the totality of possible events that are linked to an actual event by any M zig-zagging curve admitted by the pre-causal relation.[7]

The word "our" above is not merely a stylistic ornament: it reflects the construction and the fact that we have to start with the actual here-and-now. That might raise the concern that there could be many different *Our Worlds*, depending on which actual event we start with. This option is, however, excluded by our requirement that all possible courses of events are linked via a common past. The totality of possible events is therefore not relativized to an actual event. In other words, the unified structure of *Our World* can be described in indexical terms starting from *any* really possible event.[8]

---

[7] This is reminiscent of David Lewis's (1986a, p. 208) way of demarcating a possible world: its elements "stand in suitable external relations, preferably spatiotemporal". Note, however, that while Lewis's demarcation is non-modal, our characterization is modal, as indicated by the phrase "something *can really happen* in the future of something else". Furthermore, as our characterization is not spatio-temporal, BST is not threatened by current developments in physics that take space-time to be an emergent phenomenon. For weird consequences of demarcating a possible world in Lewis's way, see Wüthrich (2020).

[8] Of course, even given the uniqueness of *Our World*, agents can be *epistemically* uncertain about what is actual and what is or was really possible. Such uncertainty could be modeled via a *set* of branching models representing different epistemic alternatives. The uniqueness of *Our World* then translates into the fact that only one from among the set of these models can contain any actual and any really possible events. In this book, we are not concerned with epistemic possibilities, and so we leave the issue for another occasion.

Events. We set out our vision of *Our World* thus far in terms of (token) events without further elucidation. We are aware that talk of events is ambiguous: there are large events as well as small events, and some events have a modal multiplicity in that they can occur in different ways (e.g., faster or more slowly), whereas other events can only occur in one way. In what follows, we take idealized point-like events as the basic building blocks of our construction, in analogy to Euclidean points or to the physicists' mass points. We assume that our point-like events cannot occur in different ways, that is, they do not have the mentioned modal multiplicity. Starting from point-like token events, we then construct other varieties of events, including so-called disjunctive events, which *can* occur in different ways.

An event, as we understand it, is not a place-time or a collection of such; it is a happening. It has a time and a place, which partly describes its locus in our world—but that is not enough to confer uniqueness. An event has a concrete past and a concrete future of possibilities. Nor is the past of an event a mere array of times and places: such a past also consists in concrete events. Events, as we construe them, are as concrete as your actual eye-blinking, they have a definite relationship to this very eye-blinking and are related to all other point events that from our point of here-and-now are either actual (such as those in your past) or really possible. A consequence that we want to draw from this characterization is that token-level events cannot be repeated. In contrast, a type-event is repeatable.[9] To make one final point about events: despite the basic role they play in our theorizing, the Branching Space-Times project should not be read as a commitment to a reductionist program in ontology that aims to reduce everything to events. Our decision to focus on events is driven by our diagnosis as to which objects are known to be amenable to formal analysis: we know how to formally analyze events, but we also know that adding processes, or enduring objects, makes the task much harder. So events is what we start with: we begin with possible events, idealized as point-like, and then construct other varieties of events as well as some other kinds of objects.

Histories. The totality of all really possible events contains compatible events (i.e., events that can occur together) and incompatible events (i.e., events that cannot occur together). Of particular interest are incompatible

---

[9] In some sense, the non-repeatability of token events might still be contentious, as there might be closed causal loops—see Section 9.3.6 for discussion.

events that are local alternatives for the future, like Alice's alternative possible responses "yes" and "no" to the question about her lunch. Which events are compatible? We follow what we call the later witness intuition, according to which two events are compatible only if they belong to the past of some really possible event. Based on the criterion provided by this intuition, we then define histories as maximal sets of compatible events, where "history" is our technical notion for the informal "course of events". We consider histories a useful device for tracking local compatibilities and incompatibilities, but we tend to minimize their ontological significance. That is, we take it that local alternatives occurring in space and time are ontologically important, whereas histories merely offer a useful way of conceptualizing alternative possibilities. There are less demanding conceptualizations that better serve the local vision underlying Branching Space-Times. Two such concepts, alternative transitions (Müller, 2014; Rumberg, 2016a,b) and alternative possible continuations (Placek, 2011), have already been proposed for the branching framework. We nevertheless work here in terms of histories, mostly for reasons of simplicity. Note also that given an arbitrary criterion of (local) compatibility, one can typically define maximal sets of compatible events,[10] which means that the formal basis for defining histories is available anyway.

As we have already emphasized here, courses of events overlap, and indeed our postulates require any two histories to have a non-empty intersection. More precisely, if a structure has more than one history, it contains an object (a point event or a chain of points events) at which histories branch. But what is it, precisely, that branches? There are two ways of thinking about branching that should be avoided: (1) individual space-times branch or (2) *actual* courses of events branch. As for mistake (1), since histories are identified with concrete possible courses of events, there is more to a history than its spatio-temporal structure. Branching Space-Times in fact allows that all the different histories (all the concrete individual space-times) have the same space-time structure. For an example, see the so-called Minkowskian Branching Structures discussed in Chapter 9.1. Clearly, individual space-times do not branch in structures of this kind, as every history has the same

---

[10] Set-theoretically, the Axiom of Choice is required. In this book we freely use the Axiom of Choice and equivalent principles such as the Zorn-Kuratowski lemma or Hausdorff's maximal principle. For some details, see note 8 on p. 30.

non-branching, Minkowskian space-time structure.[11] It is spatio-temporal *histories* that branch. To address the second mistake, a point about language might help. While explaining the pre-causal relation, we used the phrase 'what can really happen later', not 'what will happen'. To illustrate, focusing on an actual event of coin tossing, the coin can really land heads up and it can really land tails up. Thus, it is alternative *possibilities* open in particular circumstances, or more generally, alternative *possible* courses of events, that branch.

Temporal directedness. We insist on the significance of alternative possibilities for the *future* in our branching framework. We assume that the temporal directedness of the pre-causal relation is objective and derivable from the modal features of our world, viz., from the distinction between a settled past and an open future. In particular, if there is no room for real possibilities in *Our World*, there is no directedness, and nothing can distinguish the past and the future of any event. In the presence of real possibilities, however, the future is modally distinguished from the past.

Given that there are real possibilities, we need to take a stance on how to represent the world-lines of point-like objects in a branching structure. Such a world-line (a trajectory) is defined in Branching Space-Times as a maximal chain of events, where a set of events is called a *chain* if any two elements of it are comparable by the pre-causal relation. With real possibilities present, some pairs of trajectories bifurcate, forming a Y-shaped figure. Two such bifurcating trajectories share a common "trunk" that is adjacent to two disjoint "arms". Given continuity (in line with the standard assumptions in physics), either (i) the trunk has a maximal element and each arm has no minimal element, or (ii) the trunk has no maximal element, but each arm has a minimal element. It turns out that both options are formally viable. For our theory, we decide the issue globally, by assuming different postulates governing lower bounded chains and upper bounded chains. A lower bounded chain has to have a (unique) greatest lower bound, whereas an upper bounded chain might have multiple minimal upper bounds, as in option (ii) for bifurcating trajectories. As a result of these diverging postulates, if we start with a branching structure and then systematically flip the direction, typically the resulting object will not be a branching

---

[11] For the record, Minkowskian Branching Structures are not the only option. Branching Space-Times also allows that different histories have vastly different space-time structures (see, e.g., the construction suggested in Exercise 2.5).

structure—it will violate the infima and suprema postulates for bounded chains. For a further discussion of temporal directedness, see Chapter 2.4, and for further discussions relating to the philosophy of time, see Chapter 10.

1.4 Branching in the landscape of possible-worlds theories: Some comments on modal metaphysics

Branching Space-Times, as we said, is not a possible-worlds theory in the standard sense: it does not posit a multiplicity of possible worlds, and thus it does not single out an actual world from among that multiplicity. The main reason for that difference is that Branching Space-Times is meant to describe a kind of possibility that is different from the usual target of possible-worlds analyses. We have argued that possible-worlds frameworks do not do justice to real possibilities, and we said that the different targets of the two approaches give rise to deep formal differences of how the two approaches represent modalities (see Section 1.2). Yet, it is useful to describe the branching approach within the larger landscape of positions in modal metaphysics. One might think that this should be straightforward. After all, the positions are differentiated by their stance with respect to just a few issues, and so, by learning the branching theorists' response to them, one should be able to locate the theory on the map of standpoints in modal metaphysics. The most important issues are the following: Is the theory intended to be reductive with respect to modalities, by reductively explaining them in non-modal terms of some sort? Are possible worlds (or histories, or any objects standing for full possible courses of events) thought of as actual, and, if not, how is the distinction between actuality and mere possibility explained? The next big question is how individuals are represented in these possible worlds or histories, and especially how the modal features of agents that pertain to exercising their agency are described. Finally, there is a question concerning the status of the laws of nature vis-à-vis possibilities: Are the possibilities dictated by the laws of nature and particular circumstances, or is it precisely the other way round, so that (real) possibilities delineate what the laws are in *Our World*?

It is true that the branching approach and the possible worlds approach do differ, more or less strongly, with respect to the mentioned issues. But the assimilation of branching with possible-worlds theory is also somewhat treacherous, and for two reasons. First, in the interest of maximal formal rigor, Branching Space-Times theory is very frugal with respect to its primitive notions. Having only a few primitive notions increases the transparency of our formal constructions. More advanced concepts are added later on, and they need to be constructed in terms of primitive objects. Now it is exactly such advanced concepts that are needed to draw comparisons between branching and possible-worlds theories. But these advanced concepts, being derived rather than primitive, are not well-suited for drawing fundamental distinctions between Branching Space-Times and possible-worlds theories.

The second and more important reason is that branching and possible-worlds theories respond to different data. They have somewhat diverging aims and different criteria of success. The theory of Branching Space-Times belongs to metaphysics. It picks out the notion of alternative possibilities for the future as its starting point and assumes that this notion is clear enough to permit a non-controversial formalization. It then uses that notion to mathematically analyze local indeterminism occurring in relativistic space-time. The hope of the branching theorist is that the theory's mathematical elegance, its broad scope, the richness of its consequences, and its applicability to the analysis of problems in metaphysics and in the foundations of science will count in its favor. These virtues should thus provide a good defense of the metaphysical position that the theory formalizes. But, crucially, accounting for ways we use modal fragments of our vernacular languages is not at the top of the list of priorities of Branching Space-Times theory. It is, for example, not the theory's business to account for our practice of using counterfactuals (i.e., to account for the linguistic fact that we intuitively take some counterfactuals to be true and others to be false). In a similar vein, our linguistic practice of using alethic modalities are not the data that Branching Space-Times theory responds to. To put our cards on the table, we share the linguistic intuition that Elizabeth II might have never become the Queen of England, but that she could not have had parents different from the parents she actually had.[12] But, to repeat, such linguistic facts are not the evidential data that Branching Space-Times is meant to accommodate. This relatively low priority of linguistic data extends to the way people speak about future possibilities and actuality. Even if it turns out that our ways of speaking favor a vision with a distinguished actual future course of events (contradicting our egalitarian concept of alternative possibilities for

---

[12] We emphasize that these are *intuitions*. With respect to the first claim, we believe that it is an objective fact whether there really were chancy events such that, if they had happened, Princess Elizabeth would not have been crowned Queen of England.

the future), we persevere in the construction of our metaphysical theory, investigating what are the consequences of this egalitarian concept (given the assumed postulates).

The relatively low priority that Branching Space-Times theory gives to linguistic data stands in stark contrast to possible-worlds theories. These theories have either emerged from semantical theories for languages with modal operators or aim at providing such a semantics. Our modal talk is likely the most important datum that influences how these theories frame their basic metaphysical concepts, such as the similarity of possible worlds, an accessibility relation between worlds, or possible worlds themselves. For an illustration, here is David Lewis explaining how our (intuitive) knowledge of counterfactuals determines which similarity relation between possible worlds is adequate. ("Analysis 2" is his possible-worlds-based analysis of counterfactuals, for which a notion of similarity between worlds is crucial.)

> [W]e must use what we know about the truth and falsity of counterfactuals to see if we can find some sort of similarity relation—not necessarily the first one that springs to mind—that combines with Analysis 2 to yield the proper truth conditions. It is this combination that can be tested against our knowledge of counterfactuals, not Analysis 2 by itself. In looking for a combination that will stand up to the test, we must use what we know about counterfactuals to find out about the appropriate similarity relation—not the other way around. (Lewis, 1986b, p. 43)

The same methodology, with a dominant role of linguistic data concerning modalities, is operative in the works of two other founding fathers of modal metaphysics, Saul Kripke and Robert Stalnaker.

Our habits of speaking might be biased toward determinism for various reasons. A possible-worlds theory could therefore be adequate in accounting for these habits, while being rather more off the mark with respect to metaphysical issues such as indeterminism. In short, what is central for a branching approach (i.e., an exercise in metaphysics to provide an analysis of local indeterminism), may be of marginal importance for possible-worlds theories. And, vice versa, the semantical enterprise, so dear to possible worlds theorists, has only limited, secondary significance for Branching Space-Times.

The semantical enterprise is also problematic because it typically brings with it a way of thinking about modalities that is foreign to real possibilities.

A particularly useful idea in modal semantics is that of an accessibility relation used to discern modalities of different grades, like logical possibilities,
metaphysical possibilities, natural (aka physical) possibilities, technological
possibilities, and so on. These distinct possibilities are thought of as differentiated by distinct accessibility relations, which in turn are explained
in terms of different kinds of laws: what is permitted by (or compatible
with) laws of a given kind, is possible in the sense related to that kind.[13]
Accordingly, we arrive at the familiar picture of increasingly smaller spaces
of possibilities. There is the largest space of logically possible propositions—
all those that are consistent with the laws of logic. There is a smaller space
of metaphysically possible propositions, all of which are consistent with the
laws of metaphysics. In a similar fashion, physical possibilities, technological
possibilities, and others complete the picture. It is debatable whether this
image of nesting possibilities is adequate even before bringing in the issue
of real possibilities (see Fine, 2005). But our concern is real possibilities,
and we are skeptical that this approach, by giving priority to laws and by
characterizing ever stricter possibilities by increasingly more demanding
laws, can capture real possibilities. Assume that we take laws of a certain
kind as fundamental for bringing modalities into our world: Are there laws
that single out real possibilities precisely? To describe a real possibility, we
need to refer to a particular concrete circumstance, a particular moment in
time, and a particular location in space. Can real possibilities be derived from
a net of laws taken together with some initial conditions? How rich would
a net of laws have to be for this to be viable? Consider Bálazs who, given
all the circumstances obtaining here-and-now, could really run to the main
station to catch the last train to Konstanz today. If he starts a minute later, he
won't make it; under slightly changed circumstances he won't make it either.
Was this feat of his really possible at the mentioned circumstance because
it is compatible with some set of laws, taken together with some initial
conditions? Our answer is that it was possible because of the particular circumstances obtaining in the relevant region of *Our World*. Perhaps one could
derive the required laws of nature from the real possibilities that obtain,
distilling the laws, so to speak, from the dispositional and modal features of
our world. But the net of laws that would be required to determine the real
possibilities that obtain for our runner in the given concrete circumstances

---

[13] In contrast, in Branching Space-Times, a notion of accessibility can be defined in terms of the
relative location of points in a branching structure, without reference to laws; see Müller (2002).

would be immense and surely beyond our comprehension. This net of laws looks very different from the laws that we know from the sciences, or from philosophical accounts of laws. Just compare the attempt at providing a law-based account of our runner's real possibilities with a smooth account concerning physical possibility (e.g., that for a photon it is physically possible to travel from the Moon to the Earth within 3 minutes because that does not contradict the relativistic limitation on the speed of light or any other known candidates for laws of nature). Branching theory thus suggests that real possibilities do not fall in the mold of law-given modalities. The suggestion, therefore, is to take real possibility as a primitive, non-reducible notion and study it by constructing a formally rigorous theory.

So much for the differences between Branching Space-Times and possible-worlds theories. Given these differences and the frugality of Branching Space-Times, we believe that it is far from helpful to attempt to locate this theory on the map of possible-worlds theories. In order to contribute to the discussions in the literature, we nevertheless end this section with some remarks on three issues that call for our particular attention: actuality, reducibility, and the meaning of "possible worlds".

Actuality and alternatives *to* vs. alternatives *for*. Branching Space-Times theory subscribes to the semantical thesis that "actually" is an indexical word, like "here". Accordingly, if we imagine a branching-world dweller, her utterance singles out a specific piece of *Our World*, namely, the event of her particular utterance. The theory idealizes this utterance to be a point-like event. Does actuality extend any further, beyond the event of utterance? The answer is relevant for the metaphysical question of how the division of actual vs. possible is drawn in Branching Space-Times. In his account of modal realism (which is the thesis that all possible worlds are equally real), David Lewis claims that actuality somehow percolates from "me and all my surroundings" (1973, p. 86) to the whole actual world. We disagree already with this starting point. Uncontroversially, an utterance is a larger affair than a point event, but it is an innocent idealization to identify it with a point event. How can one extend actuality beyond such a point-like utterance? One idea is to extend actuality beyond the actual utterance and toward its past. Given the structure of possibilities captured by the postulates of Branching Space-Times, there are alternative possibilities for the future, but no alternative possibilities for a concrete event's past. So, since the past of the actual utterance is fixed, one can extend actuality from an actual event of utterance downward, to include the whole past of this event. Can we go any further, taking Lewis's (1986a, p. 71) lead in appealing to spatio-temporal relations? On this proposal, if a possible event is spatio-temporally related to the actual event of utterance, then it is actual as well. We oppose this move for the simple reason that the alternative possibilities for the future of the actual utterance event are not necessarily distinguished via different spatio-temporal structures. For instance, there are specific structures of Branching Space-Times, such as the so-called Minkowskian Branching Structures developed in Chapter 9.1, in which all histories share the same space-time structure (in that case, the structure of Minkowski space-time). And yet these structures harbor different incompatible possibilities for the future of certain events. Therefore, no purely spatio-temporal relation involving the utterance can distinguish between its alternative possible futures.[14]

In Branching Space-Times, there is thus no actual future, and accordingly, there is no actual history. This is the expression of a basic tenet of our theory: future possibilities are alternative possibilities *for* the future and not alternative possibilities *to* an actual future. This stance contrasts with the so-called Thin Red Line doctrine, according to which there is one distinguished (actual) history and, hence, one distinguished actual future of any actual event.

One way of arguing for the metaphysics of a Thin Red Line goes via linguistic data that seem to suggest an actual future. After all, in some circumstances people utter sentences of the form "It will (actually) happen, even though it might not" (Malpass and Wawer, 2012, p. 26). One might hear in a bar, for instance, that poor Fred will actually have another beer, even though he might not. The first part of this utterance reflects on Fred's bad habits, whereas the second acknowledges an alternative following from Fred's being possibly more strong-willed. We agree that such sentences have felicitous uses, but we are skeptical of the idea that such data indicate a *metaphysical* stance about a distinguished future. It seems to us that such utterances can be accounted for in epistemic terms (e.g., by reference to the strength of expectations). We still acknowledge that at the end of the day it

---

[14] For the record, note that structures of Branching Space-Times in which histories have different topological spatio-temporal structures do not help either (for an example of this kind of structure, see Exercise 2.5). In such a structure, the event of an utterance has to lie in a region in which the alternative spatio-temporal structures coincide. To use one of the alternative spatio-temporal structures as a criterion of actuality then betrays the very idea that motivates the picture with a topology change: both spatio-temporal structures are on a par, as each can be realized as our world develops further.

may turn out that all epistemic accounts fail, leaving us with clear evidence for people's belief in an actual future. We would take that as an indication of people's deterministic preferences. Yet, in accord with our project's assignment of a low priority to linguistic data and its avowed aim to model indeterminism, we would still not accommodate such language-based evidence for a Thin Red Line by adding a preferred history to our formal theory.

Another typical argument for Thin Red Line metaphysics rests on the desire to retain the meta-semantical intuition that, given a context of evaluation, any sentence (including a sentence about future contingents) is either true simpliciter or false simpliciter. 'Simpliciter' here means that the truth-values are not relativized to possible histories. We do not share this intuition, and we note again that it relates to linguistic or semantical matters. Their role in deciding a metaphysical issue like determinism vs. indeterminism should be fairly limited. We take it that adding a distinguished "Thin Red Line" history compromises the local indeterminism that we want to model, even if the resulting structure permits true indeterministic-looking sentences such as "it is possible that it will rain tomorrow and it is possible that it will not rain tomorrow".[15]

To sum up, the theory of Branching Space-Times upholds the semantical thesis of the indexical character of "actually". With regard to metaphysics, the theory holds that actuality can be ascribed to a point event (paradigmatically, an event of utterance), and, if one likes, to the past of this event. But the theory strongly opposes ascribing actuality to histories or to future segments of histories.

Eliminative analysis or modalism . An important meta-methodological issue in modal metaphysics is the following: What is the analysis of modality meant to achieve? It is typical for philosophers in the analytic tradition to deal with philosophically problematic concepts by attempting to provide an eliminative analysis. The attempted analysis aims at reducing the problematic concept to concepts that are thought to be unproblematic, or at least significantly less problematic. A well-known example is the tripartite definition of knowledge that attempts to identify knowledge with true and justified belief. Arguably, if one knows what belief is, what truth is, and what justification is, one learns from that analysis what knowledge is (if

---

[15] Such a sentence is true at any point of evaluation because on the Thin Red Line, it either rains tomorrow, or it doesn't rain tomorrow. See, e.g., Øhrstrøm (2009) and Malpass and Wawer (2012).

one did not know it before). The controversy surrounding this example of analysis is well known. What about analyzing modality? Modal idioms might be special, in the sense that it does not seem possible to learn them by just mastering their possible-worlds analysis. Accordingly, there is the doctrine known as *modalism*, which claims that modal idioms are primitive, which implies that an eliminative analysis of modal idioms is impossible. The controversy between eliminative positions and modalism involves a number of subtleties, some related to the notions of analysis and elimination, some to drawing the line between modal and non-modal terms, and some to technical details concerning the supposed reduction. Without going into these details, we just report the consensus view that David Lewis's project of *Humean Supervenience* is intended as reductive. A non-modal analysis of laws of nature (the so-called Best System analysis) serves as its starting point.[16] This is in stark contrast to Saul Kripke's (1980, p. 19) stance, which he expresses as follows: "I do not think of 'possible worlds' as providing a *reductive* analysis in any philosophically significant sense, that is, as uncovering the ultimate nature, from either an epistemological or a metaphysical point of view, of modal operators, propositions, etc., or as 'explicating' them". Kripke's stance is likely the majority view. For instance, Stalnaker's (2012, p. 30) diagnosis is that

> [...] if by "analysis" one means an eliminative reduction, then I think most possible-worlds theorists (David Lewis aside) will agree with modalism, but one may still hold that possible-worlds semantics provides a genuine explanation, in some sense, of the meanings of modal expressions.

In this controversy, Branching Space-Times theory sides with the majority view, as it does not aim to eliminate modality, but rather to offer an elucidation of some modal and some non-modal notions. Starting with the primitive concept of real possibilities, the theory aims to describe local indeterminism as happening in relativistic space-time. On that basis, it aims to establish an analysis of causation in indeterministic settings and a theory of single case objective probabilities (propensities). It provides analyses of modal funny business and of non-local probabilistic correlations. These analyses are then used to address selected problems in the philosophy of

---

[16] See, e.g., Stalnaker (2015) for an assessment.

quantum mechanics, in the philosophy of general relativity, and in the philosophy of time.

Possible worlds or alternative states of one world. The notion of *Our World* containing multiple branching histories reminds one of the controversy as to whether one should analyze possibilities in terms alternative possible worlds or rather in terms of alternative states of one actual world. When introducing possible worlds, David Lewis (1973, p. 84) writes:

> I believe, and so do you, that things could have been different in countless ways. ...Ordinary language permits the paraphrase: there are many ways things could have been besides the way they actually are. ...I therefore believe in the existence of entities that might be called 'ways things could have been.' I prefer to call them 'possible worlds'.

Philosophers were quick to note that the passage from 'ways things could have been' to 'possible worlds' is far from innocuous (see, e.g., Stalnaker, 1976, and Kripke, 1980). As Kripke notes, the label 'possible world' is picturesque, but metaphorical, and potentially misleading. A more adequate terminology would be to call the entities posited by modal metaphysics "total 'ways the world might have been', or states or histories of the entire world" (Kripke, 1980, p. 18). A possible (total) state of the world may or may not be instantiated by the actual world. Thus, 'possible ways' suggests a picture of one actual world that is capable of taking one of possibly many alternative states, whereas Lewis's phrase invokes a multiplicity of possible worlds that includes one distinguished world, the actual one. These two pictures illustrate the distinction between two varieties of views in modal metaphysics: actualism (one world with many possible total states) and modal realism (many worlds).

Returning to Branching Space-Times, its insistence on there being just one world, *Our World*, which may comprise many histories, sounds like an actualist position. However, to make any stronger claim one needs a theory of states or properties. Since Branching Space-Times, at least at its present stage of development, does not say anything about states or properties (it is purely an event-based theory), we cannot advance any stronger claim besides noting the resemblance between Branching Space-Times and actualism. The resemblance, however, is far from perfect, marred by the fact that histories are just particular subsets of *Our World*. It sounds odd to say that a part of something, say, a part of Pittsburgh, is a state (or a property) of Pittsburgh. However, if one thinks that this case of bad English is not a major obstacle, we have no objections to understanding histories as possible ways *Our World* might be. Of course, what is then needed is a theory of states and an elaboration of histories in terms of states. The important thing which must not be lost in attempts to assimilate Branching Space-Times with actualism, is that *Our World* has a non-reducibly modal character based on multiple alternative possibilities for the future of particular events.

1.5 Outline of the book

At some point, teasing glimpses of a theory should give way to laying the groundwork for its formulation. This point has arrived and so we turn now to explaining the formal framework of Branching Space-Times.

Our book has two parts. The remaining chapters of Part I present the formal theory. Starting with Chapter 2, we introduce the Postulates of the common core of Branching Space-Times. In Chapter 3, we show that there are two options for developing this common core further, which lead to two topologically different ways for histories to branch. We introduce further defined notions in Chapter 4. The remaining three chapters of Part I introduce further formal developments of the core theory that provide the basis for applications: modal funny business (Chapter 5), causation in terms of *causae causantes* (Chapter 6), and a spatio-temporal theory of single case probabilities (Chapter 7). In Part II, we put the material of Part I to use in three concrete applications to quantum correlations (Chapter 8), to branching in (special and general) relativistic space-times (Chapter 9), and to the doctrine of presentism (Chapter 10).

1.6 Exercises to Chapter 1

Exercise 1.1. Lewis (1986a, p. 208) assumes that all elements of a possible world are to "stand in suitable external relations, preferably spatiotemporal". Somewhat similarly, in Branching Space-Times any two point events from *Our World* are linked by appropriately combined instances of the pre-causal relation $<$ (see the M property, Fact 2.4). Discuss whether the pre-causal relation (which is formally explained in Chapter 2.1) is a "suitable external relation" from Lewis's perspective.

Exercise 1.2. Branching Space-Times supplies two options of how to construe actuality as a metaphysical concept: either as a token event (typically, the event of utterance), or a token event together with its past. Discuss the pros and cons of each option.

2

The Foundations of Branching Space-Times

2.1 The underlying ideas of BST

In this chapter we guide the reader through the construction of the core theory of Branching Space-Times. This discursive approach culminates in proposing a set of postulates that a structure of the core theory of Branching Space-Times (BST) is to satisfy (see Chapter 2.6). The rigorous theory commences with Postulate 2.1. We begin with an informal gloss, explaining the main ideas of our construction as clearly as we can.

The fundamental element of the construction is the set W of all point events, ordered by a certain pre-causal relation $<$. What postulates hold for the pre-causal order? For Minkowski space-time, Mundy (1986) describes the results of Robb (1914, 1936) and gives additional results for the light-like order. That research, however, does not immediately help here because a Minkowski space-time does not contain incompatible point events. We shall need to proceed more slowly. The first postulate is so natural and vital that without it we would not know what to say next.

Postulate 2.1 (BST Strict Partial Order). *Our World $\mathscr{W}$ is a nontrivial strict partial ordering $\langle W, < \rangle$, i.e.:*

1. *Nontriviality: W is nonempty.*
2. *Nonreflexivity: For all $e \in W$, $e \not< e$.*
3. *Transitivity: For all $e_1, e_2, e_3 \in W$, if $e_1 < e_2$ and $e_2 < e_3$, then $e_1 < e_3$.*

We read $e_1 < e_2$ as "e_2 can occur after e_1". Recall that asymmetry (i.e., if $e_1 < e_2$, then $e_2 \not< e_1$) follows from nonreflexivity and transitivity. Note that asymmetry incorporates the prohibition of repeatable events. Transitivity can be motivated by reflecting on how *Our World $\mathscr{W}$* is

Branching Space-Times: Theory and Applications. Nuel Belnap, Thomas Müller, and Tomasz Placek, Oxford University Press. © Oxford University Press 2022. DOI: 10.1093/oso/9780190884314.003.0002

constructed: if e_2 is a future possibility of e_1 and e_3 is a future possibility of e_2, then e_3 is a future possibility of e_1.

For convenience, we add the following simple definition of weak partial ordering:

Definition 2.1 (BST weak companion of strict order). The symbol $\leqslant$ stands for the companion weak partial ordering: $e_1 \leqslant e_2$ if $e_1 < e_2$ or $e_1 = e_2$. We also write $\geqslant$, naturally defined as $e_1 \geqslant e_2$ iff $e_2 \leqslant e_1$.

We mark the stronger relation with 'proper', as in 'e_1 is properly earlier than e_2'.

In *Our World* possible point events are related by the pre-causal relation $<$; some such events are compatible with others. Here is an idealized illustration. There is an ideally small event, e_m, at which a certain electron's spin is measured along a certain direction. There are two possible outcomes: measured spin up or measured spin down. Take a possible point event, e_u, at which it is true to say, 'It has been measured spin up', and another, e_d, at which it is true to say 'It has been measured spin down'. The point events e_u and e_d are incompatible, though each is compatible with e_m. How precisely can two incompatible point events both fit into *Our World*? Answer: By means of the pre-causal order. In the next three paragraphs we explain the intuition that leads to a criterion of which events are compatible.[1]

Let e_1, e_2, and e_3 be three events in *Our World*, with pre-causal ordering $<$. How can these three events be ordered, and how are the resulting patterns of ordering to be interpreted? Here are three paradigms.

Causal dispersion: Causal order can hold between a given point event, e_3, and two space-like separated (space-like related) future point events, e_1 and e_2, in a single Minkowski space-time, just as one might expect: $e_3 < e_1$ and $e_3 < e_2$. The three events form an "up-fork".

Causal confluence: Causal order also can hold between two given space-like related point events, e_1 and e_2, in a single Minkowski space-time, and a single future point event, e_3, as one might equally expect: $e_1 < e_3$ and $e_2 < e_3$. The three events form a "down-fork".

---

[1] We say of two or more events that they are compatible or not; the corresponding set of events is then, accordingly, consistent or not. That is, we use "consistent" as a unary predicate for sets of entities such as events or histories, and we use "compatible with" for the relation. An alternative for "compatible with" is "co-possible with".

Causal branching: Causal order can also hold between a given e_3 and two possible future point events e_1 and e_2 that might be said to be alternative possibilities: $e_3 < e_1$ and $e_3 < e_2$. The three events form an "up-fork."

Observe that an "up-fork" allows for two interpretations: if you see just an "up-fork" (i.e., without seeing how it is related to other events), you can read its upper events either as (incompatible) modal alternatives, or as (compatible) spatially separated events. It seems to us that the analogous dichotomy is absent in the case of "down-forks", at least on the understanding of point events as concrete, non-disjunctive objects. The distinction between states and concrete events is highly relevant here: the same state can result from incompatible starting points, and a state can be repeated, whereas a concrete event, with its identity being tied to its past, is not repeatable. (At least this is a plausible and intuitive view; see a relevant caveat below.) By saying "non-disjunctive" we refer to the second relevant distinction, that between Lewis's fragile and non-fragile events (Lewis, 1986b, Ch. 23), which relates to the question of whether an event might occur in more than one alternative way. In our basic construction we rely on non-disjunctive events, that is, those events that cannot occur in more than one alternative way. In the Lewisian terminology, concrete events are therefore fragile. (In Chapter 4.1 we study other types of events, including disjunctive events.)

We thus deny *backward branching*: we deny that incompatible point events can lie in the past (i.e., that some events could have incompatible 'incomes' in the same sense that some have incompatible possible outcomes). To put it differently, we deny that two events that cannot occur together somehow combine to have a common future successor. No backward branching is part of common sense, including that of scientists when speaking of experiments, measurements, probabilities, some irreversible phenomena, and the like. We need a caveat, however: some models of the general theory of relativity (GR) allow for so-called causal loops, which are typically interpreted as involving repetitions of a concrete event (Hawking and Ellis, 1973). We return to this topic in Chapter 9.3.6, in which we suggest how to modify BST to accommodate causal loops of GR.[2]

The 'no backward branching' intuition gives rise to a semi-criterion as to which events are compatible (co-possible). The fact that a "down-fork" has a univocal meaning provides a recipe for the criterion of a later witness: if two events are seen as past from the perspective of some third event,

---

[2] An attempt to generalize BST so that it accommodates causal loops is provided in Placek (2014).

these two events are compatible. We will later strengthen this recipe to provide a criterion that will deliver possible histories (as maximal subsets of compatible events) of *Our World*.

2.2 Histories

How does one further describe the way that point events fit together in *Our World*? The crucial concept is that of a *history*, which will help us keep track of the compatibilities and incompatibilities between events in *Our World*. In BST, which adds a spatial aspect, we generalize the concept of a history as defined in Prior's theory of branching temporal histories, which is often called "Branching Time" (Prior, 1967; Thomason, 1970). The basic blocks of this theory are spatially maximal instantaneous objects, somewhat mis-leadingly called "moments".[3] The theory arranges such moments into a tree (see Figure 2.1): incompatible moments have a lower bounding moment in the tree (a feature we will call "historical connection"), but never a common upper bound (no backward branching). The formal definition of a tree gives expression to the openness of the future in contrast to the settledness of the past. A key point to always bear in mind is that in Branching Time, the entire tree is 'the world'. In addition there is the concept of a 'history', defined as a maximal chain of moments. Locate yourself at a moment in the tree, perhaps at the moment at which the spin measurement occurs. You will easily visualize that in this picture your 'world' is unique, whereas you belong to many 'histories'. Until and unless branching ceases before your expiration, there is no such thing as 'your history'. Of course in Branching

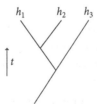

Figure 2.1 A tree-like structure of Branching Time with three histories.

[3] The construction of moments requires a frame-independent simultaneity relation, and for that reason Branching Time is not a relativity-friendly theory. The terminology "Branching Time" is doubly misleading, suggesting that moments are just instants of time and that time branches. For a discussion of what branches (viz., histories rather than time or space-time), see p. 10.

Time 'your historical past' makes perfect sense, in contrast to 'your entire history'. Branching Time takes uniqueness to fail when histories are taken as stretching into the future. Significantly, in this usage a 'world' contains incompatible possibilities, while a 'history' does not. A history represents a choice between incompatible possibilities, a resolution of all disjunctions unto the end that presumably never comes.

The present development retains Prior's idea of *Our World* as involving many possible histories. A novelty is that a history can be isomorphic to a relativistic space-time (e.g., to Minkowski space-time),[4] the latter being understood as the set of quadruples of reals, $\mathbb{R}^4$, together with Minkowskian ordering $<_M$, defined as follows:

$$x <_M y \quad \text{iff} \quad -(x^0 - y^0)^2 + \sum_{i=1}^{n-1} (x^i - y^i)^2 \leqslant 0 \text{ and } x^0 < y^0. \quad (2.1)$$

(Here $<$ and $\leqslant$ are, respectively, the strict and weak orderings of reals and it is assumed that the zero coordinate is temporal. Space-times of dimensions other than 4 are defined analogously.) Given our aim to accommodate Minkowski space-time, histories cannot be defined as maximal chains of point events; the latter are mere paths without a spatial dimension. But, as we have just seen, there is the notion of a later witness on which we may base our attempt: if two point events share a later point event, the three events must be in one history. For this to yield a viable criterion of being a history in BST, however, we had better have an opposite implication as well, in the form "if two point events do not have a common later point event, then they do not share a history". The geometry of Minkowski space-time certainly guarantees that this latter implication holds, yet it is easy to construct a truncated Minkowski space-time, or a more sophisticated space-time of general relativity, which violates this implication (cf. Müller, 2014). Nevertheless, we assume here the strengthened intuition of a later witness in the form of an equivalence: "two point events have a common later point event iff they share a history".[5]

Since the structural feature underlying the later witness intuition has a name, we will use it: a history must be a 'directed' set, defined as follows.

[4] We mean this as a promissory note; in Section 9.1 we will exhibit BST structures in which all histories are indeed isomorphic to Minkowski space-time.
[5] It is possible to work with only a one-way implication in order to properly analyze the truncated space-times that were just mentioned (Placek, 2011). This approach is, however, too complicated for the pay-off it might bring.

Definition 2.2. A subset E of *Our World* is *directed* just in case for all e_1 and e_2 in E there is a point event e_3 in E that is their common upper bound: $e_3 \in E$ and $e_1 \leqslant e_3$ and $e_2 \leqslant e_3$.

Not every directed set should be counted as a history; in line with a respectable tradition, we expect a history to be maximal.[6]

Definition 2.3. A subset h of *Our World* is a *history* just in case h is a maximal directed subset of *Our World*: h itself is a directed subset of *Our World*, and no proper superset of h has this feature.

Histories are a key conceptual tool.[7] Each history might be a Minkowski space-time; but typically *Our World* is no such thing, because a single Minkowski space-time, unlike *Our World*, fails to contain any incompatible possible point events.

Here are some elementary facts about histories.

Fact 2.1. *Let $\mathscr{W} = \langle W, < \rangle$ be a partially ordered set satisfying Postulate 2.1. Then:*

1. *Every finite set of points contained in a history, h, has an upper bound in h.*
2. *Infinite subsets of a history, for example a history itself, need not have a common upper bound.*
3. *Every directed subset of W can be extended to a history. In particular, every chain is a subset of some history, and every point event belongs to some history.*
4. *$\mathscr{W}$ has at least one history.*
5. *Histories are closed downward: if $e_1 \leqslant e_2$ and $e_2 \in h$, then $e_1 \in h$.*
6. *The complements of histories are closed upward: if $e_1 \leqslant e_2$ and $e_1 \in W \setminus h$, then $e_2 \in W \setminus h$.*
7. *No history is a subset of a distinct history.*

---

[6] Against branching theories, Barnes and Cameron (2011) object that they cannot accommodate the intuition that it "may be open whether or not reality will continue beyond tonight". The intuition, if rendered as a history possibly being a proper segment of another history, contradicts the maximality of histories. However, one may save this intuition (if one needs to) by drawing a distinction between a history and "its" space-time and appealing to an assignment of properties to spatio-temporal points. One may then consider two histories such that none is a segment of the other, yet the space-time of one is a segment of the space-time of the other, and such that they are qualitatively the same in the shared segment of space-time.

[7] Note that the ordinary use of "history" is relational, as in "the history of Pittsburgh". The monadic use appears to be technical. For example, in physics one identifies possible histories with possible evolutions. A different name for Branching-Time histories is "chronicles"; see Øhrstrøm (2009).

8. *No history, h, is a subset of the union of a finite family, H, of histories of which it is not a member. Provided H is a finite set of histories, if $h \subseteq \bigcup H$ then $h \in H$.*

9. *If e is not maximal in* Our World, *then neither is e maximal in any history to which it belongs.*

10. *If e is maximal in W, then there is exactly one history in $\mathcal{W}$ containing e, and e is the unique maximum of h.*

Proof. (1) follows from the definition of a directed subset by finite induction, starting with a common upper bound for the first two points and then consecutively constructing the upper bounds with the next points. As a witness for (2), take for instance the real line with its natural ordering. For (3), let A be a directed subset of W; note that any chain, including the singleton set of a point, is directed. To arrive at a maximal directed set (i.e., a history) containing A, we apply the Zorn-Kuratowski lemma to the family of directed supersets of A in W, ordered by set inclusion.[8] That partial ordering satisfies the premise of the lemma, as for any chain C of directed subsets of W, the union $\bigcup C$ is a directed subset of W that contains every element of C as a subset, so that $\bigcup C$ is an upper bound of C. The lemma then implies that the partial ordering of directed supersets of A contains a maximal element, which is the sought-for history. (4) follows from (3), as W is non-empty. (5) follows from the definition of history as a maximal directed subset, and (6) by taking the contraposition of (5). A maximality of histories suffices to prove (7). As for (8), let $H = \{h_1, ..., h_n\}$ and $h \neq h_i$ for every $h_i \in H$. By (7) $h \setminus h_i \neq \emptyset$ for each $i = 1, ..., n$, so for each i there is $x_i \in h \setminus h_i$. The set $\{x_i \mid i = 1, ..., n\}$ is a finite subset of h, so by (1), there is $y \in h$ upper-bounding all of the x_i. Now, it cannot be that $y \in \bigcup H$, as then $y \in h_j$ for some $h_j \in H$, and hence every $x_i \in h_j$, including $x_j \in h_j$, which contradicts the construction above. As $y \in h$, it follows that $h \not\subseteq \bigcup H$. For (9) let us suppose that $e \in h$ and that $e < e_1$. Then $\{e_2 \mid e_2 \leqslant e\} \cup \{e_1\}$ is a directed subset and a proper superset of $\{e_2 \mid e_2 \leqslant e\}$, so that the latter subset of h is not a history, so not identical with h, so a proper subset of h. Let $e_3 \in h \setminus \{e_2 \mid e_2 \leqslant e\}$. Since h is directed, there must be some element e^* in h that upper-bounds

---

[8] For future reference, here are the formulations of the versions of the Axiom of Choice that we will use: (1) The Axiom of Choice: for every set X of non-empty sets, there exists a choice function defined on X. (2) The Zorn-Kuratowski lemma: a partially ordered set containing upper bounds for every chain contains at least one maximal element. (3) Hausdorff's maximal principle: in any partially ordered set, every totally ordered subset is contained in a maximal totally ordered subset.

both e and e_3. The element e^* is distinct from e (since otherwise $e_3 < e$), hence e^* is properly later than e, and thus e is not a maximal element of h. (10) Let e be maximal in W (i.e., there is no $e' \in W$ for which $e < e'$). Then the set $h =_{df} \{x \in W \mid x \leqslant e\}$ is directed (e being an upper bound for any two of its members), and e is its unique maximum. To see that h is maximal directed, suppose for reductio that there is some directed proper superset $h' \supsetneq h$ and pick some $e' \in h' \setminus h$. Since h' is directed, there is some element e'' that upper-bounds the elements e and e'. Since e is maximal in W, we get $e = e''$. Then $e' \leqslant e$, and thus $e' \in h$ by (5), contradicting that $e' \in h' \setminus h$. Thus, h is indeed a maximal directed subset. □

Two point events evidently share some history, just in case they have a common upper bound in Our World. In contrast, two point events fail to have any history in common just in case they have no common upper bound. It would be right to mark such a fundamental matter with a definition.

Definition 2.4. Point events e_1 and e_2 are *compatible* if there is some history to which both belong, and otherwise are *incompatible*.

One may wonder if BST models, and BST histories in particular, incorporate some sense of temporal direction. Much simpler Branching Time models incorporate it, since they have a form of a tree starting from a single trunk. With respect to BST, it may have crossed the reader's mind that each Minkowski space-time appears the same upside down: each is not only directed, but also 'directed downward' in the following sense.

Definition 2.5. A subset E of *Our World* is *directed downward* just in case for all e_1 and e_2 in E there is a point event e in E that is their common lower bound: $e \in E$ and $e \leqslant e_1$ and $e \leqslant e_2$.

That each Minkowski space-time is an upside-down image of itself is of course true,[9] but this should not lead one to think that the way in which we define a 'history' makes no difference. Consider, for instance, the following. While a Minkowski space-time is indeed downward directed, it would be truly peculiar if it were maximal downward directed. For it can be proved that if a subset of a partially ordered set is maximal downward directed, then it is upward closed (see Exercise 2.1). So, if a history were maximal downward directed, it would be upward closed. And if it were upward closed, then if

---

[9] More precisely, $\langle \mathbb{R}^4, \leqslant_M \rangle$ and its image by time-reflection are order-isomorphic.

there were any incompatible possible point events in the future of any one of its members, the history would have to contain both of them, which would run counter to the idea of compatibility.

In this way, the concepts of Branching Space-Times provide a natural, unforced articulation of the 'direction of time' without complicated physics. They do so by looking beyond the properties of a single history so as to take account of how distinct histories fit together, something that becomes really clear only later in the context of further postulates.

A definition and a fact here shift our attention from single to multiple histories.

Definition 2.6. We write $\text{Hist}(W)$ for the set of all histories in $\langle W, < \rangle$,[10] and for $E \subseteq W$, we write $H_{[E]}$ for the set of those histories containing all of E (i.e.: $H_{[E]} =_{\text{df}} \{h \in \text{Hist} \mid E \subseteq h\}$). We abbreviate $H_{[\{e\}]}$ as H_e, i.e., $H_e =_{\text{df}} \{h \in \text{Hist} \mid e \in h\}$.

Fact 2.2. *(1) H_e is never empty. Also, (2) if $e_1 \leqslant e_2$, then $H_{e_2} \subseteq H_{e_1}$.*

Proof. (1) follows from Fact 2.1(3), while (2) follows from histories being downward closed; that is, Fact 2.1(5). □

One should not generally expect the converse of (2) presented earlier, as two compatible though incomparable point events may belong to exactly the same histories.

With the criterion of historicity we are able to carve out histories from *Our World* $\mathscr{W}$. Typically $\mathscr{W}$ has more than one history, and in that case, if BST is right, the phrase 'our history' or 'the actual history' is meaningless.[11] Scientists, for instance, no matter how hardheaded and downright empirical they wish to be, cannot confine their attention to 'our history' or to 'the actual history'. It is not just that they ought not. Rather it is that if Branching Space-Times is correct, scientists cannot confine their attention to 'the actual history' for precisely the same reason for which mathematicians cannot confine their attention to 'the odd prime number': there is more than one odd prime number, and there is more than one history to which we belong. On the other hand, just as a mathematician can deal with 'the odd

---

[10] We omit "(W)" if the reference is clear.

[11] This remark pertains to the debate on the doctrine that there is a distinguished history (or a distinguished future): the actual history (future), or our history (future). This doctrine is known as the "Thin Red Line" view and is defended, e.g., by Malpass and Wawer (2012). For an extended argument against this view, see Belnap et al. (2001).

prime numbers' (plural), so a scientist could manage to deal only with 'our histories' (plural); that is, with the set of all histories to which this indexically indicated context of utterance belongs.[12]

With the notion of compatibility in place, we can now define space-like separation.

Definition 2.7. If e_1 and e_2 are (i) incomparable by $\leqslant$ but (ii) compatible, then they are space-like related (SLR), written as $e_1 \, SLR \, e_2$. We may also call the events causal contemporaries (provided we bear in mind the failure of the transitivity of *SLR*).

In this definition, condition (ii) is essential. That is why it was not possible to become clear on space-like separation without the definitions of this Section. We can call events that are related by the pre-causal ordering $<$ "cause-like related". Such events are compatible because histories are closed downward. Using this terminology, we can state the following Fact:

Fact 2.3. *Incompatible points have neither a cause-like nor a space-like relation. They are thus with respect to each other neither causally future nor causally past nor causally contemporaneous.*

This fact is a trivial, albeit helpful, consequence of the definitions. Note that it leaves open the possibility that the spatio-temporal *positions* of incompatible point events are spatio-temporally related, given that a notion of spatio-temporal position of events is available.[13] Even if such a spatio-temporal concept becomes available, however, this does not imply a cause-like relation, and one cannot infer a spatio-temporal relation between the spatio-temporal positions of two point events from the mere fact that they are incompatible. Incompatibility, although defined from the pre-causal order, is not itself a spatio-temporal relation in this sense.

2.3 Historical connection

What we said thus far leaves open the question of how histories in a BST structure are to be related. That histories should somehow overlap follows from the idea of *Our World* as the totality of events accessible from a given

---

[12] Perhaps physics also considers worlds other than ours, such as those postulated by Lewis (1986a); it is important to recognize this as an entirely different question.

[13] See Chapter 2.5 for some discussion of the introduction of spatio-temporal positions in BST.

actual event by the pre-causal relation. A history that is completely severed from the rest of the model is in conflict with this construction. In the same spirit, in the theory of Branching Time, where histories are chains, one postulates that every two histories overlap; we call this property 'historical connection'.

The property should hold in BST, although the notion of 'history' now has a wider meaning:

Postulate 2.2 (Historical Connection). *Every pair of histories has a nonempty intersection.*

In the theory of Branching Time, this would be the equivalent of saying that every two moments have a lower bound. Here, where the topic is point events instead of moments, the 'common lower bound' principle is not equivalent to Historical Connection, and is not postulated. For more detail, see Fact 2.3. Note that Historical Connection, unlike Postulate 2.1, does not imply the result of replacing $<$ by its converse, and is thus sensitive to the direction of time. That is, if $\langle W, < \rangle$ satisfies Postulate 2.1, so does $\langle W, <\cdot \rangle$, where $<\cdot$ is the relation on W converse to $<$, i.e., $e <\cdot e'$ iff $e' < e$. However, since replacing $<$ by $<\cdot$ can change the set Hist of histories, it may happen that $\langle W, < \rangle$ satisfies Postulate 2.2, but $\langle W, <\cdot \rangle$ does not. A case in point is depicted in Figure 2.2.[14]

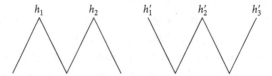

$h_1 \qquad h_2 \qquad h'_1 \qquad h'_2 \qquad h'_3$

Figure 2.2 Since $h_1 \cap h_2 \neq \emptyset$, $\langle W, < \rangle$ (on the left) satisfies Postulate 2.2, but $\langle W, <\cdot \rangle$ (on the right) does not, as $h'_1 \cap h'_3 = \emptyset$.

The following consequence of historical connection supplies a good account of Lewis's notion of a "suitable external relation": *Our World* is connected by $\leqslant$, the suitable external relation, since the trip from one point to another in *Our World* may be long, but it need not have a complicated shape:

Fact 2.4 (The M property). *Let $\mathcal{W}$ satisfy Postulates 2.1 and 2.2. Then every pair e_1, e_5 of point events can be connected by a $\leqslant / \geqslant$—path no more complex*

---

[14] Thanks to A. Barszcz for this model.

than the shape of an M, i.e., there are e_2, e_3, e_4 in $\mathcal{W}$ such that $e_1 \leqslant e_2, e_5 \leqslant e_4$ and $e_3 \leqslant e_2, e_3 \leqslant e_4$.

Proof. Left as Exercise 2.2. □

The M property gives a unity to *Our World* of really possible events, as we explained in Section 1.3.

Note that Historical Connection does not generalize to a larger number of histories. Figure 2.3 provides an illustration: There are three-point events in each history. You see that each pair of histories overlaps (historical connection), but that no point event belongs to all three. A later postulate will rule this out as a possible model, as it will require any finite number of histories to have an overlap.

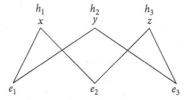

Figure 2.3 Historical Connection does not imply the existence of the overlap of more than two histories.

This observation clearly indicates that a postulate stronger than Historical Connection is needed. The core theory is too frugal to use BST to address some issues in general philosophy as well as in the philosophy of physics. In particular, a stronger postulate should govern what the overlap of two branching histories looks like. This question touches upon a contentious issue. At the bare minimum, we need a local notion of possibility, like alternative possibilities open at a junction, but to define it, we need to decide how to understand "junction" in this context: is it an event, possibly idealized as a point event, or a specific collection of point events, or some other structure? Only after deciding this can we define local alternative possibilities. These should be possibilities open at a junction, in the relevant sense, and should be defined in terms of a suitable partition of the relevant histories. The technical definition of local alternative possibilities is crucial for BST: it will later be used to analyze non-local correlations, singular causation, propensities, Bell's inequalities, tense, and the question of determinism in general relativity. Here we can note that the core of BST can be developed in two different ways. One option leads to the original theory of BST_{92}, the other to a new theory providing "new foundations" for BST, BST_{NF}. Technically, Historical Connection will be strengthened to one of two alternative Prior Choice Principles, one defining BST_{92} and the other defining BST_{NF}. We will discuss these two options in detail in Chapter 3. In the rest of this chapter, we provide a full exposition of the core theory that is common to both approaches, leading to the definition of a common BST structure (Def. 2.10).

2.4 Density and continuity

In this section we reflect further on the pre-causal relation $<$. As of now, we postulated that $\mathcal{W} = \langle W, < \rangle$ is a non-trivial strict partial ordering (Postulate 2.1). The next three postulates come from our desire to arrive at histories that are somewhat similar to space-times of physics. A world-line of a spatially non-extended object is naturally modeled in BST as a maximal chain. To be in accordance with the physics of space and time, maximal chains should be dense and continuous. Hence our next postulates:

Postulate 2.3 (Density). *If $e_1 < e_2$, then there is a point event properly between them.*

 We will discuss continuity in the sense of the existence of infima (maximal lower bounds) and suprema (suitable minimal upper bounds). Each of our respective postulates implies Dedekind continuity; see Appendix A.1. Here are the formal definitions of infima and suprema:

Definition 2.8. For $E \subseteq W$, where $\langle W, < \rangle$ is a partial order, a lower bound for E is a point e such that $e \leqslant e_1$ for every $e_1 \in E$. A maximal lower bound for E is a lower bound for E such that no lower bound for E is strictly above it. If there is a lower bound e for E such that $e_1 \leqslant e$ for every lower bound e_1 of E, it will be unique. One calls it $\inf E$, the infimum of E. Similarly for upper bound, for minimal upper bound, and for supremum, written 'sup E' when it exists.

 We postulate the existence of infima for lower bounded chains:

Postulate 2.4 (Existence of infima for chains). *Every nonempty lower bounded chain of point events has an infimum.*

 It can be shown that the infima postulate does not decide the question of suprema for upper bounded chains in BST (see Exercise 2.4). We therefore need yet another postulate. In formulating the suprema postulate, we cannot just use the mirror image of Postulate 2.4, because an upper bounded chain may end in different ways depending on the history that one considers. As a supremum has to be unique, we do not postulate the existence of suprema, but rather of history-relative suprema. This shows how temporal directedness is modally grounded: a BST structure with just one history allows for unique suprema, but once there is more than one history (i.e., once there is indeterminism), the different behavior of infima and suprema signals temporal asymmetry.

Postulate 2.5 (Existence of history-relative suprema for chains). *Each non-empty upper bounded chain l has a supremum $\sup_h l$ in each history h such that $l \subseteq h$, where $\sup_h l$ is characterized by the following three conditions: (1) $\sup_h l \in h$; (2) $e_1 \leqslant \sup_h l$ for every $e_1 \in l$, and (3) if $e_2 \in h$ and $e_1 \leqslant e_2$ for every $e_1 \in l$, then $\sup_h l \leqslant e_2$.*

Note that a corresponding reformulation of the infima postulate would change nothing, as by the downward closure of histories, any lower bound of a lower-bounded chain belongs to all histories to which the chain belongs.

The definition of history thus entails that infima of lower bounded chains exist independently of histories, while suprema of upper bounded chains exist only relative to a history. These features are essential features of Branching Space-Times. Take a 'process' as represented by a bounded causal interval without a first or last point event, and interpret the following tenses from the standpoint of a point event within it. 'How this process will end' (i.e., the supremum of the process) is historically contingent, depending as it does on (perhaps metaphorical) choices made in the neighborhood of the process. 'How this process began' (i.e., the infimum of the process) is, in contrast, independent of histories.

2.5 Weiner's postulate and spatio-temporal locations

There is one further postulate, suggested by M. Weiner, that needs to be added to the core theory of BST in order to exclude unwanted structures.[15]

---

[15] Weiner's postulate (see also Def. 2.10(6)), which was not included in early BST papers such as Belnap (1992), was added later as it was found to be necessary to develop a useful probability theory in BST$_{92}$; see Weiner and Belnap (2006) and Müller (2005).

Postulate 2.6 (Weiner's postulate). *Let $l, l' \subseteq h_1 \cap h_2$ be upper bounded chains in histories h_1 and h_2. Then the order of the suprema in these histories is the same:*

$$\sup\nolimits_{h_1} l \leqslant \sup\nolimits_{h_1} l' \quad \textit{iff} \quad \sup\nolimits_{h_2} l \leqslant \sup\nolimits_{h_2} l'.$$

Note that hereby we also have

$$\sup\nolimits_{h_1} l = \sup\nolimits_{h_1} l' \quad \textit{iff} \quad \sup\nolimits_{h_2} l = \sup\nolimits_{h_2} l'.$$

Weiner's postulate has philosophical significance as it helps to clarify which objects branch in branching-style theories. Histories represent alternative possible scenarios. In Branching Time, histories capture merely temporal aspects of a possible scenario, whereas in BST histories accommodate spatio-temporal aspects. Perhaps misled by the names, an objection has been leveled that these theories assume the branching of time or of space-time itself. But what branches, according to these theories, is spatio-temporal (temporal) histories and *not* space-time (time). Weiner's postulate provides an argument for this verdict, as it makes the following construction possible.

Having a branching model of possible histories, one might want to coordinate, in a temporal or spatio-temporal sense, events belonging to alternative histories. The motivation for this is clearly seen in a temporal case, as we often wonder what would have occurred at a given instant of time (e.g., 9 am this morning) if things had gone differently in the past. One might thus want to view some incompatible events as possibly occurring at the same instant of time. The result of introducing such temporal coordinates would be a theory of Branching Time with temporal instants, with the set of instants naturally identified with time. Since the time so constructed is common to all histories, it cannot branch, though histories do branch.[16] The same motivation, if applied to spatio-temporal histories, yields the concept of spatio-temporal (point-like) locations, with the underlying idea that events from alternative histories belong to one and the same location of this sort. The set of spatio-temporal (point-like) locations is analogously read as space-time.

The significance of Weiner's postulate is that it provides a necessary (but not sufficient) condition for spatio-temporal locations to be definable in BST structures. To prove this claim we first define space-time locations:

---

[16]   For a theory of Branching Time with Instants, see Belnap et al. (2001, Ch. 7A.5).

Definition 2.9 (BST with space-time locations). Let $\langle W, < \rangle$ satisfy Postulates 2.1–2.5. A partition S of W is a set of *spatio-temporal locations* of $\langle W, < \rangle$ iff

1. For each history h in W and for each $s \in S$, the intersection $h \cap s$ contains exactly one element.
2. S respects the ordering, i.e., for $s, s' \in S$ and $h_1, h_2 \in \text{Hist}$, $s \cap h_1 \leqslant s' \cap h_1$ iff $s \cap h_2 \leqslant s' \cap h_2$.[17]

An auxiliary result is that the history-relative suprema of upper bounded chains, guaranteed to exist by Postulate 2.5, belong to the same location—if $\langle W, < \rangle$ admits spatio-temporal locations.

Fact 2.5. *Let $\langle W, <, S \rangle$ satisfy Postulates 2.1–2.5, and let S be a set of spatio-temporal locations of $\langle W, < \rangle$. Let l be an upper-bounded chain in W such that $l \subseteq h_1 \cap h_2$, where $h_1, h_2 \in \text{Hist}$. Then for some $s \in S$, $\{\sup_{h_1} l, \sup_{h_2} l\} \subseteq s$.*

Proof. Let $c_1 = \sup_{h_1} l$ and $s \in S$ be such that $c_1 \in s$. We take the unique $c_2 \in s \cap h_2$. We need to prove that $c_2 = \sup_{h_2} l$. Since S preserves the ordering, and c_1 upper bounds l in h_1, c_2 must be an upper bound of l in h_2. Moreover, c_2 must be the least upper bound of l in h_2, for if there were a c_2' such that $l \leqslant c_2' < c_2$, then c_1 would not be the least upper bound of l in h_1, contradicting our assumption. Thus, c_2 is an h_2-relative supremum of l, i.e., $c_2 = \sup_{h_2} l$, as required. □

Then as a corollary we get that Postulate 2.6 is a necessary condition for definability of locations S on $\langle W, < \rangle$:

Corollary 2.1. *If $\langle W, < \rangle$ satisfies Postulates 2.1–2.5 and admits spatio-temporal locations S, then it satisfies Postulate 2.6.*

Proof. Let l and l' be two upper bounded chains, and let h_1, h_2 be two histories to which both l and l' belong. Then by Fact 2.5, their history-relative suprema will be at the same space-time locations, and the claim follows by order preservation of S. □

Further, the model discussed in Exercise 2.5 shows that definability of locations S is not a trivial matter, as evidenced by the following:

---

[17] We are using an extension of the ordering notation to singletons here, as $s \cap h_1$ is the singleton set of an event, not an event. The analogous claims for "=" and for "<" follow directly.

Fact 2.6. *There is a strict partial ordering* $\langle W, < \rangle$ *that satisfies Postulates 2.1–2.5, but does not admit spatio-temporal locations.*

Proof. By the model given in Exercise 2.5 and by Corollary 2.1, using contraposition.   □

2.6 Axioms of the common core of BST

We end this chapter with the "official definition" of a common BST structure, pulling together all of the above Postulates, with BST histories defined by Definition 2.3:

Definition 2.10 (Common BST structure). A *common BST structure* is a pair $\langle W, < \rangle$ that fulfills the following conditions:

1. W is a non-empty set of possible point events.
2. $<$ is a strict partial ordering denoting a pre-causal relation on W.
3. The ordering $<$ is dense;
4. The ordering contains infima for all lower bounded chains;
5. The ordering contains history-relative suprema for all upper bounded chains;
6. Weiner's postulate: Let $l, l' \subseteq h_1 \cap h_2$ be upper bounded chains in histories h_1 and h_2. Then the order of the suprema in these histories is the same:

$$\sup_{h_1} l \leqslant \sup_{h_1} l' \quad \text{iff} \quad \sup_{h_2} l \leqslant \sup_{h_2} l'.$$

7. Historical connection: Any two histories have a non-empty intersection, i.e., for $h_1, h_2 \in \text{Hist}, h_1 \cap h_2 \neq \emptyset$.

2.7 Exercises to Chapter 2

Exercise 2.1. Complete the sketch of the proof of Fact 2.1(5). Then prove that, if a subset of a partially ordered set is maximal downward directed, then it is upward closed.

Hint: Rework the proof of Fact 2.1(5) in an appropriate way.

Exercise 2.2. Prove the M property (Fact 2.4):
For every pair e_1, e_5 of point events in $\mathcal{W}$, there are e_2, e_3, e_4 in $\mathcal{W}$ such that $e_1 \leqslant e_2, e_5 \leqslant e_4$ and $e_3 \leqslant e_2, e_3 \leqslant e_4$.

Hint: Use historical connection and the directedness of histories. An explicit proof is given in Appendix B.2.

Exercise 2.3. Let $\langle W, \leqslant \rangle$ be a partially ordered set satisfying Postulates 2.1 and 2.2. Prove that if every history of W is downward directed, then so is W as a whole. (Note that the assumption is true, for example, if each history is isomorphic to Minkowski space-time.)

Hint: Pick any $e_1, e_5 \in W$. If these events share a history, we are done. If not, invoke the M property to find appropriate lower bounds. An explicit proof is given in Appendix B.2.

Exercise 2.4. Let $\langle W, \leqslant \rangle$ be a partially ordered set satisfying the infima postulate 2.4. Show that this postulate does not imply the existence of history-relative suprema for upper bounded chains (i.e., it does not imply Postulate 2.5).

Hint: Take a maximal chain in history h and its Dedekind cut $\{A, B\}$, where $A < B$. (See Appendix A.1 for the definition.) The upper sub-chain B thus has an infimum. Suppose that it belongs to B. Then this infimum is also a minimal upper bound for A. One can show, however, that the infima postulate does not prohibit the existence of another minimal upper bound for A in h. Then in h there is no unique minimal upper bound of A, and so A has no supremum in h (and hence no supremum *simpliciter*). Although a topology for BST has not yet been introduced (see Chapter 4.4), one can reasonably suspect that the structure in question violates the Hausdorff property. The postulate of history-relative suprema accordingly assures that all histories are Hausdorff.

Exercise 2.5. Show that Postulate 2.6 is independent of the remaining postulates 2.1–2.5.

Hint: For a structure that satisfies Postulates 2.1–2.6, one can pick a simple two-histories structure in which each history h_1 and h_2 is isomorphic to the two-dimensional real plane with the Minkowskian ordering $\leqslant_M$ (see Eq. 2.1).

For a structure that satisfies Postulates 2.1–2.5 but violates Postulate 2.6, consider the essentially linear two-history structure depicted in Figure 2.4.[18]

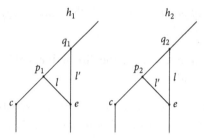

Figure 2.4 Postulate 2.6 is violated by chains l and l'.

---

[18] The figure is based on a drawing by M. Weiner. For a different example that is essentially two-dimensional, see Appendix A of Müller (2005).

3

Two Options for the Branching of Histories

3.1 Indeterminism as the branching of histories

The branching of histories in BST is intended to represent indeterminism as a feature of our world. Now, if a common BST structure contains just one history, then it is trivial from the perspective of indeterminism: all events are compatible, and the picture of a world with just one history is that of a deterministic world. Since there are multiple histories in any non-trivial BST structure, there are different ways in which these histories can interrelate. A strong intuitive principle is historical connection (Postulate 2.2): The idea is that any two histories should share some common past. As we said, we will show that historical connection is implied by stronger principles concerning the interrelation of histories. These so-called prior choice principles (Defs. 3.4 and 3.14) make specific demands on the way in which histories branch off from one another. The key decision is what the branching of histories looks like locally: What are the objects at which histories branch? BST_{92} decides for points: histories branch, or remain undivided, at points, which means that there is a maximal element, called a *choice point*, in the overlap of any two histories.

The existence of choice points has important implications for the topological properties of the resulting structures, a matter to which we will turn later in Chapter 4.4. But do choice points exist, or, more precisely, do the postulates of a common BST structure decide whether there are choice points? It turns out that the answer is in the negative: we can show that both the existence and the non-existence of choice points are live options for the branching of histories in common BST structures. While BST_{92} requires the existence of choice points, the "new foundations" theory BST_{NF} prohibits the existence of choice points and works with so-called *choice sets*. The difference is illustrated by the two common BST structures of Figure 3.1

Branching Space-Times: Theory and Applications. Nuel Belnap, Thomas Müller, and Tomasz Placek, Oxford University Press. © Oxford University Press 2022. DOI: 10.1093/oso/9780190884314.003.0003

(a) and (b).[1] These structures illustrate the two possibilities for histories to branch in common BST structures.

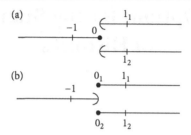

Figure 3.1 Two simple common BST structures with (a) and without (b) a choice point. Both (a) and (b) depict partial orderings in which there are two continuous histories branching at point 0. In (a), point 0 is the shared maximum in the intersection of the histories, so that 0 is a choice point. In (b), the intersection of the histories has no maximum, and points 0_1 and 0_2 are different history-relative suprema (minimal upper bounds) of the intersection, which together form a choice set.

We provide a formal definition of these structures, so that we do not rely on pictures alone. Both structures are defined as quotients of the double real line

$$L_2 =_{df} \{\langle x,i \rangle \mid x \in \mathbb{R}, i \in \{1,2\}\}$$

under the equivalence relations $\equiv_a$ and $\equiv_b$, which are defined, respectively, as

$$\langle x,i \rangle \equiv_a \langle x',i' \rangle \leftrightarrow (x = x' \wedge (i = i' \vee x \leqslant 0));$$
$$\langle x,i \rangle \equiv_b \langle x',i' \rangle \leftrightarrow (x = x' \wedge (i = i' \vee x < 0)).$$

These relations differ only in their handling of $x = 0$. The ordering on the quotient structures $M_a =_{df} L_2/\equiv_a$ and $M_b =_{df} L_2/\equiv_b$ is defined uniformly via

$$[\langle x,i \rangle] < [\langle x',i' \rangle] \Leftrightarrow_{df} (x < x' \wedge [\langle x,i \rangle] = [\langle x,i' \rangle]).$$

---

[1] These structures have just the bare minimum of complexity to fulfill the axioms of Def. 2.10 in a non-trivial way: they contain just two histories each. Furthermore, they do not include any spatial extension—so in fact they are branching time structures as well.

It is easy to check that these structures are non-empty partial orderings satisfying all the conditions of a common BST structure. The two histories h_1^a, h_2^a in M_a and h_1^b, h_2^b in M_b are, respectively (for γ one of a or b),

$$h_1^\gamma = \{[\langle x, 1\rangle] \in M_\gamma \mid x \in \mathbb{R}\}; \quad h_2^\gamma = \{[\langle x, 2\rangle] \in M_\gamma \mid x \in \mathbb{R}\}.$$

The intersections of these two histories are, respectively, the upper bounded chains

$$l_a =_{df} h_1^a \cap h_2^a = \{[\langle x, 1\rangle] \in M_a \mid x \leqslant 0\}; \; l_b =_{df} h_1^b \cap h_2^b = \{[\langle x, 1\rangle] \in M_b \mid x < 0\}.$$

The difference is the following: while the chain l_a in M_a has a maximal element, $[\langle 0, 1\rangle]$, the chain l_b in M_b has no maximal element. That latter chain instead has two different history-relative suprema:

$$\sup_{h_i^b} l_b = [\langle 0, i\rangle], \quad i = 1, 2.$$

3.2 On chains in common BST structures

As history-relative suprema of chains will play a crucial role in the chapters to come, we provide a number of pertinent definitions and facts.

Definition 3.1 (Chains and related sets). Let $\mathscr{W}$ be a common BST structure. We define the following classes of chains and related sets in $\mathscr{W}$:

- $\mathscr{C}_e$: the set of chains ending in, but not containing, e. That is:
  $l \in \mathscr{C}_e$ iff l is an upper bounded chain and there is some $h \in$ Hist for which $l \subseteq h$ and $\sup_h l = e$, but $e \notin l$.
- $\mathscr{S}(l)$: the set of history-relative suprema for an upper bounded chain l:

$$\mathscr{S}(l) =_{df} \{s \in W \mid \exists h \in \text{Hist}\, [l \subseteq h \wedge s = \sup_h l]\}.$$

- For l a chain and $e \in W$, we define initial and final segments:

$$l^{\leqslant e} =_{df} \{e' \in l \mid e' \leqslant e\}; \quad l^{\geqslant e} =_{df} \{e' \in l \mid e' \geqslant e\};$$

- $\mathscr{P}_e$: the proper past of e:

$$\mathscr{P}_e =_{\text{df}} \{e' \in W \mid e' < e\}.$$

We establish the following Facts, using the existence of history-relative suprema and Weiner's postulate (see Section 2.5):

Fact 3.1. *Let l be an upper-bounded chain in a common BST structure $\mathscr{W}$ and $l \subseteq h'$ for some $h' \in \text{Hist}(\mathscr{W})$, and $s = \sup_{h'} l$. Then for all $h \in H_s$, we have $\sup_h l = s$.*

Proof. Assume that $s \in h$. Observe that $\{s\}$ is a (trivial) chain with $\sup_{h^*} \{s\} = s$ for any $h^* \in H_s$. We can use Weiner's postulate on the chains l and $\{s\}$. As $\sup_{h'} \{s\} = s = \sup_{h'} l$, we also have to have $\sup_h l = \sup_h \{s\} = s$. □

Alternatively, the result also follows from the definition of suprema and the downward closure of histories, so Weiner's postulate does not need to be used—see Exercise 3.7.

Fact 3.2. *Let l be an upper bounded chain and $h_1, h_2 \in H_{[l]}$, and let*

$$\sup_{h_1} l = s_1 \neq s_2 = \sup_{h_2} l.$$

Then there is no history h containing both s_1 and s_2.

Proof. Assume otherwise, and let $\{s_1, s_2\} \subseteq h$ for some $h \in \text{Hist}$. We have $l \subseteq h$, since $s_1 \in h$ and $l \leqslant s_1$. By Fact 3.1, we have both $\sup_h l = s_1$ (as $s_1 \in h$) and $\sup_h l = s_2$ (as $s_2 \in h$), contradicting our assumption that $s_1 \neq s_2$. □

Here is another useful fact about the suprema of chains. If a chain l contains its history-relative supremum s, $s \in l$, then a chain obtained by removing s from l has the same history-relative supremum, s. An analogous fact holds for infima.

Fact 3.3. *The suprema and infima of maximal chains are unaffected by the removal of the supremum or infimum:*

1. *Let l be a maximal upper bounded chain, and let $h \in \text{Hist}$ such that $l \subseteq h$. Let $s =_{\text{df}} \sup_h l$. Then for $l' =_{\text{df}} l \setminus \{s\}$, we also have $\sup_h l' = s$.*
2. *Let l be a maximal lower bounded chain, and let $e =_{\text{df}} \inf l$. Then for $l' =_{\text{df}} l \setminus \{e\}$, we also have $\inf l' = e$.*

Proof. (1) If $s \notin l$, we have $l' = l$, and there is nothing to prove. Otherwise, let $s' =_{\mathrm{df}} \sup_h l'$. Clearly, $l' \leqslant s$, so $s' \leqslant s$ (by the definition of suprema). Now assume for reductio that $s \neq s'$, i.e., $s' < s$. By the construction of l', we then have

$$(*) \qquad \forall x \in l \, [x \neq s \rightarrow x \leqslant s'].$$

By density, there is some $e \in W$ for which $s' < e < s$. By $(*)$, we have $e \notin l$. But then, again by $(*)$, we have that $l^* =_{\mathrm{df}} l \cup \{e\}$ is also a chain with $\sup_h l^* = s$, and $l^* \supsetneq l$. This contradicts the maximality of l. So, we have $s = s'$.

The proof for (2) is exactly parallel to that for (1). $\qquad \Box$

Our next Fact concerns the maximality of chains in a history $h \in \mathrm{Hist}(W)$ and in W:

Fact 3.4. *Let l be a maximal chain in $h \in \mathrm{Hist}(W)$. Then l is a maximal chain in W as well.*

Proof. For reductio, let us suppose that l is a maximal chain in $h \in \mathrm{Hist}(W)$, but not in W. Since l is not a maximal chain in W, there exists some e in $W \backslash l$ such that the set $l \cup \{e\}$ forms a chain. Observe that $(\dagger)$ $l < e$, since otherwise there would be some element $x \in l$ such that $e < x$, from which it follows by downward closure of h that l is not a maximal chain in h, which is a contradiction. Thus l is an upper-bounded chain in W, so we can apply the history-relative supremum postulate to conclude that l has an h-relative supremum $s = \sup_h l \in h$. By maximality of l, $s \in l$ and (by the same assumption) s is a maximal element of h. By Fact 2.1(9), s is then a maximal element of W, so there is no $e > l$, contradicting $(\dagger)$. Thus l is a maximal chain in W. $\qquad \Box$

The next Fact shows that the proper past of an event e consists of all the chains ending in, but not containing, e.

Fact 3.5. *For $e \in W$, we have*

$$\mathscr{P}_e = \bigcup_{l \in \mathscr{C}_e} l.$$

Proof. "$\supseteq$" Let $x \in \bigcup_{l \in \mathscr{C}_e} l$, i.e., $x \in l$ for some $l \in \mathscr{C}_e$. Since $l \in \mathscr{C}_e$, there is $h \in H_e$ such that $\sup_h l = e$ and $e \notin l$, we have $l < e$, and thus, $x < e$, i.e., $x \in \mathscr{P}_e$.

"$\subseteq$" Let $x \in \mathscr{P}_e$, i.e., $x < e$. Then $\{x, e\}$ is a chain, which by the Hausdorff maximal principle[2] can be extended to a maximal chain l ending in e. By Fact 3.1, for any history h in H_e we have that $\sup_h l = e$. By Fact 3.3, for $l' =_{\mathrm{df}} l \setminus \{e\}$ we also have $\sup_h l' = e$, whereby we have some $l' \in \mathscr{C}_e$ for which $x \in l'$. $\qquad\square$

3.3 Extending common BST: two options

As suggested by Figure 3.1, there are two options for fulfilling the common BST axioms in a simple case. Which one should we choose? We could now enter into a philosophical discussion as to the correct manner in which we are to proceed—but we refrain from attempting any a priori arguments here. One can give good reasons for each option. Thus, in favor of the existence of choice points, one can argue that a causal account of indeterministic choice requires a special final element of indecision and, therefore, a maximal element of any two branching histories. On the other hand, since choice points are distinguished as maximal elements of the overlap of histories, considerations of uniformity argue against them: as we will see, it is possible to have branching without maxima in the overlap of histories in a uniform way. The further topological considerations to be discussed in Chapter 4.4 also argue against the assumption of choice points. In our view, these controversial issues provide a good motivation for investigating common BST structures both with and without choice points.

As a matter of fact, the theory of Branching Space-Times was initially developed with the requirement of the existence of choice points, and the resulting axiomatic theory, BST$_{92}$ (named after the year of publication of the original BST paper by Belnap (1992)), has proved to be fruitful for quite a number of applications, for example, to causation (Belnap, 2005b), to probability theory (Weiner and Belnap, 2006; Müller, 2005), and to physics (Placek, 2004, 2010). On the other hand, in applications of BST to physics in which an attempt is made to link BST histories to physical space-times, topological considerations argue against the existence of choice points, so that a slightly different axiomatic theory, BST$_{\mathrm{NF}}$ (with "NF" for "new foundations"), is to be preferred.

---

[2] See note 8 on p. 30.

We will proceed by introducing both developments of the common BST framework, BST_{92} and BST_{NF}. Each will be obtained by adding a (different) axiom, a prior choice principle, called PCP_{92} or PCP_{NF}, to the list of axioms given in Definition 2.10. We will investigate the consequences of the resulting theories, focusing in particular on how they define local possibilities and how histories are to branch in each of them. Finally, we will broach the larger question of whether the differences between the two kinds of structures really matter. While pointing to the significance of topological differences on the one hand, on the other hand we exhibit translatability results that lessen the importance of topology somewhat. In a nutshell, if we have a BST_{92} structure with topologically worrisome features, it can be translated into a BST_{NF} structure, in which the worrisome feature is absent; and a translation in the other direction is also available. The remainder of this chapter is divided into sections devoted to BST_{92}, to BST_{NF}, to issues of topology, and to the translatability of the two kinds of structures.

3.4 BST_{92}

BST_{92} is the original BST theory put forward by Belnap (1992). Here we approach it somewhat differently, namely by considering it as a development of the common BST framework. The development consists of the addition of just a single axiom, the Prior Choice Principle (PCP_{92}), to the set of axioms of Definition 2.10.

3.4.1 BST_{92} in formal detail

As we just said, the theory of BST_{92} posits the axioms of a common BST structure together with the so-called prior choice principle (PCP_{92}).[3] In order to motivate the idea of prior choice, we start with the notion of *undividedness*. Let two histories h_1, h_2 share some event $e \in h_1 \cap h_2$. Then they also may or may not share a later event (provided there is one at all). In the former case, we call the histories *undivided at e*:

---

[3] Belnap (1992) does not mention Weiner's postulate, which proved critical for some applications developed later, especially regarding probability theory. See note 15.

Definition 3.2 (Undividedness). Let $h_1, h_2 \in$ Hist, and let $e \in h_1 \cap h_2$. We say that h_1 *and* h_2 *are undivided at* e ($h_1 \equiv_e h_2$) iff either there is no $e' \in W$ at all for which $e < e'$, or there is some $e' \in h_1 \cap h_2$ for which $e < e'$.

There is also a relation opposite to undividedness: in case two histories h_1, h_2 share an event e but no event later than e, that event e is a maximum in the intersection of the histories $h_1 \cap h_2$. In that case (provided that e is not maximal in W to begin with), we say that the histories *split at* e:

Definition 3.3 (Splitting at a point; choice point). Let $h_1, h_2 \in$ Hist, and let $e \in h_1 \cap h_2$. We say that h_1 *and* h_2 *split at* e, and that e *is a choice point for histories* h_1 *and* h_2 ($h_1 \perp_e h_2$) iff it is not the case that $h_1 \equiv_e h_2$. We extend the "$\perp$" notation to sets, with the universal reading, i.e., for $H \subseteq$ Hist, we write $h_1 \perp_e H$ iff for any $h \in H$, we have $h_1 \perp_e h$.

It immediately emerges that this definition agrees with the informal explanation of a choice point for h_1 and h_2 as a maximal element in the intersection $h_1 \cap h_2$.

Basically, the prior choice principle PCP$_{92}$ requires that whenever an event e belongs to one history h_1 but not to another history h_2, these two histories split at a choice point c in the past of e:

$$e \in (h_1 \setminus h_2) \rightarrow \exists c \, [c \in h_1 \cap h_2 \wedge c < e \wedge h_1 \perp_c h_2].$$

A motivation for PCP$_{92}$ is that there should be a reason (however minimally understood) in the past of each point event for its being in one history rather than another.

We have written the notion of undividedness in a way that suggests that it is an equivalence relation, which seems natural enough. It turns out, however, that in order to enforce the transitivity of the relation of undividedness, PCP$_{92}$ needs to be formulated not for points, but for lower bounded chains in the difference of two histories, as follows.[4]

---

[4] Why not generalize PCP$_{92}$ to other sets of events contained in a history? Such a set can spread through much space and lack homogeneity, and so, intuitively speaking, a PCP of so large a scope seems unreasonable. Chains, however, form a distinguished category as they naturally represent (parts of) world-lines of (spatially non-extended) objects. And it is a common practice to ask questions like "where did a given particle go up, instead of going down?" We owe our thanks to J. Luc for discussing various versions of PCP.

Definition 3.4 (BST$_{92}$ prior choice principle, PCP$_{92}$). A common BST structure $\langle W, < \rangle$ fulfills the *BST$_{92}$ prior choice principle* iff it fulfills the following condition:

Let $h_1, h_2 \in$ Hist be two histories, and let $l \subseteq (h_1 \setminus h_2)$ be a lower-bounded chain that belongs fully to history h_1 but does not intersect history h_2. Then there is a choice point $c \in h_1 \cap h_2$ such that $c < l$ and $h_1 \perp_c h_2$, i.e., c lies properly below l and is a choice point for h_1 and h_2, which is maximal in the intersection of h_1 and h_2.

That definition obviously implies the point version described above, as any singleton $\{e\}$ is a lower-bounded chain.[5] It also ensures the property of historical connection independently of the explicit requirement of Def. 2.10(7): any two different histories have a non-empty difference (see Fact 2.1(7)), so that it follows that they have to share a choice point.

We can now enter the BST$_{92}$ prior choice principle in its official form as an additional item to our list of axioms for BST$_{92}$:

Definition 3.5 (BST$_{92}$ structure). A *BST$_{92}$ structure* is a common BST structure $\langle W, < \rangle$ (Definition 2.10) that also fulfills the BST$_{92}$ prior choice principle, PCP$_{92}$ (Definition 3.4).

We can prove that in BST$_{92}$ structures, there is a choice point (i.e., a maximal element in their overlap) for any two histories.

Fact 3.6. *Let $h_1, h_2 \in$ Hist be histories of a BST$_{92}$ structure, $h_1 \neq h_2$. Then there is a maximal element in $h_1 \cap h_2$.*

Proof. Left as Exercise 3.1. Note that by Def. 3.3, any maximal element in $h_1 \cap h_2$ is a choice point for h_1 and h_2. □

3.4.2 Local possibilities

One important application of PCP$_{92}$ is the construction of the concept of possibilities open at an event. For any event e, the relation $\equiv_e$ among the set H_e of all the histories containing e is obviously symmetrical (by the form of the definition) and reflexive (note that the definition takes care of e being a maximal element in W as well). It is transitive as well, as proved below.

---

[5] See Belnap (1992) for an argument that the chain version properly strengthens the point version of PCP$_{92}$.

Fact 3.7. *Suppose that a poset $\langle W, < \rangle$ satisfies density, existence of infima and PCP$_{92}$. Then for every e in W, the relation $\equiv_e$ is transitive on the set H_e.*

Proof. Fix some $e \in W$, and let $h_1, h_2, h_3 \in H_e$. Suppose toward a contradiction that (†) $h_1 \equiv_e h_2$ and $h_2 \equiv_e h_3$, but that $h_1 \not\equiv_e h_3$. By the definition of undividedness $h_1 \not\equiv_e h_3$ implies that e is not maximal in W, and hence e is not maximal in any history containing e (by Fact 2.1(9)). Consider the subset of $h_1 \cap h_2$ that is properly above e. Since $h_1 \equiv_e h_2$, and e is not maximal in h_1 and h_2, this set is nonempty. So by the Hausdorff maximal principle, there is a maximal chain l of points above e in $h_1 \cap h_2$ (i.e. a maximal chain lower-bounded by e). By Postulate 2.4, $\inf(l)$ exists, and $e \leqslant \inf(l)$. Moreover, $\inf(l) = e$ since by maximality of l there are no points properly between e and l, as this would contradict density. Further $h_1 \not\equiv_e h_3$ implies that no point later than e belongs to both histories, so $l \subseteq h_1$ but $l \cap h_3 = \emptyset$. By the construction, $l \subseteq h_2$, so $l \subseteq h_2 \setminus h_3$. Thus by PCP$_{92}$ (Def. 3.4), there must be a choice point e_1 for h_2 and h_3, (‡) $h_2 \perp_{e_1} h_3$, strictly below l, $e_1 < l$. Since $e = \inf(l)$, by the definition of infima, $e_1 \leqslant e$. But if $e_1 < e$, we contradict (‡) as $e \in h_2 \cap h_3$. And, if $e_1 = e$, then $h_2 \perp_e h_3$, contradicting (†). $\square$

Transitivity of $\equiv_e$ allows one to establish that a choice point posited by PCP splits not just two histories, but a history and a set of histories:

Fact 3.8. *Let $c < l$, $l \subseteq h_1 \setminus h_2$, and $h_1 \perp_c h_2$. Then $H_{[l]} \perp_c h_2$, where $H_{[l]} = \{h \in \text{Hist} \mid l \subseteq h\}$ is the set of histories containing all of l.*

Proof. Suppose toward a contradiction that there is some $h \in H_l$ such that $h \not\perp_c h_2$. The latter implies, since $c \in h$ (for $c < l$) and $c \in h_2$, that $h \equiv_c h_2$. Since $l \in h \cap h_1$ and $c < l$, $h \equiv_c h_1$. By the transitivity of $\equiv_c$ we get $h_1 \equiv_c h_2$, which contradicts our premise. $\square$

Given transitivity, $\equiv_e$ is an equivalence relation on H_e. We use the notation Π_e to indicate the partition of histories from H_e into equivalence classes according to $\equiv_e$, i.e.,

for $H \subseteq H_e$ with $H \neq \emptyset$, we have $H \in \Pi_e \leftrightarrow H$ is maximal with respect to the property that $\forall h_1, h_2 \in H [h_1 \equiv_e h_2]$.

It may be that in fact *all* histories from H_e are undivided at e, i.e., e is not maximal in the intersection of any two histories from H_e. In that case, we have $\Pi_e = \{H_e\}$. On the other extreme, it may be that any two histories in H_e split at e, so we have $\{h\} \in \Pi_e$ for every $h \in H_e$.

The relation of undividedness at an event is significant, as it allows us to define an entirely objective concept of 'elementary possibility at e'.

Definition 3.6 (Elementary possibility at e). Let Π_e be the partition of H_e induced by $\equiv_e$. By an *elementary possibility at e* (a *basic outcome of e*) we mean a member of Π_e.

The concept of elementary possibilities deserves a few comments.

1. An elementary possibility can be represented as a set of histories. This idea is copied from 'possible worlds' theories.
2. Which English phrase shall we use for Π_e, apart from 'the elementary possibilities (open) at e'? One might think of Π_e as representing what might happen at e, or the way things might go immediately after e, or as the possible outcomes (results) of e.
3. The range of elementary possibilities open at e is therefore not an extra primitive. It is definable from the very structure of *Our World* $\langle W, < \rangle$.
4. An elementary possibility at e is always a set of histories, all of which contain e. It may be typically or even always true in *Our World* that the unit set $\{h\}$ of a history from H_e is not an elementary possibility at any e. Thus, the competing definition of an elementary possibility as such a unit set of a history would be too fine-grained (though perhaps not too fine-grained for every purpose). There are also possibilities open at an event that are not elementary. At least any union of a set of elementary possibilities at e will need to be counted as itself a possibility at e. Some notions of less immediate possibilities, or outcomes, will be discussed in Chapter 4.1.

The uniquely determined partition Π_e is a proper locus for a ground-level theory of objective transition possibilities (or outcomes) in the single case. The significance is this: the finest partition is delivered by the causal structure of *Our World*, not by human interests, language, concepts, universals, other possible worlds, or evolutionary entrenchment. The possibility in question is conditional in form (the condition being that the point event occurs), but more than that, it has a concrete foothold in *Our World*.

With the concept of local possibilities, we are finally in a position to define a local, modal, and relativity friendly notion of determinism and indeterminism.

Definition 3.7 (Determinism and indeterminism). A point event, e, is indeterministic (is a choice point) if Π_e has more than one member. Otherwise, it is deterministic.

As a rhetorical variant, we may say that *Our World* is indeterministic at e. Note that on this account it makes perfectly good sense to locate indeterminism not metaphorically in a theory, but literally in our world, something very much in contrast with the dominant views in the philosophy of science.[6] The concept is also local: it makes sense to say that *Our World* was indeterministic in Boston yesterday, but might not be so in Austin tomorrow.

3.4.3 The pattern of branching of BST$_{92}$

Given how frugal the axioms of BST$_{92}$ are, there are many different kinds of structures of BST$_{92}$. Since our aim is that the theory can have applications in physics, it will be useful to single out BST$_{92}$ structures in which histories are isomorphic to Minkowski space-time, or to some solutions of the field equations of General Relativity. Details of these will be discussed in Chapter 9. At this stage, we will explore the pattern of branching that the BST$_{92}$ axioms impose: If two histories split at a single choice point c, at which regions they are identified, and what do the regions of difference look like? Understandably, the histories share the past light-cone of c, and have separate future light-cones of c, but are the events space-like separated from c (forming the so-called 'wings', or the 'elsewhere' of c) in the shared region, or not? As we shall see, PCP$_{92}$ provides a principled answer to this query.

To explore the matter further, it would be helpful to introduce some of the steps in our construction of Minkowskian Branching Structures, to be

---

[6] The standard approach in the philosophy of science (e.g., Earman, 2006; Butterfield, 2005) is to take determinism and indeterminism to be properties of theories, not of our world. According to that approach, roughly, a theory is called deterministic iff all of its models that have isomorphic initial segments also have isomorphic final segments. The individual models are usually taken to be separate possible worlds, so there is no notion of branching histories involved in that approach to indeterminism.

provided in Chapter 9.1. Here we put down a preliminary target definition (for the precise formulation, see Definition 9.5):[7]

Definition 3.8 (Target notion of MBS). A Minkowskian Branching Structure (MBS) is a BST structure in which each history is order-isomorphic to Minkowski space-time, with the pre-causal ordering generalizing the Minkowskian ordering $<_M$ of Definition 2.1.

To be more specific, let us set out the task of producing an MBS that models a measurement of spin with two possible outcomes (idealized as histories), measured spin up or measured spin down. How does this measurement affect the causal contemporaries of the measurement? Do they belong to the intersection of the two histories, or just to one or the other? Are they ontologically indefinite or ontologically definite (if that language helps), relatively or absolutely?

Given the problem, we can consider an MBS that satisfies the following stipulations:

1. There are exactly two histories.
2. Each history is order-isomorphic to Minkowski space-time.
3. There is precisely one choice point c.

The matter is resolved by the Prior Choice Principle of BST_{92}, as the following fact testifies:

Fact 3.9. *Let $\langle W, < \rangle$ be a BST_{92} structure that contains two histories, h_1 and h_2, and exactly one choice point, c, which is maximal in $h_1 \cap h_2$. It follows from the BST_{92} Prior Choice Principle, PCP_{92}, that the 'wings' of c, i.e. $X = \{e \in W \mid e\,SLR\,c\}$, must be in the intersection $h_1 \cap h_2$ of the two histories.*

Proof. By the definition of X, no element of X has a choice point in its past. Therefore, if an element of X failed to lie in both h_1 and h_2, PCP_{92} would be violated.   □

Thus, given PCP, the true picture of two Minkowski histories with exactly one choice point must be as it is displayed in Figure 3.2. The intersection of

[7] In Chapter 9.1 we explore MBSs that are BST_{NF} structures. A series of papers listed in Footnote 4 of Chapter 9 (p. 307) develops MBSs that are BST_{92} structures.

the two histories is shaded, and the upper borders belong 'on the light side'
in the respective differences.

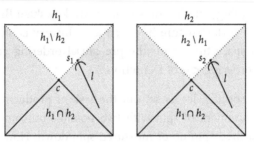

Figure 3.2 'Wings' ('elsewhere') belong to the intersection.

This formal result deserves additional comments:

(i) Observe that the difference made by the choice at c pertains only
to the future of possibilities of c. It does not pertain to the causal
contemporaries of c.

(ii) One might imagine that whenever there is a tiny indeterministic
situation such as spin up/spin down, the entire causally unrelated
universe simultaneously splits in twain. BST_{92} gives a sharp explanation of how and why this picture is generally wrong. It also offers a
competing rigorous and positive theory of what is right: splitting in
Our World fundamentally occurs at point events. A single splitting
affects only its causal future, not everything above a simultaneity
slice.

(iii) Thus, a single, local splitting (a single chancy event, idealized to be
point-like) does not give rise to a simultaneity slice that divides h_1
into $h_1 \cap h_2$ and $h_1 \setminus h_2$. BST is thus in conflict with any interpretation
of special relativity that assumes that a wave-function collapse occurring along a simultaneity surface can be effected by a single chancy
event such as a measurement.[8]

A feature specific to the BST_{92} pattern of branching is the topological
difference between what might be called "indeterminism without choice"

---

[8] In Chapter 10 we will discuss how the *additional* resources provided by BST—most prominently,
modal correlations, to be discussed in Chapter 5—can be used to define different kinds of splitting.
These, however, need coordinated choice points, and for each of them, the pattern of branching is as
described, affecting only its causal future.

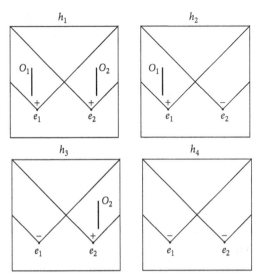

Figure 3.3 Two binary *SLR* choice points giving rise to four histories.

and "indeterminism with choice". To explain, consider again Figure 3.2 and a maximal chain, l, that traverses from the histories' overlap, $h_1 \cap h_2$, toward the histories's difference $h_1 \setminus h_2$, avoiding c. Does it have a last element in the intersection of $h_1 \cap h_2$? Note that the borders of a pair of histories that overlap do not belong to the overlap (apart from c), since they are above c. The chain l therefore has two minimal upper bounds, say, s_1 on the upper light cone of h_1 and s_2 on the upper light cone of h_2. If you are 'traveling along' this track, the situation as the track draws to a close is indeterministic: it is not determined whether you will wind up at s_1 or s_2. Still, there is no choice: the matter is entirely in the hands of your causal contemporary, c. Things are different, however, for a chain converging to a choice point, c, in the histories' overlap: this chain has a unique least upper bound (supremum), namely c. The difference between the two cases seems to be this. The only reason that l underdetermines whether s_1 or s_2 will occur is that it does not exhaust the entire past of either of these points: given the set of all proper predecessors of s_1, which includes a choice point for histories containing s_1 vs. s_2, the outcome, s_1, is uniquely determined (and analogously for s_2). In contrast, the entire past culminating in c does not suffice to decide what happens next.

Figure 3.3 illustrates the combination of choice points in the simplest case, in which two *SLR* choice points e_1 and e_2 have two outcomes each (denoted

"+" and "−"). The choices at e_1 and e_2 are uncorrelated, giving rise to a total of four histories. As we will show in Chapter 5, BST also allows for structures in which choice points are coordinated; compare Figure 5.1 on page 107.

3.4.4 Transitions

The notion of a transition is a powerful tool for discussing indeterminism. Belnap (1999) adopts the notion from von Wright (1963), who provides the basic idea of a transition as "first this, and then that", adding formal rigor. Generally, a transition is a pair $\langle I, O \rangle$, written $I \rightarrowtail O$, in which I is appropriately prior to O, and O is, in some appropriate sense, an outcome of I. There are several notions of transitions, which we discuss in Chapter 4.1. For our immediate purposes we focus here on the simplest notion of a transition, a *basic transition*, which in BST_{92} is from a point event e to one of the immediate possibilities open at e (i.e., from e to a member of the partition Π_e of the set H_e of histories containing e),

$$\tau = e \rightarrowtail H, \quad e \in W, H \in \Pi_e.$$

Observe that the kind of transitions that we consider are *modal* transitions. They are not merely *state* transitions. If at a certain moment, for example, there is a (real) possibility of motion, then 'remaining at rest' would cut off certain possibilities and thus be an (indeterministic) transition event of a kind that is the object of our investigation, even though there is no 'change of state'.

Basic transitions are divided up into those that witness local indeterminism, and those at which, so to speak, nothing happens. The formal distinction is provided by whether or not there are multiple immediate future possibilities open at e, or just one. Thus, at an indeterministic event e (at a choice point), the partition Π_e has more than one member (histories split at e; there are $h_1, h_2 \in H_e$ for which $h_1 \perp_e h_2$), whereas at a deterministic event e, there is only one immediate possibility for the future, whence for all $h_1, h_2 \in H_e$, we have $h_1 \equiv_e h_2$, and $\Pi_e = \{H_e\}$.

Definition 3.9 (Deterministic and indeterministic basic transitions). A *basic transition* is a pair $\langle e, H \rangle$, written $e \rightarrowtail H$, with $e \in W$ and $H \in \Pi_e$. For $h \in H_e$, we write $\Pi_e \langle h \rangle$ for the member of Π_e that contains h, so that the basic transition $e \rightarrowtail \Pi_e \langle h \rangle$ is from e to that (unique) basic outcome of

e that contains h. A basic transition is *indeterministic* iff Π_e has more than one member. On the other hand, if $\Pi_e = \{H_e\}$, then the transition $e \rightarrowtail H_e$ is called *deterministic* or *trivial*.

We denote the set of basic indeterministic transitions of a BST_{92} structure $\langle W, < \rangle$ by $TR(W)$, and the set of all basic transitions by $TR_{full}(W)$.

The set of basic transitions, whether deterministic or indeterministic, admits a natural ordering.

Definition 3.10 (Transition ordering). For $\tau_1 = e_1 \rightarrowtail H_1$, $\tau_2 = e_2 \rightarrowtail H_2$, we say that τ_1 *precedes* τ_2, written $\tau_1 \prec \tau_2$, iff ($e_1 < e_2$ and $H_2 \subseteq H_1$). The companion non-strict partial ordering is defined via $\tau_1 \preccurlyeq \tau_2 \leftrightarrow_{df} (\tau_1 \prec \tau_2 \vee \tau_1 = \tau_2)$.

We prove next that that $\prec$ is a partial ordering:

Fact 3.10. *Let $\langle W', <' \rangle =_{df} \langle TR_{full}(W), \prec \rangle$ be the set of all basic transitions of a BST_{92} structure $\langle W, < \rangle$ together with the transition ordering $\prec$. Then (1) $TR_{full}(W)$ is non-empty and (2) $\prec$ is a strict partial ordering on $TR_{full}(W)$.*

Proof. (1) Since W is non-empty, there is some $e \in W$ and hence $H_e \neq \emptyset$ (see Fact 2.1(4)), so there is a non-empty $H \in \Pi_e$, and hence there exists a transition $e \rightarrowtail H \in W'$.

(2) Since $<$ is irreflexive, $\prec$ is irreflexive as well. For transitivity, let $(e_1 \rightarrowtail H_1) \prec (e_2 \rightarrowtail H_2)$ and $(e_2 \rightarrowtail H_2) \prec (e_3 \rightarrowtail H_3)$. By transitivity of $<$ we have $e_1 < e_3$. Also, from $H_2 \subseteq H_1$ and $H_3 \subseteq H_2$ we have $H_3 \subseteq H_1$ by transitivity of $\subseteq$. Together this establishes $(e_1 \rightarrowtail H_1) \prec (e_3 \rightarrowtail H_3)$. $\square$

We can also prove that on the assumption that $\langle W, < \rangle$ has no maxima nor minima, $TR_{full}(W)$ also has no maxima nor minima; see Exercise 3.2.

The following facts about alternatives to the definition of the transition ordering (Def. 3.10) will be helpful later on.

Fact 3.11. *Let $\tau_1 = e_1 \rightarrowtail H_1$, $\tau_2 = e_2 \rightarrowtail H_2$ be transitions in a BST_{92} structure $\langle W, < \rangle$. (1) Generally, $\tau_1 \prec \tau_2$ iff ($e_1 < e_2$ and $H_{e_2} \subseteq H_1$) iff ($e_1 < e_2$ and for every $h \in H_{e_2}$ it is the case that $H_1 = \Pi_{e_1}\langle h \rangle$). (2) If τ_1 is deterministic, then $\tau_1 \prec \tau_2$ iff $e_1 < e_2$. (3) For the non-strict companion order, we have $\tau_1 \preccurlyeq \tau_2$ iff ($e_1 \leqslant e_2$ and $H_2 \subseteq H_1$).*

Proof. (1) We prove the first "iff", from which the second follows immediately (left as Exercise 3.3). Let $e_1 < e_2$. We have to show that $H_2 \subseteq H_1$ iff $H_{e_2} \subseteq H_1$.

As $H_2 \subseteq H_{e_2}$, the "$\Leftarrow$" direction is trivial. For "$\Rightarrow$", assume $H_2 \subseteq H_1$, and pick an arbitrary $h \in H_2$, so $h \in H_1$. Pick next some $h' \in H_{e_2}$. We have $e_2 \in h \cap h'$, which establishes $h \equiv_{e_1} h'$. It follows that $h' \in H_1 = \Pi_{e_1}\langle h \rangle$.

(2) The "$\Rightarrow$" direction is trivial. For "$\Leftarrow$", let $e_1 < e_2$, and assume that τ_1 is deterministic, so that $H_1 = H_{e_1}$. We have to show that $H_2 \subseteq H_1$. By downward closure of histories (Fact 2.1(5)), we have $H_{e_2} \subseteq H_{e_1}$, and $H_2 \subseteq H_{e_2}$ by definition, so that $H_2 \subseteq H_{e_2} \subseteq H_{e_1} = H_1$. The claim follows by transitivity of $\subseteq$.

(3) "$\Rightarrow$": Assume $\tau_1 \preccurlyeq \tau_2$ (i.e., either $\tau_1 \prec \tau_2$ or $\tau_1 = \tau_2$). In the first case, the claim follows from the definition of $\prec$, in the second case the claim is obvious as then, $e_1 = e_2$ and $H_1 = H_2$.

"$\Leftarrow$": Let $H_2 \subseteq H_1$, and let $e_1 \leqslant e_2$. Again there are two cases. If $e_1 = e_2$, then, as $H_1, H_2 \in \Pi_{e_1}$ and Π_{e_1} is a partition, $H_2 \subseteq H_1$ implies $H_2 = H_1$, whence $\tau_1 = \tau_2$, establishing the claim. The remaining case, $e_1 < e_2$ and $H_2 \subseteq H_1$, satisfies the definition of $\prec$ exactly as in Def. 3.10. $\qquad\square$

Transitions figure prominently in our account of singular causation, propensities, and many other applications of BST. Importantly, it can be shown that the set of transition $\mathrm{TR}_{\mathrm{full}}(W)$ of a BST_{92} structure $\langle W, < \rangle$ together with the ordering $\prec$ of Definition 3.10 is a common BST structure. Under the further assumption that $\langle W, < \rangle$ has no minimal elements, $\langle \mathrm{TR}_{\mathrm{full}}(W), \prec \rangle$ is a $\mathrm{BST}_{\mathrm{NF}}$ structure—a notion that we now go on to define.

3.5 Introducing BST$_{\mathrm{NF}}$

Similarly to our construction of BST_{92} structures, we begin with a common BST structure and then put forward a new prior choice principle, $\mathrm{PCP}_{\mathrm{NF}}$. With this principle in hand, we will define a $\mathrm{BST}_{\mathrm{NF}}$ structure as a common BST structure that satisfies $\mathrm{PCP}_{\mathrm{NF}}$. Having achieved this task in Chapter 3.5.1, we explore some of the basic features of the resulting theory in Chapters 3.5.2–3.5.3.

3.5.1 The new Prior Choice Principle and BST$_{\mathrm{NF}}$ structures defined

Working with a common BST structure $\langle W, < \rangle$, we define a few concepts needed to formulate $\mathrm{PCP}_{\mathrm{NF}}$. The underlying idea of $\mathrm{BST}_{\mathrm{NF}}$ is that there should be no maximal elements in the overlap of histories. Thus, for any maximal chain l traversing from the histories' overlap $h_1 \cap h_2$ to one of its differences, say $h_1 \setminus h_2$, its segment in the overlap, $l' =_{df} l \cap h_1 \cap h_2$, has at least two history-relative suprema, $s_1 = \sup_{h_1}(l')$ and $s_2 = \sup_{h_2}(l')$. Since typically such pairs of distinct suprema abound for any two histories h_1, h_2, we should not say that the two histories split at all such pairs. We had better pick distinguished pairs, analogous to the topologically distinguished choice points of BST_{92}. A relevant observation is that a chain may be such that its pair of distinct suprema s_1, s_2 cannot be avoided, because *any* chain approaching *any one* of s_1, s_2 has (at least) two history-relative suprema, s_1 and s_2. In contrast, other chains with distinct history-relative suprema s_1, s_2 may be such that these distinct suprema can be avoided, that is, there is a different chain approaching one of s_1, s_2 that fails to have them both as history-relative suprema. In fact, given PCP_{92}, there will always be a chain that has only one history-relative supremum (see Exercise 3.4). But in common BST structures without PCP_{92}, the former case of unavoidable history-relative suprema is not excluded. Accordingly, we single out the first category of sets of history-relative suprema as choice sets, at which histories split. These observations translate into the following definitions:

Definition 3.11 (Choice set). For $e \in W$, we define the *choice set based on* e, written $\ddot{e}$, to be the intersection of the sets of suprema of all chains ending in e (Def. 3.1).[9] In case e is a minimal element in W, we have $\mathscr{C}_e = \emptyset$, so we make sure that e belongs to its own choice set in this case as well.

$$\ddot{e} =_{df} \begin{cases} \{e\}, & \text{if } \mathscr{C}_e = \emptyset, \\ \bigcap_{l \in \mathscr{C}_e} \mathscr{S}(l), & \text{otherwise.} \end{cases}$$

Fact 3.12. *For any $e \in W$, $e \in \ddot{e}$.*

Proof. If $\mathscr{C}_e = \emptyset$, we have $e \in \ddot{e}$ by definition. Otherwise, for any $l \in \mathscr{C}_e$, $e \in \mathscr{S}(l)$ by definition, and hence $e \in \ddot{e}$. $\square$

We also call the choice set $\ddot{e}$ the set of *local point-wise alternatives* for e. Note that e then counts as an alternative to itself. The related notions of alternative histories and history-wise alternatives are defined via the point-wise alternatives:

---

[9] Our notation with the double dot over e is meant to be suggestive of a number of different history-relative suprema on top of a chain. Think of Figure 3.1(b) rotated counterclockwise by 90 degrees.

Definition 3.12 (Alternative histories and local history-wise alternatives). We define the set of *alternative histories at* $ë$, $H_ë$, and the set of *local history-wise alternatives for* e, $\Pi_ë$, to be

$$H_ë = \{h \in \mathrm{Hist} \mid h \cap ë \neq \emptyset\}; \qquad \Pi_ë =_{\mathrm{df}} \{H_s \mid s \in ë\}.$$

In order to spell out $\mathrm{PCP_{NF}}$, we define two new relations between histories, splitting at a choice set and being undivided at a choice set, written $h_1 \perp_ë h_2$ and $h_1 \equiv_ë h_2$, respectively, in analogy to the respective $\mathrm{BST_{92}}$ notions.

Definition 3.13. Let h_1, h_2 be histories in $\mathrm{Hist}(W)$, and let $e \in W$. We require as a presupposition for $h_1 \equiv_ë h_2$ and $h_1 \perp_ë h_2$ that $h_1, h_2 \in H_ë$ (i.e., $h_1 \cap ë \neq \emptyset$ and $h_2 \cap ë \neq \emptyset$). Then the relations are defined as follows:

$$h_1 \equiv_ë h_2 \quad \Leftrightarrow_{\mathrm{df}} \quad h_1 \cap ë = h_2 \cap ë;$$
$$h_1 \perp_ë h_2 \quad \Leftrightarrow_{\mathrm{df}} \quad h_1 \cap ë \neq h_2 \cap ë.$$

If $h_1 \perp_ë h_2$, we say that the choice set $ë$ is *a choice set for histories h_1 and h_2*.

With the required notions at hand, we put forward a new prior choice principle, $\mathrm{PCP_{NF}}$:

Definition 3.14 ($\mathrm{PCP_{NF}}$). Let $h_1, h_2 \in \mathrm{Hist}(W)$, and let l be a lower bounded chain for which $l \subseteq h_1$ but $l \cap h_2 = \emptyset$. Then there is some $e \in W$ for which $e \leqslant l$ and for which the set $ë$ of local alternatives to e satisfies $h_1 \perp_ë h_2$.

Note the weak relation $e \leqslant l$ in the formulation of $\mathrm{PCP_{NF}}$, in contradistinction to the strict relation in the formulation of $\mathrm{PCP_{92}}$ in Def. 3.4. For example, if l has just one element c (i.e., $l = \{c\}$) such that $c \in h_1 \setminus h_2$ and c is an element of a non-trivial choice set $ë \neq \{c\}$, then the choice set for $l = \{c\}$ is just $ë$ itself, and $h_1 \perp_ë h_2$. In such a case we only have the weak ordering relation, $c \leqslant l$.

Having proposed a new prior choice principle, we can now give a full definition of the "new foundations" for BST, $\mathrm{BST_{NF}}$:

Definition 3.15 ($\mathrm{BST_{NF}}$ structure). A strict partial ordering $\langle W, < \rangle$ is a *structure of $\mathrm{BST_{NF}}$* iff it is a common BST structure (Def. 2.10) for which $\mathrm{PCP_{NF}}$ (Def. 3.14) holds.

It can be shown that $\mathrm{PCP_{NF}}$ implies historical connection; see Exercise 3.5.

3.5.2 Local possibilities and the pattern of branching in BST_{NF}

In this section we ask how local possibilities are represented in BST_{NF} and how histories branch according to this theory. To this end, we first investigate some features of the relations of splitting $\perp_{\ddot{e}}$ and being undivided $\equiv_{\ddot{e}}$ at a choice set $\ddot{e}$. A handy fact that we will use below says that a history and a choice set intersect at one point event at most:

Fact 3.13. *(1) For any $h \in \text{Hist}(W)$ and for any $e \in W$, we either have $h \cap \ddot{e} = \emptyset$, or $h \cap \ddot{e} = \{e'\}$ for some $e' \in \ddot{e}$, i.e., a choice set has at most one element in common with any history. (2) The set of sets of histories $\Pi_{\ddot{e}}$ partitions $H_{\ddot{e}}$.*

Proof. (1) We have two cases here. Case 1: If e is a minimal element of W, then by Def. 3.11 we have $\ddot{e} = \{e\}$, and the result follows immediately.

Case 2: If e is not a minimal element of W, there exists some element $x < e$. We can then invoke the Hausdorff maximal principle to extend the chain $\{x, e\}$ to a maximal chain l and delete the segment of l that is above e. The remaining chain $l' =_{\text{df}} l^{<e}$ is a maximal chain upper-bounded by e, i.e., $l' \in \mathscr{C}_e$. If the set $h \cap \ddot{e} \neq \emptyset$, then $l' \subseteq h$, and thus if $h \cap \ddot{e}$ were to contain more than one element, this would contradict Fact 3.2.

(2) Exhaustiveness is immediate: by definition, $\cup \Pi_{\ddot{e}} = H_{\ddot{e}}$. To prove disjointness of the elements of $\Pi_{\ddot{e}}$, let $H_1, H_2 \in \Pi_{\ddot{e}}$ such that $H_1 \neq H_2$. Then $H_1 = H_{s_1}$ and $H_2 = H_{s_2}$ for two distinct members $s_1, s_2 \in \ddot{e}$. Let $h \in H_1$; by (1), we then have $h \notin H_2$. $\qquad\qquad\square$

The first part of this fact implies the following interrelation of the two relations, $\equiv_{\ddot{e}}$ and $\perp_{\ddot{e}}$:

Fact 3.14. *Let $e \in W$ and let $h_1, h_2 \in H_{\ddot{e}}$. Then $\equiv_{\ddot{e}}$ and $\perp_{\ddot{e}}$ are mutually exclusive and jointly exhaustive: we have $h_1 \equiv_{\ddot{e}} h_2$ iff not $h_1 \perp_{\ddot{e}} h_2$.*

Proof. Given the assumptions, we have $h_1 \cap \ddot{e} = \{s_1\}$ and $h_2 \cap \ddot{e} = \{s_2\}$ for some $s_1, s_2 \in \ddot{e}$. Now by our definitions, $h_1 \equiv_{\ddot{e}} h_2$ iff $s_1 = s_2$, and $h_1 \perp_{\ddot{e}} h_2$ iff $s_1 \neq s_2$. These are mutually exclusive and jointly exhaustive alternatives. $\qquad\square$

Finally, we can prove a much desired consequence of the above definitions:

Fact 3.15. *The relation $\equiv_{\ddot{e}}$ is an equivalence relation on the set of alternative histories at $\ddot{e}$, $H_{\ddot{e}}$.*

Proof. Let $h_1, h_2, h_3 \in H_{\ddot{e}}$. We have to establish reflexivity, symmetry and transitivity. Reflexivity and symmetry are trivial. For transitivity, assume $h_1 \equiv_{\ddot{e}} h_2$ and $h_2 \equiv_{\ddot{e}} h_3$, so $h_1 \cap \ddot{e} = h_2 \cap \ddot{e}$ and $h_2 \cap \ddot{e} = h_3 \cap \ddot{e}$. By transitivity of identity $h_1 \cap \ddot{e} = h_3 \cap \ddot{e}$, which implies $h_1 \equiv_{\ddot{e}} h_3$. □

The immediate corollary of this Fact is that $H_{\ddot{e}}/ \equiv_{\ddot{e}}$ is a partition of $H_{\ddot{e}}$. We thus define local possibilities as follows:

Definition 3.16 (Elementary possibility at $\ddot{e}$). The set of elementary possibilities open at $\ddot{e}$ (or of possible outcomes of $\ddot{e}$) is the partition $H_{\ddot{e}}/ \equiv_{\ddot{e}}$.

It is then immediately discernible that $H_{\ddot{e}}/ \equiv_{\ddot{e}}$ is identical to the partition $\Pi_{\ddot{e}}$ of Def. 3.12:

Fact 3.16. $\Pi_{\ddot{e}} = H_{\ddot{e}}/ \equiv_{\ddot{e}}$.

Proof. The claim is established by this chain of equivalences: $h_1, h_2 \in H \in (H_{\ddot{e}}/ \equiv_{\ddot{e}}) \Leftrightarrow h_1 \equiv_{\ddot{e}} h_2 \Leftrightarrow \exists e \in \ddot{e} [e \in h_1 \cap h_2] \Leftrightarrow h_1, h_2 \in H_e \in \Pi_{\ddot{e}}$. □

In what follows, we will use the succinct notation $\Pi_{\ddot{e}}$ for the possibilities open at $\ddot{e}$. We turn next to facts shedding some light on the pattern of branching in BST$_{\mathrm{NF}}$ structures. Such structures, being common BST structures, satisfy historical connection just as BST$_{92}$ structures do. The new PCP$_{\mathrm{NF}}$, however, implies that the branching of histories in BST$_{\mathrm{NF}}$ looks different from the branching in terms of choice points in BST$_{92}$: there cannot be any maximal elements in the intersection of histories in a BST$_{\mathrm{NF}}$ structure.

Fact 3.17. *Let h_1, h_2 be two histories in a BST$_{\mathrm{NF}}$ structure $\langle W, < \rangle$, $h_1 \neq h_2$. Then $h_1 \cap h_2$, which is non-empty, contains no maximal elements. Accordingly, there are no choice points in a BST$_{\mathrm{NF}}$ structure.*

Proof. By historical connection (see Exercise 3.5), $h_1 \cap h_2 \neq \emptyset$. Assume for reductio that there is a point e that is maximal in $h_1 \cap h_2$. By assumption h_1 and h_2 are distinct histories. It follows from the contrapositive of Fact 2.1(10) that the element e is not maximal in W. By Fact 2.1(9) this means that in particular, e is not maximal in h_1. Therefore there is some element x in h_1 that lies strictly above e. By the Hausdorff maximal principle, there is a maximal lower bounded chain $l \subseteq \{x \in h_1 \mid e \leqslant x\}$. As $e \in l$, we have $\inf l = e$. Since e is not a maximal element of h_1, it follows that $l' =_{\mathrm{df}} l \setminus \{e\} \neq \emptyset$. By Fact 3.3, e is also the infimum of l'. As we have $l' \subseteq h_1 \setminus h_2$, by PCP$_{\mathrm{NF}}$ there is $\ddot{c}$ with some $c_1, c_2 \in \ddot{c}$ such that $c_1 \in h_1$, $c_2 \in h_2$, $h_1 \cap \ddot{c} = c_1 \neq c_2 = h_2 \cap \ddot{c}$, and $c_1 \leqslant l'$. Then c_1 and c_2 cannot share a history (by Fact 3.13(1)).

However, since $e = \inf l'$, by the definition of inifima, $c_1 \leqslant e$, and as histories are downward closed, $c_1 \in h_2$. But we have $c_2 \in h_2$ as well, which again contradicts Fact 3.13(1). □

We also establish a fact about minimal points in BST_{NF} structures.

Fact 3.18 (Minimal elements in BST_{NF}). *A minimal point in a BST_{NF} structure belongs to all histories of that structure.*

Proof. For reductio, let e be a minimal point in a BST_{NF} structure $\langle W, < \rangle$ such that $e \in h_1 \setminus h_2$ for some histories $h_1, h_2 \in \text{Hist}(W)$. By PCP_{NF} there is a choice set $\ddot{c}$ for which $h_1 \perp_{\ddot{c}} h_2$, and so there are $c_1, c_2 \in \ddot{c}$ for which $c_i \in \ddot{c} \cap h_i$ $(i = 1, 2)$, $c_1 \leqslant e$, and $c_1 \neq c_2$. But two distinct $c_1, c_2 \in \ddot{c}$ can exist only if they are history-relative suprema of some chain $l \in \mathscr{C}_e$ (i.e., one that approaches them from below). Thus, $l < c_1 \leqslant e$, so e cannot be minimal, contradicting our assumption. □

As a consequence, in a BST_{NF} structure no two histories can branch (in the sense of Definition 3.13) at a set of minimal points. This illustrates a difference between BST_{92} structures and BST_{NF} structures: the former permit branching at minimal elements, but the latter do not.[10]

One might wonder if the BST_{NF} pattern of branching, which is different from one present in BST_{92}, affects the verdict about the set of events space-like related to a choice that was discussed above in terms of the problem of the wings. In the present context, the focus is on a single choice set $\ddot{c}$ such that two histories, h_1 and h_2 split at $\ddot{c}$, $h_1 \perp_{\ddot{c}} h_2$, and the question is whether the set of point events space-like related to an element of $\ddot{c}$, $c_i = \ddot{c} \cap h_i$, is in the histories' overlap, $h_1 \cap h_2$, or not. We invite the reader to show that in BST_{NF}, the wings are also in the overlap—see Exercise 3.6.

3.5.3 Facts about choice sets

The focus of this section are structures of BST_{NF}. We prove a few facts related to sets of local point-wise alternatives and sets of local history-wise alternatives, which will also justify our terminology. Our main result is Theorem 3.1, which states that choice sets fully capture the notion of a local alternative in BST_{NF}.

---

[10] As we will see, this difference implies a small limitation for the translatability of one kind of structure into the other kind; see Theorem 3.3.

Fact 3.19. *Let there be h_1, h_2 and $e \in W$ such that $h_1 \perp_{\ddot{e}} h_2$. Then there is no $c < e$ for which $h_1 \perp_{\ddot{c}} h_2$.*

Proof. Suppose that $e_i \in \ddot{e}$ are distinct, where $h_i \cap \ddot{e} = \{e_i\}$ $(i = 1, 2)$, and let $c \leqslant e$. Since e is a member of some history h_3, by the Hausdorff maximal principle we can extend the chain $\{c\}$ to a maximal chain l upper-bounded by e, which yields $\sup_{h_3} l = e$. This chain l is a member of $\mathscr{C}_e$, so by our supposition that $e_1 \in \ddot{e}$, it follows that there is some history h such that $\sup_h l = e_1$. By the definition of supremum, this means that $l \leqslant e_1$, and in particular, $c < e_1$. Since $e_1 \in h_1$, it follows from the downward closure of histories that $c \in h_1$, and thus we may apply Fact 3.13(1) to deduce that $h_1 \cap \ddot{c} = \{c\}$. A symmetrical argument can be made to deduce that $h_2 \cap \ddot{c} = \{c\}$, from which we observe that $h_1 \cap \ddot{c} = \{c\} = h_2 \cap \ddot{c}$, which is precisely the definition of undividedness at $\ddot{c}$. The result then follows immediately from Fact 3.14. □

Lemma 3.1. *Let $s \in \ddot{e}$ for some $e \in W$. Then we have $x < s$ iff $x < e$, i.e., $\mathscr{P}_e = \mathscr{P}_s$.*

Proof. If $e = s$, there is nothing to prove. Also, if e is a minimal element of W, then $s = e$, and there is nothing to prove either. We thus assume below that e is not a minimal element of W and $e \neq s$.

"$\Leftarrow$": Let $s \in \ddot{e}$, and let $x < e$. By the Hausdorff maximal principle, there is some chain $l \in \mathscr{C}_e$ for which $x \in l$. As $s \in \ddot{e}$, we know that there is some $h \in$ Hist for which $\sup_h l = s$. We cannot have $s \in l$: otherwise, for h' witnessing $\sup_{h'} l = e$, we would have $\{e, s\} \subseteq h'$, contradicting Fact 3.2. Thus, $l < s$, which implies $x < s$.

"$\Rightarrow$": Let $s \in \ddot{e}$, and let $x < s$. Assume for reductio that $x \not< e$. We first show that under this assumption, x and e cannot share a history. Assume otherwise, and let $h_1 \in H_e \cap H_x$. Let $h_2 \in H_s$. Take some $l \in \mathscr{C}_e$ (it exists since e is not a minimal element of W). We have $x \in h_1$, $x \in h_2$ (by downward closure of histories, as $x < s$), and $l \subseteq h_1$ (as $e \in h_1$). Now, as $s \in h_2$ and $s \in \ddot{e}$, we have $l < s$, so by downward closure of histories, $l \subseteq h_2$ as well. Noting that $\sup_{h_2} \{x\} = x < s = \sup_{h_2} l$, Weiner's postulate implies that $\sup_{h_1} \{x\} = x < e = \sup_{h_1} l$, contradicting the assumption that $x \not< e$.

Under our reductio assumption, we must thus have that x and e do not share a history. Choose some $h_1 \in H_e$ and some $h_2 \in H_s$. Since $s \in \ddot{e}$ and $e \neq s$, by Facts 3.13 and 3.14, $h_1 \perp_{\ddot{e}} h_2$. Moreover, by the downward closure of histories, we have $x \in h_2$, and as $e \in h_1$ and x and e do not share a history,

$x \notin h_1$. By PCP_{NF} applied to $x \in h_2 \setminus h_1$, there is $c \in W$ such that $h_1 \perp_{\ddot{c}} h_2$ and $c \leqslant x$, and hence $c < s$ and $c \in h_2$. And there is $c' \in \ddot{c}$ such that $c' \in h_1$. Picking $I \in \mathscr{C}_e$ and $J \in \mathscr{C}_c$, $I, J \subseteq h_1 \cap h_2$ and observing $c = \sup_{h_2} J < \sup_{h_2} I = s$, Weiner's postulate implies $c' = \sup_{h_1} J < \sup_{h_1} I = e$.

Fact 3.19 says that since $c' < e$ and $h_1 \perp_{\ddot{e}} h_2$, it cannot be the case that $h_1 \perp_{\ddot{c'}} h_2$. Thus h_1 and h_2 are undivided at c'. However, by definition of undividedness this means that $h_1 \cap \ddot{c'} = h_2 \cap \ddot{c'} = \{c'\}$, and thus $c' \in h_2$, which contradicts $h_2 \cap \ddot{c} = \{c\}$. □

With Lemma 3.1 to hand, it is easy to see that an element of a non-trivial choice set is always a minimal element in the difference of some two histories that split at this choice set.

Fact 3.20. *Let $h_1 \perp_{\ddot{c}} h_2$. Then c_1, the unique element of $h_1 \cap \ddot{c}$, is a minimal element in $h_1 \setminus h_2$, and c_2, the unique element of $h_2 \cap \ddot{c}$, is a minimal element in $h_2 \setminus h_1$.*

Proof. To prove that c_1 is a minimal element in $h_1 \setminus h_2$, let us assume for reductio that there is $e \in h_1 \setminus h_2$ such that $e < c_1$. By Lemma 3.1, $e < c_2$. But $c_2 \in h_2$, and by downward closure of histories, also $e \in h_2$, contradicting our assumption. The argument that c_2 is a minimal element in $h_2 \setminus h_1$ is exactly analogous. □

It follows, moreover, that for any two histories h_1, h_2 in a BST_{NF} structure, there is a minimal element in their difference, $h_1 \setminus h_2$.

Fact 3.21. *Let $h_1, h_2 \in Hist$, with $h_1 \neq h_2$. Then there is a minimal element in $h_1 \setminus h_2$.*

Proof. Since histories are maximal, there is $e \in h_1 \setminus h_2$. By PCP_{NF} there is $\ddot{c}$ such that $h_1 \perp_{\ddot{c}} h_2$. Then by Fact 3.20, $c_1 \in \ddot{c} \cap h_1$ is a minimal element in $h_1 \setminus h_2$. □

Another fact concerns maximal chains in the difference of two histories:

Fact 3.22. *Let $l \subseteq h \setminus h'$ be a maximal chain in the difference of histories h and h'. PCP_{NF} guarantees that there is a choice set $\ddot{c}$ such that $h \perp_{\ddot{c}} h'$, and for $c \in \ddot{c} \cap h$, we have $c \leqslant l$. We claim that in fact, $c = \inf l$.*

Proof. By PCP_{NF}, $c \leqslant l$, so by the infima postulate, $i =_{df} \inf l$ exists. As i is the greatest lower bound of l and c is a lower bound of l, we have $c \leqslant i$. We need to show that $c = i$. Assume otherwise, i.e., $c < i$. Note that this implies that $c \notin l$. We also have $c \in h$. Given that $c \in \ddot{c}$ and $h \perp_{\ddot{c}} h'$, we have that $c \notin h'$ (by Def. 3.13). But then $l \cup \{c\}$ is also a chain ($c < i \leqslant l$) that lies wholly in $h \setminus h'$, and that chain properly extends l. This contradicts the maximality of t. $\qquad\square$

Given Lemma 3.1, it is also not difficult to see that for $s \in \ddot{e}$, a chain ends in e iff it ends in s:

Fact 3.23. *Let $s \in \ddot{e}$. Then we have $l \in \mathscr{C}_s$ iff $l \in \mathscr{C}_e$.*

Proof. If $e = s$, there is nothing to prove. Also, if e is a minimal element of W, then $s = e$, and there is nothing to prove either. We thus assume below that e is not a minimal element of W and $e \neq s$. The former implies $\mathscr{C}_e \neq \emptyset$.

"$\Leftarrow$": Given $s \in \ddot{e}$ and $l \in \mathscr{C}_e$, by the definition of $\ddot{e}$ there is some history h for which $\sup_h l = s$. We argue that $s \notin l$. Suppose otherwise. Since l is upper-bounded by e, this means that $s < e$. Then for any history h containing e, we would have that $\{s, e\} \subseteq \ddot{e} \cap h$, which contradicts Fact 3.13(1). Then it must be the case that $s \notin l$, which implies that $l \in \mathscr{C}_s$.

"$\Rightarrow$": Let $s \in \ddot{e}$, and let $l \in \mathscr{C}_s$, i.e., $l < s$ and for some $h \in H_s$, $\sup_h l = s$. By Lemma 3.1, $l < e$. Take some $h' \in H_e$, and pick some $k \in \mathscr{C}_e$, so $k < e$, and hence $k \subseteq h'$. By the same Lemma, since $s \in \ddot{e}$, we have $k < s$, which gives us $k \subseteq h$. We claim that $\sup_h k = s$. For, if there were some x in h such that $k \leqslant x < s$, then Lemma 3.1 implies that $x < e$. Then $k \leqslant x < e$ in h', which means that $e \neq \sup_{h'} k$, contradicting $k \in \mathscr{C}_e$. We thus have $\sup_h l = s = \sup_h k$. Hence by Weiner's postulate, we also have $\sup_{h'} l = \sup_{h'} k = e$, and therefore, $l \in \mathscr{C}_e$. $\qquad\square$

Given the previous results, we see that the set $\ddot{e}$ is independent of the witness chosen:

Fact 3.24. *Consider a BST_{NF} structure $\langle W, < \rangle$. Let $s \in \ddot{e}$. Then $e \in \ddot{s}$.*

Proof. If e is a minimal element of W, then $e = s$ and we are done. We thus assume that e is not minimal in W, and hence $\mathscr{C}_e \neq \emptyset$. By Lemma 3.1, $\mathscr{C}_s \neq \emptyset$ as well.

Let $s \in \ddot{e}$. We have to show that $e \in \ddot{s}$, i.e., $e \in \mathscr{S}(l)$ for all $l \in \mathscr{C}_s$. Thus consider an arbitrary $l \in \mathscr{C}_s$. By Fact 3.23, $l \in \mathscr{C}_e$. Now take some $h \in H_e$; we have $\sup_h l = e$, i.e., $e \in \mathscr{S}(l)$. As l was arbitrary, we have $e \in \ddot{s}$. $\qquad\square$

Lemma 3.2. *We have $s \in \ddot{e}$ iff $\ddot{e} = \ddot{s}$.*

Proof. "⇐": Immediate, since $s \in \check{s}$ by Fact 3.12.

"⇒": Let $s \in \ddot{e}$. For $s = e$ there is nothing to prove, so suppose that $s \neq e$.

"⊆": Let $x \in \ddot{e}$. We have to show that $x \in \check{s}$, i.e., that $x \in \mathscr{S}(l)$ for all $l \in \mathscr{C}_s$. Thus, take some $l \in \mathscr{C}_s$. By Fact 3.23, $l \in \mathscr{C}_e$, and as $x \in \ddot{e}$, we have $x \in \mathscr{S}(l)$.

"⊇": Let $x \in \check{s}$. Take some $l \in \mathscr{C}_e$. As above, by Fact 3.23, $l \in \mathscr{C}_s$, and as $x \in \check{s}$, we have $x \in \mathscr{S}(l)$. □

Having prepared the groundwork, we can now finally fully justify calling the partition $\Pi_{\ddot{e}}$ the set of local history-wise alternatives: the set of sets of histories $\Pi_{\ddot{e}}$ partitions the set of histories containing $\mathscr{P}_e$. That is, any history containing the whole proper past of e ends up in exactly one of the elements of $\Pi_{\ddot{e}}$.

Theorem 3.1. *Let $e \in W$. Then $\Pi_{\ddot{e}}$ partitions $H_{[\mathscr{P}_e]}$, i.e.: (1) $\bigcup \Pi_{\ddot{e}} = H_{\ddot{e}} = H_{[\mathscr{P}_e]}$ and (2) for $H_1, H_2 \in \Pi_{\ddot{e}}$, if $H_1 \neq H_2$, then $H_1 \cap H_2 = \emptyset$.*

Proof. (1) We have to show that $\bigcup \Pi_{\ddot{e}} = H_{[\mathscr{P}_e]}$. Note that $\bigcup \Pi_{\ddot{e}} = H_{\ddot{e}}$ by Fact 3.13(2).

"⊆": Take $h \in \bigcup \Pi_{\ddot{e}}$, i.e., $h \in H_s$ for some $s \in \ddot{e}$. By the definition of $\mathscr{P}_s$, we have $\mathscr{P}_s \subseteq h$, and by Lemma 3.1, $\mathscr{P}_e \subseteq h$. Thus, $h \in H_{[\mathscr{P}_e]}$.

"⊇": We need to consider two cases.

Case 1: Event e is minimal in W. By definition this means that $\mathscr{P}_e = \emptyset$. Accordingly, $H_{[\mathscr{P}_e]} = \text{Hist}(W)$. By Fact 3.18 the minimal point e belongs to every history of W, so it is also the case that $H_{\ddot{e}} = H_e = \text{Hist}(W)$, which gives us the desired identity.

Case 2: Event e is not minimal in W, hence $\mathscr{C}_e \neq \emptyset$. Consider $h \in H_{[\mathscr{P}_e]}$, which implies $\mathscr{P}_e \subseteq h$. By Fact 3.5, for all $l \in \mathscr{C}_e$ we have $l \subseteq h$. Take some $l_0 \in \mathscr{C}_e$, and let $s =_{\text{df}} \sup_h l_0$. We show that s is the h-relative supremum of any chain from $\mathscr{C}_e$. Fix some $h' \in H_e$. Take any $l \in \mathscr{C}_e$. We have $\sup_{h'} l = e = \sup_{h'} l_0$, and thus by Weiner's postulate we also have $\sup_h l = \sup_h l_0 = s$. Thus we have $s \in \mathscr{S}(l)$ for any $l \in \mathscr{C}_e$, which implies $s \in \ddot{e}$. As $h \in H_s$, we have $h \in \bigcup \Pi_{\ddot{e}}$.

(2) This follows from Fact 3.13(2), as $H_{\ddot{e}} = H_{[\mathscr{P}_e]}$ (by item (1) of this Fact). □

The main message of the constructions studied in this section is that some $e \in W$ generate a non-trivial choice set $\ddot{e}$, in the sense that $\ddot{e} \neq \{e\}$. Such a set $\ddot{e}$ indeed consists of local point-wise alternatives to e. We can think of a choice set as a set of "indeterministic transitions", and each choice set induces a set of history-wise alternatives for e, namely $\Pi_{\ddot{e}}$. Finally, PCP$_{\text{NF}}$

requires that any two histories split at a choice set. So, in BST_{NF} the basic concepts of branching histories still apply, but in a slightly different way from BST_{92}. As we will later show (Chapter 4.4), this has some beneficial topological consequences.

3.6 BST_{92} or BST_{NF}: Does it matter?

The difference between BST_{92} and BST_{NF} amounts to the issue of whether there is a maximal element in the overlap of histories or a minimal element in the difference of histories (cf. Facts 3.6 and 3.21). This may seem to be a minor issue, and one may accordingly doubt whether the difference matters. Indeed, in many applications of BST to general philosophical problems, as well as to some problems in the philosophy of physics, the issue does not seem to have any bearing. It becomes important, however, when BST is used to model space-times of general relativity (GR), as then topological questions come to the fore. For a precise evaluation of whether BST_{92} or BST_{NF} can accommodate the topological requirements of GR space-times, we first need to describe BST structures topologically.[11] There is a natural topology on BST structures, both BST_{92} and BST_{NF}, the so-called diamond topology. We will describe this topology in technical detail when we return to topological issues in Chapter 4.4. In Chapter 3.6.1 which follows, we remain on an intuitive level, adding some promissory notes that will be substantiated later. After our overview of topology in BST, in Chapter 3.6.2 we then introduce wide-ranging translatability results for the two frameworks of BST_{92} and BST_{NF}. These can be read as showing that a choice between the two frameworks can be left a matter of pragmatic choice.

3.6.1 Topological issues: An overview

For representing space-times, physics uses so-called differential manifolds that have a number of defining topological features. In particular, differential manifolds have two properties that are hard to satisfy in branching structures.

---

[11] Topological worries about the appropriateness of BST were raised, e.g., by Earman (2008). Jeremy Butterfield asked about reasons to assume non-Hausdorff branching and its compatibility with space-time physics already in 2001 (Butterfield, personal communication with TP).

First, by definition, a differential manifold is locally Euclidean, which means that each point of the manifold has a neighborhood that can be mapped continuously onto an open subset of $\mathbb{R}^n$ (in realistic applications, $n = 4$; for the precise statement, see Def. 4.16). Local Euclidicity is standardly presupposed (often without explicitly mentioning the condition by name) when the notion of a space-time manifold is introduced. On such a manifold, local coordinates are defined via so-called charts (see, e.g., Wald, 1984, pp. 12f.): at each point of the manifold, it is possible to find a neighborhood that is homeomorphic to some open set of $\mathbb{R}^n$, and the respective mapping induces the coordinates. If a topological space is not locally Euclidean, it is not possible to assign coordinates in this way. It is hard to relax this requirement, as space-time points without some coordinates go against both common sense and the practice of physics.

Local Euclidicity is a challenge for BST. Given the frugality of the BST axioms, BST structures come in many varieties, so hoping that their topology will always be locally Euclidean is not realistic anyway. One can reasonably hope, however, that local Euclidicity should transfer from the individual histories to the whole global structure. More precisely, if for each history h of a BST_{92} structure the history-relative topology $\mathscr{T}_h$ is locally Euclidean, then the global topology $\mathscr{T}$ should be locally Euclidean as well. The underlying thought is that if we have a collection of physically reasonable space-times, each with an assignment of coordinates, then a BST analysis of indeterminism should not destroy the coordinate assignment.

Unfortunately, local Euclidicity is not preserved in BST_{92} as one moves from the history-relative topologies to the global topology. In fact, barring trivial one-history cases, BST_{92} structures are never locally Euclidean with respect to their natural topology. The reason is that a neighborhood of a maximal element in the intersection of two histories cannot be appropriately mapped onto $\mathbb{R}^n$. A case in point is the simple two-history model depicted in Figure 3.1(a) on p. 44. The two histories are h_1^a and h_2^a, and their overlap has a maximal element $[\langle 0, 1\rangle] = [\langle 0, 2\rangle]$. By the natural topology on the histories, which is the topology of the real line, the open sets of $\mathscr{T}_{h_i^a}$ ($i = 1, 2$) are either of the form $\{[\langle x, i\rangle] \mid x \in (c, d)\}$, for some open interval $(c, d) \subseteq \mathbb{R}$, or they are unions of such sets. As every element of the history h_i^a belongs to a set $\{[\langle x, i\rangle] \mid x \in (c, d)\}$, and such a set is trivially homeomorphic to an open interval of $\mathbb{R}$, $\mathscr{T}_{h_i^a}$ is locally Euclidean of dimension 1. Let us now turn to the global topology $\mathscr{T}$ on M_a and consider an open neighborhood (in $\mathscr{T}$) of the branching point $[\langle 0, 1\rangle]$. Each neighborhood of that point must extend somewhat to the trunk and to both of the arms. Accordingly, it must contain subsets $\{[\langle x, 1\rangle] \mid x \in (c, d)\}$ and $\{[\langle x, 2\rangle] \mid x \in (c, d')\}$, with $c < 0$ and $d, d' > 0$. A fork of that sort, however, cannot be homeomorphically mapped onto an open interval of the real line. Thus, the global topology of M_a is not locally Euclidean, despite the fact that each history-relative topology is.

Note that no such problem arises for the structure M_b of Figure 3.1(b), which is a BST_{NF} structure in which the intersection of the two histories does not contain a maximum. In fact, as we will show in Chapter 4.4, BST_{NF} vindicates the idea that if one starts with locally Euclidean histories (space-times) that allow for the assignment of spatio-temporal coordinates, one does not destroy that feature by analyzing indeterminism within the framework of Branching Space-Times.

The second property that individual space-times satisfy, but which is typically violated both by BST_{92} and by BST_{NF} structures, is a topological separation property known as the Hausdorff property (see, e.g., Wald, 1984, p. 12). The property requires that any two points of a topology's base set have non-overlapping open neighborhoods. In contrast to the requirement of local Euclidicity, however, a failure of the Hausdorff property in BST structures is not a troubling one. After all, in BST, individual space-times are represented by single histories, and it can be proved under modest assumptions that histories are Hausdorff both in BST_{92} and BST_{NF} (see Chapter 4.4). The non-Hausdorffness of the global topology of a BST structure simply reflects the fact that such a structure typically brings together multiple space-times, as it represents a number of alternative spatio-temporal developments.

3.6.2 Translatability results: An overview

We now present a number of theorems that show that there is a systematic way of translating branching structures of one kind into the branching structures of the other kind, and vice versa. In this sense, we can leave the question open as to what branching is really like. A key motivation for working out an ecumenical position is that there does not seem to be a really convincing argument for preferring one of the two options mentioned earlier. In the original paper developing BST_{92} (Belnap, 1992), the decision in favor of maxima in the intersection of histories is commented as follows:

> Finally, let me explicitly note that on the present theory, and in the presence of the postulates of this section, a causal origin has always 'a last point of indeterminateness' (the choice point) and never 'a first point of

determinateness'. I find the matter puzzling since it's neither clear to me how an alternate theory would work nor clear what difference it makes. (Belnap, 1992, p. 428)

This feeling of puzzlement also lies behind some of the objections to BST_{92}: the objectors ask about the reasons for assuming a specific pattern of branching, or they are skeptical whether that pattern is compatible with the physics of space-time. To such objections, the results given in this section answer that it is almost always possible to translate a branching structure with one pattern of branching into a structure with the other pattern of branching.[12] Thus, if we have a BST_{92} structure modeling some phenomena and are worried that it is non-Euclidean, we will dispense with the worry by translating it into a BST_{NF} structure. And in the opposite direction, if we prefer for some reasons (most likely, for simplicity) the BST_{92} framework, we are always able to transform a BST_{NF} structure into that framework. In order to keep our text concise, here we just give the statement of relevant theorems and facts, while putting all the required proofs in Appendix A.2.

The translatability results simplify the use of the BST structures in this book, as we need not develop the whole machinery for BST_{92} and for BST_{NF} in parallel. If a topic at hand is not related to general relativity, we always use BST_{92} structures, as they are somewhat easier to handle and also because much of our earlier work on BST and its applications applied that framework. We will provide some hints to help the reader to connect to the other framework, however. The exception is Chapter 9 which deals with the space-times of General Relativity and a related topological issue: there we rely exclusively on BST_{NF} structures.

Our first result concerns the set of transitions of a BST_{92} structure, which is needed to define a full transition structure; later we argue that it has the sought-for properties.

Definition 3.17 (The Υ transform as the full transition structure of a BST_{92} structure.). Let $\langle W, < \rangle$ be a BST_{92} structure. Then we define the transformed structure, $\Upsilon(\langle W, < \rangle)$, to be the full transition structure (including trivial transitions) together with the transition ordering $\prec$ from Def. 3.10, as follows:

$$\Upsilon(\langle W, < \rangle) =_{df} \langle W', \prec \rangle, \text{ where } W' =_{df} TR_{full}(W) = \{e \rightarrowtail H \mid e \in W, H \in \Pi_e\}.$$

---

[12] A small qualification applies if there are maxima or minima; see the Lemmas and Theorems below.

It turns out that the common BST properties of Def. 2.10 still hold for the Υ transform of a BST_{92} structure:

Lemma 3.3. *Let* $\langle W, < \rangle$ *be a* BST_{92} *structure without minima. Then its full transition structure* $\Upsilon(\langle W, < \rangle)$ *is still a common BST structure according to Definition 2.10.*

Further we have the already advertised results:

Lemma 3.4. *Let* $\langle W, < \rangle$ *be a* BST_{92} *structure without minima. Then that structure's full transition structure* $\langle W', <' \rangle =_{\mathrm{df}} \Upsilon(\langle W, < \rangle)$ *satisfies the* PCP_{NF} *as in Definition 3.14.*

Taken together, these Lemmas yield our first translatability result (note that we need to restrict to BST_{92} structures without minimal elements).

Theorem 3.2. *Let* $\langle W, < \rangle$ *be a* BST_{92} *structure without minima. Then that structure's full transition structure* $\Upsilon(\langle W, < \rangle)$ *is a* BST_{NF} *structure.*

It turns out that, in the other direction, there is also a fairly simple translation, viz., lumping together all the elements of a choice set to form a single point.[13]

Definition 3.18 (The Λ transformation from BST_{NF} to BST_{92}.). Let $\langle W, < \rangle$ be a BST_{NF} structure. Then we define the companion Λ-transformed ("collapsed") structure as follows:

$$\Lambda(\langle W, < \rangle) =_{\mathrm{df}} \langle W', <' \rangle, \quad \text{where}$$
$$W' =_{\mathrm{df}} \{\ddot{e} \mid e \in W\};$$
$$\ddot{e}_1 <' \ddot{e}_2 \quad \text{iff} \quad e'_1 < e'_2 \quad \text{for some} \quad e'_1 \in \ddot{e}_1, e'_2 \in \ddot{e}_2.$$

With the Λ transform we can prove the desired translatability results in the other direction, from BST_{NF} to BST_{92}. Mirroring the situation for the Υ transform results that required no minimal elements, here we have to work under the provision that the given BST_{NF} model contains no maximal elements.

Lemma 3.5. *Let* $\langle W, < \rangle$ *be a* BST_{NF} *structure without maxima. Then its* Λ*-transform,* $\langle W', <' \rangle =_{\mathrm{df}} \Lambda(\langle W, < \rangle)$, *is a common BST structure.*

---

[13] Graphically, we have chosen "Λ", which suggests pulling elements (of a choice set) together into one, as the reverse of "Υ", which suggests fanning out elements from a common base (viz., the choice point).

Lemma 3.6. *The Λ-transform $\Lambda(\langle W, < \rangle)$ of a BST_{NF} structure without maxima $\langle W, < \rangle$ satisfies the BST_{92} prior choice principle.*

Theorem 3.3. *The Λ-transform $\Lambda(\langle W, < \rangle)$ of a BST_{NF} structure without maxima $\langle W, < \rangle$ is a BST_{92} structure.*

We can even go full circle. As there is a way to get from BST_{92} structures without minimal elements to BST_{NF} structures, and a way to get from BST_{NF} structures without maximal elements to BST_{92} structures, the question arises as to where we end up when we concatenate these transformations. We can show that, as hoped, we return to where we started: the resulting structures are order-isomorphic to the ones we started with. For simplicity's sake, we work with structures without maximal or minimal elements.

Consider first the direction from BST_{92} to BST_{NF} to BST_{92}.

Theorem 3.4. *The function $\Lambda \circ \Upsilon$ is an order isomorphism of BST_{92} structures without maximal or minimal elements: Let $\langle W_1, <_1 \rangle$ be a BST_{92} structure without maximal or minimal elements, let $\langle W_2, <_2 \rangle =_{df} \Upsilon(\langle W_1, <_1 \rangle)$, and let $\langle W_3, <_3 \rangle =_{df} \Lambda(\langle W_2, <_2 \rangle)$. Then there is an order isomorphism φ between $\langle W_1, <_1 \rangle$ and $\langle W_3, <_3 \rangle$, i.e., a bijection between W_1 and W_3 that preserves the ordering. Accordingly, $\langle W_3, <_3 \rangle$ has no minima and no maxima.*

The result in the opposite direction, that is, from BST_{NF} to BST_{92} to BST_{NF}, also holds:

Theorem 3.5. *The function $\Upsilon \circ \Lambda$ is an order isomorphism of BST_{NF} structures without maximal or minimal elements: Let $\langle W_1, <_1 \rangle$ be a BST_{NF} structure without maximal or minimal elements, let $\langle W_2, <_2 \rangle =_{df} \Lambda(\langle W_1, <_1 \rangle)$, and let $\langle W_3, <_3 \rangle =_{df} \Upsilon(\langle W_2, <_2 \rangle)$. Then there is an order isomorphism φ between $\langle W_1, <_1 \rangle$ and $\langle W_3, <_3 \rangle$, i.e., a bijection between W_1 and W_3 that preserves the ordering. Accordingly, $\langle W_3, <_3 \rangle$ has no minima and no maxima.*

3.7 Exercises to Chapter 3

Exercise 3.1. Prove that there is a choice point for any two histories in a BST_{92} structure (i.e., a maximal element in their overlap).

Hint: Use the fact that histories are maximal and apply PCP_{92}.

Exercise 3.2. Prove the following extension of Fact 3.10: For a BST_{92} structure $\langle W, < \rangle$ that has neither maximal nor minimal elements, its full transition structure $\langle W', <' \rangle =_{df} \langle TR_{full}(W), \prec \rangle$ has no maxima nor minima either.

Hint: Take an appropriate initial from W and a fitting history to define the outcome of a witnessing transition. (A full proof is given in Appendix B.3.)

Exercise 3.3. Prove the second "iff" in Fact 3.11(1).

Exercise 3.4. Prove that there are no non-trivial choice sets in BST_{92} structures.

Hint: For reductio, assume $\ddot{e} \subseteq W$, with $e, e' \in \ddot{e}$ and $e \neq e'$. There are histories h, h' with $e \in h, e' \in h'$. Since $e \notin h'$ (why?), we have $e \in h \setminus h'$, to which we apply PCP of BST_{92} to get some $c < e$ such that $h \perp_c h'$. Pick next a chain $l \in \mathscr{C}_e$ such that $c \in l$. Observe then that $\sup_{h'}(l) \neq e'$ (why?). This proves $e' \notin \ddot{e}$ (why?) which contradicts the reductio assumption.

Exercise 3.5. Prove that the BST_{NF} prior choice principle (Def. 3.14) implies historical connection.

Hint: By maximality of histories, for any two histories h_1, h_2 there is e such that $e \in h_1 \setminus h_2$. Apply then PCP_{NF} to e to obtain a choice set $\ddot{c}$, at which h_1 and h_2 split. There must then be two different $c_1, c_2 \in \ddot{c}$. By definition, they are history h_1- and h_2-relative suprema of chains from $\mathscr{C}_{c_1}$, where these chains are contained in $h_1 \cap h_2$.

Exercise 3.6. Discuss the problem of wings in BST_{NF} structures (cf. Figure 3.2). Show that the wings are in the histories' overlap.

Hint: produce a proof analogous the proof of Fact 3.9, using PCP_{NF} at some appropriate stage.

Exercise 3.7. Let $\langle W, < \rangle$ satisfy Postulates 2.1–2.5. Let l be an upper-bounded chain, and let $e =_{df} \sup_{h'}(l)$. Then for every history h of W containing the chain l, if e lies in h, then $e = \sup_h(l)$.

Hint: Derive a contradiction from the assumption that there is an upper bound of l below e. (A full proof is given in Appendix B.3.)

4

Building upon the Foundations of Branching Space-Times

In this chapter we introduce a variety of events that are definable in BST and discuss in which histories these events occur. This will give rise to the concept of the occurrence proposition for events of various kinds. We also hint at how BST structures may be used to build semantic models for languages with temporal and modal operators. In this way we provide the machinery for our BST theories of causation and of propensities, and we prepare the ground for a number of further applications.

Perhaps surprisingly, a large portion of this material is independent of the choice between BST_{92} and BST_{NF}, so in Chapter 4.1 we remain in the common BST framework. The situation changes once we discuss basic transitions in Chapter 4.2, as these objects are sensitive to topology (which dictates the pattern of branching) and need to be defined differently in the two frameworks. We already introduced the notion of a transition in Chapter 3.4.4, working in BST_{92}. Here we will provide a number of additional definitions, mostly for sets of transitions which will be used extensively in later chapters. We also provide some details about basic transitions in BST_{NF}.

4.1 A variety of events and their occurrence propositions

We begin with the notion of a proposition. In line with a prominent tradition, propositions are identified with sets of histories:

Definition 4.1 (Basics of propositions). H is a proposition $\Leftrightarrow_{df} H \subseteq \text{Hist}$. H is defined as true or false in a history h according to whether or not $h \in H$. H is consistent $\Leftrightarrow_{df} H \neq \emptyset$. H is universal $\Leftrightarrow_{df} H = \text{Hist}$, and H is contingent $\Leftrightarrow_{df} H$ is consistent but not universal.

Branching Space-Times: Theory and Applications. Nuel Belnap, Thomas Müller, and Tomasz Placek, Oxford University Press. © Oxford University Press 2022. DOI: 10.1093/oso/9780190884314.003.0004

We define the notion of a necessary and of a sufficient condition in the standard way.[1]

Definition 4.2 (Necessary and sufficient conditions). For propositions X and Y, we say that X is a sufficient condition for $Y \Leftrightarrow_{df} X \subseteq Y$; we also say that X implies Y, or that Y is a necessary condition for X.

Although BST starts with the meagre primitive notion of a point event, it permits the introduction of a number of more complex event-like concepts. We define these together with their respective occurrence propositions. We start with initial and outcome events. The distinction is especially important for the representation of indeterministic processes. Typically, such a process can be conceived as containing an initial event and one of its multiple possible outcomes. A radioactive particle's decay or a measurement process can serve as illustrations. Our definition below reflects the intuition that for an initial event to have occurred, it needs to have come to an end. Thus, in a measurement, before an outcome occurs, the whole measurement initial event (e.g., the preparation of the apparatus in the 'ready' state) needs to have come to completion. In contrast, for an outcome to occur it is enough that it has just begun: to say that a particle has decayed, we need to witness just some arbitrarily small part of the world after the decay. Although the intuition is perhaps not crystal clear, the definitions below turn out to be fruitful in many later developments.

Definition 4.3 (Initial and outcome events and occurrence propositions).

1. I is an *initial event* $\Leftrightarrow_{df} I$ is a consistent nonempty set of point events (i.e., a set of point events all of which are members of some one history). The occurrence proposition for I is $H_{[I]} =_{df} \{h \in \text{Hist} \mid I \subseteq h\}$. Equivalently, $H_{[I]} = \bigcap_{e \in I} H_e$.
2. O is an *outcome chain* $\Leftrightarrow_{df} O$ is a non-empty and lower-bounded chain. The occurrence proposition for O is $H_{\langle O \rangle} =_{df} \{h \in \text{Hist} \mid h \cap O \neq \emptyset\}$.
3. $\hat{O}$ is a *scattered outcome event* $\Leftrightarrow_{df} \hat{O}$ is a set of outcome chains all of which overlap some one history (i.e., there is one history that contains an initial segment of each of the chains). The occurrence proposition for $\hat{O}$ is $H_{\langle \hat{O} \rangle} =_{df} \bigcap_{O \in \hat{O}} H_{\langle O \rangle}$.

---

[1] As we proceed to develop the BST account of causation in Chapter 6, we will need the related notion of a non-redundant part of a sufficient condition for a proposition. As it will turn out, however, this notion is subtle and not univocal. We will therefore discuss it only after we develop the framework within which this subtlety plays a role.

4. $\check{\mathbf{O}}$ is a *disjunctive outcome event* $\Leftrightarrow_{df} \check{\mathbf{O}}$ is a set of pairwise incompatible scattered outcomes (a set of sets of sets), where 'pairwise incompatible' means that for any $\hat{O}_1, \hat{O}_2 \in \check{\mathbf{O}}$, if $\hat{O}_1 \neq \hat{O}_2$, then $H_{\langle \hat{O}_1 \rangle} \cap H_{\langle \hat{O}_2 \rangle} = \emptyset$.

The occurrence proposition for $\check{\mathbf{O}}$ is $H_{\langle \check{\mathbf{O}} \rangle} =_{df} \bigcup_{\hat{O} \in \check{\mathbf{O}}} H_{\langle \hat{O} \rangle}$.

According to this definition, there is a hierarchy of outcome events, which differ in their complexity. Starting with an outcome chain, it is a part of a world-line that is bounded from below. A scattered outcome is a more complex spatiotemporal affair. A given result, say 'side 1 up', $\boxed{1}$, of a particular rolling of a die, is a scattered outcome. That outcome is composed of a huge number of lower bounded segments of the world-lines of the particles involved. It does not matter that, realistically speaking, no single history will contain sizable segments of all these world-lines. To constitute a scattered outcome, however, there should be at least one history in which all these segments *begin*. Next in the hierarchy come disjunctive outcomes, which are more complex, as they combine different possibilities. As an illustration, consider a set of two possible results of a particular rolling of a die, say $\{\boxed{1}, \boxed{3}\}$. Clearly, this disjunctive outcome occurs if $\boxed{1}$ occurs or $\boxed{3}$ occurs. Thus, a disjunctive outcome, in contrast to an outcome chain and a scattered outcome, can be realized in a number of different ways, as its elements belong to alternative possibilities. We thus say that disjunctive outcomes are multiply realizable. To return to our die, note that the set of *all* possible results of rolling the die, $\{\boxed{1}, \boxed{2}, \ldots, \boxed{6}\}$, counts as a disjunctive outcome as well. This disjunctive outcome has the peculiar feature that it is bound to occur once the rolling of the die occurs. Later, we will capture such cases via the concept of a deterministic transition to a disjunctive outcome: although the world is as indeterministic as you like, a deterministic transition to the disjunctive outcome including all possible outcomes is bound to occur once the initial event has occurred.

One may wonder what the initial of a particular process of rolling a die consists of. To describe a concrete happening: you put a particular die in a dice cup, give it a shake, and roll it onto a flat surface. That complex affair needs to be over before a particular result shows up. As this example makes clear, an initial event I may be extended in space and time—there are no restrictions except for consistency. As $I \subseteq h$ for some history h, $H_{[I]} \neq \emptyset$. In a similar vein, occurrence propositions for outcomes are never the empty set. Since a chain O is a directed subset of $\mathcal{W}$, it can be extended to a full history, hence $O \subseteq h$ for some $h \in$ Hist, and thus $H_{\langle O \rangle}$ is non-empty. With this observation it is easy to see that $H_{\langle \hat{O} \rangle}$ and $H_{\langle \breve{O} \rangle}$ are non-empty as well, and thus consistent in the sense of Def. 4.1.

We next turn to transition events, to be understood as a liberalized notion of change. Following von Wright (1963), we take it that a transition is of the form "something and then something", but not necessarily "...something else". "Then" has to be spelled out to mean that the beginning of a transition is appropriately below its final part. We insist, moreover, that the parts of a transition are categorically different: its beginning is to be an initial event, while its final part—an outcome event.

A paradigm example of a transition event is a choice. Before the choice there is no choice, and after the choice there is no choice. So when is the choice? Bad question: a choice, like any transition event, has no 'simple location' (Whitehead, 1925, Ch. 3). You can locate its initial in the causal order, and you can locate its outcome in the causal order; and having done that, you have done all that you can do. When a choice is made, something happens, but 'when' it happens can only be described by giving separate 'whens' to its initial and to its outcome. Exactly the same holds for any other transition event. (This thought will be applied to measurements in quantum mechanics in Chapter 8.)

In what follows, we will use a generic notation for outcomes to define transitions: we will write O^* for an outcome event from Definition 4.3, that is, an outcome chain O, a scattered outcome $\hat{O}$, or a disjunctive outcome $\breve{O}$. We obtain thus three kinds of transitions; we refer to them all by the generic notation, $I \rightarrowtail O^*$. Clearly, a singleton $\{e\}$ of a point event is an initial. We simplify the unwieldy $\{e\} \rightarrowtail O^*$ as $e \rightarrowtail O^*$.

In Chapter 3.4.4 we already introduced a basic transition in BST$_{92}$ as a pair $\langle e, H \rangle$ with $H \in \Pi_e$, written $e \rightarrowtail H$. In this representation, one element of a transition is an event and the other is a proposition. In contexts in which the spatio-temporal location of outcome events is irrelevant, we will often use such a "quasi-propositional notation", writing $e \rightarrowtail H_{\langle O \rangle}$ instead of $e \rightarrowtail O$, and analogously for transitions to scattered outcomes and to disjunctive outcomes. While such hybrid objects are often handy, it is also natural to represent transitions in terms of events only, as in the following definition. It will turn out that for basic transitions, both representations are equivalent; see Fact 4.3 in Chapter 4.2. Here is how we spell out that the initial of a transition of one of the kinds we consider is *appropriately below* the outcome:

Definition 4.4 (Transition events). For I and $\mathcal{O}^*$ an initial event and a generic outcome event, respectively, a transition is the pair $\langle I, \mathcal{O}^* \rangle$, written $I \rightarrowtail \mathcal{O}^*$, where I is appropriately below $\mathcal{O}^*$, $I <_i \mathcal{O}^*$. "Appropriately below" is defined as follows:

$$e <_1 O \Leftrightarrow_{df} \forall e'[e' \in O \to e < e']$$
$$I <_2 O \Leftrightarrow_{df} \forall e[e \in I \to e <_1 O]$$
$$e <_3 \hat{O} \Leftrightarrow_{df} \exists O[O \in \hat{O} \wedge e <_1 O]$$
$$I <_4 \hat{O} \Leftrightarrow_{df} \forall e[e \in I \to e <_3 \hat{O}]$$
$$e <_5 \check{O} \Leftrightarrow_{df} \forall \hat{O}[\hat{O} \in \check{O} \to e <_3 \hat{O}]$$
$$I <_6 \check{O} \Leftrightarrow_{df} \forall e[e \in I \to e <_5 \check{O}].$$

A welcome consequence of our definitions is that for any transition $I \rightarrowtail \mathcal{O}^*$, the occurrence of the initial is a necessary condition for the occurrence of the outcome:

Fact 4.1. *For a generic transition* $I \rightarrowtail \mathcal{O}^*$, $H_{\langle \mathcal{O}^* \rangle} \subseteq H_{[I]}$.

Proof. We consider the six cases of Def. 4.4 in turn. (1) Since histories are downward closed, $e <_1 O$ implies $H_{\langle O \rangle} \subseteq H_e$. (2) $I <_2 O$ implies that for every $e \in I$: $H_{\langle O \rangle} \subseteq H_e$, and hence $H_{\langle O \rangle} \subseteq \bigcap_{e \in I} H_e = H_{[I]}$. (3) Next, $e <_3 \hat{O}$ implies $H_{\langle O \rangle} \subseteq H_e$ for some $O \in \hat{O}$, and hence $H_{\langle \hat{O} \rangle} = \bigcap_{O \in \hat{O}} H_{\langle O \rangle} \subseteq H_e$. (4) For $I <_4 \hat{O}$, since $H_{\langle \hat{O} \rangle} \subseteq H_e$ for every $e \in I$, we get $H_{\langle \hat{O} \rangle} \subseteq \bigcap_{e \in I} H_e = H_{[I]}$. (5) And, if $e <_5 \check{O}$, then for all $\hat{O} \in \check{O}$: $H_{\langle \hat{O} \rangle} \subseteq H_e$, and hence $H_{\langle \check{O} \rangle} = \bigcup_{\hat{O} \in \check{O}} H_{\langle \hat{O} \rangle} \subseteq H_e$; thus, (6) for $I <_6 \check{O}$ we have $H_{\langle \check{O} \rangle} \subseteq H_e$ for every $e \in I$, which entails $H_{\langle \check{O} \rangle} \subseteq \bigcap_{e \in I} H_e = H_{[I]}$. $\square$

A transition event, like any event, can occur or not occur. What, then, is the occurrence proposition for a transition event? A good guess would be that it should be an *and then* proposition: first the initial occurs, and then the outcome occurs. It turns out, however, that in BST, it is more appropriate to take the occurrence proposition for a transition event to be the material implication: *if* the initial occurs, *then* the outcome occurs.

Definition 4.5. Let $I \rightarrowtail \mathcal{O}^*$ be a transition event of one of the types allowed by Definition 4.4, and let $H_{[I]}$ and $H_{\mathcal{O}^*}$ be the occurrence propositions defined for I and $\mathcal{O}^*$ respectively. Then $H_{I \rightarrowtail \mathcal{O}^*} \Leftrightarrow_{df} (\text{Hist} \setminus H_{[I]}) \cup H_{\mathcal{O}^*}$ is the occurrence proposition for $I \rightarrowtail \mathcal{O}^*$, true in h iff $h \in H_{I \rightarrowtail \mathcal{O}^*}$, hence iff, if $h \in H_{[I]}$, then $h \in H_{\mathcal{O}^*}$.

The final 'if – then' must be truth-functional. Usually, in ordinary language applications, the negation of a material implication 'if A then B' seems wrong; this is of course one of the motivations for various theories of counterfactual conditionals, to say nothing of relevance logic. Here, however, there is a better fit: for the transition $I \rightarrowtail \mathcal{O}^*$ *not* to occur is for the initial to occur and then for *some other outcome* of I to occur instead. It is not merely for the outcome $\mathcal{O}^*$ not to occur. The non-occurrence proposition of $\mathcal{O}^*$ is simply Hist $\setminus H_{\mathcal{O}^*}$; the non-occurrence of the transition $I \rightarrowtail \mathcal{O}^*$ is more specific. For instance, if you understand a particular choice as a transition from a particular occasion of indecision to a settled state of having selected the tuna sandwich, then for that transition event not to occur is for the chooser to have chosen otherwise from that very same occasion of indecision. For the non-occurrence of the transition event, it does not suffice that the chooser was never born—although that would certainly be sufficient for the non-occurrence of the tuna-selection outcome. Furthermore, we naturally say that a transition $I \rightarrowtail \mathcal{O}^*$ is (historically) 'noncontingent' when the initial already deterministically guarantees the outcome; that is, when $H_{\langle \mathcal{O}^* \rangle}$ is not merely a subset of $H_{[I]}$ (as must always be the case, see Fact 4.1), but identical to $H_{[I]}$. In that case, the transition-event occurrence proposition rightly turns out to be the universal proposition: $(\text{Hist} \setminus H_{[I]}) \cup H_{\langle \mathcal{O}^* \rangle} = (\text{Hist} \setminus H_{\langle \mathcal{O}^* \rangle}) \cup H_{\langle \mathcal{O}^* \rangle} = \text{Hist}$, which signals historical noncontingency. One should not be deeply interested in transition events whose occurrence in h is merely a matter of the initial not occurring in h, and so it is good to mark this by saying that the transition event occurs *vacuously* in h if $h \notin H_{[I]}$.

As we mentioned earlier, there are deterministic (historically noncontingent) transitions to disjunctive outcomes even in an indeterministic context. To return to our example of die casting, the set of all possible results, $\check{\mathbf{O}} =_{df} \{\boxed{1}, \boxed{2} \ldots \boxed{6}\}$, of a particular act of die casting is the exhaustive disjunctive outcome. (We exclude weird cases, such as the die landing on its edge.) The particular act of casting the die is an initial event I. Given their location, I and $\check{\mathbf{O}}$ form the transition $I \rightarrowtail \check{\mathbf{O}}$. By Fact 4.1, we have $H_{\langle \check{\mathbf{O}} \rangle} \subseteq H_{[I]}$, and by exhaustiveness, $H_{[I]} \subseteq H_{\langle \check{\mathbf{O}} \rangle}$, i.e., $H_{[I]} = H_{\langle \check{\mathbf{O}} \rangle}$, which means that if I occurs, so does $\check{\mathbf{O}}$. Moreover, by Definition 4.5, $I \rightarrowtail \check{\mathbf{O}}$ occurs in every history, as $H_{I \rightarrowtail \check{\mathbf{O}}} = (\text{Hist} \setminus H_{[I]}) \cup H_{\langle \check{\mathbf{O}} \rangle} = (\text{Hist} \setminus H_{[I]}) \cup H_{[I]} = \text{Hist}$.

For the record, here we define transitions to deterministic disjunctive outcomes:

Definition 4.6. Let I be an initial event, Γ some index set, $|\Gamma| > 1$. We call $\mathbf{1}_I = \{\hat{O}_\gamma \mid \gamma \in \Gamma\}$ a *deterministic disjunctive outcome of I* iff (1) each $\hat{O}_\gamma$ is above I in the sense of Def. 4.4, (2) for $\gamma, \gamma' \in \Gamma$, $H_{\langle \hat{O}_\gamma \rangle} \cap H_{\langle \hat{O}_{\gamma'} \rangle} = \emptyset$ if $\gamma \neq \gamma'$, and (3) $\bigcup_{\gamma \in \Gamma} H_{\langle \hat{O}_\gamma \rangle} = H_{[I]}$. We call $I \rightarrowtail \mathbf{1}_I$ a *transition to a deterministic disjunctive outcome.*

By this definition, if I occurs, some $\hat{O}_\gamma$ occurs, and hence $\mathbf{1}_I$ occurs as well. Despite indeterminism, witnessed by multiple transitions to different scattered outcomes, $I \rightarrowtail \mathbf{1}_I$ is a deterministic transition.

We end this section by noting the following simple relations between the occurrence propositions of different types of transitions:

Fact 4.2. *(1)* $H_{I \rightarrow \hat{O}} = \bigcap_{O \in \hat{O}} H_{I \rightarrow O}$. *(2)* $H_{I \rightarrow \check{O}} = \bigcup_{\hat{O} \in \check{O}} H_{I \rightarrow \hat{O}}$.

Proof. (1) This follows by Def. 4.3(3), noting that $(\text{Hist} \setminus H_{[I]}) \cap H_{\langle O \rangle} = \emptyset$ by Fact 4.1. (2) Observe that $(\text{Hist} \setminus H_{[I]}) \cup \bigcup_{\hat{O} \in \check{O}} H_{\langle \hat{O} \rangle} = \bigcup_{\hat{O} \in \check{O}}((\text{Hist} \setminus H_{[I]}) \cup H_{\langle \hat{O} \rangle})$. $\qquad\square$

4.2 Basic transitions

In this section we develop further the theory of basic transitions that we started in Chapter 3.4.4. Our discussion in Chapter 4.1 has been phrased in the framework of common BST structures, but basic transitions look different in BST_{92} and BST_{NF}, as they are sensitive to the topology of branching. Moreover, the proof of the interchangeability of two representations of basic transition below (Fact 4.3) appeals to the prior choice principle, which works differently in BST_{92} than in BST_{NF}. So in this section we cannot work in the common BST framework. In line with our general approach of working with BST_{92} in the main text in cases where the choice matters, we will discuss basic transitions in BST_{92} structures. We will, however, also provide some details that lead to the definition of two representations of basic transitions in BST_{NF} (see Chapter 4.2.2).

4.2.1 Basic transitions in BST_{92}

Basic transitions are the irreducible local elements of indeterminism in a BST_{92} structure, consisting of a point event e and one of its immediate possible outcomes. We introduce two alternative views of basic transitions in BST_{92}, based on the observation that the outcome of a basic transition can be represented in either of two (equivalent) ways: as a proposition defined in terms of undividedness, or as a scattered outcome event consisting of outcome chains all of which begin immediately after e:

1. In Chapter 3.4.4, we introduced a basic transition as a transition from a single point event e to one of its elementary possibilities understood propositionally. The outcome of a basic transition is then a proposition $H \in \Pi_e$, where Π_e is the partition of the set of histories containing e, H_e, induced by the equivalence relation of undividedness-at-e, $\equiv_e$. We write $e \rightarrowtail H$ for a basic transition, so understood.

2. In line with Def. 4.4, we can understand a basic transition as a transition from an initial point event e to a particular scattered outcome event $\hat{O}$ that we call an immediate (basic scattered) outcome of e. What makes the scattered outcome event $\hat{O}$ an immediate outcome of e is that for every outcome chain $O \in \hat{O}$, $\inf O = e$ and $e \notin O$. There are many such chains. In the definition below, we consider maximal chains, and we divide them up to form a particular scattered outcome event via a given history, as follows:

Definition 4.7 (Basic scattered outcomes of e). Let $h \in$ Hist, and let $e \in h$. We define $\Omega_e\langle h\rangle =_{df} \{O \mid O$ is a chain maximal with respect to $\inf O = e \wedge e \notin O \wedge h \cap O \neq \emptyset\}$. $\Omega_e =_{df} \{\Omega_e\langle h\rangle \mid h \in H_e\}$. Each member $\Omega_e\langle h\rangle$ of Ω_e is a *basic scattered outcome* of e.

The members of $\Omega_e\langle h\rangle$ evidently begin in the immediate future of e, so that between e and members of $\Omega_e\langle h\rangle$ there is no room for influences from the past. Since $\Omega_e\langle h\rangle$ is a scattered outcome event, which can occur or not occur, $H_{\langle\Omega_e\langle h\rangle\rangle}$ makes sense as a proposition. That proposition equals the propositional basic outcome of e determined by h:

Fact 4.3 (Interchangeability of $\Omega_e\langle h\rangle$ and $\Pi_e\langle h\rangle$). *The occurrence proposition $H_{\langle\Omega_e\langle h\rangle\rangle}$ for $\Omega_e\langle h\rangle$ is the same proposition as $\Pi_e\langle h\rangle$.*

Proof. Let $h' \in H_{\langle\Omega_e\langle h\rangle\rangle}$. This implies that every $O \in \Omega_e\langle h\rangle$ intersects nonemptily with h', $O \cap h' \neq \emptyset$. Since $O \in \Omega_e\langle h\rangle$, we get $O \cap h \neq \emptyset$ as well. Since $\inf O = e$ and $e \notin O$, there is some $e' \in h \cap h'$ such that $e < e'$, so we have $h \equiv_e h'$, hence $h' \in \Pi_e\langle h\rangle$.

In the opposite direction, let $h' \in \Pi_e\langle h \rangle$ and suppose for reductio that $h' \notin H_{\langle \Omega_e \langle h \rangle \rangle}$, which implies that for some $O \in \Omega_e\langle h \rangle$, $h' \cap O = \emptyset$. Hence there would be an initial segment O' of O such that $O' \subseteq h \setminus h'$. Since $e = \inf O = \inf O'$, by PCP$_{92}$ we have that for some $c < O'$: $h \perp_c h'$. This contradicts $h \equiv_e h'$, as $c \leqslant e$ by the definition of the infimum. $\qquad\square$

Occurrence propositions do not in general determine outcome events. For the special outcome events of the form $\Omega_e\langle h \rangle$, however, when we are not only given the proposition but also e, we can recover the event from the proposition:

Fact 4.4. *Let $e \in W$, and let $h_1, h_2 \in H_e$. We have $H_{\langle \Omega_e \langle h_1 \rangle \rangle} = H_{\langle \Omega_e \langle h_2 \rangle \rangle}$ iff $\Omega_e\langle h_1 \rangle = \Omega_e\langle h_2 \rangle$.*

Proof. The direction from right to left is trivial. For the other direction, assume that $H_{\langle \Omega_e \langle h_1 \rangle \rangle} = H_{\langle \Omega_e \langle h_2 \rangle \rangle}$. By Fact 4.3, $H_{\langle \Omega_e \langle h_i \rangle \rangle} = \Pi_e\langle h_i \rangle$ ($i = 1, 2$), so in particular, $h_1 \equiv_e h_2$. Now assume for reductio that there is $O \in \Omega_e\langle h_1 \rangle$ while $O \notin \Omega_e\langle h_2 \rangle$ (the case with h_1 and h_2 reversed is exactly analogous). Let $h \in$ Hist be such that $O \subseteq h$. By Def. 4.7, $O \in \Omega_e\langle h_1 \rangle$ implies that $O \cap h_1 \neq \emptyset$, and as $e < O$, we have $h \equiv_e h_1$. By transitivity of $\equiv_e$, $h \equiv_e h_2$ as well. On the other hand, $O \notin \Omega_e\langle h_2 \rangle$ implies that $O \subseteq h \setminus h_2$, so that by PCP$_{92}$, there is $c < O$ for which $h \perp_c h_2$. As $c \leqslant e$ by $\inf O = e$, this contradicts $h \equiv_e h_2$. $\qquad\square$

So there is a natural one-to-one correspondence between the set of basic scattered outcomes Ω_e of e and the set of basic propositional outcomes Π_e of e.[2] As a further consequence of Fact 4.3, in the same way in which Π_e partitions the set H_e of histories containing e, Ω_e partitions the future of possibilities of e:

Fact 4.5. *Let $e \in W$, and let $F_e =_{df} \{e' \in W \mid e < e'\}$. Then Ω_e is a partition of F_e: (1) the union of all the chains that make up all the basic scattered outcomes of e cover the whole future of possibilities of e, i.e., $\bigcup \bigcup \Omega_e = F_e$, and (2) the basic scattered outcomes of e do not overlap, i.e., for $\hat{O}_1, \hat{O}_2 \in \Omega_e$, if $\hat{O}_1 \neq \hat{O}_2$, then $\hat{O}_1 \cap \hat{O}_2 = \emptyset$.*

---

[2] It by no means follows that for two different e_1 and e_2, if $H_{\langle \Omega_{e_1} \langle h_1 \rangle \rangle} = H_{\langle \Omega_{e_2} \langle h_2 \rangle \rangle}$, then $\Omega_{e_1}\langle h_1 \rangle = \Omega_{e_2}\langle h_2 \rangle$. You must hold e constant.

Proof. If e is a maximal element of W, there is nothing to prove. So we assume that e is not maximal.

(1) "$\subseteq$": Let $e' \in \bigcup\bigcup\Omega_e$, i.e., $e' \in \bigcup\Omega_e\langle h \rangle$ for some $h \in H_e$. As $\Omega_e\langle h \rangle$ is a set of outcome chains, there must be some outcome chain $O \in \Omega_e\langle h \rangle$ for which $e' \in O$. By Def. 4.7, $e < O$, so that $e < e'$, i.e., $e' \in F_e$.

"$\supseteq$": Let $e' \in F_e$, i.e., $e < e'$. The set $\{e, e'\}$ can be extended to a maximal chain l that begins at e and contains e'. As l is directed, there is a history $h \supseteq l$. Now by Fact 3.3, $O =_{df} l \setminus \{e\}$ is an outcome chain with infimum e, so that by Def. 4.7, $O \in \Omega_e\langle h \rangle$. As $e' \in O$, this means $e' \in \bigcup\Omega_e\langle h \rangle$, i.e., $e' \in \bigcup\bigcup\Omega_e$.

(2) Let $\hat{O}_i = \Omega_e\langle h_i \rangle$ $(i = 1, 2)$, and assume that $\hat{O}_1 \neq \hat{O}_2$. By Fact 4.4 and 4.3, this implies $\Pi_e\langle h_1 \rangle \neq \Pi_e\langle h_2 \rangle$. Assume for reductio that there is some outcome chain O with infimum e for which $O \in \hat{O}_1 \cap \hat{O}_2$, and let h be a history containing O. Then, as in the proof of Fact 4.4, we have $h_1 \equiv_e h \equiv_e h_2$, which contradicts $\Pi_e\langle h_1 \rangle \neq \Pi_e\langle h_2 \rangle$. $\qquad\square$

The results above show that it makes sense to extend Def. 3.9 and to define two varieties of basic transitions, Π_e-based and Ω_e-based, which are equivalent:

Definition 4.8 (Basic transitions in BST$_{92}$). For $e \in h$, $e \rightarrowtail \Omega_e\langle h \rangle$ is a basic transition event, and $e \rightarrowtail \Pi_e\langle h \rangle$ is a basic propositional transition. Both $e \rightarrowtail \Omega_e\langle h \rangle$ and $e \rightarrowtail \Pi_e\langle h \rangle$ may be called *basic transitions*.

Fact 4.6. *Let $e \in W$ and $h \in H_e$. The basic transitions $e \rightarrowtail \Omega_e\langle h \rangle$ and $e \rightarrowtail \Pi_e\langle h \rangle$ are equivalent in the sense of having the same occurrence proposition, i.e.,*

$$H_{e \rightarrowtail \Omega_e\langle h \rangle} = H_{e \rightarrowtail \Pi_e\langle h \rangle}.$$

Proof. Obviously $e \rightarrowtail \Omega_e\langle h \rangle$ and $e \rightarrowtail \Pi_e\langle h \rangle$ have the same initials, and therefore Fact 4.3 implies that

$$(\text{Hist} \setminus H_e) \cup H_{\langle \Omega_e\langle h \rangle \rangle} = (\text{Hist} \setminus H_e) \cup \Pi_e\langle h \rangle. \qquad\square$$

We can extend the outcome selection notation for propositions given a history, $\Pi_e\langle h \rangle$, to point events and to outcome chains: Given some initial e_1, an outcome $H \in \Pi_{e_1}$ of e_1 is not just uniquely determined by some $h \in H_{e_1}$ (which motivates the notation $\Pi_{e_1}\langle h \rangle$), but also by any later event e_2 in the future of possibilities of e_1, or by any outcome chain O for which $e_1 < O$. Therefore, we can introduce the notation "$\Pi_{e_1}\langle e_2 \rangle$" and "$\Pi_{e_1}\langle O \rangle$":

Fact 4.7. *(1) Let $e_1 < e_2$. Then there is exactly one basic outcome of e_1 that is compatible with e_2, which we denote by $\Pi_{e_1}\langle e_2 \rangle$. (2) Let $e_1 < O$ for an outcome chain O. Then there is exactly one basic outcome of e_1 that is compatible with O, which we denote by $\Pi_{e_1}\langle O \rangle$. (3) Let $e < O$, $h_1 \perp_e h_2$, and $h_2 \in H_{\langle O \rangle}$. Then $h_1 \perp_e H_{\langle O \rangle}$.*

Proof. (1) Let $h \in H_{e_2}$. By the downward closure of histories, $e_1 \in h$, so that h determines the element $\Pi_{e_1}\langle h \rangle$ of the partition Π_{e_1} of H_{e_1}. And for any $h' \in H_{e_2}$, we have $h \equiv_{e_1} h'$, as witnessed by $e_2 \in h \cap h'$. So we can set $\Pi_{e_1}\langle e_2 \rangle =_{\mathrm{df}} \Pi_{e_1}\langle h \rangle$.

The proof for (2) is exactly parallel, and is left as Exercise 4.2.

For (3), as $e < O$, for all $h, h' \in H_{\langle O \rangle}$, we have $h \equiv_e h'$. The claim follows from the transitivity of $\equiv_e$ on the set H_e. $\qquad\square$

A similar construction for extending the notation $\Omega_{e_1}\langle h \rangle$ for basic scattered outcomes to $\Omega_{e_1}\langle e_2 \rangle$ and $\Omega_{e_1}\langle O \rangle$ is left as Exercise 4.3.

In order to extend our results to disjunctive outcomes, we first need to extend our propositional notation somewhat. In analogy with the notation $\breve{O}$ for disjunctive outcome events, we write $\breve{H}$ for disjunctive propositional events. Typically, such disjunctive outcomes should have at least two elements. For technical reasons (see Chapter 6.4), it is useful to be more general and allow for one-element disjunctions as well, so that in the following, we only require that the disjunction be non-empty.

Definition 4.9. Let $e \in W$. A *basic disjunctive outcome event of e*, generically written $\breve{O}$, is any non-empty subset of Ω_e (i.e., a set of some basic scattered outcome events of e). A *basic propositional disjunctive outcome of e*, generically written $\breve{H}$, is any non-empty subset of Π_e, (i.e., a set of some basic propositional outcomes of e). The occurrence proposition for $\breve{H}$ is

$$H_{\breve{H}} =_{\mathrm{df}} \bigcup \breve{H}.$$

Given this definition, $\breve{H}$ occurs in precisely those histories in which one of its members (one of the disjuncts) occurs. In full analogy with Def. 4.5, the occurrence proposition for the transition from e to a basic propositional disjunctive outcome $\breve{\mathbf{H}}$ is

$$H_{e \rightarrowtail \breve{\mathbf{H}}} =_{df} (\text{Hist} \setminus H_e) \cup H_{\breve{\mathbf{H}}}.$$

If e is an indeterministic event, both Π_e and Ω_e are disjunctive outcomes of e, but their occurrence propositions exhaust all of H_e, and, accordingly, the occurrence propositions of the respective transitions from e are universal:

Fact 4.8. *Let* $e \in W$. *Then* $H_{\Pi_e} = H_{\langle \Omega_e \rangle} = H_e$, *and* $H_{e \rightarrowtail \Pi_e} = H_{e \rightarrowtail \Omega_e} = \text{Hist}$.

Proof. Let $h \in H_e$, then $h \in \Pi_e \langle h \rangle$, and so $h \in H_{\Pi_e}$. In the other direction, let $h \in H_{\Pi_e}$, so that $h \in \Pi_e \langle h' \rangle$ for some $h' \in H_e$, which implies $h \in H_e$. The claim for $H_{\langle \Omega_e \rangle}$ follows by Fact 4.3, and the claim about the occurrence propositions of the respective transitions is an immediate consequence of the definitions.   □

In a somewhat idealized fashion, we can represent the die rolling example from the end of Chapter 4.1 as follows: the rolling of the die corresponds to a point event e with six immediate basic scattered outcome events, $\hat{O}_i = \Omega_e \langle h_i \rangle = \boxed{i}$, $i = 1, \ldots, 6$. Exhaustiveness then means that $\Omega_e = \{\hat{O}_i \mid i = 1, \ldots, 6\}$. In this representation, it is immediately clear that $e \rightarrowtail \Omega_e$ is a deterministic transition to a disjunctive outcome, as $H_e = H_{\langle \Omega_e \rangle}$.

By Fact 4.3, the equivalence between basic disjunctive and basic propositional disjunctive outcomes also holds in non-extremal cases. We therefore have two equivalent representations of basic outcomes, single or disjunctive, and consequently, of basic transitions: in terms of propositions or sets of propositions ($\Pi_e \langle h \rangle$ or $\breve{\mathbf{H}}$) and in terms of scattered outcomes or disjunctive outcomes ($\Omega_e \langle h \rangle$ or $\breve{\mathbf{O}}$). We will use these two equivalent representations almost interchangeably, mostly giving preference to the propositional version in proofs and theorems. The chief place in which we rely on the interchange is in the idea of a *causa causans*, which will be discussed extensively in Chapter 6.

In Chapter 4.3, we will discuss sets of transitions. With a view to that discussion, a note of caution may be useful: transitions to disjunctive outcomes are *not* equivalent to sets of transitions to the disjuncts. Take, for example, some indeterministic $e \in W$ and two $H_1, H_2 \in \Pi_e$, $H_1 \neq H_2$. Then $e \rightarrowtail \{H_1, H_2\}$ is a transition to a basic disjunctive outcome of e, and that outcome occurs on $H_{\{H_1,H_2\}} = H_1 \cup H_2$. On the other hand, $\{e \rightarrowtail H_1, e \rightarrowtail H_2\}$ is a set of two transitions that are incompatible local alternatives and which, therefore, cannot occur together.

4.2.2 A note on basic transitions in BST$_{\text{NF}}$

We end this section with an observation concerning basic transitions in the BST$_{\text{NF}}$ framework. Recall that by Def. 3.12, $H_{\ddot{e}} = \bigcup_{e \in \ddot{e}} H_e$. Read propositionally, the occurrence proposition for a choice set $\ddot{e}$ is therefore $H_{\ddot{e}} = H_{[\mathscr{P}_e]}$ (see Theorem 3.1). Since in a BST$_{\text{NF}}$ structure there is a minimal element in the difference of any two histories, an element of a choice set uniquely singles out a choice. It is therefore tempting to identify a BST$_{\text{NF}}$ basic transition simply with an element of a choice set, $c \in \ddot{c}$, where trivial choice sets $\ddot{c} = \{c\}$ would obviously be allowed as well and would give rise to trivial basic transitions. Thus, in this proposal, any point event of *Our World* counts as a transition. Care is, however, needed when passing to propositional basic outcomes, since the set H_c of histories by itself carries no information about the relation of c to other point events of the structure. That is, there will generally also be point events $e \neq c$ for which $H_c = H_e$. To avoid ambiguity, we could associate the history set H_e with the point event e, and take the propositional basic transition corresponding to e to be the pair of e and H_e, which we could write $e \rightarrowtail H_e$.

That is still troublesome, however, as the occurrence proposition of $e \rightarrowtail H_e$ is universal. Recall that via Def. 4.5, we opted for an implication-like reading of occurrence propositions for transitions. But if e occurs, then H_e occurs as well. So $H_{e \rightarrowtail H_e} = \text{Hist}$, providing no information at all.

To resolve this difficulty, we take as the initial not e, but the choice set $\ddot{e}$, which is still weakly before e in the sense that $e \in \ddot{e}$. We define two kinds of BST$_{\text{NF}}$ basic transitions in analogy to the BST$_{92}$ case, as follows:

Definition 4.10 (Basic transitions in BST$_{\text{NF}}$). For $\langle W, < \rangle$ a BST$_{\text{NF}}$ structure and for $e \in W$, any pair $\langle \ddot{e}, e \rangle$ with $e \in \ddot{e}$, written as $\ddot{e} \rightarrowtail e$, is a *basic transition event*. The pair $\langle \ddot{e}, H_e \rangle$, written as $\ddot{e} \rightarrowtail H_e$, is a *basic propositional transition*. Both $\ddot{e} \rightarrowtail e$ and $\ddot{e} \rightarrowtail H_e$ are called *basic transitions* in BST$_{\text{NF}}$.

Basic transition events are ordered by $(\ddot{e}_1 \rightarrowtail e_1) \prec (\ddot{e}_2 \rightarrowtail e_2)$ iff $e_1 < e_2$, and basic propositional transitions are ordered by $(\ddot{e}_1 \rightarrowtail H_{e_1}) \prec (\ddot{e}_2 \rightarrowtail H_{e_2})$ iff $e_1 < e_2$.

The occurrence proposition for a basic transition $\ddot{e} \rightarrowtail e$ (as well as for a basic propositional transition $\ddot{e} \rightarrowtail H_e$) is $(\text{Hist} \setminus H_{\ddot{e}}) \cup H_e$, which provides an analogon of Fact 4.3 about the interchangeability of basic transition events and basic propositional transitions in BST_{NF}.

Note that unless $\ddot{e} = \{e\}$ (i.e., $\ddot{e} \rightarrowtail e$ is deterministic), the occurrence proposition of $\ddot{e} \rightarrowtail e$, $(\text{Hist} \setminus H_{\ddot{e}}) \cup H_e$, is contingent, not universal. And clearly, the ordering $\prec$ of basic transitions is a strict partial ordering.

It turns out that a fact analogous to Fact 4.7 also holds for basic transitions in BST_{NF}:

Fact 4.9. *(1) Let $e_1 < e_2$. Then there is exactly one basic outcome of $\ddot{e}_1$ that is compatible with e_2. We write the corresponding basic transition as $\ddot{e}_1 \rightarrowtail \Pi_{\ddot{e}_1}\langle e_2 \rangle$. (2) Let $e_1 < O$ for an outcome chain O. Then there is exactly one basic outcome of $\ddot{e}_1$ that is compatible with O, which we denote by $\Pi_{\ddot{e}_1}\langle O \rangle$. (3) Let $e < O$, $h_1 \perp_{\ddot{e}} h_2$, and $h_2 \in H_{\langle O \rangle}$. Then for every $h \in H_{\langle O \rangle}$: $h_1 \perp_{\ddot{e}} h$.*

Proof. (1) Since $H_{e_2} \cap H_{e_1} = H_{e_2} \neq \emptyset$, e_2 and $\ddot{e}_1 \rightarrowtail e_1$ are compatible: their occurrence propositions intersect non-emptily. Since distinct elements of $\ddot{e}_1$ must be incompatible and histories are downward closed, no other basic transition from $\ddot{e}_1$ is compatible with e_2. We have $\Pi_{\ddot{e}_1}\langle e_2 \rangle = H_{e_1}$. (2) and (3) The arguments for these claims are analogous to the one just given.   □

4.3 Sets of basic transitions

In later chapters we will continue to employ a dual view of transitions as either BST events of a spatio-temporal-modal kind, as introduced via Def. 4.4, or as proposition-like objects as discussed in Chapter 4.2 (see Def. 4.8). In the latter approach, the notion of a basic transition is best generalized via *sets of basic transitions*. We introduce a number of relevant notions here, retaining the BST_{92} framework for concreteness. Our definitions easily transfer to the BST_{NF} framework.

As stated in Chapter 3.4.4 (Def. 3.9), there are two kinds of basic transitions, deterministic and indeterministic ones. Deterministic basic transitions are trivial, from a deterministic point event e (a point event that is not a choice point) to the only immediate outcome of e; indeterministic transitions are from an indeterministic point event (a choice point) e to one of the several immediate outcomes of e. For some applications, it is useful to look at all transitions, deterministic and indeterministic alike—and this approach was taken to define $\mathrm{BST_{NF}}$ in Chapter 3.5. In many other contexts, however, it is most useful to disregard the trivial deterministic transitions, and to focus exclusively on indeterministic ones. This is the route we will follow in our discussion of modal funny business (Chapter 5), of causation (Chapter 6), of probabilities (Chapter 7), and in many applications. Accordingly, we develop our notation here with a view to later uses of sets of indeterministic transitions.

Working in terms of propositions, an indeterministic basic transition is of the form

$$\tau = e \rightarrowtail H, \quad \text{where } e \in W, H \in \Pi_e, \text{ and } \Pi_e \neq \{H_e\},$$

where the last clause implies that e, and thereby τ, is indeed indeterministic. In order to be able to identify initials and outcomes easily, we often write sets of such non-trivial basic transitions as

$$T = \{\tau_\gamma = e_\gamma \rightarrowtail H_\gamma \mid \gamma \in \Gamma\}, \quad \text{where } \Gamma \text{ is some index set.}$$

Here are some pertinent definitions.

Definition 4.11 (Notation for sets of transitions). Let $\langle W, < \rangle$ be a $\mathrm{BST_{92}}$ structure.

- We denote the set of all basic indeterministic transitions in W by $\mathrm{TR}(W)$, as already announced on p. 59.
- For $h \in \mathrm{Hist}(W)$, we write $\mathrm{TR}(h)$ for those basic indeterministic transitions that occur non-vacuously in h. That is, we have

$$\mathrm{TR}(h) =_{\mathrm{df}} \{\tau = e \rightarrowtail H \in \mathrm{TR}(W) \mid h \in H\}.$$

- We write $H(T)$ for the set of histories admitted by the outcomes of a set of transitions T. That is, for $T = \{\tau_\gamma = e_\gamma \rightarrowtail H_\gamma \mid \gamma \in \Gamma\}$ (where Γ is some index set), we set

$$H(T) =_{\mathrm{df}} \bigcap_{\gamma \in \Gamma} H_\gamma.$$

We extend this notation to single transitions, writing $H(\tau)$ in place of $H(\{\tau\})$. That is, for $\tau = e \rightarrowtail H$, we have $H(\tau) = H$.

Given that $H(T)$ is a set of histories (i.e., a proposition), the notion of consistency of Def. 4.1 naturally applies. We extend that notion to sets of transitions in the obvious way:

Definition 4.12 (Consistency of a set of transitions). We call a set of transitions T *consistent* iff $H(T)$ is consistent (i.e., iff $H(T) \neq \emptyset$). A consistent set of transitions thus admits at least one history. If $H(T) = \emptyset$, we call T *inconsistent*.

The above notation, as well as Def. 4.12, naturally extends to sets of transitions in BST_{NF}.

4.4 Topological aspects of BST

In the following section we describe the natural topology for common BST structures, and we comment on some of the topological features of BST_{92} and of BST_{NF}. To recall, a topology on a set X is given by specifying a family of subsets of X, known as "open sets", that is closed under finite intersection and arbitrary union, and which contains X as well as the empty set. (See, e.g., Munkres, 2000, for an overview.)

4.4.1 General idea of the diamond topology

BST admits a natural topology, introduced by Paul Bartha,[3] which we call the diamond topology. The topology is defined either for W, the base set of a BST structure, or for a given history $h \in \text{Hist}(W)$. In the definitions below, $MC(e)$ stands for the set of maximal chains in W that contain e, whereas $MC_h(e)$ stands for the set of maximal chains in history h that contain e.

Definition 4.13 (Diamond topology $\mathscr{T}$ on W). Z is an open subset of W, $Z \in \mathscr{T}$, iff $Z = W$ or for every $e \in Z$ and for every $t \in MC(e)$, there are $e_1, e_2 \in t$ such that $e_1 < e < e_2$ and the diamond $D_{e_1,e_2} \subseteq Z$, where

$$D_{e_1,e_2} =_{df} \{e' \in W \mid e_1 \leqslant e' \leqslant e_2\}.[4]$$

---

[3] Cf. note 26 of Belnap (2003b), the "postprint" of the original BST paper (Belnap, 1992).
[4] Note that the diamonds themselves are *not* open sets. It is possible to introduce borderless diamonds, which are in fact open sets in the topology defined here, but they are harder to work

Definition 4.14 (History-relative diamond topologies $\mathscr{T}_h$ on W). For $h \in$ Hist, Z is an open subset of h, $Z \in \mathscr{T}_h$, iff $Z = h$ or for every $e \in Z$ and for every $t \in MC_h(e)$, there are $e_1, e_2 \in t$ such that $e_1 < e < e_2$ and the diamond $D_{e_1, e_2} \subseteq Z$.

It is not too difficult to check that $\mathscr{T}$ and $\mathscr{T}_h$ are indeed topologies; that is, both the empty set and the base set (W or h, respectively) are open, the intersection of two open sets is open, and the union of countably many open sets is open (see Exercise 4.4). The claim of the naturalness of the diamond topology is based on the observation that for an important class of BST structures, this topology coincides with the standard open-ball topology on $\mathscr{T}(\mathbb{R}^n)$ (to be described in Def. 4.16 on p. 95) and that the notion of convergence it induces coincides with the order-theoretic notions of infima and suprema.[5] As one can see from the definition, the history-relative topologies are the so-called subspace topologies induced by the diamond topology on W, by taking a history as a subspace of W. This means that $A \in \mathscr{T}_h$ iff there is $A' \in \mathscr{T}$ such that $A = A' \cap h$.

In BST$_{92}$, the global topology and the history-relative topologies have different features. As we will show, this fact reflects a problem with local Euclidicity.

4.4.2 Properties of the diamond topology in BST$_{92}$

We review here some facts about the diamond topology in BST$_{92}$. The first observation is that unless $\langle W, < \rangle$ is a one-history structure, the history-relative and the global topologies disagree with respect to a topological separation property called the Hausdorff property. That property is defined as follows:

Definition 4.15 (Hausdorff property). A topological space $\langle X, \mathscr{T}(X) \rangle$ is *Hausdorff* iff for any distinct $x, y \in X$ there are disjoint open neighborhoods of x and of y (i.e., there are $O_x, O_y \in \mathscr{T}(X)$ for which $O_x \cap O_y = \emptyset$).

Putting aside BST$_{92}$ structures that are pathological in the sense that they prohibit the construction of light-cones, it can be proved that the

---

with technically (Placek et al., 2014, Def. 23). Simply removing e_1 and e_2 from the definition does not help, as the borders of the respective space-time region will then be retained.

[5] For a discussion of the naturalness of the diamond topology, see Placek et al. (2014, §6). See also Fact 9.13 in Chapter 9.

history-relative topologies $\mathscr{T}_h$ on a BST_{92} structure have the Hausdorff property.[6] This fact stands in sharp contrast with the properties of the global diamond topology $\mathscr{T}$: if a BST_{92} structure has more than one history, its global topology is non-Hausdorff (again, ignoring pathological structures); see Figure 3.1. In fact, the non-Hausdorffness of the global topology is related to the existence of upper-bounded chains that have more than one history-relative supremum. As one might expect, a pair of distinct history-relative suprema of a chain provides a witness for non-Hausdorffness: if any two open sets in $\mathscr{T}$ each contain a distinct supremum, they must overlap because they share some final segment of the chain in question.

These results about Hausdorffness in the diamond topology in BST_{92} appear encouraging as far as the relation to physics in concerned. In physics it is standardly required that individual space-times be Hausdorff (see, e.g., Wald, 1984, p. 12). As individual space-times are represented by single histories in a BST_{92} structure, we take it that BST_{92} structures are not in conflict with the Hausdorffness requirement of space-time physics. The non-Hausdorffness of the global topology of a BST_{92} structure simply reflects the fact that such a structure brings together more than one history (space-time), explicitly representing a number of alternative spatio-temporal developments.

There is, however, another difference between the history-relative and the global topologies in BST_{92} that is more problematic: again, putting aside trivial one-history structures, a history h is not open in the global topology $\mathscr{T}$, whereas it is open by definition in its own history-relative topology $\mathscr{T}_h$. Generally, if an open set A from a history-relative topology $\mathscr{T}_h$ contains a choice point, then $A \notin \mathscr{T}$:

Fact 4.10. *Let* $\langle W, < \rangle$, $h \in \mathrm{Hist}(W)$ *and* $A \in \mathscr{T}_h$. *Then, if* A *contains a choice point,* $A \notin \mathscr{T}$. *This implies that unless* Hist *has only one member, for any* $h \in \mathrm{Hist}$, $h \in \mathscr{T}_h$, *but* $h \notin \mathscr{T}$.

Proof. Let $A \in \mathscr{T}_h$, and let $e \in A$ be a choice point, so that $h \perp_e h'$ for some $h' \in H_e$. Thus, e is not maximal in W, and hence, not maximal in h' (by Fact 2.1(9)). Now pick a maximal chain $t \in MC_{h'}(e)$, so that $e \in t$ and $t \subseteq h'$. By Fact 3.4, $t \in MC(e)$. As e is not maximal in h' and t is a maximal chain in h', t extends above e in h'. For A to be open in $\mathscr{T}$, by Def. 4.13 there needs to be $e_2 \in t$, $e < e_2$, such that $e_2 \in A \subseteq h$. But, since e is a choice point for h

---

[6] For the proofs, see Placek et al. (2014). The mentioned pathological BST_{92} structures violate one of the conditions C1–C4 discussed in that paper.

and h', which is maximal in the intersection of h and h', there is no such e_2. Thus, $A \notin \mathscr{T}$. Note that in a BST_{92} structure with more than one history, by PCP_{92}, any history contains at least one choice point.   □

There is thus a systematic discrepancy between the global and the history-relative notions of openness. This result spells trouble for an important topological property called local Euclidicity. Technically, this property is defined as follows:

Definition 4.16 (Local Euclidicity). *A topological space $\langle X, \mathscr{T}(X) \rangle$ is locally Euclidean of dimension n iff for every $x \in X$ there is an open neighborhood $O_x \in \mathscr{T}(X)$ and a homeomorphism φ_x that maps O_x onto an open set $R_x \in \mathscr{T}(\mathbb{R}^n)$. Here, $\mathscr{T}(\mathbb{R}^n)$ is the standard so-called open ball topology of $\mathbb{R}^n$, which has as a basis open balls of the form $B(x, \varepsilon) =_{df} \{y \in \mathbb{R}^n \mid d(x,y) < \varepsilon\}$ according to the standard Euclidean distance d.*

In Chapter 3.6.1 we already noted that local Euclidicity is standardly presupposed, often without mentioning the condition by name, when the notion of a space-time manifold is introduced. On such a manifold, local coordinates are defined via so-called charts (see, e.g., Wald, 1984, pp. 12f.), and the existence of charts is guaranteed by local Euclidicity: at each point of the manifold, one can find a neighborhood that is homeomorphic to some open set of $\mathbb{R}^n$. In this way (additionally assuming some compatibility requirements between charts), coordinates can be introduced. If a topological space is not locally Euclidean, it is not possible to assign coordinates in this way.

We also noted that given the frugality of the BST_{92} postulates, BST_{92} structures can differ widely. It would not be realistic to hope that their global topology will always be locally Euclidean—but one can reasonably require that local Euclidicity should transfer from the individual histories to the whole global structure. More precisely, if for each history h of $\langle W, < \rangle$, $\mathscr{T}_h$ is locally Euclidean, then the global topology $\mathscr{T}$ should be locally Euclidean as well. If we have some collection of physically reasonable space-times, each with an assignment of coordinates, then a BST analysis of indeterminism should not destroy the coordinate assignment. Unfortunately, in BST_{92} local Euclidicity is not preserved as one moves from the history-relative topologies to the global topology. As a case in point, in Chapter 3.6.1 we already discussed the simple example from Figure 3.1 (p. 44).

4.4.3 The diamond topology in BST_{NF}

The situation of BST_{92} is unfortunate with respect to local Euclidicity. It would be better if we could have a BST framework for local indeterminism that preserves local Euclidicity: if each history (space-time) is locally Euclidean of dimension n, then the global topology should be locally Euclidean of dimension n as well. As we saw in §4.4.2, the diamond topology on BST_{92} structures does not preserve local Euclidicity when moving from the history-relative topologies to the global topology. In contrast, we can prove that the diamond topology on BST_{NF} structures preserves local Euclidicity. Working toward Theorem 4.1 about the preservation of local Euclidicity, we first need an auxiliary Lemma, which is also of interest on its own. Recall the disturbing feature of BST_{92} discussed as Fact 4.10 in Chapter 4.4.2: a set that is open in a history-relative topology need not be open in the corresponding global topology. The Lemma below states that this problem cannot occur in the diamond topology on BST_{NF} structures:

Lemma 4.1. *Let a BST_{NF} structure $\langle W, < \rangle$ be given, let $h \in \mathrm{Hist}(W)$, and let $Z \subseteq W$ be such that $Z \in \mathscr{T}_h$, i.e., Z is an open set with respect to the history-relative topology $\mathscr{T}_h$. Then $Z \in \mathscr{T}$, i.e., Z is also open with respect to the global topology on W.*

Proof. Let $Z \in \mathscr{T}_h$ for some $h \in \mathrm{Hist}$. Let $e \in Z$, and let $t \in MC(e)$. In order to establish the openness of Z with respect to $\mathscr{T}$, we need to show that there is an e-centered diamond with vertices on t wholly contained in Z. The openness of Z with respect to $\mathscr{T}_h$ gives us such a diamond for any $t_h \in MC_h(e)$, but not necessarily for our given $t \in MC(e)$.

We show that the given t has a segment both below and above e that is contained in some $t_h \in MC_h(e)$. The segment below e is contained in h by downward closure of histories. For the segment above e, we proceed in two steps. First, we claim that t contains some $e' \in h$ for which $e' > e$. Assume otherwise, i.e., the chain $t^+ =_{df} \{e^* \in t \mid e^* > e\}$ contains no element of h. Note that by construction, $\inf t^+ = e$. As t^+ is a chain, it is directed, and thus wholly contained in some history h_2. Pulling these facts together, $t^+ \subseteq h_2 \setminus h$, and by the maximality of t and the construction of t^+, we have that t^+ is a maximal chain in $h_2 \setminus h$. The PCP_{NF} gives us a choice set $\ddot{c}$ such that (†) $h \perp_{\ddot{c}} h_2$, and for the unique $c' \in \ddot{c} \cap h_2$, we have $c' \leqslant t^+$. We observe next that from the fact that t^+ is a maximal chain in $h_2 \setminus h$, it follows that $c' = \inf t^+$.

Otherwise, for $i = \inf t^+$ we would have $c' < i \leqslant t^+$. By (†) we have $c' \notin h$, so $\{c'\} \cup t^+ \subseteq h_2 \setminus h$. As this chain extends t^+, it contradicts the maximality of t^+ in $h_2 \setminus h$. Thus, $c' = \inf t^+$, whence $c' = e$. It follows that $e \in h_2 \setminus h$, which contradicts our initial assumption that $e \in h$. So indeed, t contains some $e' \in h$ for which $e' > e$.

Second, we construct t_h by starting with an initial segment of the given t, as follows: Let $t^- =_{\mathrm{df}} \{e^* \in t \mid e^* \leqslant e'\}$; we have $t^- \subseteq h$ and $e \in t^-$. By the Hausdorff maximal principle we can extend t^- with elements of h to form a chain t_h that is maximal in h, so that $t_h \in MC_h(e)$. The chains t and t_h share the initial segment t^-. We can now invoke the openness of Z with respect to $\mathcal{T}_h$ for e and t_h, which gives us a diamond $D_{e_1^h, e_2^h} \subseteq Z$ for which $e_1^h, e_2^h \in t_h$ and $e_1^h < e < e_2^h$. We set $e_1 =_{\mathrm{df}} e_1^h$ and $e_2 =_{\mathrm{df}} \min\{e', e_2^h\}$. We thus have $e_1 < e < e_2$ with $e_1 = e_1^h \in t$, and also $e_2 \in t$ because $e' \in t$. And as the diamond $D_{e_1, e_2} \subseteq D_{e_1^h, e_2^h}$, we have $D_{e_1, e_2} \subseteq Z$. So we have found the witnessing e-centered diamond with vertices on t, which establishes the openness of Z with respect to $\mathcal{T}$. □

The above Lemma immediately implies a fact about histories in $\mathrm{BST_{NF}}$ that shows that the consequences of Fact 4.10 are avoided in $\mathrm{BST_{NF}}$:

Fact 4.11. *Let $\langle W, < \rangle$ be a $\mathrm{BST_{NF}}$ structures. Then for every $h \in \mathrm{Hist}(W)$: $h \in \mathcal{T}$.*

Proof. By Lemma 4.1, since $h \in \mathcal{T}_h$. □

It is easy to see that the converse of Lemma 4.1 holds as well, that is, if $A \in \mathcal{T}$, then $A \cap h \in \mathcal{T}_h$ (see Exercise 4.5). More interestingly, $\mathcal{T}$ has a handy basis, the elements of which are subsets of histories:

Lemma 4.2. *Let $\langle W, < \rangle$ be a $\mathrm{BST_{NF}}$ structure with diamond topology $\mathcal{T}$. Then the set B, defined as*

$$B =_{\mathrm{df}} \cup_{h \in \mathrm{Hist}(W)} \{O \cap h \mid O \in \mathcal{T}\},$$

is a basis of $\mathcal{T}$.

Proof. We have to show two things: (1) Any element $O \in \mathcal{T}$ is a union of elements of B, and (2) the elements $b \in B$ are in fact open in $\mathcal{T}$.

For (1), let $O \in \mathcal{T}$. Then, since $\cup_{h \in \mathrm{Hist}} h = W$, we have $O = \cup_{h \in \mathrm{Hist}} O \cap h$, and any $O \cap h \in B$ by construction of B. For (2), let $b \in B$ be given, so that there is $O \in \mathcal{T}$ and $h \in \text{Hist}(W)$ for which $b = O \cap h$. By Fact 4.11, $h \in \mathcal{T}$, so that the openness of b follows by the finite intersection property of $\mathcal{T}$. $\square$

With Lemma 4.1 in hand, we can prove the sought-for theorem about the transfer of local Euclidicity from histories to the whole structure in BST_{NF}:

Theorem 4.1. *Let $\langle W, < \rangle$ be a BST_{NF} structure. If there is an $n \in \mathbb{N}$ such that for every $h \in \text{Hist}(W)$, the topological space $\langle h, \mathcal{T}_h \rangle$ is locally Euclidean of dimension n, then the topological space $\langle W, \mathcal{T} \rangle$ is also locally Euclidean of dimension n.*

Proof. We need to show that each $e \in W$ has a neighborhood $O_e \in \mathcal{T}$ that is mapped by some homeomorphism φ_e to an open set $R_e \in \mathcal{T}(\mathbb{R}^n)$. Let $e \in W$, and pick some $h \in H_e$. Since h is locally Euclidean with respect to $\mathcal{T}_h$, there is a $\mathcal{T}_h$-open neighborhood $O_e^h \subseteq h$ of e, an open set of $\mathbb{R}^n$, $R_e^h \in \mathcal{T}(\mathbb{R}^n)$, and a homeomorphism φ_e^h such that $\varphi_e^h[O_e^h] = R_e^h$. By Lemma 4.1, from $O_e^h \in \mathcal{T}_h$ it follows that $O_e^h \in \mathcal{T}$. We thus let $O_e =_{\text{df}} O_e^h$, $R_e =_{\text{df}} R_e^h$, and we can use $\varphi_e =_{\text{df}} \varphi_e^h$ as our homeomorphism between the $\mathcal{T}$-open neighborhood O_e of e and the open set $R_e \in \mathcal{T}(\mathbb{R}^n)$. $\square$

BST_{NF} thus vindicates the idea that if one starts with locally Euclidean histories (space-times) that allow for the assignment of spatio-temporal coordinates, one does not destroy that feature by analyzing indeterminism within the framework of Branching Space-Times.

4.5 A note on branching-style semantics

As we have shown in Chapters 4.1 and 4.2, BST—both in the form of BST_{92} and in the form of BST_{NF}—can be developed to provide a theory of events as well as a theory of propositions, and both approaches will be used in later chapters. When we introduce the notion of a cause-like locus for an outcome (Chapter 5.3, Def. 5.10) and, more generally, for a transition (Chapter 6.3.1, Def. 6.1), we will describe a cause-like locus as a risky juncture for the occurrence of some event. We will motivate the notion of a cause-like locus in part through some claims about the truth at such a locus of certain sentences with temporal and modal operators. We will say, for instance, that a cause-like locus for an outcome O is the last event at which both the sentences "it is possible that O will occur" and "it is possible that O will not occur" are true. These claims need not be left on a merely intuitive level, as branching structures can be used to provide formal semantics for languages with temporal and modal operators. Here we recall a few definitions and facts pertaining to how semantic models on branching structures can be constructed and how temporal-modal sentences are evaluated in the resulting models. Where the difference matters, we stick to BST_{92} for simplicity's sake.

Historically, branching structures for the combination of temporal and modal information and for the formal analysis of indeterminism were first discussed in a now famous exchange between Saul Kripke and Arthur Prior in the late 1950s (see Ploug and Øhrstrøm, 2012). Roughly, such branching structures, which became known under the name 'Branching Time' (BT),[7] use a tree-like ordering to depict the difference between an open future of possibilities and a fixed past: such an ordering is backwards-linear, thus allowing for branching toward the future, but not toward the past. Branching Time provides the formal basis for many applications in computer science, in logic, and in philosophy, including the *stit* ('seeing to it that') formal theory of agency in BT (Belnap et al., 2001). In a BT structure, a history is a maximal chain (a maximal linear subset). There is a direct connection to the notion of a history as a maximal directed set in BST; see Belnap (2012). In fact, BT structures are BST structures of a particularly simple kind, namely, BST structures without SLR elements.

It is not easy to fulfill all our intuitive requirements for the notion of an open future as incorporated in natural language expressions, let alone for the subtle natural-language interaction between tenses and modals. Faced with these problems, Prior developed two different BT-based approaches to the formal semantics for a temporal-modal language, which he called 'Peircean' and 'Ockhamist' (Prior, 1967, pp. 126ff.). It is generally acknowledged that the Peircean approach, which is less expressive than the Ockhamist one, faces such serious difficulties that the Ockhamist approach is usually taken as default. That approach was formally precisified by Thomason (1970). For a comprehensive introduction to Ockhamist semantics for BT, see Belnap

---

[7] As already mentioned in note 3 on p. 27, the terminology is unfortunate, suggesting perhaps that time itself is branching, while the theory clearly pictures branching *histories* before the background of a linear order of temporal instants (see also Belnap et al., 2001, p. 29).

et al. (2001, Ch. 8). In what follows, we will stick to the Ockhamist approach.[8]
The Ockhamist language $\mathscr{L}$ based on BT has temporal operators for the past
("it was the case that", P) and for the future ("it will be the case that", F) as well
as the dual modal operators of settledness ("it is settled that", Sett), which is
sometimes also called "historical necessity", and real possibility ("it is really
possible that", Poss).

Note that in generalizing from BT to BST, the addition of space-like
related events in BST permits the introduction of additional spatio-temporal
operators. It is challenging to work out how these new operators should be
defined and especially how they should interact with the mentioned temporal and modal operators. This is a large topic that will not be discussed in this
book.[9] Our explicit motivational claims that refer to semantics concern only
the combination of tenses and possibilities, and therefore we focus on the
BT-based Ockhamist language only. That is, we work with BST structures,
but we assume that the formal language $\mathscr{L}$ that we are dealing with contains
only the usual propositional connectives (written $\land$, $\lor$, $\neg$, $\rightarrow$, $\leftrightarrow$) and the
temporal and modal operators mentioned above, P, F, Sett, and Poss. In later
applications we will need to consider sentences of the form "At st-location
x it is φ", like "The value of electromagnetic field at x is such-and-such." The
truth conditions for such sentences (i.e., with At_x as the main operator) can
only be formulated with respect to a BST model with set S of spatio-temporal
locations (see Definition 2.9), so that $x \in S$.

The most salient feature of Ockhamist logic is that sentences are evaluated
as true or false at an index of evaluation that specifies an event and a history
containing that event. To have a handy notation, we will write e/h to stand
for a pair of exactly this kind, i.e., $e \in W$, $h \in \text{Hist}(W)$, and $e \in h$. A semantic
model $\mathscr{M}$ based on a BST structure $\mathscr{W} = \langle W, < \rangle$ for our language $\mathscr{L}$ is a
pair $\mathscr{M} = \langle \mathscr{W}, \Psi \rangle$, where Ψ is an interpretation function from the set $Sent$
of sentences of $\mathscr{L}$ to the set of sets of indexes of evaluation, i.e., $\Psi : Sent \rightarrow$
$\wp(E/\text{Hist})$, where $E/\text{Hist} =_{\text{df}} \{e/h \mid e \in W \land h \in H_e\}$.

---

[8] Recently, a more general framework for BT semantics, the so-called transition semantics, has
been proposed by Rumberg (2016a), building upon, but going far beyond earlier work of Placek
(2011) and Müller (2014). For a more thorough investigation of formal semantics based on BST, the
transition framework would be the ideal starting place, and we strongly encourage its use. Here we
only provide a brief introduction of simple BT semantics for purely motivational purposes, so that
a detailed introduction to transition semantics seems unwarranted at this point.

[9] For an approach to alternative space-times that centers on operators, see Strobach (2007).

The interpretation function Ψ is required to satisfy the following semantic clauses, where $e/h \models \varphi$ means that $e/h \in \Psi(\varphi)$, to be read as "φ is true at e/h".

$e/h \models \varphi$ iff $e/h \in \Psi(\varphi)$, for φ an atomic sentence of $\mathcal{L}$;

$e/h \models \neg\varphi$ iff it is not the case that $e/h \models \varphi$;

$e/h \models (\varphi \vee \Psi)$ iff $e/h \models \varphi$ or $e/h \models \psi$

(and similarly for the other propositional connectives);

$e/h \models \mathrm{F} : \varphi$ iff for some $e' \in h$ such that $e < e' : e'/h \models \varphi$;

$e/h \models \mathrm{P} : \varphi$ iff for some $e' \in h$ such that $e' < e : e'/h \models \varphi$;

$e/h \models \mathrm{Poss} : \varphi$ iff for some $h' \in H_e : e/h' \models \varphi$;

$e/h \models \mathrm{Sett} : \varphi$ iff for all $h' \in H_e : e/h' \models \varphi$;

$e/h \models At_x : \varphi$ iff $\exists e' : e' \in h \cap x \wedge e'/h \models \varphi$, where $x \in S$.

Note that, generally, the truth of a sentence depends on both parameters, the event e and the history $h \in H_e$. However, since in the clauses for modal operators one quantifies over histories, a sentence beginning with a modal operator is evaluated the same on any history $h \in H_e$, so that we may set:

$$e \models \mathrm{Sett} : \varphi \Leftrightarrow_{\mathrm{df}} \exists h \in H_e \, [e/h \models \mathrm{Sett} : \varphi];$$
$$e \models \mathrm{Poss} : \varphi \Leftrightarrow_{\mathrm{df}} \exists h \in H_e \, [e/h \models \mathrm{Poss} : \varphi].$$

Having provided the background for the semantics, let us return to the motivational use of the semantics in the definition of a cause-like locus later on. Let $e \in W$, and let O be an outcome chain for which $e < O$. We need to explain why e appears to be decisive for the occurrence of O, given that at e, all histories h on which O remains possible ($h \in H_{\langle O \rangle}$) split from some history h' on which O does not occur. Given such e and O, by the clauses above, we have both

$$e \models \mathrm{Poss} : F : (O \text{ is occurring}) \quad \text{and}$$
$$e \models \mathrm{Poss} : \neg F : (O \text{ is occurring}),$$

where the proposition "O is occurring" is formally represented as $H_{\langle O \rangle}$. Furthermore, these two sentences are true at any e' below e. In contrast, for any $e' > e$, we have:

$$e' \models \text{Sett} : \neg F : (O \text{ is occurring}) \quad \text{or}$$
$$e' \models \text{Sett} : F : (O \text{ is occurring}).$$

This justifies our later claim that a cause-like locus e for O such that $e < O$ is decisive for the occurrence of O in the following sense: e is the last event at which both the future occurrence as well as the future non-occurrence of O is possible. At each event in the future of possibilities of e, depending on its location, it is either settled that O will occur, or it is settled that O will not occur.

4.6 Exercises to Chapter 4

Exercise 4.1. Prove that $H_{[I]} =_{\text{df}} \{h \in \text{Hist} \mid I \subseteq h\} = \bigcap_{e \in I} H_e$.

Exercise 4.2. Provide an explicit proof of Fact 4.7(2).

Exercise 4.3. Prove a variant of Fact 4.7 for basic scattered outcomes; that is, show that given $e_1 < e_2$ [$e_1 < O$], there is exactly one basic scattered outcome of e_1 that is compatible with e_2 [with O], which we therefore denote by $\Omega_{e_1} \langle e_2 \rangle$ [$\Omega_{e_1} \langle O \rangle$].

Exercise 4.4. Prove that the diamond topology of Def. 4.13 and the history-relative diamond topologies of Def. 4.14 are indeed topologies for both BST_{92} and BST_{NF}, i.e., prove that both (1) the base set (W or h, respectively) and (2) the empty set are open, (3) arbitrary unions of open sets are open, and (4) finite intersections of open sets are open.

Hint: For intersection (4), identify the relevant maximal chains and appropriate limits on them. (A full proof is given in Appendix B.4.)

Exercise 4.5. Let $\langle W, < \rangle$ be a BST_{92} or BST_{NF} structure. Then for every $h \in \text{Hist}(W)$, if $A \in \mathcal{T}$, then $A \cap h \in \mathcal{T}_h$.

Hint: This establishes that $\mathcal{T}_h$ is a subset topology of $\mathcal{T}$. For the proof, consider the respective maximal chains in W and in h.

5

Modal Funny Business

In this chapter we deal with a type of correlation that is not much discussed and which only becomes properly analyzable by means of BST's focus on possibilities that are localized in both space and time.

There is much interest in correlations, and for good reasons. Correlations are often a good guide to causal dependencies, and we are naturally inclined to look for explanations behind observed correlations. To give a trivial example, the light is on in this room 20% of the time, and the light switch in this room is thrown 20% of the time. And these events are correlated perfectly: the probability of the switch being thrown, given that the light is on, is 100%, as is the probability of the light being on given that the switch is thrown. The explanation for this correlation is a simple direct causal connection: the throwing of the switch causes the light to turn on. Famously, not all correlations are due to direct causation between the correlated events. Sometimes we can explain a correlation conceptually. For example, it is day 50% of the time, and it is night 50% of the time, but whenever it's day it's not night, and vice versa, whereas if day and night were uncorrelated, it should be day half of the time that it is night. This correlation is just due to the fact that day *is* when it's not night, by the very concept of day. No causal link is involved here. In other cases, there is a causal link, but not a direct one. The standard example is a barometer falling and the coming of a storm: these events are highly correlated, but neither causes the other. Rather, there is a common cause, low atmospheric pressure, which causes both the barometer's fall and the storm.

Such probabilistic correlations have been much discussed. BST allows us to dig one level deeper and to unearth and analyze an unspoken assumption behind most talk of probabilistic correlations. As we will discuss at length in Chapter 7, probabilities need to be analyzed on the basis of possibilities; probabilities are graded possibilities. So to understand to what to assign a probability in the first place, we need to understand the underlying possibilities. Standard expositions of probability theory usually assume

Branching Space-Times: Theory and Applications. Nuel Belnap, Thomas Müller, and Tomasz Placek,
Oxford University Press. © Oxford University Press 2022. DOI: 10.1093/oso/9780190884314.003.0005

that one can combine probability spaces smoothly by forming Cartesian products. Behind this construction there is the assumption that the underlying possibilities combine smoothly such as to form Cartesian products—but this is not always the case. BST has the resources to analyze this issue: The spatio-temporal anchoring of possibilities in BST structures allows us to make sense of what we will call *modal correlations*. Roughly, a modal correlation is present whenever possibilities do not combine in the simplest imaginable way. It turns out, interestingly, that BST can distinguish two types of modal correlations. Many modal correlations are to be expected upon a simple reflection on the fact that possiblities in BST are concrete possibilities in space and time. Paying attention to these modal correlations is crucial for the theory of causation and of probabilities that we will build up in Chapters 6 and 7. But these modal correlations are not by themselves strange at all. Pointing to their existence is just highlighting an important lacuna in the general discussion of combining possibilities. The other type of modal correlations for which BST makes conceptual room *is* strange, and we do not take it to be a settled matter that such modal correlations exist in our world. BST does, however, provide a formally precise picture of what our world would have to be like in order for such strange modal correlations, which we call *modal funny business*, to exist.

We will use the term "modal correlation" in exact analogy with the notion of a probabilistic correlation: some joint occurrences are not independent. In probabilistic terms, dependence is expressed in terms of the probability of the joint occurrence vis-à-vis the individual occurrences. Thus, given events A and B in a probability space with probability measure pr, a correlation between A and B means that

$$pr(A \cap B) \neq pr(A) \cdot pr(B).$$

Such correlated individual events do not combine smoothly in the probabilistic sense of independence.

In the modal case, there are no probabilities assigned to the occurrences (yet), so a correlation has to be expressed solely in terms of the presence and absence of combined possibilities. Our main idea is the following: a modal correlation is present whenever two individually possible outcomes do not combine to yield a possible joint outcome. In terms of transitions, the simplest case of a modal correlation consists of two basic transitions that are individually possible, but not jointly possible. That is:

Definition 5.1. Two basic transitions $\tau_1 = e_1 \rightarrowtail H_1$ and $\tau_2 = e_2 \rightarrowtail H_2$ constitute a case of *modal correlation of the simplest kind* iff they are individually possible but not jointly so; that is, iff $H_1 \neq \emptyset$ and $H_2 \neq \emptyset$, but $H_1 \cap H_2 = \emptyset$. In terms of the histories admitted by (sets of) transitions (see Def. 4.11), we can also write this as: $H(\tau_1) \neq \emptyset, H(\tau_2) \neq \emptyset, H(\{\tau_1, \tau_2\}) = \emptyset$.

5.1 Motivation for being interested in modal correlations

Earlier we said that some modal correlations are to be expected, while others would be strange. It is "scientifically natural" to be puzzled by modally correlated transitions whose initials are space-like related—in a way, that is the gist of the famous "EPR" argument against the completeness of quantum mechanics (Einstein et al., 1935, see Chapters 5.2 and 8). On the other hand, we can show that if the initials of two transitions are *not* space-like related, then there is no deeper interest in their modal correlation.

As usual, we work in BST_{92} for concreteness, deferring a discussion of BST_{NF} to Chapter 5.4. Consider then two basic transitions $e_1 \rightarrowtail H_1$ and $e_2 \rightarrowtail H_2$, where $H_i \in \Pi_{e_i}$ ($i = 1, 2$). There are three ways (1)–(3) in which their initials can fail to be space-like related. In each of these cases, we indicate why the question of modal correlation is obviously uninteresting.

1. If the initials e_1 and e_2 are incompatible, there is an inevitable and indeed rampant modal correlation, since every member of H_1 must contain e_1, whereas in virtue of the inconsistency of e_1 and e_2, no member of H_2 can contain e_1. So in this case, modal correlation is trivially inescapable. Since the existence of incompatible point events is a direct consequence of indeterminism (that is, it follows from the bare existence of more than one history), such modal correlations do not by themselves warrant our interest.

2. If $e_1 = e_2$, then intuitively we might not even speak of "correlation". But it is illuminating to spell out the two equally uninteresting cases. Case (a). If $H_1 \neq H_2$ (where $H_1, H_2 \in \Pi_{e_1}$ because $\Pi_{e_1} = \Pi_{e_2}$), then $H_1 \cap H_2 = \emptyset$ as Π_{e_1} partitions H_{e_1}, and modal correlation cannot be avoided. It is a conceptual truth that different immediate outcomes of the same event are incompatible and cannot occur together. Case (b). On the other hand, if $H_1 = H_2$, then the absence of modal correlation is vacuous and of equal lack of interest.

3. If e_1 is in the causal past of e_2 ($e_1 < e_2$), then according to BST_{92} theory, the very occurrence of e_2 is consistent with one and only one basic outcome of e_1, viz., $\Pi_{e_1} \langle e_2 \rangle$—see Fact 4.7. There are two cases. Case (a): Perhaps e_1 has H_{e_1} as its single vacuous basic outcome. This is evidently a case of uninteresting absence of modal correlation. Case (b): e_1 has more than one basic outcome. We know that e_2 is compatible with only one of them, so that *each* of the *other* outcomes of e_1 is incompatible with *all* of the outcomes of e_2; which is equally uninteresting. So if $e_1 < e_2$, "modal correlation" is in either case uninteresting. And, of course, the case is the same if e_2 lies in the causal past of e_1.

What happens when one eliminates these three uninteresting cases? In BST_{92} that is *exactly* to say that e_1 and e_2 are "space-like related": A modal correlation between transitions $e_1 \rightarrowtail H_1$ and $e_2 \rightarrowtail H_2$ is puzzling only if e_1 is space-like related to e_2. In our view, such correlations are interesting in precisely the same way that EPR-like phenomena are, which is why we call such modal correlations a kind of "funny business", an opinion built into the wording of the following *definiendum*.

Definition 5.2. Two basic transitions $\tau_1 = e_1 \rightarrowtail H_1$ and $\tau_2 = e_2 \rightarrowtail H_2$ constitute a case of *space-like-related modal funny business of the simplest kind* $\Leftrightarrow_{df} e_1 \, SLR \, e_2$ and $H_1 \cap H_2 = \emptyset$.

5.2 Modal funny business

The basic message of our discussion of modal correlations so far is this: Quite a number of modal correlations are to be expected, but given the BST framework, there is a class of modal correlations that constitute funny business. Empirically, such funny business appears to be present (at least ideally) in EPR-like scenarios,[1] in which an entangled two-partite quantum

---

[1] So-called because such a scenario figures prominently in the famous 1935 article by Einstein, Podolsky, and Rosen (EPR), "Can quantum-mechanical description of reality be considered complete?" (Einstein et al., 1935). In that paper the authors argue that space-like correlations need an explanation in terms of an expanded description of reality that goes beyond the quantum mechanical formalism. John Bell (1964) showed a way of deriving empirical predictions from this assumption of (certain forms of) quantum-mechanical incompleteness (the existence of so-called hidden variables), and the predictions of a number of hidden variable theories have been found to be empirically violated, providing an argument for the completeness of quantum mechanics *pace* Einstein et al. These issues will be discussed in more detail in Chapter 8.

system shows perfect correlations of outcomes for space-like separated measurements. For example, the maximally entangled singlet state of two spin-1/2 particles, $|\psi\rangle$ (written in the basis $|\pm,\pm\rangle = |\pm\rangle_1 \otimes |\pm\rangle_2$)

$$|\psi\rangle = \frac{1}{\sqrt{2}}(|+,-\rangle - |-,+\rangle)$$

is such that upon measuring the first particle in the $\pm$ basis, the result of a measurement on the second particle in that basis is already determined, even if the measurement events are space-like separated. In modal terms, while for the measurement on the first particle, both the outcomes $+$ and $-$ are possible, and the same holds for the measurement on the second particle, it is impossible that the measurements give the same results, and the only possible joint outcomes are $(+,-)$ and $(-,+)$. The outcomes $(+,+)$ and $(-,-)$, which appear to be combinatorially possible given the individual possibilities, are missing from the set of joint possibilities. The large literature on quantum correlations and their possible causal interpretation is witness to the fact that people find such phenomena strange or funny, or even "spooky" (to use the common translation of Einstein's phrase, "spukhaft").[2]

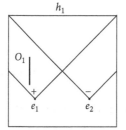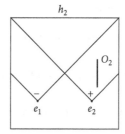

Figure 5.1 BST diagram for the EPR scenario. Choice points e_1 and e_2 represent the measurement events. O_1 and O_2 are outcome chains that represent measurement results. Note that there are only two rather than four histories, even though there are two binary choice points.

The scenario discussed so far has just two binary choice points, thus constituting the simplest possible case in which modal funny business can occur. There are a number of ways of generalizing the notion of modal funny business from this base case to arrive at more encompassing notions of

---

[2] The phrase occurs in a letter from Einstein to Max Born dated March 3, 1947; see Einstein et al. (1971, p. 158). We defer a discussion of the literature to Chapter 8.

modal funny business. We present the two most general definitions in what follows.[3] With a view to the prominence of possibilities described via sets of basic transitions, esp. in Chapters 6 and 7, we phrase our definitions in terms of sets of basic transitions.

5.2.1 Expected inconsistencies in sets of basic transitions

Modal correlations, generally speaking, are present whenever there is a set of transitions that are individually possible but jointly inconsistent in the sense of Def. 4.12 (i.e., not admitting a joint outcome). The individual consistency of a single basic transition is guaranteed as a matter of definition: A basic transition in BST_{92} is of the form

$$\tau = e \rightarrowtail H, \quad H \in \Pi_e,$$

and as e is always a member of some histories ($H_e \neq \emptyset$) and Π_e partitions H_e into non-empty subsets, we have that $H(\tau) = H \neq \emptyset$.

Accordingly, we link the notion of modal correlation to the inconsistency of a set of transitions:

Definition 5.3. A set of transitions

$$T = \{\tau_i = e_i \rightarrowtail H_i \mid i \in \Gamma\}, \text{ where } \Gamma \text{ is some index set}$$

constitutes a case of *modal correlation* iff it is inconsistent, i.e., iff $H(T) = \bigcap_{i \in \Gamma} H_i = \emptyset$.

Our preceding discussion has shown that certain cases of modal correlations are to be expected once one acknowledges the spatio-temporal nature of individual possibilities (transitions). Thus, to mention the simplest case, a set T containing two different transitions from the same initial has to be inconsistent. Such an inconsistency is of the clearest and most easily discernible variety: the set T runs together different local alternatives. We call such an inconsistency "blatant inconsistency":

---

[3] Four different notions of modal funny business were initially introduced by Belnap (2002, 2003c). They all turn out to be special cases of the definitions given here, as shown in Müller et al. (2008). Note that while Theorem 2 in the latter paper is indeed correct as stated, its proof is faulty, as kindly pointed out to us by Leszek Wroński. See Theorem 5.1 for a corrected version.

Definition 5.4 (Blatant inconsistency). A set $T = \{\tau_i = e_i \rightarrowtail H_i \mid i \in \Gamma\}$ of transitions is *blatantly inconsistent* iff there are $\tau_1, \tau_2 \in T$ such that $e_1 = e_2$, but $H_1 \neq H_2$.

Apart from blatant inconsistency, there are two other forms of inconsistency that are not surprising, as the above discussion in Chapter 5.1 has shown. The first form is that the initials of two transitions $\tau_1, \tau_2 \in T$, e_1 and e_2, are incompatible to begin with, that is, that these initials are incomparable and do not belong to any one history. In that case, no outcomes of e_1 and e_2 can share a history as the initials do not share a history to begin with (i.e., $H_1 \cap H_2 = \emptyset$ because $H_{e_1} \cap H_{e_2} = \emptyset$). This is not surprising. The second form of unproblematic inconsistency can occur when the initials e_1 and e_2 of two transitions τ_1 and τ_2 are consistent and in fact comparable (let us assume $e_1 < e_2$ for concreteness). In that case, there is exactly one outcome of e_1 that is compatible with the occurrence of e_2, namely, $\Pi_{e_1}\langle e_2 \rangle$ (see Fact 4.7(1)). So if we have $\tau_1 = e_1 \rightarrowtail H_1 \in T$, where $H_1 \neq \Pi_{e_1}\langle e_2 \rangle$, the transition τ_1 excludes the occurrence of τ_2 by already excluding the occurrence of its initial e_2. Again, it is obvious why the whole set T is inconsistent; the blame is on τ_1 selecting the wrong outcome. Interestingly, this case can be linked to blatant inconsistency in the following way, and which we will make use of later in Chapter 5.2.3. In BST, the occurrence of any point event implies the previous occurrence of its complete spatio-temporal past; histories are closed downward. We thus do not add anything modally substantial (we do not add new choices) if we complete a transition set toward the past. But given that $\tau_2 \in T$ and $e_1 < e_2$, any history $h \in H_2$ contains e_1 *and* fixes the outcome of e_1 to be $\Pi_{e_1}\langle h \rangle = \Pi_{e_1}\langle e_2 \rangle$, a superset of H_2. So adding the transition $\tau_1' =_{df} e_1 \rightarrowtail \Pi_{e_1}\langle e_2 \rangle$ to T should be an innocuous downward extension that just adds a part of the story that was implied by the downward closure of histories anyway.[4] Yet it turns out that the set of transitions $T' =_{df} T \cup \{\tau_1'\}$ is not just inconsistent (as was T), but even blatantly inconsistent, containing now two different transitions τ_1 and τ_1' with the same initial e_1. In this way, we see that inconsistency due to consistent, order-related initials with incompatible outcomes is quite close to blatant inconsistency.[5]

Pulling together the various strands of the discussion so far, we can repeat the observation from Chapter 5.1 that the only surprising cases of modal

---

[4] The case for the term "downward extension" is further strengthened by noting that in terms of the ordering of transitions (Def. 3.10), $\tau_1' \prec \tau_2$.

[5] See Chapter 5.2.3 for more details on downward extensions and the related idea of explanatory funny business as absence of blatant inconsistency in all downward extensions.

correlations—the ones whose presence has no immediate explanation—are those linked to space-like related initials which have incompatible outcomes.

5.2.2 Combinatorial funny business

Summing up the discussion of the expected inconsistencies of sets of basic transitions above, we can say that transition sets of the following form are well-behaved in the sense of not containing a direct case of obvious inconsistency. We call such sets combinatorially consistent:

Definition 5.5 (Combinatorial consistency). A set $T = \{\tau_i = e_i \rightarrowtail H_i \mid i \in \Gamma\}$ of basic transitions is *combinatorially consistent* iff for any $\tau_i, \tau_j \in T$:

1. if $e_i = e_j$, then $H_i = H_j$ (i.e., $\tau_i = \tau_j$);
2. if $e_i < e_j$, then $H_{e_j} \subseteq H_i$ (i.e., $\tau_i \prec \tau_j$);
3. if $e_j < e_i$, then $H_{e_i} \subseteq H_j$ (i.e., $\tau_j \prec \tau_i$);
4. if e_i and e_j are incomparable, then $e_i\,SLR\,e_j$.

Thus, in a combinatorially consistent set of transitions, there is no blatant inconsistency (1), there is no order-related inconsistency (2, 3), and there is no inconsistency related to inconsistent initials (4). It is indeed the last clause (4) that relies on combinatorics, as it suggests that the compatibility of two SLR initials should be enough for a pair of transitions starting with these initials to be consistent.

A combinatorially consistent set looks well behaved. One aspect of this well-behavedness is that any two initials from transitions from such a set T share some history—so T is, in some sense, almost consistent:

Fact 5.1. *If $T = \{\tau_i = e_i \rightarrowtail H_i \mid i \in \Gamma\}$ is combinatorially consistent, then for any two initials e_i and e_j $(i, j \in \Gamma)$, there is a history h containing them both.*

Proof. Let $i, j \in \Gamma$. We argue by cases. (1) If $e_i = e_j$, then any history h from H_{e_i} serves as a witness. (2) If $e_i < e_j$, any history h from H_{e_j} serves as a witness (note that $h \in H_i$, and thus in particular, $e_i \in h$). (3) If $e_j < e_i$, then similarly, take any history h from H_{e_i}. (4) The fact that $e_i\,SLR\,e_j$ implies, by definition, that there is a history h containing them both. ☐

So there is no apparent reason why a combinatorially consistent set of transitions should not be consistent. To support this idea, we can note that sets of transitions that are in fact consistent are also well-behaved according to the definition:

Lemma 5.1. *If a set of basic transitions T is consistent, then it is also combinatorially consistent.*

Proof. Assume that T is not combinatorially consistent. Thus, there are $\tau_1, \tau_2 \in T$ ($\tau_i = e_i \rightarrowtail H_i$, $i = 1, 2$) constituting a counterexample to one of the four clauses from Def. 5.5. For each of these cases, the discussion of Section 5.2.1 has shown that $H_1 \cap H_2 = \emptyset$, so that T is not consistent. $\square$

The other direction does not hold in general but, if it fails, something at least mildly counterintuitive is going on: The set T is well-behaved, but the combinatorics do not work out as expected. Some histories that should witness the consistency of T are, as it were, missing. Thus we define:

Definition 5.6 (Combinatorial funny business). A set of basic transitions T constitutes a case of *combinatorial funny business* (CFB) iff T is combinatorially consistent (Def. 5.5), but inconsistent ($H(T) = \emptyset$).

It should be reassuring that the simplest EPR-like case of funny business modal correlations, discussed earlier, falls under this definition.

Fact 5.2 (The EPR scenario exhibits CFB). *The EPR scenario of Figure 5.1, discussed at the beginning of Section 5.2, constitutes a case of combinatorial funny business according to Definition 5.6.*

Proof. Let e_1, e_2 denote the two binary choice points in the EPR scenario (measurement in the left and right wing of the experiment, respectively). Each of these choice points has two basic scattered outcomes, $\Omega_{e_i}^+$ and $\Omega_{e_i}^-$, so that for $i = 1, 2$, there are two transitions each, $\tau_i^+ = e_i \rightarrowtail \Omega_{e_i}^+$ and $\tau_i^- = e_i \rightarrowtail \Omega_{e_i}^-$. The scenario thus contains four basic indeterministic transitions in total, τ_1^+, τ_1^-, τ_2^+, and τ_2^-. The measurement outcomes $\Omega_{e_i}^\pm$ are correlated to allow only the combination of one $+$ outcome with one $-$ outcome. Thus, out of the four (2×2) histories one would combinatorially expect, only two, $h_1 = h^{+-}$ (containing the transitions τ_1^+ and τ_2^-) and $h_2 = h^{-+}$ (containing τ_1^- and τ_2^+), are possible. Thus, in particular, the joint $++$ outcome cannot happen, meaning that the set of transitions $T =_{df} \{\tau_1^+, \tau_2^+\}$ is inconsistent. This set is, however, combinatorially consistent, as the initials of the two transitions, e_1 and e_2, are *SLR*: they are incomparable, and their compatibility is witnessed, for example, by the history h^{+-} featuring the $+-$ joint outcome. $\square$

5.2.3 Explanatory funny business

According to the preceding discussion, an instance of modal correlations constitutes funny business iff the spatio-temporal layout of the possibilities (transitions) that are combined does not provide a reason for the inconsistency of the whole set of transitions. On the face of it, such a set T looks consistent, and it is a raw additional fact that it is nevertheless inconsistent.

Another line of looking at inconsistency and funny business is opened up by not just looking at the transition set T as given, but by attempting to come up with a satisfactory explanatory account of the inconsistency. As we said earlier, an immediately understandable form of inconsistency is blatant inconsistency (i.e., the running together of incompatible local alternatives). This form of incompatibility is also acknowledged in standard probability theory: If we have two members A, B of the event algebra of a probability space and $A \cap B = \emptyset$, then we know that $pr(A \cap B) = 0$ even if $pr(A) \neq 0$ and $pr(B) \neq 0$.

In probabilistic contexts, correlations are often interesting, and an account for a correlation can often be found by looking at circumstances in the past of the correlated events. Thus, in the case of the correlation between the falling barometer and the impeding storm described at the beginning of this chapter, we find a previous event, the advent of a low pressure weather system, that explains the correlation. Technically, we often assume that given two correlated contemporaneous events A and B,

$$pr(A \cap B) \neq pr(A) \cdot pr(B),$$

there is an additional past event C such that conditional on the occurrence of C, A and B become probabilistically independent, i.e.,

$$pr(A \cap B \mid C) = pr(A \mid C) \cdot pr(B \mid C).$$

In this spirit, Reichenbach (1956) proposed his common cause principle, which has been the subject of much recent research and discussion.[6] When transferred to the context of modal correlations in sets of transitions in BST, we cannot, of course, expect the whole idea of the common cause principle to carry over, but the idea of "explaining a surprising phenomenon

---

[6] See, e.g., Hofer-Szabó et al. (2013) and Wroński (2014).

by looking in the past" generalizes in a useful way. Earlier, in Section 5.2.2, we already pointed out that in case a set of transitions $T = \{\tau_1, \tau_2\}$ (with $\tau_i = e_i \rightarrowtail H_i$, $i = 1, 2$) is inconsistent ($H(T) = \emptyset$) and $e_1 < e_2$, we can provide a local account of that inconsistency by adding in the seemingly innocuous downward extension $\tau_1' =_{df} e_1 \rightarrowtail \Pi_{e_1}\langle e_2 \rangle$, which, lying in the past of e_2, has to have occurred for the initial of τ_2 to occur. The extended set $T' =_{df} T \cup \{\tau_1'\}$ is blatantly inconsistent, which readily explains why it is inconsistent.

It is interesting to inquire as to whether we can always explanatorily extend a given inconsistent transition set such that the blame for the inconsistency is ultimately on some blatant inconsistency (i.e., such that the extended set is blatantly inconsistent). It turns out that the answer is no. The EPR-like example of Figure 5.1 that was discussed in the context of Fact 5.2 already suffices as an illustration: In this structure, there are only two choice points, e_1 and e_2, so that the inconsistent set of transitions $T =_{df} \{\tau_1^+, \tau_2^+\}$ cannot be extended to the past in any way. And that set is inconsistent, but not blatantly inconsistent. So we can have cases of modal correlations that do not have a (local) explanation. We call such cases of modal correlations, accordingly, *explanatory funny business*.

To make this notion formally precise, we start with the notion of downward extension. The idea is that, in searching for the explanation of the inconsistency of a given set of transitions T, we may add transitions in the past, as these cannot get in the way modally speaking (their occurrence is implied by the occurrence of a later transition), and they may add explanatory detail.

Definition 5.7 (Downward extension). *The set of basic transitions T^* is a downward extension of T iff (1) $T \subseteq T^*$ and (2) for any (new) $\tau^* \in (T^* \setminus T)$, there is some $\tau \in T$ for which $\tau^* \prec \tau$.*

We can prove the following facts about downward extensions:

Lemma 5.2. *Let T^* be a downward extension of a given set of basic transitions T. Then the following holds: (1) T^* is consistent iff T is consistent. (2) T^* is combinatorially consistent iff T is combinatorially consistent.*

Proof. (1) "$\Rightarrow$" Let T be inconsistent, i.e., $H(T) = \cap_{\tau \in T} H(\tau) = \emptyset$. Then we have

$$H(T^*) = \cap_{\tau \in T^*} H(\tau) = H(T) \cap (\cap_{\tau^* \in T^* \setminus T} H(\tau^*)) = \emptyset,$$

so that T^* is also inconsistent.

"⇐" Let T be consistent, i.e., $H(T) = \bigcap_{\tau \in T} H(\tau) \neq \emptyset$. Again we can write

$$H(T^*) = \bigcap_{\tau \in T^*} H(\tau) = H(T) \cap (\bigcap_{\tau^* \in T^* \setminus T} H(\tau^*)).$$

Now take some $\tau^* \in T^* \setminus T$. By definition of downward extension, there is some $\tau \in T$ for which $\tau^* \prec \tau$, and this implies that $H(\tau^*) \supseteq H(\tau)$. So (as $H(T) \subseteq H(\tau)$) we have for any of the new τ^* that $H(\tau^*) \supseteq H(T)$. This implies that

$$H(T^*) = \bigcap_{\tau \in T^*} H(\tau) = H(T) \cap (\bigcap_{\tau^* \in T^* \setminus T} H(\tau^*)) = H(T) \neq \emptyset,$$

establishing the consistency of T^*.

(2) "⇒" Let T^* be combinatorially consistent, which means that for any two $\tau_1 = e_1 \rightarrowtail H_1, \tau_2 = e_2 \rightarrowtail H_2 \in T^*$, one of the four clauses from Definition 5.5 applies. This implies in particular that for any two transitions from $T \subseteq T^*$, one of the four clauses applies, so that T is also combinatorially consistent.

"⇐" Let T be combinatorially consistent, which means that for any two transitions from T, one of the four clauses from Definition 5.5 applies. We have to show that that feature transfers to T^*, which is a downward extension of T. Thus, take some $\tau_1 = e_1 \rightarrowtail H_1, \tau_2 = e_2 \rightarrowtail H_2 \in T^*$. There are three cases, depending on whether or not τ_1 and τ_2 already belong to T.

Case 1. If both $\tau_1, \tau_2 \in T$, then the combinatorial consistency of T alone suffices to show that one of the four clauses holds for τ_1 and τ_2.

Case 2. Assume that $\tau_1 \in T$, but $\tau_2 \in T^* \setminus T$. (The case with τ_1 and τ_2 reversed is exactly analogous.) As τ_2 is part of a downward extension of T, there is some $\tau_3 = e_3 \rightarrowtail H_3 \in T$ for which $(*)$ $\tau_2 \prec \tau_3$ (i.e., $e_2 < e_3$ and $H_{e_3} \subseteq H_2$; see Fact 3.11). As $\tau_1, \tau_3 \in T$ and T is combinatorially consistent, by Fact 5.1 there is $h \in \text{Hist}$ such that $e_1, e_3 \in h$, and since $e_2 < e_3$, $(\dagger)$ $h \in H_{e_1} \cap H_{e_2} \cap H_{e_3}$. We show that one of four clauses applies for τ_1 and τ_2. (1) It cannot be that $e_1 = e_2$. Assume otherwise, then $\tau_2 \in T^* \setminus T$ implies that $H_1 \neq H_2$. Now $e_1 < e_3$ (as $e_1 = e_2$), so the combinatorial consistency of T implies $H_{e_3} \subseteq H_1$. But by $\tau_2 \prec \tau_3$, we also have $H_{e_3} \subseteq H_2$. This is a contradiction, because H_1 and H_2 are by assumption different elements of the partition Π_{e_1}, and $H_{e_3} \neq \emptyset$. (2) If $e_1 < e_2$, we have to show that $H_{e_2} \subseteq H_1$. By combinatorial consistency of T, we have $\tau_1 \prec \tau_3$ as $e_1 < e_3$, so for the h witnessing $(\dagger)$, we have $H_1 = \Pi_{e_1} \langle e_3 \rangle = \Pi_{e_1} \langle h \rangle$. Pick an arbitrary $h' \in H_{e_2}$.

As $e_2 > e_1$, we have $h' \equiv_{e_1} h$, i.e., $h' \in H_1$. (3) If $e_2 < e_1$, we have to show that $H_{e_1} \subseteq H_2$. Analogously to case (2), we have $H_2 = \Pi_{e_2}\langle e_3 \rangle = \Pi_{e_2}\langle h \rangle$. Pick an arbitrary $h' \in H_{e_1}$. Since $h, h' \in H_{e_1}$ and $e_2 < e_1$, we have $h \equiv_{e_2} h'$, and hence $h' \in H_2$. Finally, (4) if e_1 and e_2 are incomparable, then $e_1 \, SLR \, e_2$ by (†).

Case 3. Assume that $\tau_1, \tau_2 \in T^* \setminus T$. Then there are $\tau_1', \tau_2' \in T$, $\tau_i' = e_i' \rightarrowtail H_i'$ such that (i) $\tau_i \prec \tau_i'$ $(i = 1, 2)$. Hence by Fact 5.1, there is $h \in$ Hist such that $e_1', e_2' \in h$, and hence (ii) $h \in H_1 \cap H_2$ by (i). We show that τ_1, τ_2 satisfy one of the four clauses (1)–(4). (1) If $e_1 = e_2$, then (ii) implies that $H_1 = H_2$. (2) If $e_1 < e_2$, then for any $h' \in H_{e_2}$ we have $h \equiv_{e_1} h'$ since $e_2 \in h$ and $e_1 < e_2$, so $h' \in H_1$. Case (3) follows analogously. And (4) if e_1 and e_2 are incomparable, since by (ii) $e_1, e_2 \in h$, it follows that $e_1 \, SLR \, e_2$. $\qquad \square$

If a given set T is inconsistent, one can hope that it will be possible to find a downward extension of T that is blatantly inconsistent. This would plainly make the inconsistency intelligible. If that hope is frustrated, something funny is going on. Thus we define:

Definition 5.8 (Explanatory funny business). A set T of transitions is a case of *explanatory funny business* (EFB) iff (1) T is inconsistent ($H(T) = \emptyset$) and (2) there is no downward extension T^* of T that is blatantly inconsistent.

Note that the verdict of explanatory funny business can be checked by considering a single structure, viz., the maximal downward extension of a given transition set. The maximal downward extension of T is the set

$$T^{*max} =_{\mathrm{df}} \{\tau' \in \mathrm{TR}(W) \mid \exists \tau \in T \, [\tau' \preccurlyeq \tau] \},$$

where $\mathrm{TR}(W)$ is the set of basic indeterministic transitions introduced in Chapter 3.4.4, $\mathrm{TR}(W) = \{e \rightarrowtail H \mid e \in W, \Pi_e \neq \{H_e\}, H \in \Pi_e \}$.

Obviously, if some downward extension T^* of T is blatantly inconsistent, then so is T^{*max}, which is a superset of any T^*, and if no downward extension of T is blatantly inconsistent, then this excludes T^{*max} from being blatantly inconsistent as well.

5.2.4 On the interrelation of combinatorial and explanatory funny business

Given the result about the preservation of (combinatorial) consistency and inconsistency from Lemma 5.2, it turns out that in any case in which there is combinatorial funny business, there is also explanatory funny business:

Lemma 5.3 (Combinatorial funny business implies explanatory funny business). *If a set of transitions T is an instance of combinatorial funny business (i.e., if T is inconsistent, but combinatorially consistent), then T is also an instance of explanatory funny business (i.e., no downward extension of T is blatantly inconsistent).*

Proof. Let T be an instance of combinatorial funny business (i.e., inconsistent but combinatorially consistent), and let T^* be some downward extension of T. By Lemma 5.2(1), T^* is also inconsistent. By part (2) of that Lemma, T^* is also combinatorially consistent. Thus, for any $\tau_1 = e_1 \rightarrowtail H_1, \tau_2 = e_2 \rightarrowtail h_2 \in T^*$, if $e_1 = e_2$, then by clause (1) of Definition 5.5, we have $H_1 = H_2$, i.e., $\tau_1 = \tau_2$. So T^* is not blatantly inconsistent. As T^* was an arbitrary downward extension of T, we have established that T is an instance of explanatory funny business.   □

The other direction of Lemma 5.3, however, fails to hold. The reason is that the notion of combinatorial funny business cannot detect funny business in cases in which transitions with inconsistent initials are present: by Def. 5.6, a transition set that includes inconsistent initials cannot be combinatorially consistent, and therefore cannot be a case of combinatorial funny business. For a relevant example, consider the two structures shown in Figure 5.2.

In each of these two BST_{92} structures, the possible point events e_3 and e_4 are incompatible; they are not order related and do not share a history. Therefore, the set of transitions $T =_{\mathrm{df}} \{e_3 \rightarrowtail \Omega_{e_3}^+, e_4 \rightarrowtail \Omega_{e_4}^+\}$ is inconsistent, but it is also combinatorially inconsistent—the initials e_3 and e_4 fail clause (4) of Definition 5.5. So the definition of combinatorial funny business in both cases gives the verdict that there is nothing funny going on. The notion of explanatory funny business, however, requires us to consider the downward extensions of the transition set T, and these lead to different verdicts. In case (a), the maximal downward extension of T is $T_a = \{e_3 \rightarrowtail \Omega_{e_3}^+, e_4 \rightarrowtail \Omega_{e_4}^+, e_1 \rightarrowtail \Omega_{e_1}^+, e_1 \rightarrowtail \Omega_{e_1}^-\}$, and this set is blatantly inconsistent as it combines different basic outcomes of e_1. So for case (a), the verdict of the definitions of explanatory and combinatorial funny business agree—nothing funny going on. In case (b), however, the maximal downward extension of T is $T_b = \{e_3 \rightarrowtail \Omega_{e_3}^+, e_4 \rightarrowtail \Omega_{e_4}^+, e_1 \rightarrowtail \Omega_{e_1}^+, e_2 \rightarrowtail \Omega_{e_2}^-\}$, and this set is not blatantly inconsistent. So the definition of explanatory funny business signals, correctly, that something funny is going on—the outcomes of the *SLR* initials e_1 and e_2 are, after all, correlated. The definition of combinatorial funny business does not signal anything funny, however, as the initials e_3

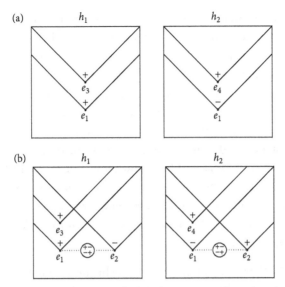

Figure 5.2 Two BST structures with four histories each and with inconsistent initials e_3 and e_4. (Histories h_3 and h_4 are not shown; they correspond to copies of h_1 and h_2 with the "−" outcome of e_3 and e_4, respectively.) Upper panel (a): no funny business, lower panel (b): explanatory funny business.

and e_4 are incompatible, which is enough of a combinatorial reason for the inconsistency of T. We can describe the upshot of this discussion as a useful fact:

Fact 5.3. *In the BST structure of Figure 5.2(b), the transition set $T =_{df} \{e_3 \rightarrowtail \Omega^+_{e_3}, e_4 \rightarrowtail \Omega^+_{e_4}\}$ exhibits explanatory funny business, but no combinatorial funny business.*

We sum up our results about the interrelation of combinatorial and explanatory funny business so far:

Lemma 5.4. *The notion of explanatory funny business properly extends the notion of combinatorial funny business. That is: (1) If a set of transitions T is an instance of combinatorial funny business, then T is also an instance of explanatory funny business. (2) There are instances of transition sets T exhibiting explanatory funny business that are not instances of combinatorial funny business.*

Proof. (1) is the content of Lemma 5.3. (2) has been shown via Fact 5.3. □

Does this mean that the notion of combinatorial funny business is too narrow and therefore inadequate? It would be good if we could salvage the notion of combinatorial funny business in some way, as it captures an important intuition behind the notion of funny business, viz., puzzlement over SLR correlations. In the example of Figure 5.2 (b), starting with the given transition set T, we can exhibit such underlying SLR correlations by downward extending T and then stripping away unnecessary transitions with inconsistent initials. That is: in the given transition set T, there is only explanatory, not combinatorial funny business, but in the BST_{92} structure from which T is taken, there *is* an instance of combinatorial funny business to be found. We will show that this idea generalizes in a useful way: we can show that *at the level of BST structures rather than transition sets*, our two notions of funny business agree.

Theorem 5.1 (There is combinatorial funny business iff there is explanatory funny business). *Let $\langle W, < \rangle$ be a BST_{92} structure. For its set of basic indeterministic transitions, $\text{TR}(W)$, the following holds: There is a subset $T_1 \subseteq \text{TR}(W)$ exhibiting combinatorial funny business iff there is a subset $T_2 \subseteq \text{TR}(W)$ exhibiting explanatory funny business.*

Proof. The "$\Rightarrow$" direction has been established via Lemma 5.3: we can take $T_2 = T_1$. The proof of the "$\Leftarrow$" direction is rather lengthy, so we provide full details in Appendix A.3. The main idea of that proof direction is to work under the assumption that there is no combinatorial funny business in the given structure. We then assume for reductio that there is a set of transitions T that witnesses explanatory funny business. Its maximal downward extension T^* thus contains no instance of blatant inconsistency. Given that there is no combinatorial funny business by assumption, T^* must be inconsistent, but combinatorially inconsistent. The bulk of the proof consists in using T^* to construct a transition set T_C that, contrary to our main assumption, exhibits combinatorial funny business. This proves that our reductio assumption (the existence of explanatory funny business in T) must be false, so that from the assumption of no combinatorial funny business in a given BST_{92} structure, we can show that there can be no explanatory funny business either.   □

The two notions of explanatory and combinatorial funny business are thus equivalent at the level of BST_{92} structures. We take this result to support the view that we have thereby provided a stable explication of the notion of a puzzling modal correlation, or modal funny business, in branching space-times.[7] In the rest of the book, we will therefore always speak of "modal funny business" as a unified concept. Here is our official definition, for which we use CFB—which can be replaced by EFB if one so wishes.

Definition 5.9. A BST_{92} structure $\langle W, < \rangle$ exhibits *modal funny business (MFB)* iff among its set of basic indeterministic transitions $TR(W)$ there is a set $T \subseteq TR(W)$ that constitutes combinatorial funny business according to Def. 5.6.

5.3 Some consequences of modal funny business

In this section, we prove some facts related to the BST_{92} prior choice principle that will be used later on. Recall that PCP_{92} guarantees that for any two histories h_1, h_2 and for any lower bounded chain O, if $O \subseteq h_1 \setminus h_2$, then there is at least one choice point c for h_1 and h_2 that is in the past of O, $c < O$. Such a choice point c is maximal in $h_1 \cap h_2$ (i.e., $h_1 \perp_c h_2$). There can be other choice points for h_1 and h_2, and these may not lie in the past of O. The existence of such choice points not in the past of O is a common feature, independent of whether there is modal funny business in a structure or not. Compare Figure 3.3 (p. 57), in which there are two uncorrelated *SLR* choice points. In the structure depicted in that figure, the outcome chain O_1 belongs to h_2 but not h_3, and these histories split both at the choice point e_1 in the past of O_1 and at the choice point e_2, which is not in the past of O_1. It is not the case, however, that *all* histories in which O_1 begins split off from h_2 at e_2 (the counterexample is h_1). This is different in the EPR scenario of Figure 5.1 (p. 107), which exhibits a related, slightly different notion that is of greater interest here: we may look at choice points at which *all* histories from $H_{\langle O \rangle}$ (all histories in which the chain O begins to occur) split from a given history h. As $H_{\langle O \rangle}$ is the occurrence proposition for O, a choice point of this kind constitutes a decisive event as discussed in our introduction to semantic notions in Chapter 4.5: at such a choice point, the occurrence of O is still contingent (it is possible, but not settled, that O will occur), but immediately after this event, in some histories such as h, the occurrence of O is prohibited (see Fact 4.7). We call such choice points *cause-like loci* for

---

[7] This attitude is further strengthened by noting that EFB and CFB extend the previous notions of funny business due to Belnap (2002, 2003c). See also note 3.

O, written $cll(O)$. Cause-like loci will figure prominently in our discussion of causation and probabilities and in related applications.

Definition 5.10 (Cause-like locus). Let O be a lower bounded chain in a BST_{92} structure $\langle W, < \rangle$. The set of cause-like loci for O, written as $cll(O)$, is defined as:

$$cll(O) = \{e \in W \mid \exists h\, [h \in \text{Hist} \wedge h \perp_e H_{\langle O \rangle}]\},$$

where $h \perp_e H_{\langle O \rangle} \Leftrightarrow_{df} \forall h'[h' \in H_{\langle O \rangle} \rightarrow h \perp_e h']$. We simplify $cll(\{e\})$ to $cll(e)$.

It turns out that whether or not all elements of $cll(O)$ are in the past of O depends on the presence of modal funny business. Indeed, look at the EPR scenario from Figure 5.1, which was already formally discussed in the proof of Fact 5.2. There are two SLR choice points, e_1 and e_2, with two immediate outcomes $\Omega_{e_i}^+$ and $\Omega_{e_i}^-$ each ($i = 1, 2$), and thus the scenario contains the four indeterministic transitions τ_1^+, τ_1^-, τ_2^+, and τ_2^-. As the initials are SLR, one would expect there to be four histories ($h^{++}, h^{+-}, h^{-+}, h^{--}$), but due to modal funny business, there are only the two histories h^{+-} and h^{-+}. That is, the only consistent two-element sets of transitions are $\{\tau_1^+, \tau_2^-\}$ and $\{\tau_1^-, \tau_2^+\}$. Let now O_1 be a chain in $h_1 = h^{+-}$ above e_1 (in $\Omega_{e_1}^+$) that does not lie above e_2, as indicated in Figure 5.1. Intuitively, the occurrence of this chain means that in the left wing of the apparatus, the outcome "+" has been registered. The chain O_1 does not occur in $h_2 = h^{-+}$, in which the left measurement registers outcome "−". It is obvious that $h^{+-} \perp_{e_1} h^{-+}$, and as $H_{\langle O_1 \rangle} = \{h^{+-}\}$, we have $e_1 \in cll(O_1)$, as is to be expected: e_1 is a decisive event for the occurrence of O_1. Yet, as one can easily verify, we have $e_2 \in cll(O_1)$ as well, even though $e_2 \not< O_1$. Thus, something strange is going on.[8] Since the notion of $cll(O_1)$ has causal connotations, it appears that some causal factors responsible for keeping O_1 possible are not in the past of O_1. For further analysis of this weird feature, see Chapter 8. Luckily, however, we can show that this weirdness will not occur if there is no modal funny business in a BST_{92} structure. The following Fact has variants for

---

[8] In Belnap (2002), scenarios like this are called "Some-cause-like-locus-not-in-past funny business". As we said earlier (note 3), our notion of MFB generalizes this notion. Examples proving that MFB properly extends the earlier notions are quite involved; see Müller et al. (2008) for a pertinent construction with infinitely many SLR choice points.

scattered outcomes and disjunctive outcomes, which are left as Exercises 5.3 and 5.4. Here we state and prove the relevant Fact for outcome chains.

Fact 5.4. *Assume that in a BST_{92} structure $\langle W, < \rangle$ there is no modal funny business. For $O \subseteq W$ a lower bounded chain, $h \in$ Hist, and $e \in W$: if $h \perp_e H_{\langle O \rangle}$, then $e < O$. Thus, for every $e \in cll(O)$, we have $e < O$.*

Proof. Let (i) $h \perp_e h_O$, where $h_O \in H_{\langle O \rangle}$ and $h \notin H_{\langle O \rangle}$. Since O is by definition lower bounded, there is $i = \inf(O)$, and we have $i \in h_O$. We will consider the possible ordering relations between e and i.

Consider first (ii) $e \leqslant i$; in that case it is impossible that both $i \in O$ and $e = i$ since then $e \in O$, $e \in h$, and hence $h \in H_{\langle O \rangle}$; then (i) implies $h \perp_e h$, which contradicts the reflexivity of $\equiv_e$. Accordingly, $[i \notin O$ or $e \neq i]$, which together with (ii) implies $e < O$, so we are done. We next show that all other order relations between e and i lead to a contradiction.

Let us then suppose that (iv) $i < e$. Then we claim that $\forall h \, [h \in H_e \to h \in H_{\langle O \rangle}]$, which contradicts (i). For reductio, let (v) $h' \in H_e$ but $h' \notin H_{\langle O \rangle}$, hence $h' \cap O = \emptyset$. On the other hand, there is a non-empty chain $O' = O \cap h_O$ for some $h_O \in H_{\langle O \rangle}$, so $O' \subseteq h_O \setminus h'$. By PCP of BST_{92} there is $c < O'$, hence (vi) $c < O$, such that (vii) $h_O \perp_c h'$. By (vi), since i is the infimum of O, $c \leqslant i$. Together with (iv) this implies $c < e$, and we also have $e \in h_O \cap h'$ from (i) and (v), which contradicts (vii). We thus established the above claim, which proves that the option (iv) is not possible.

The remaining case is that i and e are incomparable. Since $e, i \in h_O$, we get (viii) $i \, SLR \, e$. Consider transitions $\tau_1 = i \rightarrowtail \Pi_i \langle h_O \rangle$ and $\tau_2 = e \rightarrowtail \Pi_e \langle h \rangle$. Since by (viii) the set $\{\tau_1, \tau_2\}$ is combinatorially consistent, and by assumption there is no MFB, the set cannot exhibit CFB, so is consistent as well; that is, $\Pi_i \langle h_O \rangle \cap \Pi_e \langle h \rangle \neq \emptyset$. Let (†) $h^* \in \Pi_i \langle h_O \rangle \cap \Pi_e \langle h \rangle$, so (ix) $h^* \equiv_e h$. We claim now that (x) $h^* \in H_{\langle O \rangle}$. Otherwise there would be $c' < O$ such that (xi) $h^* \perp_{c'} H_{\langle O \rangle}$, by PCP_{92} and an argument like for the case above. By the definition of infimum, $c' \leqslant i$; yet if $c' = i$, then $h^* \perp_i H_{\langle O \rangle}$, hence $h^* \perp_i h_O$, which contradicts $h^* \in \Pi_i \langle h_O \rangle$, where the last follows from (†). And if $c' < i$, then since $i \in h^*$ and $i \in h_O$, we have $h^* \equiv_{c'} h_O$, in contradiction to (xi). We thus established (x), which given (i) implies $h \perp_e h^*$, whereas (ix) says $h^* \equiv_e h$: a contradiction. We have thus shown that incomparability of i and e is not a possible option either, and the only possible ordering is $e < i$, which implies $e < O$, as required. $\square$

We mention two helpful consequences of $cll(O)$ being in the past of O, which follow directly from Fact 4.7(2) and (3): First, given $e < O$, there is a unique element of Π_e that is consistent with $H_{\langle O \rangle}$, for which we write $\Pi_e \langle O \rangle$.

Second, if every cause-like locus $e \in cll(O)$ is in the past of O, every history from $H_{(O)}$ selects the same (unique) outcome $\Pi_e \langle h \rangle = \Pi_e \langle O \rangle$ of e. One may wonder what happens to these two results if for some $e \in cll(O)$, $e \not< O$. Very briefly, in such a case O might be compatible with more than one basic outcome of e, thus singling out a basic *disjunctive* outcome of e. We will return to this topic in our analysis of causation in BST in the presence of MFB (Chapter 6.4).

Here is a further consideration that illustrates the effects of modal funny business. Assume there is a set of point events E that consists of pairwise compatible points (i.e., any two elements of E share some history). Does it follow that E itself is consistent, that there is a history containing all of E? The following Fact shows that this depends on the presence or absence of MFB.

Fact 5.5. *Let $E \subseteq W$ be a set of events that are pairwise compatible, i.e., for any $e_1, e_2 \in E$, there is some $h \in$ Hist for which $e_1, e_2 \in h$. (For example, such a set E could consist of pairwise SLR events.) If there is no history h for which $E \subseteq h$, then W exhibits MFB.*

Proof. Let $E \subseteq W$ be as above, and assume for reductio that no history contains all of E while there is no MFB. Let $C =_{df} \bigcup_{e \in E} cll(e)$ be the set of cause-like loci for all members of E. As by assumption no history contains all of E, the set C is non-empty: Let $E = E_1 \cup E_2$ for non-empty subsets $E_1, E_2 \subseteq E$ such that $E_1 \cap E_2 = \emptyset$ and such that for some history h_1, $E_1 \subseteq h_1$, while $E_2 \cap h_1 = \emptyset$. Then for any $e_2 \in E_2$, H_{e_2} splits off from h_1 at some choice point $c < e_2$, and $c \in cll(e_2) \subseteq C$ (see Fact 3.8).

Consider now the set of transitions

$$T =_{df} \{c \rightarrowtail \Pi_c \langle e \rangle \mid e \in E \wedge c \in cll(e) \wedge c < e\}.$$

We can show that this set is combinatorially consistent: Let $\tau_i = c_i \rightarrowtail \Pi_{c_i} \langle e_i \rangle \in T$ $(i = 1, 2)$. As E consists of pairwise compatible events, there is some history h_{12} containing both e_1 and e_2. It is easy to check that that history serves as a witness for the relevant clause from the definition of combinatorial consistency (Def. 5.5): (1) If $c_1 = c_2$, then $h_{12} \in \Pi_{c_1} \langle e_1 \rangle \cap \Pi_{c_1} \langle e_2 \rangle$, which implies $\Pi_{c_1} \langle e_1 \rangle = \Pi_{c_1} \langle e_2 \rangle$. (2, 3) If $c_1 < c_2$, we have to show that $H_{c_2} \subseteq \Pi_{c_1} \langle e_1 \rangle$. We have $h_{12} \in \Pi_{c_1} \langle e_1 \rangle \cap \Pi_{c_2} \langle e_2 \rangle$. Let now $h' \in H_{c_2}$; as $c_2 > c_1$, we have $h' \equiv_{c_1} h_{12}$, so $h' \in \Pi_{c_1} \langle e_1 \rangle$, which implies $\tau_1 \prec \tau_2$. (4) In case c_1 and c_2 are incomparable, h_{12} witnesses their being SLR.

So, given our assumptions, T is combinatorially consistent. We can now use our assumption that there is no MFB to show that there must be some history containing all of E, which is a contradiction, and thus finishing our proof.

As there is no MFB, the combinatorially consistent set T is in fact consistent: there is some history $h \in H(T)$. We claim that $E \subseteq h$. Assume for reductio that for some $e \in E, e \notin h$, and let $h_e \in H_e$. Then by PCP$_{92}$, there is a choice point $c < e$ for which $h \perp_c h_e$, and indeed $h \perp_c H_e$ (Fact 3.8). So $c \in cll(e)$, and by the definition of T, we have $c \rightarrowtail \Pi_c\langle e \rangle \in T$. But this implies $h \in \Pi_c\langle e \rangle \subseteq H(T)$, contradicting $h \perp_c H_e$. □

5.4 On MFB in BST$_{\mathrm{NF}}$

In this section we outline an account of MFB in BST$_{\mathrm{NF}}$ structures. A full account of this kind requires one to formulate counterparts of the proofs from Section 5.2 in BST$_{\mathrm{NF}}$. By focusing on BST$_{92}$ structures without maximal elements, however, we can use our powerful translatability results between BST$_{\mathrm{NF}}$ and BST$_{92}$ (see Chapter 3.6.2) to prove formal results concerning MFB-related notions in BST$_{\mathrm{NF}}$ much more easily.

Let us thus begin by recalling the format of basic indeterministic transitions in BST$_{\mathrm{NF}}$: an event-type basic indeterministic transition is a pair $\langle \ddot{e}, e \rangle$, written as $\ddot{e} \rightarrowtail e$, where $\ddot{e}$ is a choice set and $e \in \ddot{e}$. On the other hand, a proposition-type basic indeterministic transition is a pair $\langle \ddot{e}, H_e \rangle$, written as $\ddot{e} \rightarrowtail H_e$, where again $\ddot{e}$ is a choice set and $e \in \ddot{e}$. The set $H_{\ddot{e}}$ is defined as $H_{\ddot{e}} = \bigcup_{e \in \ddot{e}} H_e$, and $\Pi_{\ddot{e}}$ is the partition of $H_{\ddot{e}}$ whose elements are the individual $H_e, e \in \ddot{e}$, so that $H_e \in \Pi_{\ddot{e}}$. We define an ordering of choice sets by

$$\ddot{e}_1 < \ddot{e}_2 \text{ iff } \exists e_1' \in \ddot{e}_1 \exists e_2' \in \ddot{e}_2 \, [e_1 < e_2],$$

so that we need only small modifications to formulate a parallel to our Definition 5.5 of combinatorial consistency:

Definition 5.11 (Combinatorial consistency in BST$_{\mathrm{NF}}$). A set $T = \{\tau_i = \ddot{e}_i \rightarrowtail H_i \mid i \in \Gamma\}$ of basic transitions is *combinatorially consistent* iff for any $\tau_i, \tau_j \in T$:

1. if $\ddot{e}_i = \ddot{e}_j$, then $H_i = H_j$ (i.e., $\tau_i = \tau_j$);
2. if $\ddot{e}_i < \ddot{e}_j$, then $H_{\ddot{e}_j} \subseteq H_i$ (i.e., $\tau_i \prec \tau_j$);

124 BRANCHING SPACE-TIMES

3. if $\ddot{e}_j < \ddot{e}_i$, then $H_{\ddot{e}_i} \subseteq H_j$ (i.e., $\tau_j \prec \tau_i$);
4. if $\ddot{e}_i$ and $\ddot{e}_j$ are incomparable (i.e., if neither $\ddot{e}_i = \ddot{e}_j$ nor $\ddot{e}_i < \ddot{e}_j$ nor $\ddot{e}_i > \ddot{e}_j$), then $H_{\ddot{e}_i} \cap H_{\ddot{e}_j} \neq \emptyset$ (i.e., for some $e_1 \in \ddot{e}_1$ and some $e_2 \in \ddot{e}_2$, $e_1 \, SLR \, e_2$).

Thus, in a combinatorially consistent set of transitions, there is no blatant inconsistency (1), there is no order-related inconsistency (2, 3), and there is no inconsistency related to inconsistent initials (4). Having modified the definition of combinatorial consistency for BST$_{\mathrm{NF}}$, the remaining definitions of this chapter remain intact. That is, combinatorial funny business (CFB) is defined as in Def. 5.6, downward extensions are defined as in Def. 5.7, explanatory funny business (EFB) is defined as in Def. 5.8, and at the level of BST$_{\mathrm{NF}}$ structures, MFB is defined exactly as in Def. 5.9.

Importantly, thanks to our translatability results (see Chapter 3.6.2 and Appendix A.2, esp. A.2.5), the BST$_{\mathrm{NF}}$ analogues of all MFB-related results investigated in Section 5.2 hold. As the use of the translatability results is rather mechanical, we illustrate their working by proving just the BST$_{\mathrm{NF}}$ analogon of Theorem 5.1. This should illustrate how to produce counterparts of the remaining (and less difficult) proofs.

Theorem 5.2. *Let $\langle W, < \rangle$ be a BST$_{\mathrm{NF}}$ structure without maximal elements. For its set of basic indeterministic transitions, $\mathrm{TR}(W)$, the following holds: There is a subset $T_1 \subseteq \mathrm{TR}(W)$ exhibiting combinatorial funny business iff there is a subset $T_2 \subseteq \mathrm{TR}(W)$ exhibiting explanatory funny business.*

Proof. "$\Rightarrow$" For reductio, assume that there is a $T_1 \subseteq \mathrm{TR}(W)$ that is a case of CFB, but no set of transitions is a case of EFB. By Lemma A.3(4–5), in $\mathscr{W}' =_{\mathrm{df}} \Lambda(\mathscr{W})$, which is a BST$_{92}$ structure, there is some $T_1' = \tilde{\Lambda}(T_1)$ that is a case of EFB, but no set of transitions is a case of CFB ($\tilde{\Lambda}$ is introduced in Def. A.1). This is, however, impossible by Theorem 5.1, which proves the claim.

"$\Leftarrow$" For reductio, we assume that there is a $T_2 \subseteq \mathrm{TR}(W)$ that is a case of EFB, but no set of transitions is a case of CFB. By Lemma A.3(4–5), in $\mathscr{W}' =_{\mathrm{df}} \Lambda(\mathscr{W})$, which is a BST$_{92}$ structure, there is some $T_2' = \tilde{\Lambda}(T_2)$ that is a case of EFB, but no set of transitions is a case of CFB. This is impossible by Theorem 5.1, so the claim is proved. $\square$

Turning to the implications of MFB, the absence of MFB in a BST$_{\mathrm{NF}}$ structure again has a welcome consequence for the location of cause-like loci. These are defined as follows:

Definition 5.12 (Cause-like locus in BST_{NF}). Let O be a lower bounded chain in a BST_{NF} structure $\langle W, < \rangle$. The set $cll(O)$ is:

$$cll(O) = \{\ddot{e} \subseteq W \mid \exists h\, [h \in \text{Hist} \wedge h \perp_{\ddot{e}} H_{\langle O \rangle}]\},$$

where $h \perp_{\ddot{e}} H_{\langle O \rangle} \Leftrightarrow_{df} \forall h'[h' \in H_{\langle O \rangle} \rightarrow h \perp_{\ddot{e}} h']$.

Note that in BST_{NF}, a chain O may begin at a member of $\ddot{e} \in cll(O)$, so that in place of the strict ordering relation $<$, we have to work with the weak ordering relation, $\leqslant$, in what follows. As in the BST_{92} case, PCP_{NF} guarantees that *some* elements of $cll(O)$ are (weakly) below O, but this does not necessarily hold for all of them—consider again the EPR scenario depicted in Figure 5.1, which indicates a cause-like locus e_2 that is not in the past of the chain O_1. If the EPR scenario is reconstructed as a BST_{NF} structure, it is clearly an example of MFB, now witnessing Def. 5.11. Again, in all cases in which there is no MFB, each element of $cll(O)$ lies in the (weak) past of O, so we have Fact 5.7, a parallel of Fact 5.4. For the proof, we need an additional Fact about outcome chains in BST_{NF}:

Fact 5.6. *Let O be a lower bounded chain and let $i =_{df} \inf O$. In BST_{NF}, for any history $h \in \text{Hist}$, if $i \in h$, then $h \in H_{\langle O \rangle}$.*

Proof. Assume for reductio that $i \in h$, but $h \notin H_{\langle O \rangle}$. Let $h_O \in H_{\langle O \rangle}$, and consider $O' =_{df} O \cap h_O$, which is non-empty by the definition of $H_{\langle O \rangle}$. As O' is an initial segment of O, we have $i = \inf O'$ as well. By our assumption, $O' \subseteq h_O \setminus h$, so by PCP_{NF}, there is some choice set $\ddot{c}$ with $c \in \ddot{c}$, $c \leqslant O'$, and for which $h_O \perp_{\ddot{e}} h$. We have $\ddot{c} \cap h_O = \{c\} \neq \ddot{c} \cap h = \{c'\}$. As $c \leqslant O'$, it must be that $c \leqslant i$ (i being the infimum of O'), so as $i \in h$, we have $c \in h$. But then $\{c, c'\} \subseteq h$, which contradicts Fact 3.13(1) about the unique intersection of choice sets with histories. $\square$

Fact 5.7. *Assume that in a BST_{NF} structure $\langle W, < \rangle$ there is no MFB. Let O be a lower-bounded chain. Then for every $\ddot{e} \in cll(O)$, we have $e \leqslant O$ for some unique $e \in \ddot{e}$.*

Proof. If $cll(O) = \emptyset$, the claim holds vacuously. Let us thus assume $cll(O) \neq \emptyset$, and pick some $\ddot{e} \in cll(O)$. Then there is $h^* \in \text{Hist}$ such that $(*)$ $h^* \perp_{\ddot{e}} H_{\langle O \rangle}$, which means that $h^* \cap \ddot{e} = \{e^*\}$ while for any $h \in H_{\langle O \rangle}$, $h \cap \ddot{e} \neq \emptyset$ and $h \cap \ddot{e} \neq \{e^*\}$.

Since O is lower bounded, it has an infimum $i \leqslant O$. If there is $e \in \ddot{e}$ for which $e \leqslant i$, then $e \leqslant O$, and that member of $\ddot{e}$ is unique (take some $h \in H_i$ and apply Fact 3.13(1)), and so we are done.

We can show that given no MFB, this is in fact the only consistent option. So, assume for reductio that there is no such $e \in \ddot{e}$. It cannot be that *all* $e \in \ddot{e}$ are incompatible with i, due to (*), which implies that any $h \in H_{\langle O \rangle}$ (which contains i) is compatible with some member of $\ddot{e}$. So there has to be some $e \in \ddot{e}$ and some history h such that $\{e, i\} \subseteq h$, and as $e \not\leqslant i$ by our assumption, either (1) $e > i$ or (2) $e \, SLR \, i$.

Case (1) leads to a contradiction: Given that $e > i$, there is a chain $l \in \mathscr{C}_e$ for which $i \in l$, and so by the definition of a choice set, $i < e^*$ as $e^* \in \ddot{e}$. Thus for the history h^* witnessing (*), $i \in h^*$, whence by Fact 5.6, $h^* \in H_{\langle O \rangle}$, which contradicts (*).

Case (2) is also excluded. Given that $e \, SLR \, i$, consider the transitions $\tau_1 =_{df} \ddot{e} \rightarrowtail H_{e^*}$ and $\tau_2 =_{df} i \rightarrowtail H_i$. The set $T =_{df} \{\tau_1, \tau_2\}$ is combinatorially consistent (see clause 4 of Def. 5.11), and so, by the assumption that there is no MFB, T is consistent. This means in particular that there is a history $h' \supseteq \{e^*, i\}$. Again, by Fact 5.6, $h' \in H_{\langle O \rangle}$, which contradicts (*), because as $e^* \in h'$, it must be that $h' \cap \ddot{e} = \{e^*\}$. $\qquad\square$

An analogous fact is true for scattered outcomes—see Exercise 5.5.

5.5 Exercises to Chapter 5

Exercise 5.1. Let O be an outcome chain. Prove that if for all $e \in cll(O)$, we have $e < O$, then $H_{\langle O \rangle} = \bigcap_{e \in cll(O)} \Pi_e \langle O \rangle$.

Hint: The "$\subseteq$" direction is simple. For the other direction, argue indirectly. An explicit proof is given in Appendix B.5.

Exercise 5.2. Suppose that not all elements of $cll(O)$ are in the past of O. Will there still be some formula analogous to the identity established in Exercise 5.1?

Hint: Assume that $e \in cll(O)$. Instead of a basic propositional outcome $\Pi_e \langle O \rangle$, use the basic disjunctive propositional outcome of e that is consistent with O, i.e., the set $\check{\mathbf{H}} =_{df} \{H \in \Pi_e \mid H \cap H_{\langle O \rangle} \neq \emptyset\}$. Note the extra set-theoretic layer.

Exercise 5.3. Prove a version of Facts 5.4 and 4.7(2) for $\hat{O}$ a scattered outcome. That is, prove the following facts:

Let $\hat{O}$ be a scattered outcome. (1) If there is no MFB, then for all $e \in cll(\hat{O})$, we have $e < \hat{O}$. (2) If $e < \hat{O}$, then there is a unique basic outcome of e that is consistent with $H_{\langle\hat{O}\rangle}$, which we denote $\Pi_e\langle\hat{O}\rangle$.

Hint: For (1), use Fact 5.4 and Exercise 5.1. For (2), invoke Fact 4.7(2) for $O \in \hat{O}$. An explicit proof for (1) is given in Appendix B.5.

Exercise 5.4. Formulate and prove a version of Facts 5.4 and 4.7(2) for $\check{O}$ a disjunctive outcome.

Hint: Use Exercise 5.3 for the disjuncts $\hat{O} \in \check{O}$.

Exercise 5.5. Prove the following parallel to Exercise 5.3 in BST_{NF}:

Let $\hat{O}$ be a scattered outcome in a BST_{NF} structure $\mathscr{W}$. (1) If there is no MFB in $\mathscr{W}$, then [if $h \perp_{\ddot{e}} H_{\langle\hat{O}\rangle}$, then $e \leqslant \hat{O}$ for a unique $e \in \ddot{e}$]. (2) If $e \leqslant \hat{O}$ for some $e \in \ddot{e}$, then no other $e' \in \ddot{e}$ is consistent with $H_{\langle\hat{O}\rangle}$.

Hint: Use the BST_{NF} versions of the Facts that are useful for Exercise 5.3, (i.e., Fact 5.7 for (1) and Fact 4.9(2) for (2)).

6

Causation in Terms of *causae causantes*

We now turn to the task of analyzing causation in Branching Space-Times. Since BST has a 'pre-causal order' $<$ as one of its primitives, one might be tempted to pick $<$ as the causal relation. Yet this relation obviously falls short of being a fully fledged causal notion. Consider the pre-causal order of either Minkowski space-time or branching time. No one thinks that in all those cases in which the pre-causal order relation $<$ holds between two events, and in which therefore the first is in the 'causal past' of the second, the earlier event 'is a cause of' the later event. The sun rose six hours and four minutes ago, and a red car is passing the house right now, but the sun's rising is not a cause of the car's passing in any meaningful way. A useful BST theory of causation will have to be more elaborate.

6.1 Causation: Causes and effects as BST transitions

Working toward a more sensible account of causation that will still be based on BST's pre-causal relation, we begin with the uncontroversial assumption that causation is a two-place relation between cause and effect. Controversial, but crucial for any theory of causation, are responses to the following two questions.

Question 6.1. What is caused, that is, which entities are *effects*?

Question 6.2. What causes, that is, which entities are *causes*?

As is well known, there are numerous different answers to these questions. One main divide is between singular and generic causes and effects. Obviously, both notions are important—consider 'Bob's sudden move caused the boat to rock' vs. 'smoking causes cancer'. It seems both ontologically most promising and most in line with our framework to address questions of singular causation first. Thus we look for singular, concrete causes and effects that are anchored in the branching histories of a BST structure. What sorts

Branching Space-Times: Theory and Applications. Nuel Belnap, Thomas Müller, and Tomasz Placek,
Oxford University Press. © Oxford University Press 2022. DOI: 10.1093/oso/9780190884314.003.0006

of entities are causes and effects? A common intuition has it that causation involves change, but it is notoriously difficult to say what precisely constitutes a change. The notion of a transition, developed in Chapter 4.1, is a liberalized notion of change. Following this idea, in causation in Branching Space-Times the crucial causal notion is therefore that of a transition.

The answer to question 6.1, from the point of view of causation in Branching Space-Times, is accordingly:

Answer to Question 6.1: *Transitions* are caused.

Given this decision, one needs to ask for causes of a transition from an initial event to an outcome event. In other words, an effect always has to involve not just an outcome event, but also an initial event. This decision appears to be well aligned with our practice of responding to questions about causes. Typically, a context of uttering a causal question delineates how far into the past we are supposed to look for causes. We hardly ever mention the Big Bang or the deeds of our remote ancestors when addressing questions concerning the causes of, say, the Tunguska event, or Bob's sudden motion in the boat. In practice, we ignore most remote happenings, although they do come up as causes on some accounts of causation. On our BST approach, the context of a causal question excludes remote causal factors by providing an initial event, which is part of the transition constituting the effect that we are after. Given our formal machinery to be developed soon, it is provable that no BST cause of a transition can occur before any element of the initial defining the transition—see Fact 6.1(1).[1]

The crucial idea of causation in Branching Space-Times is that, in sharp contrast to other theories of causation, *non-trivial causation depends on indeterminism*. A deterministic transition $I \rightarrowtail O^*$, whose outcome O^* is bound to occur given that I occurs, has no causes. In terms of occurrence propositions, for a deterministic transition, $H_{[I]} = H_{\langle O^* \rangle}$. Such a transition does not need any causes, since it happens anyway. Causes in BST are thus understood as indeterministic originating causes (*causae causantes*). This seems to capture our actual everyday usage of the notion of causation better than other accounts.

In philosophy, however, the view that non-trivial causation needs indeterminism is a minority view, even though important philosophers such

---

[1] To link this idea to some existing literature, the initial of a transition plays the role of Spohn's (1983) concept of background or *Seinsgrund*, or of the background circumstances discussed in Xu (1997).

as Anscombe (1971) and von Wright (1974) have held that position. In contrast, there are—as far as we know—only a few philosophers who object to the intelligibility of causation in a deterministic universe, which suggests that the combination of causation with determinism is not widely perceived to be controversial.[2] It is, therefore, the view that causation requires indeterminism that needs to be defended.

The theory of causation that we will present here is a formal development of the view that objective causation as objective difference-making requires objective indeterminism. It may sound like a platitude to say that causes make a difference to how the world develops. Yet, how to understand the notion of difference-making, and consequently, how to capture it formally, remains contentious. Different accounts of causation employ relations like deducibility, counterfactual dependence, or manipulability in order to draw a distinction between factors that make a difference, and factors that do not make a difference. Now, if the aim is to give an account of objective causation, the notions used to capture the notion of difference-making need to be shown to be objective. This means to argue, for instance, that counterfactual dependence is relevantly independent of our linguistic practices, or that the notion of manipulation is not invariably tied to the technological and scientific development achieved in a given society at a given time. We do not claim that such arguments cannot be given, but we want to stress that they are genuinely required for accounts of objective causation. In contrast, local objective indeterminism, such as represented in BST, provides an inherent and clear-cut notion of objective difference-making, as there are critical junctions at which things can go differently (see Chapter 4.5). More precisely, at a critical junction, two or more alternative continuations are still possible, but immediately after that junction, some continuations are prohibited, while others are still possible. In other words, what happens at a critical junction makes an objective difference: some developments stop being possible, while other remain possible. Local objective indeterminism thus provides the sought-for notion of objective difference-making. In BST, we can spell this out formally in terms of choice points and non-trivial basic transitions.

---

[2] Philosophers who object to the combination of non-trivial causation and determinism include, as already mentioned, Anscombe and von Wright. A critical view of causation under determinism is also common in some approaches to agency theory. This is important, as an action is certainly a prime example of a causal happening. It is typical in theories of agency to assume that an outcome that is determined (or 'settled') cannot be the outcome of an action. Nobody, that is, sees to what is happening anyway. See Belnap et al. (2001) for a formal statement in terms of the *stit* ('seeing to it that') account of agency.

To continue our defense of the combination of causation and indeterminism, we observe that under determinism it is hard to ensure that a given concept of difference-making picks intuitively adequate candidates for causes: not too many, nor too few, just the right ones. A deterministic world simply does not provide enough useful resources for an objective notion of difference-making.

By way of illustration, we will focus on the concrete example of a deterministic Newtonian pendulum that is completely isolated from its surroundings. Consider a particular event of the bob being in a given position at a given instant. On a determinism-friendly account of objective causation as difference-making, that event has innumerably many causes. After all, the given position is derivable from any instantaneous dynamical state of the pendulum (a pair of the bob's position and momentum at a given instant) and Newton's law of gravitation. Thus, any pair of the bob's instantaneous position and momentum, at any instant—earlier or later than the one under consideration—counts as a cause of the bob's position in question, because any instantaneous position and momentum different from the actual one leads to a different current position of the bob.

One might think that the popular counterfactual approach to causation fares better under determinism, but we are also doubtful. On the counterfactual approach, a cause is an event whose non-occurrence would have prevented the occurrence of the effect. This is taken to mean, roughly, that from among all scenarios (possible worlds) in which the cause-event does not occur, in those that are most similar to actuality, the effect-event does not occur either. It appears that again, this account is overly generous: after all, under determinism, almost any purported change to an earlier state would have resulted in a change to a later state, so that (almost) any earlier state of a deterministic system counts as a cause of its present state. Returning to our Newtonian pendulum, it is plausible to keep Newton's law of gravitation fixed, as sameness of laws trumps other entries on the list of criteria of similarity for possible worlds. But then one has to assent to counterfactuals like "if the bob had had an instantaneous state different from the actual one, it would not have landed in the position that it has now", and so all the previous states count as causes. Hopefully, not every feature of our Newtonian bob is classified as a cause on the counterfactual analysis. Supposing that our bob has a certain color, say blue, we can show that its being blue is *not* a cause of its current position and momentum. Arguably, it is false to say that if the bob had not been painted blue, it would have had a different instantaneous position and momentum than it actually has.[3] Thus, being painted a particular color does not count as a cause of the bob's having a particular instantaneous position and momentum. This is a welcome result, but we do not think that it is satisfactory. The illustration shows that the counterfactual analysis provides a distinction between factors that are relevant for the bob's motion and factors that are irrelevant. But the factors thus relevant to a system's behavior still seem to form a much larger category than the category of properly causal factors. In physics, the family of factors relevant to a system's evolution is subsumed under the heading of 'dynamical states', so each dynamical state should be a cause. This, however, goes against our causal intuitions. We would not say each dynamical state is a cause of a system's behavior, while only irrelevant factors that are not mentioned in respectable physics can count as non-causes. In reaction, one may attempt to define a more discerning category of causal factors by giving a more detailed description of the pendulum, so that one could narrow down the candidates for causes via a relation of similarity. But this move raises the worry that the extra criteria that play a role are too epistemic.

We do not mean to dismiss accounts of deterministic causation out of hand. Our analysis shows, however, that in deterministic contexts, we will either have no causes (this will be the verdict of our own approach), or far too many, or we will have to take recourse to some non-objective criteria. We are after an objective, non-trivializing account in which the only resources that play a role are provided by the objective pre-causal ordering of our world. Such an account will accord with our assumption of indeterminism and with the fact that local transitions are the basic building blocks for objective difference-making in BST. The following answer to our second question, "what are the causes?", therefore seems reasonable:

Answer to Question 6.2: Causes are sets of (especially simple) transitions.

It is important to note that our analysis of causation is linked to indeterminism of a local kind which is captured by BST. If we have a non-trivial, indeterministic transition $I \rightarrowtail \mathscr{O}^*$, by which we mean that I can occur with or without the occurrence of $\mathscr{O}^*$, then even if I occurs, there is at least one local risky junction at which things can go wrong for $\mathscr{O}^*$, that is, at which $\mathscr{O}^*$

---

[3] But note that assenting to this counterfactual requires a specific context. As the counterfactual analysis proceeds in terms of events, we need to bring in an event of painting the bob, plausibly before the bob is set in motion. Then, to assent to the counterfactual, we need to picture the event of painting as not disturbing the bob's later motion in any way. It is unclear whether that is compatible with determinism.

could be prohibited from occurring. But, similarly, there is a development at this junction that keeps the occurrence of $\mathcal{O}^*$ possible. The causes of a non-trivial transition $I \rightarrowtail \mathcal{O}^*$ are, therefore, those developments at risky junctions that keep the occurrence of $\mathcal{O}^*$ possible.

To translate this idea into the formal framework of BST_{92}, a risky junction for a transition $I \rightarrowtail \mathcal{O}^*$ will be identified with a choice event consistent with the initial I and at which the bundle of histories $H_{\langle\mathcal{O}^*\rangle}$ splits off from a history in which $\mathcal{O}^*$ does not occur. To recall Def. 5.10, the set of such crucial choice points for an outcome event $\mathcal{O}^*$ we called the *cause-like loci* for the outcome, $cll(\mathcal{O}^*)$. The developments at a cause-like locus that keep $\mathcal{O}^*$ possible are identified as those basic outcomes of the cause-like locus that are consistent with the occurrence of $\mathcal{O}^*$. In what follows we will consider cause-like loci not just for outcome events, but for transitions, to be written $cll(I \rightarrowtail \mathcal{O}^*)$. The details will be provided via Def. 6.1. Given no MFB, a cause-like locus $e \in cll(I \rightarrowtail \mathcal{O}^*)$ must be in the past of $\mathcal{O}^*$ (see Fact 5.4 and Exercise 5.3(1)). In this case there is then a unique basic outcome $H \in \Pi_e$ of e that is consistent with $H_{\langle\mathcal{O}^*\rangle}$, namely, $H = \Pi_e\langle\mathcal{O}^*\rangle$ (see Fact 4.7(2) and Exercise 5.3(2)). Thus, in the absence of MFB, the individual local causes for $I \rightarrowtail \mathcal{O}^*$ will be the basic transitions $e \rightarrowtail \Pi_e\langle\mathcal{O}^*\rangle$ with $e \in cll(I \rightarrowtail \mathcal{O}^*)$. Each of these basic transitions will be called a *causa causans*, and together they will be called the *causae causantes*, of $I \rightarrowtail \mathcal{O}^*$ (see Def. 6.2).[4]

Our approach implies that a basic transition from a point event e to one of its outcomes, $e \rightarrowtail H$ with $H \in \Pi_e$, is its own cause. This is as it should be: basic indeterministic transitions constitute the irreducibly indeterministic building blocks of our indeterministic world, so that there can be no further account of why they occur. If we ask why such an indeterministic transition occurred, the only answer that we can give is that that is what happened. And if we ask why the specific outcome H of e occurred rather than one of e's other possible outcomes, there can be no answer—it is a conceptual truth that there can be no contrastive explanation of the occurrence of basic indeterminism.[5]

In the probability theory that we develop in Chapter 7, causes understood as basic transitions will form the building blocks for the construction of

---

[4] If there is MFB in the BST_{92} structure under consideration, a certain modification is required; see Chapter 6.4.

[5] There is a tendency, which comes forcefully to the fore in the literature on free will, to keep insisting on contrastive explanations even after it has been acknowledged that the indeterminism in question is basic. See Ometto (2016) and Müller and Briegel (2018) for some discussion.

probability spaces, thus providing a close link between causation and probabilities and between indeterministic and probabilistic causation.

6.2 At least an inus condition

The idea behind the theory we develop in detail below is about tracing causality back to its beginnings in objectively indeterministic originating causes or *causae causantes* (we use these as synonyms).[6] The payoff will be a technical result: we will identify certain objects in a BST structure as *causae causantes* of a transition. Needless to say, these objects will be set-theoretical constructions. This raises the question of what reasons we have to believe that these objects are causes, or represent causes working in our indeterministic world. The question is reminiscent of Tarski's problem of how to ensure that a predicate T, as formally defined by him, is indeed a truth predicate that captures the meaning of the classical concept of truth, which is truth as correspondence. Tarski came up with his criterion of adequacy, namely, a definition of T adequately captures the classical concept of truth if every instance of the so-called truth schema is deducible from the definition of T.[7] Do we have any similar tool to ascertain that particular set-theoretical constructions, *causae causantes*, represent causes indeed? Well, in the BST framework, our constructions follow what we believe is a persuasive idea: in indeterministic settings, causes are those developments at risky junctions that keep the effect's occurrence possible. But surely this appeal to persuasiveness falls short of the rigor of Tarski's muster, as not everybody will be persuaded by our standards for persuasiveness. Fortunately, there is semi-formal support for the claim that our particular set-theoretic constructions represent causes, which comes from Mackie's (1974) analysis of causes as *inus* conditions for the occurrence of their effects: a *causa causans* of a transition is *at least an inus* condition for the occurrence of that transition.

---

[6] In this respect, our theory is indebted to the account of agency of Belnap, Perloff and Xu (2001), which shares an outline with the much earlier theory of von Kutschera (1986).

[7] Tarski discusses the adequacy issue in his famous paper on the concept of truth in formalized languages, which went through a number of versions in different languages: Polish (Tarski, 1933), German (Tarski, 1935), and English (Tarski, 1956). An instance of the truth schema, sometimes called "partial definition of truth", is a sentence like:

"Schnee ist weiß" is T in German iff snow is white.

Let us recall Mackie's idea:

Quasi-definition of inus condition. An inus condition of an event type is 'an insufficient but non-redundant part of an unnecessary but sufficient condition'. (Mackie, 1974, p. 62)

The surrounding text makes it clear that Mackie has in mind a disjunction of conjunctions such as '*ABC* or *DEF* or *JKL*' (this is his example), such that the whole disjunction is a necessary condition of some P (this feature is implicit in Mackie's formula), and each disjunct is sufficient for P, and further each element, such as A, of a disjunct is a non-redundant part of 'its' conjunction. That is, if A is omitted from its conjunction ABC, then the remaining conjunction BC is no longer sufficient for P. Even so, since each B and C is non-redundant as well, A alone is insufficient to bring about P — to this end it needs to be combined with BC. Under these circumstances, A is an 'inus condition' of P, that is, an insufficient but non-redundant part of an unnecessary but sufficient condition for the occurrence of P. In an alternative mouthful, A is a non-redundant though insufficient conjunctive part of a sufficient condition for P that is a disjunctive part of a necessary condition for P. It is worth observing that all the concepts required for stating inus conditions are well-defined in terms of the relations between occurrence propositions in BST (see Definition 4.2), at least once we settle on a reading of 'non-redundant', which will be discussed in Chapter 6.3.2.1.

The central—and perhaps surprising—result of this chapter is that the *causae causantes* of BST have precisely this 'inus' structure. In cases similar to those considered by Mackie, that is, if an outcome part of a transition can occur in a number of alternative ways (i.e., it is a disjunctive outcome), the transition's *causae causantes* are typically inus conditions. In a less complex case *causae causantes* turn out to be *at least* an inus condition.[8] The notion of *at least* an inus condition is motivated by the observation that inus is a weak condition, which can be strengthened in the appropriate circumstances.

Causal circumstances can be stricter in a variety of ways. To explain this idea schematically, suppose that a necessary condition for P has the form 'A or *DEF* or *JKL*' (that is, there is only A in the first disjunct). Since A has no companion, it is trivially a non-redundant part of itself; and it is *not* an insufficient part; in short, A is thus a nus and at least an inus condition for P.

---

[8] We owe this term to Paul Bartha.

A further case is if the condition has the form 'ABC' (so DEF and JKL are missing). Since now the single disjunct ABC is both sufficient and necessary for P, in this case we say that A is an inns condition for P, and again we call A at least inus condition for P.

Yet another case is with 'BC' and 'DEF or JKL' missing (i.e., A stands alone). Then A is a necessary and sufficient, ns, and at least an inus, condition for P. As it is also non redundant, for trivial reasons, it serves as a case of an nns condition as well, and as that is more informative, we drop the abbreviation ns.

To sum up, by "at least an inus condition for P" we mean inus, inns, nus, or nns conditions for P.

One might object that by accommodating stricter causal settings and consequently moving from inus conditions to at least inus conditions, we betray Mackie's idea. But we believe that there are various settings for asking causal questions, and that strict causal settings are natural for singular event causation. For Mackie, both effects and inus conditions are types of events, types that may have instances (Mackie, 1974, p. 262). In the present development, however, at least inus conditions are concrete possible events. They are neither types, nor are they considered here as instances of types.[9] Now, a type event like a forest fire (Mackie's example) can be produced along different causal routes, but not so *that* particular forest fire, as the indexical phrase denotes a particular event with a fixed past history. Thus, 'u', abbreviating 'unnecessary' and occurring in 'inus', is to be expected only in a context of multiply realizable objects, which in BST are disjunctive outcomes. Turning to the 'i' in 'inus', its presence seems to be a contingent matter. In everyday examples it is natural to think that a single factor, say the lighting of a match, is *insufficient* for bringing about the fire (along a given causal route). On the other hand, in indeterministic contexts, where one focuses on risky junctions (i.e., cause-like loci) at which the occurrence of a putative effect can become jeopardized, a single risky junction is perfectly intelligible. Just think of a particle's radioactive decay that makes Schrödinger's cat well-fed.[10] The transition from the non-decayed to the decayed particle turns out, as it should, to be the single necessary and sufficient (ns) condition for that

---

[9] Of course every event is an instance of arbitrarily many types. The point is that we consider the concrete possible events themselves, not the types that they may instantiate.

[10] For our love of cats, we prefer the rendering of Schrödinger's cat paradox attributed to John Bell, with the cat well-fed or hungry, rather than the original story with the cat dead or alive.

particular event of feeding the cat, which is thereby also non-redundant (nns).

We note for fairness that a BST analysis of transitions to disjunctive outcomes Ŏ still falls short of delivering full-fledged generic, type-level causation. The reason is that in such a transition $I \rightarrowtail \check{O}$, all the alternative component outcomes $\hat{O} \in \check{O}$ still need to arise from a common initial I. That is, all these scattered outcomes are in the future of possibilities of the initial event I, which fully occurs in one history (see Def. 4.3(1)). So, to handle Mackie's example of the causes of a forest fire (type-event), we need to fix some initial event—which can be as large as the past of the world before the 20th century. This initial event then delimits the forest fires to those that occurred, or might really have occurred, in the 20th century. Clearly, that is not a fully generic notion of the type-event 'forest fire'. Ultimately, the difference between the type-event approach and our BST approach is due to the frugality of the causal structures represented in BST: they are specified merely by their spatio-temporal and modal aspects, without recourse to content. Such content would have to be added, for instance, via appropriate descriptions, implying a much richer formal machinery in the background.

Having explained what the main differences are between our approach and that of Mackie's, it is still worth being clear about three further similarities and dissimilarities, since our account is supposed to gain support from his analysis of causation. First, Mackie (1974, p. 265) distinguishes between 'explaining causes' (facts) and 'producing causes' (events). Given this dichotomy, we shall only concern ourselves here with ontology (i.e., with producing causes and with produced effects or results).

Second, Mackie notes that the 'occurrence' of an event makes no sense if an event is taken to be merely a chunk of space-time: "Causation is not something between events in a spatio-temporal sense" (Mackie, 1974, p. 296). In a crucially important shift, BST theory considers causation as a relation between concrete possible events. The causal relation between such events has a spatio-temporal component, but also a modal one, so that the occurrence of a BST cause or effect does not reduce to merely a spatio-temporal matter. The smallest BST objects that have these two components are non-trivial basic transitions.

Third and finally, Mackie's theory permits the possibility of backward causation. Various analytical shifts make it difficult to compare Mackie on this point directly with BST theory, but the following is true and may shed light on the issue. Central to our theory, as explained in Definition 6.1, is the notion of a 'cause-like locus' for a transition. It follows from the postulates of BST that no cause-like locus for a transition can lie in the future of any part of the outcome-part of that transition (see Fact 6.1(3) for outcome chains). This is perhaps a difference from Mackie. BST theory, however, leaves open whether or not every cause-like locus must occur in the past of the outcome-part. Some such cause-like loci might, as far as BST theory goes, be space-like related to the outcome part, and so neither past nor future. This seems to happen, for example, in the strange case of EPR-like quantum-mechanical correlations of Figure 5.1 (p. 107), which we discussed under the heading of "modal funny business" (MFB) in Chapter 5 and which we will consider in detail in Chapter 8. Our theory works both with and without MFB, but as it is more intuitive to develop it under the assumption of no MFB, we first assume this simplification. Later on, in Chapter 6.4, we then discuss the general case.

6.3   *Causae causantes* in BST$_{92}$ in formal detail

We now turn to the formal theory of *causae causantes*, which we eventually show to be at least inus conditions. The leading idea is to identify *causae causantes* neither with initial events nor with outcome events, but instead with basic transitions. Since basic transitions are different in BST$_{92}$ and BST$_{NF}$, we need to decide upon one of these frameworks. As before, we develop the theory in BST$_{92}$ structures; we defer remarks about the corresponding BST$_{NF}$ definitions to Chapter 6.5. As stated earlier, in this section we assume that there is no modal funny business.

6.3.1   Defining *causae causantes* in BST$_{92}$

A transition event is "where" something happens; it is "where" there is a transition from (to use Mackie's language) unfixity to fixity. (The shudder quotes remind us that for transition events there is no 'simple location'.) If Alice voluntarily sends a letter to Bob, there is in her personal life a decisive event after which (but not before which) it is settled that the letter is on its way. If Bob receives the letter that, by choice or chance, might or might not have been sent, then there is for him somewhere on his world line a transition from 'might not receive letter' to 'settled that he will have received letter', but that transition event is purely passive, a mere effect. The *causa causans* in this example is along the world line of Alice, not of Bob.

We have spoken of a single *causa causans*, but of course realistically, a great many *causae causantes* must cooperate in order to produce the receive-letter effect. Each *causa causans* of a given transition keeps the occurrence of that transition possible. We do not take deterministic transition events to be *causae causantes*. (Recall that a transition is deterministic if there is no transition which is locally alternative to it.)

As a first step in our formal construction, we return to the concept of a cause-like locus. In Def. 5.10 we introduced the set of cause-like loci for an outcome O, $cll(O)$. A cause-like locus e for O is a point event that marks a critical juncture for the occurrence of O: at e, the occurrence of O is still possible, but immediately after e, the occurrence of O may be impossible. Here we define cause-like loci not for outcomes, but for general transition events $I \rightarrowtail \mathcal{O}^*$. These may be from an initial I to an outcome chain O, to a scattered outcome $\hat{O}$, or to a disjunctive outcome $\breve{O}$. The basic idea is that a cause-like locus for a transition is a critical juncture for the outcome that is compatible with the occurrence of the initial.

There is one complication: For disjunctive outcomes, which are multiply realizable, it may turn out that some critical junctures for the disjuncts are not critical at all for the occurrence of the whole disjunction. Recall the example of rolling a die from the end of Chapter 4.1, and consider the transition from the die-rolling initial I to the disjunctive outcome $\breve{O} =_{df} \{\boxed{1}, \ldots, \boxed{6}\}$ of *any* outcome occurring. In this example, each individual (scattered) outcome making up $\breve{O}$, such as $\boxed{3}$, certainly depends on what happens at some critical juncture, but the transition to the complete disjunctive outcome $\breve{O}$ is guaranteed to occur once the initial is finished—the transition $I \rightarrowtail \breve{O}$ is *deterministic* even though it is, in the relevant sense, a disjunction of individually *indeterministic* parts such as $I \rightarrowtail \boxed{3}$. Technically, this means that for a transition to a disjunctive outcome, we cannot simply identify the cause-like loci of that transition with the union of the cause-like loci for the individual transitions to the scattered outcomes that make up the disjunctive outcome. Rather, we need to remove all those cause-like loci that guarantee $\breve{O}$ to occur and whose occurrence therefore makes no difference for the occurrence of the disjunction. The following definition takes care of this complication.

Definition 6.1 (Cause-like loci of a transition). Let $I \rightarrowtail \mathscr{O}^*$ be a transition from an initial to an outcome chain or a scattered outcome. The set of cause-like loci for this transition is

$$cll(I \rightarrowtail \mathscr{O}^*) =_{df} \{e \in W \mid \exists h \in H_{[I]}[h \perp_e H_{\langle\mathscr{O}^*\rangle}]\}.$$

For $\mathscr{O}^*$ a disjunctive outcome $\check{\mathbf{O}} = \{\hat{O}_\gamma\}_{\gamma\in\Gamma}$, we take care of the mentioned complication by successively defining deterministic points for $\check{\mathbf{O}}$, reduced cause like loci (*cllr*) for $I \rightarrowtail \hat{O}_\gamma$, and finally $cll(I \rightarrowtail \check{\mathbf{O}})$. We rely on the definition of *cll* for scattered outcomes that we just gave.

$$DET_{I\rightarrowtail\check{o}} =_{df} \{e \in W \mid H_e \cap H_{[I]} \subseteq H_{\langle\check{o}\rangle}\};$$
$$cllr(I \rightarrowtail \hat{O}_\gamma) =_{df} cll(I \rightarrowtail \hat{O}_\gamma) \setminus DET_{I\rightarrowtail\check{o}};$$
$$cll(I \rightarrowtail \check{\mathbf{O}}) =_{df} \bigcup_{\gamma\in\Gamma} cllr(I \rightarrowtail \hat{O}_\gamma).$$

To illustrate this definition, let us consider fully deterministic transitions to an exhaustive disjunctive outcome that occur in an indeterministic context. The simplest case is an exhaustive disjunctive transition $e \rightarrowtail \Omega_e$ from a single indeterministic event e to the set Ω_e of all its basic scattered outcomes (see Fact 4.8). In this case, $H_{\langle\Omega_e\rangle} = H_e$, $e \in DET_{e\rightarrowtail\Omega_e}$, and for any $\hat{O}_\gamma \in \Omega_e$, we have $cll(e \rightarrowtail \hat{O}_\gamma) = \{e\}$ and $cllr(e \rightarrowtail \hat{O}_\gamma) = \emptyset$. Accordingly, $cll(e \rightarrowtail \Omega_e) = \emptyset$.

More generally, we consider $I \rightarrowtail \mathbf{1}_I$, where $\mathbf{1}_I = \{\hat{O}_\gamma\}_{\gamma\in\Gamma}$ with $\bigcup H_{\langle\hat{O}_\gamma\rangle} = H_{[I]}$. (Our die rolling example can serve as an example: for the die-rolling-initial I, we have $\mathbf{1}_I = \{\boxed{1},\dots,\boxed{6}\}$.) By saying that the context is indeterministic, we mean that for an individual $\hat{O}_\gamma$, $cll(I \rightarrowtail \hat{O}_\gamma) \neq \emptyset$. However, for any $e \in cll(I \rightarrowtail \hat{O}_\gamma)$, if $h \in H_{[I]}$ splits from $H_{\langle\hat{O}_\gamma\rangle}$ at e, then $h \in H_{\langle\hat{O}_\delta\rangle}$ for some $\delta \in \Gamma$ distinct from γ. Thus, $h \in H_{\langle\check{o}\rangle}$, so $e \in DET_{I\rightarrowtail\check{o}}$ by the definition above. Accordingly, again for the reduced sets of cause-like loci, $cllr(I \rightarrowtail \hat{O}_\gamma) = \emptyset$, and hence $cll(I \rightarrowtail \check{\mathbf{O}}) = \emptyset$.

Before we proceed to define *causae causantes*, we state some useful general facts about *cll*. These are fairly simple in the no-MFB case. We will write them out anyway, with a view to the parallel, more complicated statement of Fact 6.4 for the MFB case to be discussed in Chapter 6.4.

Fact 6.1. *We assume no MFB. Consider a generic transition $I \rightarrowtail \mathscr{O}^*$, i.e., $\mathscr{O}^*$ is an outcome chain O, a scattered outcome $\hat{O}$, or a disjunctive outcome $\check{\mathbf{O}}$.*

(1) Let $e \in cll(I \rightarrowtail \mathcal{O}^*)$. Then e does not causally precede I, i.e., it is not the case that there is some $e' \in I$ for which $e < e'$.

(2) Let $\hat{O}$ be a scattered outcome. Then $cll(I \rightarrowtail \hat{O}) = \bigcup_{O \in \hat{O}} cll(I \rightarrowtail O)$.

(3) Let $e \in cll(I \rightarrowtail \mathcal{O}^*)$ and $\mathcal{O}^*$ be an outcome chain O. Then $e < O$, i.e., for every $e' \in O$: $e \leqslant e'$.

(4) Let $e \in cll(I \rightarrowtail \mathcal{O}^*)$ and $\mathcal{O}^*$ be a scattered outcome $\hat{O}$. Then there is $O \in \hat{O}$ such that $e < O$.

(5) Let $e \in cll(I \rightarrowtail \mathcal{O}^*)$ and $\mathcal{O}^*$ be a disjunctive outcome $\check{O}$. Then there is $\hat{O} \in \check{O}$ with an $O \in \hat{O}$ such that $e < O$.

Proof. (1) For reductio, assume that $e < e'$ for some $e' \in I$. Then for any $h \in H_{[I]}$: $e' \in h$; and by Fact 4.1, for any $h' \in H_{\langle \mathcal{O}^* \rangle}$: $e' \in h'$. Since $e < e'$ we have $h \equiv_e h'$ for any $h \in H_{[I]}$ and any $h' \in H_{\langle \mathcal{O}^* \rangle}$, so $e \notin cll(I \rightarrowtail \mathcal{O}^*)$.

(2) Left as Exercise 6.1.

(3, 4) By Fact 5.4 in Chapter 5 and Exercise 5.3 to that chapter and by noting that $cll(I \rightarrowtail \mathcal{O}^*) \subseteq cll(\mathcal{O}^*)$ for $\mathcal{O}^*$ an outcome chain or a scattered outcome.

(5) follows from the above as an immediate consequence. $\square$

In what follows, we will appeal extensively to a useful consequence of items (3) and (4) of Fact 6.1: If a BST_{92} structure contains no MFB and $\mathcal{O}^*$ is an outcome chain or a scattered outcome for initial I, then all cause-like loci e of the transition $I \rightarrowtail \mathcal{O}^*$ are in the past of $\mathcal{O}^*$, and the outcome $\mathcal{O}^*$ determines a unique basic propositional outcome of e, written $\Pi_e \langle \mathcal{O}^* \rangle$.

Fact 6.2. *Assume no MFB, and let $I \rightarrowtail \mathcal{O}^*$ be a transition to an outcome chain or to a scattered outcome. Then for $e \in cll(I \rightarrowtail \mathcal{O}^*)$, (1) there is a unique basic outcome of e that is compatible with the occurrence of $\mathcal{O}^*$, which we denote by $\Pi_e \langle \mathcal{O}^* \rangle$, and (2) we have $H_{\langle \mathcal{O}^* \rangle} \subseteq \Pi_e \langle \mathcal{O}^* \rangle$.*

Proof. (1) By Fact 6.1(3) and (4), $e < \mathcal{O}^*$. The claim for outcome chains then follows from Fact 4.7, and the claim for scattered outcomes is the subject of Exercise 5.3(2) in Chapter 5 (for which a solution is given in Appendix B.5).

(2) We consider the case of $\mathcal{O}^* = \hat{O}$ a scattered outcome; the case for outcome chains is a straightforward simplification. Let $h \in H_{\langle \hat{O} \rangle}$, i.e., $h \in H_{\langle O \rangle}$ for all $O \in \hat{O}$, and let $e \in cll(I \rightarrowtail \hat{O})$. By Fact 6.1(4), $e < \hat{O}$, i.e., there is some $O \in \hat{O}$ for which $e < O$. As $h \in H_{\langle O \rangle}$, there is some $e' \in O \cap h$, and as $e < e'$, we have $h \in H_e$. Now $\Pi_e \langle \hat{O} \rangle$ is defined to be that unique outcome of e that is compatible with the occurrence of $\hat{O}$, i.e., $\Pi_e \langle \hat{O} \rangle \cap H_{\langle \hat{O} \rangle} \neq \emptyset$. Let $h' \in \Pi_e \langle \hat{O} \rangle \cap H_{\langle \hat{O} \rangle}$; by the above reasoning, there is $e'' \in O \cap h'$. Let $e^* =_{\mathrm{df}} \min\{e', e''\}$, we have $e^* \in O \cap h \cap h'$, and so e^* witnesses $h \equiv_e h'$, proving that $h \in \Pi_e \langle \hat{O} \rangle$. $\square$

We arrive at the following definition of *causae causantes*. For disjunctive outcomes, we need to keep the disjuncts separate because the union of their *causae causantes* is generally inconsistent; we invoke the notion of a reduced set of cause-like loci, *cllr*, as in Def. 6.1.

Definition 6.2 (*Causae causantes* in no-MFB contexts). Let $I \rightarrowtail \mathscr{O}^*$ be a transition from an initial I to an outcome chain or to a scattered outcome. The set of *causae causantes* for this transition is

$$CC(I \rightarrowtail \mathscr{O}^*) =_{\text{df}} \{e \mapsto \Pi_e\langle \mathscr{O}^* \rangle \mid e \in cll(I \rightarrowtail \mathscr{O}^*)\}.$$

If $\mathscr{O}^*$ is a disjunctive outcome $\breve{\mathbf{O}} = \{\hat{O}_\gamma\}_{\gamma \in \Gamma}$ (where Γ is an index set), we define first the reduced set of *causae causantes* $CCr(I \rightarrowtail \hat{O}_\gamma)$ and then the set $CC(I \rightarrowtail \breve{\mathbf{O}})$ as the family of *causae causantes* of the disjuncts:

$$CCr(I \rightarrowtail \hat{O}_\gamma) =_{\text{df}} \{e \mapsto \Pi_e\langle \hat{O}_\gamma \rangle \mid e \in cllr(I \rightarrowtail \hat{O}_\gamma)\};$$
$$CC(I \rightarrowtail \breve{\mathbf{O}}) =_{\text{df}} \{CCr(I \rightarrowtail \hat{O}_\gamma)\}_{\gamma \in \Gamma}.$$

Observe that for $\mathscr{O}^*$ an outcome chain or a scattered outcome, the set $CC(I \rightarrowtail \mathscr{O}^*)$ is consistent since $H_{\langle \mathscr{O}^* \rangle} \subseteq \Pi_e\langle \mathscr{O}^* \rangle$ for every $e \in cll(I \rightarrowtail \mathscr{O}^*)$ (Fact 6.2), and $H_{\langle \mathscr{O}^* \rangle} \neq \emptyset$. Accordingly, the set $CC(I \rightarrowtail \breve{\mathbf{O}})$, where $\breve{\mathbf{O}}$ is a disjunctive outcome, is a *family of consistent sets* of basic transitions, which keeps track of the individual causes of the disjuncts. Note that for a deterministic transition to a disjunctive outcome, $I \rightarrowtail \mathbf{1}_I$, the family $CC(I \rightarrowtail \mathbf{1}_I)$ just contains the empty set.

6.3.2 *Causae causantes* are at least inus conditions

Our goal now is to show that each *causa causans* for a transition is at least an inus condition for that transition. We must be clear on what form exactly these conditions take. Earlier, in Chapter 6.2, we noted that the multi-realizability of outcomes is essential for obtaining the "unnecessary" qualification in in*us*, so we split our discussion into the case of uniquely realizable outcomes (outcome chains and scattered outcomes; Chapter 6.3.2.1), and the case of multiply realizable outcomes (disjunctive outcomes; Chapter 6.3.2.2). For the following discussion, we assume that there is no modal funny business.

6.3.2.1 Transitions to outcome chains or scattered outcomes

Our main Theorem 6.1 is stated for both outcome chain events and scattered outcomes alike. For these types of transitions, we first prove joint sufficiency: if every single *causa causans* of the transition $I \rightarrowtail \mathscr{O}^*$ occurs in a history h, then $I \rightarrowtail \mathscr{O}^*$ occurs in h as well. Joint necessity goes in the opposite direction: if the initial I and the transition $I \rightarrowtail \mathscr{O}^*$ occur, then all the *causae causantes* of $I \rightarrowtail \mathscr{O}^*$ occur as well. (Note that we have to require the occurrence of the initial, given the implication-like reading of the occurrence proposition for the transition.) Observe also that joint necessity entails necessity of each individual *causa causans* (i.e., it entails $H_{[I]} \cap H_{I \rightarrowtail \mathscr{O}^*} \subseteq \Pi_e\langle \mathscr{O}^* \rangle$ for each $e \in cll(I \rightarrowtail \mathscr{O}^*)$). In other words, each individual *causa causans* is necessary for the occurrence of $\mathscr{O}^*$.

It is the non-redundancy clause that raises some subtle issues. Is each individual $\tau \in CC(I \rightarrowtail \mathscr{O}^*)$ a non-redundant part of the set, in the set's capacity of bringing about $\mathscr{O}^*$? Note that if, for example, the set of *causae causantes* is finite and linearly ordered, then in some sense the last transition is sufficient for $\mathscr{O}^*$; in this sense all the remaining *causae causantes* are redundant. More generally, if there are $\tau_1, \tau_2 \in CC(I \rightarrowtail \mathscr{O}^*)$ with $\tau_1 \prec \tau_2$, then τ_1 is redundant—but only because its occurrence is already implied by τ_2 occurring. In concrete terms, that sense of redundancy is spurious, however, since if the concrete transition τ_2 occurs, all the preceding concrete transitions in the past of τ_2 must occur as well—so their specification is redundant, but their occurrence is not. There is another, more useful sense of non-redundancy, which comes from our implication-style reading of the occurrence proposition for a transition (Def. 4.5). Given that definition, we read the *non-occurrence* of a given basic transition $\tau = e_0 \rightarrowtail H$ as the occurrence of a basic transition that is a local alternative to it, $\tau' = e_0 \rightarrowtail H'$, with $H \neq H'$ and $H, H' \in \Pi_{e_0}$. To ask whether a given $\tau \in CC(I \rightarrowtail \mathscr{O}^*)$ is redundant, we thus produce a tweaked set S of basic transitions by taking $CC(I \rightarrowtail \mathscr{O}^*)$ and replacing its member τ by a local alternative τ', and we then ask if the tweaked set S is adequate for bringing about $\mathscr{O}^*$. This may fail to be so for two reasons: first, S may turn out to be inconsistent, or, second, it may happen to be consistent but insufficient for the occurrence of $\mathscr{O}^*$.[11]

---

[11] Given no MFB, the first case occurs if e_0 is a non-maximal element of $cll(I \rightarrowtail \mathscr{O}^*)$, and the second case occurs if e_0 is a maximal element of that set. In the presence of MFB, things are more complicated: we may still obtain an inconsistent set by replacing an outcome of a maximal element of $cll(I \rightarrowtail \mathscr{O}^*)$, which will thus fall under the first case.

With these explanations provided, we can state our Theorem:

Theorem 6.1 (nns for transitions to outcome chains or scattered outcomes in BST_{92} without MFB). *Let $\langle W, < \rangle$ be a BST_{92} structure in which there is no MFB. Let $I \rightarrowtail \mathcal{O}^*$ be a transition to an outcome chain or a scattered outcome. Then the causae causantes of $I \rightarrowtail \mathcal{O}^*$ satisfy the following inus-related conditions:*

1. *joint sufficiency – nns:* $\bigcap_{e \in cll(I \rightarrowtail \mathcal{O}^*)} H_{e \rightarrowtail \Pi_e \langle \mathcal{O}^* \rangle} \subseteq H_{I \rightarrowtail \mathcal{O}^*}$;
2. *joint necessity – nns:* $H_{\langle \mathcal{O}^* \rangle} = H_{[I]} \cap H_{I \rightarrowtail \mathcal{O}^*} \subseteq \bigcap_{e \in cll(I \rightarrowtail \mathcal{O}^*)} H_{e \rightarrowtail \Pi_e \langle \mathcal{O}^* \rangle}$;
3. *non-redundancy – nns: for every $(e_0 \rightarrowtail H) \in CC(I \rightarrowtail \mathcal{O}^*)$ and every $H' \in \Pi_{e_0}$ such that $H' \cap H = \emptyset$:*

$$\text{either} \quad H' \cap \bigcap_{e \in cll(I \rightarrowtail \mathcal{O}^*) \setminus \{e_0\}} \Pi_e \langle \mathcal{O}^* \rangle = \emptyset, \tag{6.1}$$

$$\text{or} \quad H' \cap \bigcap_{e \in cll(I \rightarrowtail \mathcal{O}^*) \setminus \{e_0\}} \Pi_e \langle \mathcal{O}^* \rangle \not\subseteq H_{\langle \mathcal{O}^* \rangle} = H_{[I]} \cap H_{I \rightarrowtail \mathcal{O}^*}. \tag{6.2}$$

Proof. (1) If $H_{[I]} = H_{\langle \mathcal{O}^* \rangle}$, then $H_{I \rightarrowtail \mathcal{O}^*} = \mathrm{Hist}$, so the inclusion is satisfied. Otherwise we argue for the contraposition of (1), considering a scattered outcome $\hat{O}$. Let us suppose that for some $h \in \mathrm{Hist}$, $h \in H_{[I]}$, but $h \notin H_{\langle \hat{O} \rangle}$, so there is $O \in \hat{O}$ such that $h \notin H_{\langle O \rangle}$, i.e., $h \cap O = \emptyset$. On the other hand, there is some $h' \in H_{\langle \hat{O} \rangle}$, hence $h' \in H_{\langle O \rangle}$, so $h \cap O \neq \emptyset$. By Fact 4.1, $h' \in H_{[I]}$. Then $O' =_{df} O \cap h'$ is an initial segment of O for which we have $O' \subseteq h' \setminus h$. By PCP of BST_{92} and Fact 3.8, there is $c < O'$ such that $h \perp_c H_{\langle O' \rangle}$, which implies $c < \hat{O}$ and $h \perp_c H_{\langle O \rangle}$ (as O' is an initial segment of O). Since $H_{\langle \hat{O} \rangle} \subseteq H_{\langle O \rangle}$, we get $h \perp_c H_{\langle \hat{O} \rangle}$, which means that $c \in cll(I \rightarrowtail \hat{O})$. Since $c < \hat{O}$, $\Pi_c \langle \hat{O} \rangle$ is well-defined (see Fact 6.2(1)). Accordingly $h \perp_c \Pi_c \langle \hat{O} \rangle$. As $c \in h$, but $h \notin \Pi_c \langle \hat{O} \rangle$, we get $h \notin H_{c \rightarrowtail \Pi_c \langle \hat{O} \rangle}$, and hence $h \notin \bigcap_{e \in cll(I \rightarrowtail \hat{O})} H_{e \rightarrowtail \Pi_e \langle \hat{O} \rangle}$, which proves the contraposition of (1). By simplifying this argument, we can prove our claim for transitions to outcome chains.

(2) By no-MFB and by Fact 6.1(4), $e < \mathcal{O}^*$ for every $e \in cll(I \rightarrowtail \mathcal{O}^*)$, and so by Fact 6.2, $\Pi_e \langle \mathcal{O}^* \rangle$ is well-defined, and we have $H_{[I]} \cap H_{I \rightarrowtail \mathcal{O}^*} = H_{\langle \mathcal{O}^* \rangle} \subseteq \Pi_e \langle \mathcal{O}^* \rangle \subseteq (\mathrm{Hist} \setminus H_e) \cup \Pi_e \langle \mathcal{O}^* \rangle = H_{e \rightarrowtail \Pi_e \langle \mathcal{O}^* \rangle}$. So $H_{[I]} \cap H_{I \rightarrowtail \mathcal{O}^*} \subseteq \bigcap_{e \in cll(I \rightarrowtail \mathcal{O}^*)} H_{e \rightarrowtail \Pi_e \langle \mathcal{O}^* \rangle}$.

(3) Pick an arbitrary $(e_0 \rightarrowtail H) \in CC(I \rightarrowtail \mathcal{O}^*)$. Recall that by no-MFB, $H =_{df} \Pi_{e_0} \langle \mathcal{O}^* \rangle$ is well-defined. Pick then an arbitrary $H' \in \Pi_{e_0}$ such that

$H' \cap H = \emptyset$. (Such H' exists as e_0 must be a choice point.) Since by Fact 6.2, $H_{\langle \mathscr{O}^* \rangle} \subseteq \Pi_{e_0}\langle \mathscr{O}^* \rangle = H$, (*) $H' \cap H_{\langle \mathscr{O}^* \rangle} = \emptyset$. Let us abbreviate $H^- =_{\mathrm{df}}$ $\bigcap_{e \in cll(I \rightarrowtail \mathscr{O}^*) \setminus \{e_0\}} \Pi_e \langle \mathscr{O}^* \rangle$. Consider then two cases: (i) $H^- \cap H' = \emptyset$ and (ii) $H^- \cap H' \neq \emptyset$. Case (i) is identical to Eq. (6.1), and so we are done. In case (ii), by (*) we have $H^- \cap H' \cap H_{\langle \mathscr{O}^* \rangle} = \emptyset$. Since $H_{\langle \mathscr{O}^* \rangle} \neq \emptyset$, it follows that $H^- \cap H' \nsubseteq H_{\langle \mathscr{O}^* \rangle}$, which is Eq. (6.2).[12]   □

We thus obtain, for transitions to outcome chains or scattered outcomes, a result very much in line with Mackie's inus analysis, namely: each *causa causans* is an nns condition. The divergence from Mackie's account is due to our focus on singular causation rather than type causation. Since outcome chains and scattered outcomes are uniquely realizable, there is only one disjunctive part in the condition, and hence that part is necessary, not unnecessary, as Mackie has it. Also, we do not have insufficiency of a *causa causans* in general: if the set *cll* consists of exactly one element, there is also exactly one *causa causans*, and this one is then already sufficient for the occurrence of the transition effect in question.

We turn next to analyzing the causes of transitions to disjunctive outcomes.

6.3.2.2 Transitions to disjunctive outcomes

In what follows, we consider a transition $I \rightarrowtail \check{\mathbf{O}}$ to a non-trivial disjunctive outcome $\check{\mathbf{O}}$, by which we mean a disjunctive outcome consisting of two or more mutually incompatible scattered outcomes. (If there is only one disjunct, we are effectively considering a transition to a single scattered outcome, which has already been dealt with.) According to Def. 6.2, the set of *causae causantes* for a transition to a disjunctive outcome $I \rightarrowtail \check{\mathbf{O}}$ is the family of the *reduced* sets of *causae causantes* for the component transitions, $CCr(I \rightarrowtail \hat{O}_\gamma)$, $\hat{O}_\gamma \in \check{\mathbf{O}}$. There is, however, a subtle interplay between $CCr(I \rightarrowtail \hat{O}_\gamma)$ and $CC(I \rightarrowtail \hat{O}_\gamma)$ with respect to their role in our inus-related conditions. Beginning with sufficiency (the "s" in "nus"), we can show that each $CCr(I \rightarrowtail \hat{O}_\gamma)$ is sufficient for the occurrence of the transition

---

[12] One might be tempted to try and make the statements of the three parts of the Theorem more symmetrical, perhaps rewriting (2) as $H_{I \rightarrowtail \mathscr{O}^*} \subseteq \bigcap_{e \in cll(I \rightarrowtail \mathscr{O}^*)} H_{e \rightarrowtail \Pi_e \langle \mathscr{O}^* \rangle}$, or formulating Eq. (6.2) in condition (3) as $H_{e_0 \rightarrow H'} \cap \bigcap_{e \in cll(I \rightarrowtail \mathscr{O}^*) \setminus \{e_0\}} H_{e \rightarrow \Pi_e \langle \mathscr{O}^* \rangle} \nsubseteq H_{I \rightarrowtail \mathscr{O}^*}$. It can be shown, however, that both these variants are false, and philosophically dubious, which is the topic of Exercises 6.3 and 6.4.

to the disjunctive outcome. Clearly, $CC(I \rightarrowtail \hat{O}_\gamma)$, which is a superset of the former, is then sufficient for this occurrence as well. For non-redundancy (the "n" in "nus"), we can provide a reading that is fully in line with Mackie's idea of a "non-redundant part" (see the quote on p. 135): it turns out that each element of $CCr(I \rightarrowtail \hat{O}_\gamma)$ is non-redundant, in the sense that if it is removed from $CCr(I \rightarrowtail \hat{O}_\gamma)$, the resulting set is not sufficient for bringing about $I \rightarrowtail \check{O}$ via $I \rightarrowtail \hat{O}_\gamma$. On the other hand, there might be a redundant element in the possibly larger set $CC(I \rightarrowtail \hat{O}_\gamma)$. Finally, the unnecessary aspect (the "u" in "nus") comes from the fact that there is more than one $\hat{O}_\gamma$ in $\check{O}$, so a particular $I \rightarrowtail \hat{O}_\gamma$, and hence $CC(I \rightarrowtail \hat{O}_\gamma)$, is not necessary for the occurrence of $I \rightarrowtail \check{O}$.[13] Thus, the unnecessary aspect concerns $CC(I \rightarrowtail \hat{O}_\gamma)$, not the reduced set $CCr(I \rightarrowtail \hat{O}_\gamma)$. Having explained the various occurrences of CC vs. CCr, we state our Theorem:

Theorem 6.2. (nus for transitions to disjunctive outcomes in BST_{92} without MFB) *Let $\check{O} = \{\hat{O}_\gamma\}_{\gamma \in \Gamma}$ be a disjunctive outcome consisting of more than one scattered outcome. The family $CC(I \rightarrowtail \check{O}) = \{CCr(I \rightarrowtail \hat{O}_\gamma)\}_{\gamma \in \Gamma}$ of causae causantes of $I \rightarrowtail \check{O}$ and the component causae causantes $CC(I \rightarrowtail \hat{O}_\gamma)$ satisfy the following inus-related conditions:*

1. *each $CCr(I \rightarrowtail \hat{O}_\gamma)$ is sufficient – n<u>u</u>s: for every $\gamma \in \Gamma$:*
$$\bigcap_{e \in cllr(I \rightarrowtail \hat{O}_\gamma)} H_{e \rightarrowtail \Pi_e \langle \hat{O}_\gamma \rangle} \subseteq H_{I \rightarrowtail \check{O}}$$
2. *each $CC(I \rightarrowtail \hat{O}_\gamma)$ is unnecessary – n<u>u</u>s: for every $\gamma \in \Gamma$:*
$$H_{\langle \check{O} \rangle} = H_{[I]} \cap H_{I \rightarrowtail \check{O}} \not\subseteq \bigcap_{e \in cll(I \rightarrowtail \hat{O}_\gamma)} H_{e \rightarrowtail \Pi_e \langle \hat{O}_\gamma \rangle}.$$
3. *for each $\gamma \in \Gamma$, each $\tau_0 = (e_0 \rightarrowtail H) \in CCr(I \rightarrowtail \hat{O}_\gamma)$ is non-redundant – n<u>u</u>s. That is, for every $H' \in \Pi_{e_0}$ such that $H \cap H' = \emptyset$:*

$$either \quad H' \cap \bigcap_{e \in cllr(I \rightarrowtail \hat{O}_\gamma) \setminus \{e_0\}} \Pi_e \langle \hat{O}_\gamma \rangle = \emptyset, \tag{6.3}$$

$$or \quad H' \cap \bigcap_{e \in cllr(I \rightarrowtail \hat{O}_\gamma) \setminus \{e_0\}} \Pi_e \langle \hat{O}_\gamma \rangle \not\subseteq H_{\langle \hat{O}_\gamma \rangle} = H_{[I]} \cap H_{I \rightarrowtail \hat{O}_\gamma}. \tag{6.4}$$

[13] One might be tempted to think that, in contrast, $CCr(I \rightarrowtail \hat{O}_\gamma)$ is necessary for the occurrence of $I \rightarrowtail \check{O}$; but this is not the case. It is enough to consider a BST_{92} structure with one choice point c with three basic outcomes, such that $\hat{O}_1$ and $\hat{O}_2$ occur in the first and the second outcome, resp., whereas a history from the third outcome belongs to $H_{[I]}$. Then $c \in cllr(I \rightarrowtail \hat{O}_i)$, but $H_{\langle \check{O} \rangle} \not\subseteq \bigcap_{c \in cllr(I \rightarrowtail \hat{O}_i)} H_{c \rightarrowtail \Pi_c \langle \hat{O}_i \rangle}$ for each $i = 1, 2$.

Proof. (1) We consider two cases. If $I \rightarrowtail \breve{\mathbf{O}}$ is deterministic, i.e., if $H_{[I]} = H_{\langle \breve{\mathbf{O}} \rangle}$, then $H_{I \rightarrowtail \breve{\mathbf{O}}} = \text{Hist}$, and the inclusion holds trivially. Otherwise we argue for the contraposition. Consider some $h \in H_{[I]} \setminus H_{\langle \breve{\mathbf{O}} \rangle}$, which must exist since $I \rightarrowtail \breve{\mathbf{O}}$ is not deterministic. As $H_{\langle \breve{\mathbf{O}} \rangle} = \cup_{\hat{O} \in \breve{\mathbf{O}}} H_{\langle \hat{O} \rangle}$ and $h \notin H_{\langle \breve{\mathbf{O}} \rangle}$, it follows that $h \notin H_{\langle \hat{O}_\gamma \rangle}$ for every $\gamma \in \Gamma$. Exactly like in the proof of Theorem 6.1(1), by PCP92, we arrive at some $e \in cll(I \rightarrowtail \hat{O}_\gamma)$ with $h \perp_e \Pi_e \langle \hat{O}_\gamma \rangle$. Since $h \in H_e \cap (H_{[I]} \setminus H_{\langle \breve{\mathbf{O}} \rangle})$, we have that $e \notin DET_{I \rightarrowtail \breve{\mathbf{O}}}$, and hence $e \in cllr(I \rightarrowtail \hat{O}_\gamma)$. Accordingly, since $h \in H_e$ but $h \notin \Pi_e \langle \hat{O}_\gamma \rangle$, we get $h \notin H_{e \rightarrowtail \Pi_e \langle \hat{O}_\gamma \rangle}$, which proves the contraposition.

(2) Since $\breve{\mathbf{O}}$ consists of more than one scattered outcome, for every $\hat{O}_\gamma \in \breve{\mathbf{O}}$, there is $h \in H_{\langle \breve{\mathbf{O}} \rangle} \setminus H_{\langle \hat{O}_\gamma \rangle}$. Hence $h \notin H_{I \rightarrowtail \hat{O}_\gamma}$. By the contraposition of Theorem 6.1(1), $h \notin \bigcap_{e \in cll(I \rightarrowtail \hat{O}_\gamma)} H_{e \rightarrowtail \Pi_e \langle \hat{O}_\gamma \rangle}$, providing a witness for the non-inclusion claim (2).

(3) Pick some $\gamma \in \Gamma$. If $CCr(I \rightarrowtail \hat{O}_\gamma) = \emptyset$, the theorem trivially holds. Let us thus assume that $CCr(I \rightarrowtail \hat{O}_\gamma) \neq \emptyset$. Then $cllr(I \rightarrowtail \hat{O}_\gamma) \neq \emptyset$ as well. Pick an arbitrary $(e_0 \rightarrowtail H) \in CCr(I \rightarrowtail \hat{O}_\gamma)$. Recall that by no-MFB, $H =_{df} \Pi_{e_0} \langle \hat{O}_\gamma \rangle$ is well-defined. Pick then an arbitrary $H' \in \Pi_{e_0}$ such that $H' \cap H = \emptyset$. Since by Fact 6.2, $H_{\langle \hat{O}_\gamma \rangle} \subseteq \Pi_{e_0} \langle \hat{O}_\gamma \rangle$, (*) $H' \cap H_{\langle \hat{O}_\gamma \rangle} = \emptyset$. Let us abbreviate $H^- =_{df} \bigcap_{e \in cllr(I \rightarrowtail \hat{O}_\gamma) \setminus \{e_0\}} \Pi_e \langle \hat{O}_\gamma \rangle$. Consider then two cases: (i) $H^- \cap H' = \emptyset$ and (ii) $H^- \cap H' \neq \emptyset$. Case (i) is identical to Eq. (6.3), and so we are done. In case (ii), by (*) we have $H^- \cap H' \cap H_{\langle \hat{O}_\gamma \rangle} = \emptyset$. Since $H_{\langle \hat{O}_\gamma \rangle} \neq \emptyset$, it follows that $H^- \cap H' \not\subseteq H_{\langle \hat{O}_\gamma \rangle}$, which is Eq. (6.4). $\qquad \square$

6.4 Causation in the presence of modal funny business

We have seen that the proofs of two clauses, (2) and (3), in our two theorems above go through on the assumption of no MFB. If we want to adapt our theory to allow for MFB, we need to reflect on what changes this brings. Observe that in the proofs we used Fact 6.1(3)–(5), which depends on no-MFB, and according to which every cause-like locus of a transition is below the outcome part of that transition (in the appropriate sense of 'below' from Def. 4.4). This further entails that each cause-like locus has exactly one basic outcome consistent with the outcome part of the transition, as noted in Fact 6.2(1).

In MFB contexts, however, it may happen that a cause-like locus e for a transition $I \rightarrowtail \mathcal{O}^*$ is *not* in the past of $\mathcal{O}^*$ and that e has *more than one* basic

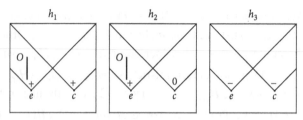

Figure 6.1 MFB with three histories, a binary choice point e (outcomes $+$ and $-$) and a ternary choice point c (outcomes $+$, 0, and $-$).

outcome that is consistent with $H_{\langle \mathcal{O}^* \rangle}$, the occurrence proposition for $\mathcal{O}^*$. In other words, there can be more than one basic outcome of a cause-like locus that keeps the possibility of $\mathcal{O}^*$ open. For illustration, consider Figure 6.1. That figure depicts a BST_{92} structure with three histories, h^{++}, h^{+0}, and h^{--}. The outcome chain O occurs in h^{++} and h^{+0}, but not in h^{--}, whereas the initial I occurs in all these histories. The structure contains two choice points, e with double splitting $(+,-)$ and c with triple splitting $(+,-,0)$, and these choice points constitute the set of cause-like loci $cll(I \rightarrowtail O)$ (h^{--} is the witness). One of these events, e, is below O, but the other, c, is not.

The sets of basic outcomes of e and c are $\Pi_e = \{\{h^{++}, h^{+0}\}, \{h^{--}\}\}$ and $\Pi_c = \{\{h^{++}\}, \{h^{+0}\}, \{h^{--}\}\}$, respectively. Since $H_{\langle O \rangle} = \{h^{++}, h^{+0}\}$, there are two basic outcomes of c, $\{h^{++}\}$ and $\{h^{+0}\}$, that are consistent with $H_{\langle O \rangle}$. Note that e and c are SLR and the structure contains sets of transitions that are combinatorially consistent, but not consistent (see Exercise 6.5), thus exhibiting combinatorial funny business (Def. 5.6), for example, in the set $T = \{e \rightarrowtail \{h^{++}, h^{+0}\}, c \rightarrowtail \{h^{--}\}\}$.

This example suggests a natural generalization of our no-MFB definition of *causae causantes*. By this definition (Def. 6.2), a *causa causans* of $I \rightarrowtail \mathcal{O}^*$ (where $\mathcal{O}^*$ is either a scattered outcome or an outcome chain) is a *basic transition* from $e \in cll(I \rightarrowtail \mathcal{O}^*)$ to the *unique basic* outcome of e consistent with $H_{\langle \mathcal{O}^* \rangle}$, i.e., to $\Pi_e \langle \mathcal{O}^* \rangle \in \Pi_e$. However, as we have just seen, in the presence of MFB we lose the guarantee that there is always a unique basic outcome consistent with the outcome part $\mathcal{O}^*$ of $I \rightarrowtail \mathcal{O}^*$. Nevertheless, there is always a *non-empty set of basic outcomes* $H_1, H_2 \ldots \in \Pi_e$ each of which is consistent with the occurrence proposition $H_{\langle \mathcal{O}^* \rangle}$ for $\mathcal{O}^*$ (see Fact 6.3). Hence the set of those basic outcomes of e, which is a *basic disjunctive* propositional outcome of e (see Def. 4.9), is consistent with $H_{\langle \mathcal{O}^* \rangle}$ as well. Thus, if MFB is allowed, a *causa causans* of the transition

$I \rightarrowtail \mathscr{O}^*$ should be a transition from $e \in cll(I \rightarrowtail \mathscr{O}^*)$ to a basic *disjunctive* outcome of e.[14]

To make these observations more precise, we can use the following handy notation.

Definition 6.3. For $A \subseteq$ Hist, $\check{\mathbf{H}}_e\langle A \rangle$ is the basic disjunctive outcome of e that is consistent with A, i.e.,

$$\check{\mathbf{H}}_e\langle A \rangle = \{H \in \Pi_e \mid H \cap A \neq \emptyset\}.$$

Taking $H_{\langle \mathscr{O}^* \rangle}$ for A, we get $\check{\mathbf{H}}_e\langle H_{\langle \mathscr{O}^* \rangle} \rangle$, which is just what we need: a basic disjunctive outcome of e that is the set of exactly those basic outcomes of e that are consistent with $H_{\langle \mathscr{O}^* \rangle}$. Of course, for the case that $e < \mathscr{O}^*$, the respective outcome of e is unique, so that in that case, we just acquire an extra layer of set-theoretic wrapping—that is the price we pay for a unified formal framework in the MFB case.

As the following fact shows, for $e \in cll(I \rightarrowtail \mathscr{O}^*)$, the basic disjunctive outcome $\check{\mathbf{H}}_e\langle H_{\langle \mathscr{O}^* \rangle} \rangle$ captures the notion of "admitting $\mathscr{O}^*$" in the relevant sense.

Fact 6.3. *Let $\mathscr{O}^*$ be an outcome chain or a scattered outcome. For $e \in cll(I \rightarrowtail \mathscr{O}^*)$, $H_{\langle \mathscr{O}^* \rangle} \subseteq \bigcup \check{\mathbf{H}}_e\langle H_{\langle \mathscr{O}^* \rangle} \rangle$.*

Proof. Let $e \in cll(I \rightarrowtail \mathscr{O}^*)$ and let $h \in H_{\langle \mathscr{O}^* \rangle}$. Then $h' \perp_e h$ for some $h' \in H_{[I]}$, so $e \in h$, and hence there is $H \in \Pi_e$ such that $h \in H$, hence $h \in H \cap H_{\langle \mathscr{O}^* \rangle}$. Accordingly, $h \in \bigcup\{H' \in \Pi_e \mid H' \cap H_{\langle \mathscr{O}^* \rangle} \neq \emptyset\}$, i.e., $h \in \bigcup \check{\mathbf{H}}_e\langle H_{\langle \mathscr{O}^* \rangle} \rangle$. $\qquad\square$

We can now state an analogue of Fact 6.1 in the general case.

Fact 6.4. *We make no assumptions about MFB. Consider a generic transition $I \rightarrowtail \mathscr{O}^*$, i.e., $\mathscr{O}^*$ is an outcome chain O, a scattered outcome $\hat{O}$, or a disjunctive outcome $\check{\mathbf{O}}$.*

(1) *Let $e \in cll(I \rightarrowtail \mathscr{O}^*)$. Then e does not causally precede I, i.e., it is not the case that there is some $e' \in I$ for which $e < e'$.*

(2) *Let $\hat{O}$ be a scattered outcome. Then $\bigcup_{O \in \hat{O}} cll(I \rightarrowtail O) \subseteq cll(I \rightarrowtail \hat{O})$.*

(3) *Let $e \in cll(I \rightarrowtail \mathscr{O}^*)$ and $\mathscr{O}^*$ be an outcome chain O. Then for every $e' \in O$: $e \leqslant e'$ or $e\,SLR\,e'$.*

(4) *Let $e \in cll(I \rightarrowtail \mathscr{O}^*)$ and $\mathscr{O}^*$ be a scattered outcome $\hat{O}$. Then there is an initial segment O' of some $O \in \hat{O}$ such that for every $e' \in O': e \leqslant e'$ or $e\,SLR\,e'$.*

(5) *Let $e \in cll(I \rightarrowtail \mathscr{O}^*)$ and $\mathscr{O}^*$ be a disjunctive outcome $\check{O}$. Then there is $\hat{O} \in \check{O}$ with an initial segment O' of some $O \in \hat{O}$ such that for every $e' \in O': e \leqslant e'$ or $e\,SLR\,e'$.*

Proof. (1) The proof is exactly as for Fact 6.1(1).

(2) This direction follows as in the no-MFB case (see Exercise 6.1).

(3) Let $e \in cll(I \rightarrowtail \mathscr{O}^*)$, and let h be a history containing the whole chain O. This implies $h \in H_{\langle O \rangle}$. By the definition of cll, there must be some history $h_I \in H_{[I]}$ for which $h_I \perp_e H_{\langle O \rangle}$, implying $h_I \perp_e h$. Let now $e' \in O$. As $\{e, e'\} \subseteq h$, it must be that $e \leqslant e'$ or $e\,SLR\,e'$ or $e > e'$. But the latter case is ruled out: as $e \in h_I$, by the downward closure of histories we would have $e' \in h_I$ as well. Now $e' \in O$, so if $e' \in h_I$, then $h_I \in H_{\langle O \rangle}$. This, however, contradicts $h_I \perp_e H_{\langle O \rangle}$.

(4) and (5) are proved by an argument analogous to that for (3). We leave these proofs as Exercise 6.2. $\qquad\qquad\qquad\qquad\qquad\qquad\qquad\square$

Having introduced the notion of a general (basic disjunctive) outcome $\check{\mathbf{H}}_e \langle H_{\langle \mathscr{O}^* \rangle} \rangle$ of e consistent with $\mathscr{O}^*$, we generalize Definition 6.2 to both MFB and no-MFB contexts:

Definition 6.4 (*Causae causantes* generally). Let $I \rightarrowtail \mathscr{O}^*$ be a transition from initial I to a scattered outcome or outcome chain $\mathscr{O}^*$. The set of *causae causantes* for this transition is

$$CC(I \rightarrowtail \mathscr{O}^*) = \{e \rightarrowtail \check{\mathbf{H}}_e \langle H_{\langle \mathscr{O}^* \rangle} \rangle \mid e \in cll(I \rightarrowtail \mathscr{O}^*)\}.$$

If $\mathscr{O}^*$ is a disjunctive outcome $\check{O} = \{\hat{O}_\gamma\}_{\gamma \in \Gamma}$, where Γ is an index set, we define first the reduced set of *causae causantes* $CCr(I \rightarrowtail \hat{O}_\gamma)$ and then the set $CC(I \rightarrowtail \check{O})$:

$$CCr(I \rightarrowtail \hat{O}_\gamma) =_{\mathrm{df}} \{e \rightarrowtail \check{\mathbf{H}}_e \langle H_{\langle \hat{O}_\gamma \rangle} \rangle \mid e \in cllr(I \rightarrowtail \hat{O}_\gamma)\};$$

$$CC(I \rightarrowtail \check{O}) =_{\mathrm{df}} \{CCr(I \rightarrowtail \hat{O}_\gamma)\}_{\gamma \in \Gamma}.$$

Observe that if no-MFB is assumed, this definition almost agrees with Def. 6.2, since no-MFB implies that there is always a unique basic outcome

$\Pi_e\langle\mathcal{O}^*\rangle$ of e that is consistent with the occurrence of $\mathcal{O}^*$, so $\check{\mathbf{H}}_e\langle H_{\langle\mathcal{O}^*\rangle}\rangle = \{\Pi_e\langle\mathcal{O}^*\rangle\}$. Thus in the no-MFB case, even according to our generalized definition, *causae causantes* turn out to be (singleton sets of) basic transitions whose initials are cause-like loci.

We can illustrate the generalized concept by returning to Figure 6.1. There the transition $I \rightarrowtail O$ has two cause-like loci: e and c. The *causa causans* associated with e is a basic transition $e \rightarrowtail \{\{h^{++}, h^{+0}\}\}$, as $\{h^{++}, h^{+0}\}$ is a basic outcome of e. However, c has two basic outcomes consistent with O, $\{h^{++}\}$ and $\{h^{+0}\}$, so the *causa causans* associated with c is the transition from c to the disjunctive basic outcome $\{\{h^{++}\}, \{h^{+0}\}\}$; that is, $c \rightarrowtail \{\{h^{++}\}, \{h^{+0}\}\}$.

Having given a general definition of *causae causantes*, we now provide analogues of Theorems 6.1 and 6.2 that prove that even in MFB cases, the *causae causantes* fulfill inus-like conditions. The two general Theorems read as follows:

Theorem 6.3. (nns for transitions to outcome chains or scattered outcomes in BST$_{92}$ with MFB) *Let $\mathcal{O}^*$ be an outcome chain or a scattered outcome. Then the* causae causantes *of $I \rightarrowtail \mathcal{O}^*$ satisfy the following inus-related conditions:*

1. *joint sufficiency – nns:* $\bigcap_{e \in cll(I \rightarrowtail \mathcal{O}^*)} H_{e \rightarrowtail \check{\mathbf{H}}_e\langle H_{\langle\mathcal{O}^*\rangle}\rangle} \subseteq H_{I \rightarrowtail \mathcal{O}^*}$;

2. *joint necessity–nns:* $H_{\langle\mathcal{O}^*\rangle} = H_{[I]} \cap H_{I \rightarrowtail \mathcal{O}^*} \subseteq \bigcap_{e \in cll(I \rightarrowtail \mathcal{O}^*)} H_{e \rightarrowtail \check{\mathbf{H}}_e\langle H_{\langle\mathcal{O}^*\rangle}\rangle}$;

3. *non-redundancy – nns: for every $(e_0 \rightarrowtail \check{\mathbf{H}}) \in CC(I \rightarrowtail \mathcal{O}^*)$ and every $\check{\mathbf{H}}'$ such that $\check{\mathbf{H}} \cap \check{\mathbf{H}}' = \emptyset$, where $\check{\mathbf{H}}, \check{\mathbf{H}}' \subseteq \Pi_{e_0}$*

$$either \quad \bigcup\check{\mathbf{H}}' \cap \bigcap_{e \in cll(I \rightarrowtail \mathcal{O}^*)\setminus\{e_0\}} (H_e \cap H_{e \rightarrowtail \check{\mathbf{H}}_e\langle H_{\langle\mathcal{O}^*\rangle}\rangle}) = \emptyset, \quad or \quad (6.5)$$

$$\bigcup\check{\mathbf{H}}' \cap \bigcap_{e \in cll(I \rightarrowtail \mathcal{O}^*)\setminus\{e_0\}} (H_e \cap H_{e \rightarrowtail \check{\mathbf{H}}_e\langle H_{\langle\mathcal{O}^*\rangle}\rangle}) \not\subseteq H_{[I]} \cap H_{I \rightarrowtail \mathcal{O}^*}. \quad (6.6)$$

Note the difference in the statement of the non-redundancy clause between this Theorem and Theorems 6.1–6.2. The difference is due to the extra set-theoretical level in the definition of (proposition-style) basic disjunctive outcomes (Def. 6.3). The two conditions of non-redundancy concern inter-sections of sets of histories, and both $H \in \Pi_e$ and $\Pi_e\langle\mathcal{O}^*\rangle$ are sets of histories. $\check{\mathbf{H}}' \subseteq \Pi_e$ and $\check{\mathbf{H}}_e\langle H_{\langle\mathcal{O}^*\rangle}\rangle \subseteq \Pi_e$, however, are *sets* of sets of histories. We thus use the union $\bigcup\check{\mathbf{H}}'$ and the occurrence proposition $H_{e \rightarrowtail \check{\mathbf{H}}_e\langle H_{\langle\mathcal{O}^*\rangle}\rangle}$ (see Def. 4.9).

Theorem 6.4. (nus for transitions to disjunctive outcomes in BST_{92} with MFB) *Let* $\breve{O} = \{\hat{O}_\gamma \mid \gamma \in \Gamma\}$ *be a disjunctive outcome consisting of more than one scattered outcome. The set of causae causantes of* $I \rightarrowtail \breve{O}$, *i.e.,* $\{CCr(I \rightarrowtail \hat{O}_\gamma)\}_{\gamma \in \Gamma}$ *as well as each* $CC(I \rightarrowtail \hat{O}_\gamma)$, *satisfy the following inus-related conditions:*

1. *each* $CCr(I \rightarrowtail \hat{O}_\gamma)$ *is sufficient – <u>nus</u>: for every* $\gamma \in \Gamma$:
   $$\bigcap\nolimits_{e \in cllr(I \rightarrowtail \hat{O}_\gamma)} H_{e \rightarrowtail \breve{H}_e \langle H_{\langle \hat{O}_\gamma \rangle} \rangle} \subseteq H_{I \rightarrowtail \breve{O}};$$

2. *each* $CC(I \rightarrowtail \hat{O}_\gamma)$ *is unnecessary – <u>nus</u>: for every* $\gamma \in \Gamma$:
   $$H_{\langle \breve{O} \rangle} = H_{[I]} \cap H_{I \rightarrowtail \breve{O}} \not\subseteq \bigcap\nolimits_{e \in cll(I \rightarrowtail \hat{O}_\gamma)} H_{e \rightarrowtail \breve{H}_e \langle H_{\langle \hat{O}_\gamma \rangle} \rangle}.$$

3. *for each* $\gamma \in \Gamma$, *each* $\tau_0 = (e_0 \rightarrowtail \breve{H}) \in CCr(I \rightarrowtail \hat{O}_\gamma)$ *is non-redundant – <u>nus</u>. That is, for every* $\breve{H}' \subseteq \Pi_{e_0}$ *such that* $\breve{H} \cap \breve{H}' = \emptyset$:

$$either \quad \bigcup \breve{H}' \cap \bigcap_{e \in cllr(I \rightarrowtail \hat{O}_\gamma) \setminus \{e_0\}} (H_e \cap H_{e \rightarrowtail \breve{H}_e \langle H_{\langle \hat{O}_\gamma \rangle} \rangle}) = \emptyset, \qquad (6.7)$$

$$or \quad \bigcup \breve{H}' \cap \bigcap_{e \in cllr(I \rightarrowtail \hat{O}_\gamma) \setminus \{e_0\}} (H_e \cap H_{e \rightarrowtail \breve{H}_e \langle H_{\langle \hat{O}_\gamma \rangle} \rangle}) \not\subseteq H_{[I]} \cap H_{I \rightarrowtail \hat{O}_\gamma}. \quad (6.8)$$

The proofs of these theorems, which parallel the proofs of Theorems 6.1 and 6.2, are left as Exercises 6.6 and 6.7.

Two comments about the BST analysis of causation in MFB contexts may be in order at this point. First, by PCP_{92}, there is always a *causa causans* in $cll(I \rightarrowtail \mathcal{O}^*)$ that is in the past of $\mathcal{O}^*$, in the relevant sense of "past", depending on the kind of $\mathcal{O}^*$ (see Def. 4.4). But as shown by Fact 6.4(3)–(5), a *causa causans* may also be space-like related to $\mathcal{O}^*$ (again, in the relevant sense). So the general situation is as depicted in Figure 6.1: there is a kosher *causa causans* for $I \rightarrowtail \mathcal{O}^*$, which is a basic transition whose initial lies below $\mathcal{O}^*$, but there is also a weird companion *causa causans*, which is a non-trivial basic disjunctive transition and whose initial is not in the past of $\mathcal{O}^*$. One might perhaps think that the weird companion is superfluous. But by the non-redundancy results of Theorems 6.3 and 6.4, the weird companion does some real work: if it were replaced by one of its alternatives, $I \rightarrowtail \mathcal{O}^*$ would not occur. Our second comment is related to the fact that in MFB contexts, a *causa causans* is a transition to a basic disjunctive outcome; the disjunctive outcome is a family of mutually inconsistent basic outcomes. One might worry whether such a transition can really occur. This worry is dispelled by our formal machinery, as the (non-trivial) occurrence proposition for such a transition is well-defined (see Def. 6.3 and Fact 6.3). We can dismiss this worry for non-technical reasons as well. Recall that the underlying idea of causation in BST is that causing X is making a difference by keeping the occurrence of X possible at a critical junction. Now, a transition from a cause-like locus e of $I \rightarrowtail \mathcal{O}^*$ to a basic disjunctive outcome $\breve{\mathbf{H}}_e \langle H_{\langle \mathcal{O}^* \rangle} \rangle$ keeps the occurrence of $I \rightarrowtail \mathcal{O}^*$ possible, and it also makes a difference, since an alternative transition from e precludes the occurrence of that transition. There is thus nothing objectionable to the notion of a transition to a basic disjunctive outcome keeping some outcome possible.

To finish this chapter, we provide some observations regarding how to reformulate our theory of causation so that it applies to $\mathrm{BST}_{\mathrm{NF}}$ structures as well.

6.5 *Causae causantes* in $\mathrm{BST}_{\mathrm{NF}}$ structures

Recall that a basic propositional transition in a $\mathrm{BST}_{\mathrm{NF}}$ structure $\langle W, < \rangle$ is a pair $\langle \ddot{e}, H_e \rangle$, written as $\ddot{e} \rightarrowtail H_e$, where $e \in W$ and where $\ddot{e}$ is the choice set associated with e. The occurrence proposition for $\ddot{e} \rightarrowtail H_e$ is $H_{\ddot{e} \rightarrow H_e} = (\text{Hist} \setminus H_{\ddot{e}}) \cup H_e$, where $H_{\ddot{e}} = \bigcup_{e \in \ddot{e}} H_e$. The definition of cause-like loci for transitions generalize Definition 5.12 for outcomes:

Definition 6.5 (Cause-like loci in $\mathrm{BST}_{\mathrm{NF}}$). Let $\mathcal{O}^*$ be an outcome chain or a scattered outcome in a $\mathrm{BST}_{\mathrm{NF}}$ structure $\langle W, < \rangle$. The set of cause-like loci for a transition $I \rightarrowtail \mathcal{O}^*$ is the following set:

$$cll(I \rightarrowtail \mathcal{O}^*) =_{\mathrm{df}} \{\ddot{e} \subseteq W \mid \exists h \in H_{[I]} \, [h \perp_{\ddot{e}} H_{\langle \mathcal{O}^* \rangle}]\}.$$

If $\mathcal{O}^*$ is a disjunctive outcome $\breve{\mathbf{O}} = \{\hat{O}_\gamma\}_{\gamma \in \Gamma}$, where Γ is an index set,

$$DET_{I \rightarrowtail \breve{\mathbf{O}}} =_{\mathrm{df}} \{\ddot{e} \subseteq W \mid H_{[I]} \cap H_{\ddot{e}} \subseteq H_{\langle \breve{\mathbf{O}} \rangle}\},$$
$$cllr(I \rightarrowtail \hat{O}_\gamma) =_{\mathrm{df}} cll(I \rightarrowtail \hat{O}_\gamma) \setminus DET_{I \rightarrowtail \breve{\mathbf{O}}},$$
$$cll(I \rightarrowtail \breve{\mathbf{O}}) =_{\mathrm{df}} \bigcup_{\gamma \in \Gamma} cllr(I \rightarrowtail \hat{O}_\gamma).$$

Assuming no MFB, it can be proved[15] that for every $\ddot{e} \in cll(I \rightarrowtail \mathcal{O}^*)$, where $\mathcal{O}^*$ is an outcome chain or a scattered outcome, there is a single

---

[15] See the Exercises to Chapter 5, esp. Exercise 5.5.

$e' \in \ddot{e}$ such that e' is appropriately below $\mathcal{O}^*$, where 'appropriately' indicates different senses of 'below', depending on the kind of outcome $\mathcal{O}^*$ (see Def. 4.4). It follows that there is a single outcome $H_{e'} \in \Pi_{\ddot{e}}$ consistent with $H_{\langle \mathcal{O}^* \rangle}$; we may refer to this outcome as $\Pi_{\ddot{e}}\langle \mathcal{O}^* \rangle$. In no-MFB contexts we thus define *causae causantes* as follows:

Definition 6.6 (*Causae causantes* with no MFB). Let $I \rightarrowtail \mathcal{O}^*$ be a transition from an initial I to a scattered outcome or to an outcome chain. The set of *causae causantes* for this transition is

$$CC(I \rightarrowtail \mathcal{O}^*) = \{\ddot{e} \rightarrowtail \Pi_{\ddot{e}}\langle \mathcal{O}^* \rangle \mid \ddot{e} \in cll(I \rightarrowtail \mathcal{O}^*)\}.$$

If $\mathcal{O}^*$ is a disjunctive outcome $\check{\mathbf{O}} = \{\hat{O}_\gamma\}_{\gamma \in \Gamma}$, where Γ is an index set, we first define the reduced set of *causae causantes* $CCr(I \rightarrowtail \hat{O}_\gamma)$ and then the set $CC(I \rightarrowtail \check{\mathbf{O}})$:

$$CCr(I \rightarrowtail \hat{O}_\gamma) =_{df} \{\ddot{e} \rightarrowtail \Pi_{\ddot{e}}\langle \hat{O}_\gamma \rangle \mid \ddot{e} \in cllr(I \rightarrowtail \hat{O}_\gamma)\};$$
$$CC(I \rightarrowtail \check{\mathbf{O}}) =_{df} \{CCr(I \rightarrowtail \hat{O}_\gamma) \mid \gamma \in \Gamma\}.$$

This definition can again be generalized to MFB contexts by identifying *causae causantes* with transitions from cause-like loci to basic disjunctive outcomes.

Definition 6.7 (General *causae causantes*). Let $I \rightarrowtail \mathcal{O}^*$ be a transition from initial I to a scattered outcome or outcome chain $\mathcal{O}^*$. The set of *causae causantes* for this transition is

$$CC(I \rightarrowtail \mathcal{O}^*) = \{\ddot{e} \rightarrowtail \check{\mathbf{H}}_{\ddot{e}}\langle H_{\langle \mathcal{O}^* \rangle} \rangle \mid \ddot{e} \in cll(I \rightarrowtail \mathcal{O}^*)\}.$$

If $\mathcal{O}^*$ is a disjunctive outcome $\check{\mathbf{O}} = \{\hat{O}_\gamma\}_{\gamma \in \Gamma}$, where Γ is an index set, we define first the reduced set of *causae causantes* $CCr(I \rightarrowtail \hat{O}_\gamma)$ and then the set $CC(I \rightarrowtail \check{\mathbf{O}})$:

$$CCr(I \rightarrowtail \hat{O}_\gamma) =_{df} \{\ddot{e} \rightarrowtail \check{\mathbf{H}}_{\ddot{e}}\langle H_{\langle \hat{O}_\gamma \rangle} \rangle \mid \ddot{e} \in cllr(I \rightarrowtail \hat{O}_\gamma)\};$$
$$CC(I \rightarrowtail \check{\mathbf{O}}) =_{df} \{CCr(I \rightarrowtail \hat{O}_\gamma) \mid \gamma \in \Gamma\}.$$

The challenge is then to prove, using these definitions, theorems analogous to Theorems 6.1, 6.2, 6.3, and 6.4. These tasks we also leave to the reader. As a hint, we provide here the statement of one of the theorems, and give its proof in the Appendix:

Theorem 6.5. *(nns for transitions to outcome chains or scattered outcomes in* BST_{NF} *with no MFB) Let* $\mathscr{O}^*$ *be an outcome chain or a scattered outcome. The causae causantes of* $I \rightarrowtail \mathscr{O}^*$ *satisfy the following inus-related conditions:*

1. *joint sufficiency –* <u>nn</u>s: $\bigcap_{\ddot{e} \in cll(I \rightarrowtail \mathscr{O}^*)} H_{\ddot{e} \rightarrow \Pi_{\ddot{e}} \langle \mathscr{O}^* \rangle} \subseteq H_{I \rightarrowtail \mathscr{O}^*}$;
2. *joint necessity – n<u>n</u>s:* $H_{\langle \mathscr{O}^* \rangle} = H_{[I]} \cap H_{I \rightarrowtail \mathscr{O}^*} \subseteq \bigcap_{\ddot{e} \in cll(I \rightarrowtail \mathscr{O}^*)} H_{\ddot{e} \rightarrow \Pi_{\ddot{e}} \langle \mathscr{O}^* \rangle}$;
3. *non-redundancy – <u>n</u>ns: for every* $(\ddot{e}_0 \rightarrowtail H) \in CC(I \rightarrowtail \mathscr{O}^*)$ *and every* $H' \in \Pi_{\ddot{e}_0}$ *such that* $H' \cap H = \emptyset$:

$$\text{either } H' \cap \bigcap_{\ddot{e} \in cll(I \rightarrowtail \mathscr{O}^*) \setminus \{\ddot{e}_0\}} \Pi_{\ddot{e}} \langle \mathscr{O}^* \rangle = \emptyset, \qquad (6.9)$$

$$\text{or } H' \cap \bigcap_{\ddot{e} \in cll(I \rightarrowtail \mathscr{O}^*) \setminus \{\ddot{e}_0\}} \Pi_{\ddot{e}} \langle \mathscr{O}^* \rangle \nsubseteq H_{[I]} \cap H_{I \rightarrowtail \mathscr{O}^*}. \qquad (6.10)$$

6.6 Conclusions

We have proposed a non-probabilistic theory of indeterministic singular causation in which both causes and effects are given via transitions. A crucial definition is that of a cause-like locus for the effect. Such a cause-like locus should be thought of as the locus of a risky junction for the effect, such that the effect may cease to be possible right there. Our main idea is that an originating cause, or a *causa causans*, for an effect is a local indeterministic transition from a cause-like locus to that local outcome that keeps the possibility of the effect open (i.e., a local transition that does not prohibit the occurrence of the effect). The full cause of a given effect is then the set of its *causae causantes*.

We argued that our theory gains philosophical support from its agreement in spirit with the 'inus' analysis of type-level causation developed by Mackie. We proved that each *causa causans* of a given transition is at least an inus condition for its occurrence. We traced the differences between Mackie's 'inus' and our 'at least inus' conditions to Mackie's interest in types of events and our focus on particular, concrete events.

On a formal level, we defined cause-like loci for transitions of various types, $cll(I \rightarrowtail \mathscr{O}^*)$, and the resulting sets of *causae causantes*. Our theory can be made to work in contexts reminiscent of quantum non-locality as well, such as in cases in which there is (non-probabilistic) modal funny business as analyzed in Chapter 5. A *causa causans* $\tau \in CC(I \rightarrowtail \mathscr{O}^*)$ is formally defined as the transition $\tau = e \rightarrowtail H$ from a cause-like locus e for $I \rightarrowtail \mathcal{O}^*$ to that outcome of e that keeps $I \rightarrowtail \mathcal{O}^*$ possible. That outcome will be a basic one if no-MFB is assumed, and a basic disjunctive one in the presence of MFB.

6.7 Exercises to Chapter 6

Exercise 6.1. Prove clause (2) of Fact 6.1.

Hint: The argument from right to left is simple. For left to right, consider a witnessing chain O for $e < \hat{O}$ (guaranteed to exist by no-MFB), and extend the splitting-off claim from $H_{\langle \hat{O} \rangle}$ to $H_{\langle O \rangle}$. A full proof is given in Appendix B.6.

Exercise 6.2. Prove clauses (4) and (5) of Fact 6.4.

Hint: For the cases covered by clause (2), we can invoke clause (3). Otherwise, pick some O and find an additional *cll*, then use the fact that the *cll* are *SLR* in the appropriate way. A full proof is given in Appendix B.6.

Exercise 6.3. Exhibit a BST$_{92}$ structure that falsifies a candidate for the joint necessity condition in Theorem 6.1, $H_{I \rightarrowtail \mathcal{O}^*} \subseteq \bigcap_{e \in cll(I \rightarrowtail \mathcal{O}^*)} H_{e \rightarrowtail \Pi_e \langle \mathcal{O}^* \rangle}$, which was mentioned in Footnote 12. Discuss why, philosophically speaking, this candidate is not to be expected to hold.

Hint: Consider the fact that a transition can occur vacuously, by its initial failing to occur.

Exercise 6.4. Exhibit a BST$_{92}$ structure that falsifies an alternative formulation of Eq. (6.2) for the non-redundancy condition in Theorem 6.1, also mentioned in Footnote 12, namely:

$$H_{e_0 \rightarrowtail H'} \cap \bigcap_{e \in cll(I \rightarrowtail \mathcal{O}^*) \setminus \{e_0\}} H_{e \rightarrowtail \Pi_e \langle \mathcal{O}^* \rangle} \not\subseteq H_{I \rightarrowtail \mathcal{O}^*}.$$

Discuss why, philosophically speaking, this candidate is not to be expected to hold.

Hint: Again, consider the possibility of vacuous occurrence.

Exercise 6.5. Show explicitly that MFB is present in the BST$_{92}$ structure depicted in Figure 6.1.

Exercise 6.6. Prove Theorem 6.3.

Exercise 6.7. Prove Theorem 6.4.

Exercise 6.8. Prove Theorem 6.5

Exercise 6.9. Formulate and prove the BST_{NF} versions of Theorems 6.2–6.4.

7

Probabilities

In the preceding chapters we laid down a comprehensive formal framework
for describing concrete spatio-temporal events in an indeterministic setting.
We defined a number of different types of events and introduced their occurrence propositions. These, in turn, allowed us to define algebraic operations
on events, such as union, intersection, or complement. We showed how
transition-events can be defined on the basis of initial and outcome events of
different types. Transitions are a handy concept to discuss indeterminism, as
local alternatives to a transition represent indeterminism in a concrete way.
There is a particularly simple type of transition, the so-called basic transitions. These are the basic indeterministic building blocks of BST structures.
For a transition $I \rightarrowtail \mathscr{O}^*$ to a singular, non-disjunctive outcome, we defined
the *causae causantes* to be a certain set of basic transitions, $CC(I \rightarrowtail \mathscr{O}^*)$,
and we argued that these *causae causantes* play the causal role of bringing
about the transition in question.[1] To support this claim, we showed that
the *causae causantes* for a given transition satisfy Mackie's inus (or some
inus-like) conditions: At a junction at which an outcome could be rendered
impossible, a *causa causans* keeps the occurrence of the outcome possible,
but the *causa causans* need not necessitate the outcome. *Causae causantes*
thus represent an objective notion of causation under indeterminism

Given these features and defined notions, in this chapter we will show that
BST also provides a promising background for a theory of propensities (i.e.,
of objective single-case probabilities).

7.1 Two conditions of adequacy and two crucial questions

The key ingredients for a theory of objective single-case probabilities in
BST have already been supplied in the previous chapters: BST combines

---

[1] As detailed in Chapter 6.3 (see Def. 6.2), a transition to a disjunctive outcome calls for an
additional set-theoretical layer: $CC(I \rightarrowtail \breve{\mathbf{O}})$ is identified with the family of the reduced sets of *causae
causantes* to the "ingredient" transitions, $I \rightarrowtail \hat{O}_\gamma$, where $\breve{\mathbf{O}} = \{\hat{O}_\gamma \mid \gamma \in \Gamma\}$.

Branching Space-Times: Theory and Applications. Nuel Belnap, Thomas Müller, and Tomasz Placek,
Oxford University Press. © Oxford University Press 2022. DOI: 10.1093/oso/9780190884314.003.0007

possibilities with space and time, it provides a way of defining algebraic operations on concrete events, and it allows for a formal analysis of causation in indeterministic settings. Our aim in this chapter is to use these resources to lay down a general framework for probabilities that does justice to the indeterministic and spatio-temporal features of our world. We will propose a rigorous formal theory of objective single-case probabilities (propensities), in which we define probabilities as graded possibilities. Probability, in other words, codes the degree to which a given event is possible.[2]

In parallel to the development of our theory of causation in Chapter 6, we motivate the overall features of our theory of probabilities through considerations of conditions of adequacy and via the answer to two crucial questions about the representation of probability structures in BST.

7.1.1 Two conditions of adequacy

We impose two conditions of adequacy for our approach. The first is a purely formal condition, one which will ensure that our approach stays within a strictly mainstream notion of probability theory: we require that our theory should follow the axioms of standard (Kolmogorovian) probability theory. We provide the standard definition of a probability space here for later reference.

Definition 7.1 (Probability space). A *probability space* is a triple $\langle S, \mathscr{A}, p \rangle$, where S is the countable base set, or sample space, $\mathscr{A}$ is a σ-algebra of subsets of S (i.e., a set of subsets of S with the operations of union, intersection, and complement defined on $\mathscr{A}$, and which is closed under countable union), and p is a normalized, σ-additive measure; that is, a function $p : \mathscr{A} \mapsto [0,1]$ such that $p(S) = 1$ and for a family $\{a_i \mid i \in \Gamma\}$ of disjoint elements of $\mathscr{A}$, we have $p(\cup_{i \in \Gamma} a_i) = \sum_{i \in \Gamma} p(a_i)$.

In what follows, we will mostly focus on *finite probability spaces*, in which S is a finite base set. In that case, $\mathscr{A}$ can be taken to be the power-set (the set of *all* subsets) of S, and the requirement of σ-additivity boils down to finite

---

[2] This idea is advocated, e.g., by Van Fraassen (1980, p. 180), who says that "probability is a modality, it is a kind of graded possibility", or by Popper (1982, p. 70). See Gigerenzer et al. (1989, pp. 7f.) for some historical material on early 18th-century references to graded possibilities (e.g., in Leibniz).

additivity, which already holds iff for any $a_1, a_2 \in \mathscr{A}$ for which $a_1 \cap a_2 = \emptyset$, we have $p(a_1 \cup a_2) = p(a_1) + p(a_2)$.

Our second condition of adequacy is not formal but rather philosophical, although it has formal aspects: we require that our theory be able to make good sense of objective single-case probabilities (propensities) vis-à-vis known objections. We by no means argue that *all* uses of probability theory are concerned with objective single-case probabilities. There are, for example, completely adequate uses of subjective probabilities. But we aim at providing a BST analysis of the notion of ontologically basic, objective single-case probabilities, and so we are affirming the sensibility of the notion of propensities as probabilities. That idea has, however, been attacked as both philosophically ill-motivated and, perhaps more alarmingly, as formally untenable. The latter charge, which is often raised via an argument known as Humphreys's paradox, is connected with the alleged inapplicability of Bayes's theorem, a simple basic result of standard (Kolmogorovian) probability theory, to any notion of propensities. As we aim at a formally tenable and well-motivated theory, our second condition of adequacy is that our approach has to provide an answer to the challenge posed by Humphreys's paradox.

We take on our first condition of adequacy immediately, in Chapter 7.2, where we will work out a BST-based causal probability theory whose probability structures are probability spaces fulfilling Def. 7.1. A discussion of the second condition of adequacy, in the form of a defense of BST probabilities as propensities and of a thorough analysis of Humphreys's paradox, we defer to Chapter 7.3.

7.1.2 Two crucial questions

The following two questions are fundamental for any approach to probability, and they guide the development of our formal theory.

Question 7.1. What are the entities to which probabilities are assigned?

Question 7.2. What are the formal structures in which these entities are assigned their probabilities? In other words, what are the probability spaces?

There are several options for answering Question 7.1 in BST. For example, one could assign probabilities fundamentally to the individual histories in a BST structure, perhaps even in such a way that each history is assigned the same probability. One could then derive probabilities for other entities such as events, which typically occur in many histories, by measuring the respective set of histories, which might amount to simply counting their number. The idea of equiprobable histories and history counting has a certain intuitive appeal, but it faces severe technical challenges in the general case.[3] In what follows, we will sometimes appeal to the idea of probabilities attaching to histories for illustration, and in fact, our theory allows one to define the causal probability of a history as a derivative concept.[4] We will, however, assign probabilities fundamentally to other, more local objects. In parallel with our approach to objective causation, our answer to the question of which entities are assigned objective probabilities refers to the notion of a transition in BST.

Answer to Question 7.1: The entities to which probabilities are assigned are *indeterministic transitions* of any of the types definable in BST (Def. 4.4).

To motivate this answer, note that indeterministic transitions are a convenient, albeit abstract, means of representing various "chance set-ups", and objective probabilities arguably attach fundamentally to concrete chance set-ups such as a concrete toss of a concrete die. A chance set-up, understood as a singular entity, involves an initial (for example, a concrete measurement process) and a collection of possible outcomes together with a probability distribution on those outcomes conditional on the initial. Clearly, a chance set-up need not be human-made. A distant collision of two asteroids, together with a number of different possible outcomes of the collision and with a probability distribution on these outcomes conditional on the collision, counts as a chance set-up as well.

We do not claim that *every* BST transition can be assigned a probability. There are technical reasons for leaving the (anyway, uninteresting) case of trivial deterministic transitions out of the picture (see footnote 9), and there may be philosophical reasons in non-trivial cases as well—we believe that that issue is best left to further metaphysical investigation.[5] But if a

---

[3] McCall (1994, Ch. 5) bases his probabilistic theory of Branching Space-Times on counting histories that are taken to be equiprobable, providing a number of suggestive illustrations in support of his idea. McCall clearly recognizes the technical challenges of general real-valued probabilities, but in our view, his approach does not address these challenges in a fully satisfactory way, so that it remains open to formal and philosophical criticism (see, e.g. Briggs and Forbes, 2019).

[4] In the manner described in Section 7.2.4, one can define a causal probability space based on $TR(h)$, the set of transitions fully characterizing the history h (see Def. 4.11), and then look at $p(TR(h))$, the probability of that set of transitions, in that space.

[5] One might, for example, be hesitant to assign a probability to a particular free action of an agent. We do not take a stance on this matter, we merely flag it as providing a reason for perhaps not requiring *all* transitions to have a probability.

probability is assigned to some entity in BST, then that entity has to be a transition involving indeterminism, or a set of such transitions.

In what follows, we will make a terminological distinction between two notions of probability: causal probabilities or propensities defined on BST transitions, which we will denote by μ and which will be given directly via probabilistic BST structures (Def. 7.3), and measures p on mathematical probability spaces, which will be defined in accordance with Def. 7.1. It is important to keep these two notions separate, as they play different roles in our theory. Our discussion below can be viewed as providing an interface between them.

7.1.3 Propensities μ and probability measures p

Our aim is to develop the notion of the causal probability, or propensity, of a concrete BST transition $I \rightarrowtail \mathcal{O}^*$. We will follow our analysis of causation in Chapter 6, which has highlighted the crucial role of a transition's *causae causantes*, $CC(I \rightarrowtail \mathcal{O}^*)$. We will assume that the causal probability of an arbitrary transition should be provided via the causal probability of its set of *causae causantes*. As the set of $CC(I \rightarrowtail \mathcal{O}^*)$ is a subset of $TR(W)$ (or, for disjunctive outcomes, a family of such sets), it makes sense to assign propensities to the relevant subsets of $TR(W)$. For non-disjunctive outcomes, we will thus write the propensity as $\mu(CC(I \rightarrowtail \mathcal{O}^*))$, and we will interpret it as a measure of the causal strength—as the grade of the possibility—of the *causae causantes* in question. For some transitions, however, a propensity might not be defined. As a consequence, μ should be a partial function on the powerset of $TR(W)$. Its range is the unit interval $[0,1]$ of real numbers.

Having decided to let μ be a partial function on the powerset of $TR(W)$, the identification of propensities of transitions with propensities of sets of *causae causantes* for these transitions could be handled, formally speaking, by introducing another function, say μ', defined generally on transition events rather than on subsets of $TR(W)$. Our identification thesis would then be rendered as $\mu(CC(I \rightarrowtail \mathcal{O}^*)) = \mu'(I \rightarrowtail \mathcal{O}^*)$. However, in order not to introduce too much symbolism, we will use the same symbol, μ, and write $\mu(CC(I \rightarrowtail \mathcal{O}^*)) = \mu(I \rightarrowtail \mathcal{O}^*)$, as the domain of μ will always be clear from context.

In what follows, we first discuss a few metaphysical constraints on μ that follow from the fact that $\mu(I \rightarrowtail \mathcal{O}^*)$ is a propensity, that is, a measure of the grade of possibility of a concrete transition in BST. Then we discuss how propensities μ can be reflected via mathematically well-defined probability measures p. The first metaphysical constraint concerns the propensity μ of transitions to disjunctive outcomes. To begin, consider a non-trivial basic disjunctive transition $\tau =_{df} e \rightarrowtail \breve{I}_e$ from a choice point e to the exhaustive disjunctive outcome $\breve{I}_e =_{df} \{H \mid H \in \Pi_e\} = \Pi_e$ that collects together all of its (two or more) possible immediate outcomes.[6] The transition τ is deterministic in the sense that it happens anyway: its occurrence proposition is universal, $H_{e \rightarrowtail \breve{I}_e} = \text{Hist}$. So τ has to have propensity one:

$$\mu(\tau) = \mu(e \rightarrowtail \breve{I}_e) = 1. \tag{7.1}$$

Given that the initial e is indeterministic, the transition $e \rightarrowtail H$ to any individual possible outcome $H \in \Pi_e$ of e has a cause-like locus at which it could be prevented from occurring, namely, e, and therefore, the grade of possibility of $e \rightarrowtail H$ will normally be strictly less than one. The complete disjunctive outcome $\breve{I}_e$, however, exhausts all the possibilities. Given that e occurs, one of the outcomes of e has to be realized, and this is what is expressed by Eq. (7.1).[7]

This constraint on the propensity function μ is of a logico-metaphysical nature: it follows from the way in which concrete transitions are defined in BST. The constraint will become important when we discuss marginal probabilities below. In a similar vein, we also assume a logico-metaphysically motivated constraint for other disjunctive outcomes of a point event: for a disjunctive outcome $\breve{H} \subseteq \Pi_e$ (i.e., for $\breve{H}$ a set of immediate basic outcomes of a choice point e, which are mutually incompatible by definition) we postulate

$$\mu(e \rightarrowtail \breve{H}) = \sum_{H \in \breve{H}} \mu(e \rightarrowtail H), \tag{7.2}$$

---

[6] In this chapter, we rewrite the definition of $\breve{I}_e$ (Def. 4.6) in terms of propositional outcomes rather than in terms of scattered outcomes. As shown in our discussion in Chapter 4.2, the two representations are equivalent for initials consisting of single events e.

[7] In a mathematical probability space that represents $\tau = e \rightarrowtail \breve{I}_e$, the full disjunction of all possible outcomes of e is represented via the unit element of the algebra, which has probability 1 by normalization. And the zero element has probability 0, of course. We remain impartial on the issue of so-called faithfulness, i.e., on the question of whether *only* the zero element of the algebra can have probability zero. By allowing for the assignment of zero probability to other elements of the algebra, we can, for example, simulate modal funny business in a probabilistic BST structure that harbors no real modal funny business. We will make use of this option in Chapter 8.4.

that is, the propensity of the transition from e to its disjunctive outcome $\breve{H}$ is the sum of the propensities of these basic outcomes taken individually. For example, considering a concrete throw of a fair die, for which all immediate outcomes $\boxed{1}, \ldots, \boxed{6}$ have propensity $1/6$, the disjunctive outcome "even number" (immediate outcome $\boxed{2}$, $\boxed{4}$, or $\boxed{6}$) has propensity $1/6 + 1/6 + 1/6 = 1/2$. In fact, Eq. (7.1) can be viewed as a special case of this additivity constraint, given the logico-metaphysical rule that unavoidable transitions have propensity one.

These considerations show that the set of all basic transitions from a given choice point naturally gives rise to a probability space. In the finite case on which we focus here, that is, when e has finitely many immediate basic outcomes, one can simply use the Boolean algebra $\mathscr{A}$ of the power-set of all the basic transitions, and a measure on the algebra can be induced by the propensities of the individual basic transitions.

At this juncture, our second notion of probability, the mathematical probability measure p, becomes important. Even if we can introduce some more constraints on the propensity function μ, we are not working toward a theory in which the propensity function μ itself fulfills the axioms of probability theory of Def. 7.1. On the one hand, this is due to our choice of answer to Question 7.1: We are working toward a theory in which concrete local BST transitions, which can represent concrete chance set-ups, have causal probabilities assigned, and therefore, we target the definition of local probability spaces with local measures p, whereas μ is defined globally. In addition, there are several technical hurdles that stand in the way of interpreting μ as a probability measure. For starters, note that we have allowed μ to be a partial function, which directly contradicts the assumption that a probability measure be defined for all elements of the algebra, (i.e., for all possible events). Second, and more alarmingly, consider the fact that, as we just said, all unavoidable transitions have propensity one. How should this be accommodated in a probability space defined on all transitions? Consider two transitions from choice points e_i to their exhaustive disjunctive outcome, $\tau_i =_{df} e_i \rightarrowtail \breve{I}_{e_i}$ ($i = 1, 2$), in incompatible possible futures (basic scattered outcomes) of another choice point e, so that $e < e_1$ and $e < e_2$. The two unavoidable transitions τ_1 and τ_2 both have propensity one, but their outcome-parts are incompatible. So it seems that in a probability space representation, τ_1 and τ_2 have to belong to disjoint elements of the algebra. But then we have to have two disjoint elements of the algebra that each have probability 1, and by the additivity of the measure, their union has to have probability 2, violating the normalization of the measure. Something is badly amiss here. Apart from extremely simple cases such as BST structures with just one choice point, it seems hopeless to try and define a *global* probability space in which the propensity function μ is really a probability measure.[8,9]

We do not claim that the issue cannot be resolved by some clever way of defining a global probability space based on a whole BST structure and deriving propensities of transitions in some other way. The propensities of concrete transitions in the BST structure would most likely have to be defined as conditional probabilities of some sort (something like the probability of the occurrence of the outcome conditional on the occurrence of the initial). We will not go down that route, however, as on the one hand it leads to unmanageably large probability spaces while, on the other, it goes against a basic principle of our approach to possibilities: a main advantage of BST is that we work with local notions of possibilities and transitions.[10] We look for a corresponding, local way of handling propensities as well.

So, given that we want to talk about the propensity $\mu(I \rightarrowtail \mathcal{O}^*)$ of some concrete transition $I \rightarrowtail \mathcal{O}^*$ *as a probability*, and a globally defined probability space is not promising in this respect, we have to find some locally defined mathematical probability space $\langle S, \mathscr{A}, p \rangle$ in which the given transition (together with some other transitions) is represented and in which it is assigned a probability that can be read as a propensity. This is the question of providing an interface between the causal BST notion μ and mathematical probability theory. As we will see, logico-metaphysical and causal constraints on μ (such as the one about unavoidable transitions having propensity one) provide useful guidance as to which mathematical structures are appropriate. To provide an indication of our approach: for a given indeterministic transition $I \rightarrowtail \mathcal{O}^*$, an adequate causal probability

---

[8] In this sense, therefore, propensities and probabilities come apart—but they are intimately linked. Our answer to Humphreys's challenge that "propensities cannot be probabilities" will stress that link; see Chapter 7.3.3.

[9] A further problem is how to handle trivial deterministic transitions, such as a transition from a deterministic event e to its only outcome H_e. For such transitions, the set of *causae causantes* is empty. The only sensible propensity one can assign for $\mu(\emptyset)$ is one, as such a transition is inevitable, having the universal occurrence proposition. But it is not possible to build a probability space in which the event algebra has only one element, the empty set. (The problem for fulfilling Def. 7.1 is with the measure, not with the algebra.) This is our technical reason for not considering causal probability spaces (see Def. 7.4) for transitions that have no indeterministic causes: there are no causal probability spaces without causation.

[10] Defining probabilities on the histories, as discussed in note 3, would be one such global approach; another would be to define a joint probability space for *all* indeterministic transitions in a given BST structure $\langle W, < \rangle$, i.e., on $TR(W)$. Our approach naturally extends to accommodate these two ideas, but it does not presuppose them. It is, therefore, both more local and more general.

space, which is able to represent that transition and its propensity, will contain $CC(I \rightarrowtail \mathscr{O}^*)$ together with some appropriate representation of causal alternatives to $I \rightarrowtail \mathscr{O}^*$ (see Defs. 7.2 and 7.4). As we said, we will make use of the assumption that the propensity of a transition may only depend on the propensities of its set of *causae causantes*. The latter set is consistent, and accordingly, a causal probability space will have a sample space consisting of consistent sets of transitions.[11] Once the details are in place, a number of mathematical constraints on the measure in a given causal probability space in relation to measures in other causal probability spaces can be motivated as either logico-metaphysical postulates (e.g., involving marginal probabilities), or as causal postulates (e.g., involving a form of the Markov property).

We sum up the partial answer to our Question 7.2 which is implicit in our discussion above.[12]

Partial answer to Question 7.2: When $\mu(I \rightarrowtail \mathscr{O}^*)$ is defined, its value must be fully determined via the propensities of the basic transitions from the set $CC(I \rightarrowtail \mathscr{O}^*)$. This emphatically includes the possibility that one may need to take into account not only the propensities of the individual *causae causantes*, but also propensities of certain sets of them, taken as operating jointly.

This answer guarantees that for BST transitions that have causal probabilities assigned, if they have the same set of *causae causantes*, then their causal probability also has to be the same. Note that we made a caveat about sets of *causae causantes* possibly working together. This option is needed in order to make room for probabilistic correlations. Correlations are often scientifically important. In Chapter 5.1 we motivated our account of modal funny business via the phenomenon of modal correlations, and here, similarly, we also have to make room for the phenomenon of probabilistic correlations. We work toward the notion of probabilistic correlations, to be provided in Chapter 7.2.6, by considering a number of scenarios that will help to anchor our general ideas about representing propensities μ as Kolmogorovian probabilities p.

---

[11] Note that by Def. 6.4 the set of *causae causantes* for a BST transition to an outcome chain or to a scattered outcome is consistent, and that the *causae causantes* for a transition to a disjunctive outcome are a family of consistent sets of transitions.

[12] A full answer to Question 7.2 will be provided via our definition of causal probability spaces in Chapter 7.2.4; see p. 193.

7.2 Causal probability spaces in BST

With a view to our first condition of adequacy, we will start by developing our theory of BST probabilities for indeterministic transitions, which will lead to the definition of a *causal probability space*. This will help provide a full answer to Question 7.2 about the formal structures in which transitions are assigned probabilities that can be read as propensities.

7.2.1 Probabilities for transitions: The simplest case

We illustrate the idea of defining probabilities for transitions by starting with the simplest case, which we have already outlined above. The simplest case of a transition is a basic indeterministic transition, a notion that we discussed in Chapter 4.2. Consider Alice's throwing of a fair die, and assume for simplicity that this die-throwing is a localized indeterministic event that has exactly six possible immediate outcomes, namely, the numbers 1 through 6.[13] We thus represent the die-throwing by a set of six mutually incompatible basic transitions $\tau_1 = e \rightarrowtail \boxed{1}, \ldots, \tau_6 = e \rightarrowtail \boxed{6}$ with the same initial e, whose set of basic outcomes we write as $\Pi_e = \{\boxed{1}, \ldots, \boxed{6}\}$. Given that the die is fair, each of the concrete transitions τ_i has the same propensity, $\mu(\tau_i) = 1/6$ $(i = 1, \ldots, 6)$. In this case, it is perfectly natural to take the six basic transitions as the elementary events constituting a sample space $S = \{\tau_1, \ldots, \tau_6\} = \{e \rightarrowtail H \mid H \in \Pi_e\}$. We can then define a finite probability space $\langle S, \mathscr{A}, p \rangle$ using the Boolean algebra $\mathscr{A}$ of subsets of S, and assigning the fully symmetrical probability measure p in accord with the transitions' propensities: the value of p on the elements of the sample

---

[13] It is not easy to say whether a concrete transition of Alice's throwing a die, with the initial of Alice prepared to throw in a concrete situation and with the outcome of the die showing, for example, outcome 1, is indeed indeterministic. The issue may in fact depend on details of Alice's current physiological state. As to the mechanics of throwing leading to a specific number, given a concrete die thrown in a concrete way (with concrete speed and angular momentum etc.), the transition to the outcome is perhaps even deterministic. (See Diaconis et al., 2007, for a study of the related problem of coin-tossing.) In what follows, we idealize Alice's die-throwing, as well as other chance set-ups (Bob's tossing a coin and Eve's throwing an octahedron), to be indeterministic. The most realistic examples of truly indeterministic transitions with precisely specified probabilities are, for all we know, quantum experiments, which will be discussed in Chapter 8. We stick to everyday examples in this chapter so as not to bring up too many complications at once. And, historically, probability theory was in fact developed initially for use in the context of simple games of chance such as games with dice (see Hacking, 2006).

space (the atoms of the algebra) is $p(\tau_i) = \mu(\tau_i) = 1/6$ $(i = 1, \ldots, 6)$.[14] Apart from the BST notion of a transition, this is all perfectly standard. Note that non-atomic elements of the Boolean algebra can be viewed as disjunctive outcomes of the initial e, in full accordance with Def. 4.4, as indicated above. To repeat, the event "throwing an even number", which is represented as the set $\{\tau_2, \tau_4, \tau_6\} \in \mathscr{A}$, corresponds to the transition from e to the disjunctive outcome $\{\boxed{2}, \boxed{4}, \boxed{6}\}$ of e. This will become important below in connection with marginal probabilities.

As transitions represent causal happenings in BST, our mathematically defined probabilities p for transitions represent causal probabilities, or propensities, μ. The causal probability $\mu(e \rightarrowtail \boxed{1})$ of $\tau_1 = e \rightarrowtail \boxed{1}$ can be viewed as the grade of possibility of the concrete possible causal process that leads from the initial event e to the concrete outcome-event $\boxed{1}$. In our example, using our probability space $\langle S, \mathscr{A}, p \rangle$ as the mathematical background representation, that probability is given as $\mu(e \rightarrowtail \boxed{1}) = p(\tau_1) = 1/6$. So, in this simple case, there is an immediate correspondence between the causal probabilities μ and the mathematical probabilities p.

Earlier, we discussed a logico-metaphysical constraint on causal probabilities: the causal probabilities for incompatible basic outcomes of the same initial event have to add up. Given the probability space of our example, the additivity of the measure secures that this constraint is satisfied. Take two basic outcomes $\boxed{i}$ and $\boxed{j}$ of e $(i \neq j)$. The concrete transition $e \rightarrowtail \{\boxed{i}, \boxed{j}\}$ to the disjunctive outcome $\{\boxed{i}, \boxed{j}\}$ corresponds to the element $\{\tau_i, \tau_j\} \in \mathscr{A}$, and as $\{\tau_i\} \cap \{\tau_j\} = \emptyset$, by additivity we have

$$\mu(e \rightarrowtail \{\boxed{i}, \boxed{j}\}) = p(\{\tau_i, \tau_j\}) = p(\tau_i) + p(\tau_j) = \mu(e \rightarrowtail \boxed{i}) + \mu(e \rightarrowtail \boxed{j}).$$

Similarly, it follows by normalization that

$$\mu(e \rightarrowtail \breve{\mathrm{I}}_e) = p(\mathbf{1}_{\mathscr{A}}) = p(S) = 1.$$

Note that the probability space in which $p(\tau_1)$ is assigned a probability contains all the immediate causal alternatives to τ_1, namely, all the other transitions from e to one of its immediate outcomes, as well. This idea

---

[14] We follow the standard practice in abusing the notation slightly: given an elementary outcome $a \in S$, the probability is officially defined only for the corresponding element of the algebra, $\{a\} \in \mathscr{A}$, but we allow ourselves to write $p(a)$ instead of the more cumbersome $p(\{a\})$. We take the same liberty for μ.

will guide our general construction below. We work toward that general construction by considering a number of further simple stories.

7.2.2 Two BST transitions, one basic transition

Recall the story about Alice sending a letter to Bob from Chapter 6.3.1, which we used to make a distinction between an active and a passive causal contribution. In order to stay close to our story, assume that Alice is undecided as to whether she wants to send a letter to Bob (initial A-undec), and has chosen to leave the matter to chance. She will throw a fair die, and she will send a letter (outcome A-sends) exactly if the die shows 1 (outcome $\boxed{1^A}$ of die-throwing event e^A). This chance set-up is mathematically represented by the probability space for the die-throwing discussed above, which we now write $\langle S^A, \mathscr{A}^A, p^A \rangle$, where the sample space $S^A = \{\tau_1^A, \ldots, \tau_6^A\}$. Bob, on the other hand, is far away, and he is facing a (passive) transition from having no information in the morning as to whether he will receive a letter (initial B-noinf) to either having personally received Alice's letter (outcome B-receives) or not having received a letter (outcome B-noletter). Given our minimal story, we have

$$CC(\text{A-undec} \rightarrowtail \text{A-sends}) = \{e^A \rightarrowtail \boxed{1^A}\} = CC(\text{B-noinf} \rightarrowtail \text{B-receives}).$$

The scenario is pictured in Figure 7.1. Given our reliance on *causae causantes*, we already have

$$\mu(\text{A-undec} \rightarrowtail \text{A-sends}) = \mu(e^A \rightarrowtail \boxed{1^A}) = \mu(\text{B-noinf} \rightarrowtail \text{B-receives}),$$

and by our representation of Alice's die-throwing event via the probability space $\langle S^A, \mathscr{A}^A, p^A \rangle$, we have

$$\mu(e^A \rightarrowtail \boxed{1^A}) = p^A(\tau_1^A) = 1/6.$$

That is, our decision to make single-case probabilities depend exclusively on the *causae causantes* of a transition already has a substantial consequence for our simple example: given that there is only one indeterministic transition doing any causal work, it must be that all concrete transitions that have the

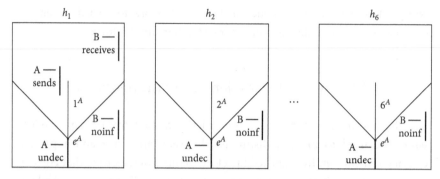

Figure 7.1 Alice throws a fair die to decide whether or not to send a letter to Bob. The BST structure contains six histories, one for each outcome of the throw. Alice sends the letter, and Bob receives it, exactly in history h_1.

same *causae causantes* also have the same causal probability. That causal probability is represented in a suitable probability space.

It is crucial that here we are dealing with causal probabilities of transitions, rather than the causal probabilities of outcomes in isolation. It arguably makes little sense to ask about the objective probability of the occurrence of an outcome event such as B-receives alone. The event of Bob's receiving the letter may be taken to have probability $1/6$, given the background of our story, but it may also be taken to be exceedingly improbable, considering, for example, the fact that Bob's parents only met via an unlikely coincidence and that it is therefore highly improbable that Bob was even ever born. A causal probability always depends on how far back in time one looks for alternatives.[15] As discussed in Chapter 6, a BST transition supplies all the details of the causal background before which the causal influences bringing about an event are assessed. Likewise, a transition supplies all the details of the causal background (the initial) before which the probability of an outcome is determined.

---

[15] It is possible to look back all the way, to the beginning of time, in which case no initial has to be given. Before we introduced the notion of *causae causantes* for a transition $I \rightarrowtail \mathcal{O}^*$ via its set of cause-like loci $cll(I \rightarrowtail \mathcal{O}^*)$ (Defs. 6.2 and 6.4), we in fact defined the set of cause-like loci $cll(O)$ of an outcome chain (Def. 5.10). This set includes *all* the risky junctions at which the occurrence of O could ever have been prevented. That set will normally be far too large to be of interest, and the objective probability of the occurrence event given no initial will normally be vanishingly small. (If we had not ruled this out by requiring an initial event to be non-empty (Def. 4.3), we could indicate the occurrence of the outcome $\mathcal{O}^*$ alone via the transition $\emptyset \rightarrowtail \mathcal{O}^*$.) To repeat, we aim at providing a *locally* anchored BST probability theory, and so we avoid the idea of looking back all the way.

7.2.3 Two or more transitions and some complications

On a more realistic account of the letter sending and receiving example, both concrete transitions in question—Alice's transition from being undecided to having sent the letter and Bob's transition from having no information to having personally received the letter—of course have many more *causae causantes*.

We first provide two different variants of our simplest story in which we add one extra indeterministic event, first on Bob's side and then on Alice's. We later combine the two variants into a more complex story, which already contains most of the ingredients needed for motivating our general account.

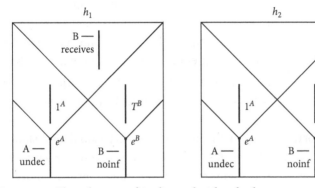

Figure 7.2 Alice throws a fair die to decide whether or not to send a letter to Bob, and Bob throws a fair coin to decide whether or not to go out. Bob receives the letter if and only if Alice's die has outcome 1 and his coin lands tails. The BST structure contains twelve histories, one for each pairing of an outcome of the throw of the die and the toss of the coin. Only histories h_1 and h_2 are shown.

Variant 1: Bob goes out. Consider, first, that Bob's personally receiving the letter depends causally on his being at home. In the morning, Bob is undecided whether to stay at home or to go out, and similar to Alice, he leaves the matter to chance: he tosses a fair coin, and if the coin shows heads (H^B), he stays at home; if not, if the coin shows tails (T^B), he takes a long walk (see Figure 7.2). Bob's coin-tossing can be represented in exact analogy to the formal account of Alice's die-throwing: the relevant mathematical probability space is based, in this case, on an underlying set of two mutually incompatible basic transitions leading from the choice point e^B to the outcomes T^B ($\tau_1^B = e^B \rightarrowtail T^B$) or H^B ($\tau_2^B = e^B \rightarrowtail H^B$). The symmetrical probability measure p^B for Bob's coin-toss assigns the values that we know to be the causal probabilities of the chance set-up of the concrete tossing of the fair coin, one half:

$$p^B(\tau_1^B) = \mu(e^B \rightarrowtail T^B) = p^B(\tau_2^B) = \mu(e^B \rightarrowtail H^B) = 1/2.$$

Let us assume that, as Alice and Bob are far away, their respective chance events are space-like related (for the initials, $e^A \, SLR \, e^B$): these events cannot causally influence one another. The transition A-undec $\rightarrowtail$ A-sends from Alice's being undecided to her having sent the letter, due to her die having shown outcome 1, still has a single *causa causans*, namely, the transition $\tau_1^A = e^A \rightarrowtail \boxed{1^A}$. The transition B-noinf $\rightarrowtail$ B-receives from Bob's having no information about the letter and being undecided whether to go out to his personally receiving the letter, on the other hand, now has two *causae causantes*: The originating causes of that transition are the two basic transitions $e^A \rightarrowtail \boxed{1^A}$ (Alice's die showing 1), which triggers Alice's sending the letter, and $e^B \rightarrowtail H^B$ (Bob's coin showing heads), which triggers Bob's staying at home.

This causal story, together with our decision that causal probabilities have to be based exclusively on *causae causantes*, affirms our earlier verdict about the probability of Alice's sending-the-letter-transition: μ(A-undec $\rightarrowtail$ A-sends) $= 1/6$. For Bob's transition from no information to having received the letter personally, however, our account gives only a partial answer:

$$\mu(\text{B-noinf} \rightarrowtail \text{B-receives}) = \mu(\{e^A \rightarrowtail \boxed{1_A}, e^B \rightarrowtail T^B\}) =?$$

The causal probability of Bob's transition has to be the causal probability of its two-transition set of *causae causantes*, but that latter probability is not necessarily determined by the individual causal probabilities of the basic transitions alone. And so far it is also not clear in which mathematical probability space that causal probability can be adequately represented.

To make some headway, we provide some relevant logico-metaphysical observations about causal probabilities. Given that in our structure, e^A and e^B are the only choice points and these choice points are SLR, it follows that any history $h \in$ Hist contains both e^A and e^B. (For our reasoning below it is enough that for any history h, $e^A \in h$ iff $e^B \in h$.) The set $I = \{e^A, e^B\}$ makes sense as a BST initial event, as it is consistent. Now we can look at its outcomes. Assuming that there is no modal funny business (see p. 175 for a brief discussion), any basic outcome of e^A is compatible with any basic outcome of e^B, so that there are $6 \cdot 2 = 12$ possible basic (non-disjunctive) joint outcomes, and $2^{12} - 1 = 4095$ disjunctive outcomes (sets of basic outcomes; we have to disregard the empty set). We represent the elementary joint outcomes via outcome chains O_{ij}, where $i \in \{1, \ldots, 6\}$ and $j \in \{1, 2\}$. Spatio-temporally, the outcome chains O_{ij} begin in the joint future of possibilities of e^A and e^B. Epistemologically speaking, if you are situated at one of these chains, you have information about the outcome of both e^A and of e^B. Each joint outcome corresponds to one of the 12 histories in the structure, which are accordingly denoted as h_{ij}, so that $O_{ij} \subseteq h_{ij}$.

With a view to calculating marginal probabilities, consider the disjunctive outcome $\check{\mathbf{O}}_i =_{df} \{O_{i1}, O_{i2}\}$, so $H_{\langle \check{\mathbf{O}}_i \rangle} = \{h_{i1}, h_{i2}\}$ $(i \in \{1, \ldots, 6\})$. The transition $I \rightarrowtail \check{\mathbf{O}}_i$ fulfills the requirement of the appropriate spatio-temporal location of I and $\check{\mathbf{O}}_i$ of Def. 4.4, as is easy to check. We then calculate

$$CC(I \rightarrowtail \check{\mathbf{O}}_i) = \{\{e^A \rightarrowtail \boxed{i}, e^B \rightarrowtail H^B\}, \{e^A \rightarrowtail \boxed{i}, e^B \rightarrowtail T^B\}\}. \quad (7.3)$$

We can now employ the logico-metaphysical principles about causal probabilities being based solely on *causae causantes* and about the probabilities of disjoint outcomes adding up. Since the elements of $CC(I \rightarrowtail \check{\mathbf{O}}_i)$ are incompatible, the probability of $CC(I \rightarrowtail \check{\mathbf{O}}_i)$ should be the sum of the probabilities of these elements. Further, since every history passing through $\boxed{i}$ contains either H^B or T^B, for our particular setup, in which e^A and e^B occur in each history, we have that:

$$H_{\{e^A \rightarrowtail \boxed{i}, e^B \rightarrowtail H^B\}} \cup H_{\{e^A \rightarrowtail \boxed{i}, e^B \rightarrowtail T^B\}} = H_{\{e^A \rightarrowtail \boxed{i}\}}. \quad (7.4)$$

These observations lie behind the following summation formula, where we use O_i for an outcome chain witnessing the outcome $\boxed{i}$ of Alice's die throwing:

$$\mu(I \rightarrowtail \check{\mathbf{O}}_i) = \mu(CC(I \rightarrowtail \check{\mathbf{O}}_i)) =$$
$$\mu(\{\{e^A \rightarrowtail \boxed{i}, e^B \rightarrowtail H^B\}, \{e^A \rightarrowtail \boxed{i}, e^B \rightarrowtail T^B\}\}) =$$
$$\mu(\{e^A \rightarrowtail \boxed{i}, e^B \rightarrowtail H^B\}) + \mu(\{e^A \rightarrowtail \boxed{i}, e^B \rightarrowtail T^B\}) =$$
$$\mu(\{e^A \rightarrowtail \boxed{i}\}) = \mu(CC(I \rightarrowtail O_i)) = \mu(I \rightarrowtail O_i).$$

The first transformation comes from Eq. 7.3, the second results from the summation of probabilities of incompatible elements of *causae causantes*, the third is based on Eq. 7.4, and the remaining transformations invoke our convention about the domain of μ. This calculation establishes that the causal probabilities yield sensible marginal probabilities: The causal probability of Alice's die showing outcome i and Bob's coin turning up either way is just the causal probability of Alice's die showing outcome i, full stop.

Given these facts about the causal probabilities, it is natural to try to represent them by combining the two given probability spaces for Alice's and for Bob's chance set-ups via their Cartesian product. That operation leads to the sample space $S^C = S^A \times S^B$. That is, a basic outcome $\langle \tau_i^A, \tau_j^B \rangle \in S^C$ in that probability space specifies one transition τ_i^A of Alice's die-throwing and one transition τ_j^B of Bob's coin-tossing, in natural correspondence to the outcome events O_{ij}. The algebra $\mathscr{A}^C$ will be the algebra of subsets of S^C, as usual. Furthermore, given the *SLR* layout, it seems natural to require that the distant outcomes be probabilistically uncorrelated. In that case, the joint measure p^C is the product measure, for which $p^C(\langle \tau_i^A, \tau_j^B \rangle) = p^A(\tau_i^A) \cdot p^B(\tau_j^B)$. In particular, we then have

$$p^C(\langle \tau_1^A, \tau_2^B \rangle) = p^A(\tau_1^A) \cdot p^B(\tau_2^B) = 1/6 \cdot 1/2 = 1/12.$$

If we assume that the probability space $\langle S^C, \mathscr{A}^C, p^C \rangle$ provides an adequate representation of the causal probabilities, we thus have $\mu(\text{B-noinf} \longmapsto \text{B-receives}) = p^C(\langle \tau_1^A, \tau_2^B \rangle) = 1/12$.

The just mentioned construction proceeds in two steps: first, combine the small (local) probability spaces by taking their Cartesian product, and then define a measure on the joint space via the product of the local measures. This construction is quite useful in the case at hand but, emphatically, we do *not* propose any of these two steps as a general basic recipe: we do not require in general that causal probability spaces combine via Cartesian products, and we do not require that in all cases, *SLR* initials give rise to probabilistically uncorrelated transitions.

We will soon turn to a case in which the Cartesian product construction is inappropriate; see our next scenario, Variant 2 (p. 177). There are, however, a number of important issues that we can discuss in relation to the present case (Variant 1). First, in the case of *SLR* initials, the question of whether the individual local possibilities combine so as to yield the Cartesian product is exactly the question of whether there is modal funny business (MFB) as discussed in Chapter 5; see clause 4 of Def. 5.5 of combinatorial consistency and the following Def. 5.6 of combinatorial funny business. Given the simplifying assumption of no MFB, the sample space for the combination of the possibilities in Alice's and Bob's chance set-ups is indeed adequately represented by the Cartesian product $S^C = S^A \times S^B$ as defined above.

Second, the Cartesian product construction does not uniquely determine the joint probability measure p^C, but it constrains the options. That is, again, in full accord with the situation regarding the causal probabilities. It might be, for all we know, that the transitions are fully correlated, so that in any case in which Alice's die shows 1, Bob's coin shows tails. In that case,

$$\mu(\text{B-noinf} \rightarrowtail \text{B-receives}) = \mu\left(\{e^A \rightarrowtail \boxed{1^A}, e^B \rightarrowtail T^B\}\right)$$
$$= \mu(e^A \rightarrowtail \boxed{1^A}) = 1/6,$$

which in terms of the joint measure would correspond to $p^C(\langle \tau_1^A, \tau_2^B \rangle) = p^A(\tau_1^A) = 1/6$, rather than $1/12$ on the product measure. On the other hand, it might also be that the transitions are fully anti-correlated (either one can occur only if the other does not occur), so that

$$\mu(\text{B-noinf} \rightarrowtail \text{B-receives}) = \mu(\{e^A \rightarrowtail \boxed{1^A}, e^B \rightarrowtail T^B\}) = 0,$$

corresponding to $p^C(\langle \tau_1^A, \tau_2^B \rangle) = 0$. In terms of possible probability measures p^C based on p^A and p^B, all these numbers make sense, and empirically, in terms of observed causal probabilities, the well-confirmed phenomenon of EPR-like probabilistic quantum correlations arguably shows that the full range of these options can in fact be produced.[16] This means that given our causal set-up of $e^A \, SLR \, e^B$, there is quite some freedom for assigning a numerical value to the joint probability $p^C(\langle \tau_1^A, \tau_2^B \rangle)$.

Even so, there are still some constraints—one cannot just assign *any* number between 0 and 1 to $p^C(\langle \tau_1^A, \tau_2^B \rangle)$. Technically, we require that the probability of the single-transition events can be recovered as marginal probabilities from the joint probability measure. That is, $p^A(\tau_1^A)$ has to be recovered from the joint probabilities of all two-transition pairs including τ_1^A and some τ_j^B, and similarly for $p^B(\tau_2^B)$:

$$p^A(\tau_1^A) = \sum_{j=1}^{2} p^C(\langle \tau_1^A, \tau_j^B \rangle); \quad p^B(\tau_2^B) = \sum_{i=1}^{6} p^C(\langle \tau_i^A, \tau_2^B \}). \qquad (MP)$$

---

[16] We will discuss these issues in more detail in Chapter 8.4.

These equations simply follow from representing probabilities of transitions via the causal probabilities of the sets of their *causae causantes* and from our account of *causae causantes*.

Given the Cartesian product construction of joint probability spaces, the marginal property follows as a mathematical consequence. Our reason for requiring the marginal property is, however, not purely mathematical, but properly causal. Consider any history h in which e^A occurs, so that by our assumptions $e^B \in h$ as well, and $e^A SLR e^B$. As e^B is the initial of a set of incompatible basic transitions, it follows that on h, exactly one of the possible outcomes occurs, viz., $\Pi_{e^B}\langle h \rangle$. So, any history on which τ_1^A occurs must be a history on which one of the τ_j^B occurs as well: for τ_1^A to occur *is* for τ_1^A to occur and for some τ_j^B to occur. Thus, the probability for τ_1^A to occur must be divided up, so to speak, between the different ways for τ_1^A to occur and some τ_j^B to occur. Each of the pairs $\langle \tau_1^A, \tau_j^B \rangle$ has to have a non-negative probability with respect to the measure p^C, and together, these probabilities have to add up to the probability $p^A(\tau_1^A)$, as Eq. (*MP*) specifies.

To put this matter in explicitly causal terms, we can say that given our scenario, the transition from Alice's initial e^A to the outcome $\boxed{1^A}$ is causally equivalent to (has the same *causae causantes* as) the transition from the initial $\{e^A, e^B\}$ to the disjunctive outcome $\{\{\boxed{1^A}, T^B\}, \{\boxed{1^A}, H^B\}\}$, and therefore, as a matter of causal probability irrespective of the representation of these probabilities in any mathematical structures, we have to have

$$\mu(\tau_1^A) = \mu(\{\tau_1^A, \tau_1^B\}) + \mu(\{\tau_1^A, \tau_2^B\}).$$

So the marginal property has a proper metaphysical foundation. It follows in particular that, as probabilities are non-negative, for $j \in \{1,2\}$,

$$0 \leqslant \mu(\{\tau_1^A, \tau_j^B\}) \leqslant \mu(\tau_1^A) = 1/6,$$

and a parallel consideration establishes the bound that for $i \in \{1, \ldots, 6\}$,

$$0 \leqslant \mu(\{\tau_i^A, \tau_2^B\}) \leqslant \mu(\tau_2^B) = 1/2.$$

So our requirement of basing probabilities solely on probabilities of sets of *causae causantes* (possibly working together), supplemented by considerations of the causal consequences of the *SLR* relation of the initials e^A and e^B, yields the following constraint:

$$0 \leqslant \mu(\text{B-noinf} \rightarrowtail \text{B-receives}) = \mu(\{\tau_1^A, \tau_2^B\}) \leqslant \min(\mu(\tau_1^A), \mu(\tau_2^B)) = 1/6.$$

And, to repeat, in the case of no correlations, which for our example is exceedingly plausible, we have the simple "just multiply" result,

$$\mu(\text{B-noinf} \rightarrowtail \text{B-receives}) = \mu(e^A \rightarrowtail \boxed{1}) \cdot \mu(e^B \rightarrowtail H^B) = 1/6 \cdot 1/2 = 1/12.$$

Variant 2: Mail gets lost. We return to our basic story of Alice throwing a die to determine whether or not to send a letter to Bob. This time round, we let Bob stay at home deterministically, canceling the indeterministic transition event on his side. However, we introduce a different complication, namely, there is an eavesdropper, Eve, who indeterministically blocks some letters from being delivered. If Alice's die has shown outcome 1 and she has sent her letter, there is a further indeterministic transition, from Eve seeing the letter (E-sees) to one of two possible outcomes: taking the letter, so that it will not arrive at Bob's place (E-takes), or letting it pass (E-passes). Given that BST represents concrete events, the initial E-sees can occur only after Alice sends the letter. The concrete event E-sees does not occur in any of the other, incompatible outcomes of A-undec. Now we assume again that the indeterminism of Eve's taking or not taking the letter is due to a chance set-up, this time the throw of an octahedron by Eve, which is represented by a set of eight basic outcomes $\boxed{i^E}$ of the choice point e^E at which the octahedron is thrown, corresponding to the eight basic transitions $\tau_i^E = e^E \rightarrowtail \boxed{i^E}$ ($i = 1, \ldots, 8$). We assume that Eve lets the letter pass iff her octahedron shows 3, and that she takes the letter on all other outcomes. We also assume that the octahedron is fair, so that the causal probabilities are $\mu(e^E \rightarrowtail \boxed{i^E}) = 1/8$ ($i = 1, \ldots, 8$). An adequate probability space for representing Eve's throw can be based on the sample space $S^E = \{\tau_1^E, \ldots, \tau_8^E\}$ and the symmetrical measure p^E that assigns the value $p^E(\tau_i^E) = 1/8$ for each of the τ_i^E ($i = 1, \ldots, 8$).

As in Variant 1, the transition A-undec $\rightarrowtail$ A-sends, from Alice's being undecided to her having sent the letter, still has a single *causa causans*, namely, the transition $\tau_1^A = e^A \rightarrowtail \boxed{1^A}$. The transition B-noinf $\rightarrowtail$ B-receives from Bob's having no information about the letter to his personally receiving the letter again has two *causae causantes*, but this time, these are the basic transitions $\tau_1^A = e^A \rightarrowtail \boxed{1^A}$, which triggers Alice's sending the letter, and $\tau_3^E = e^E \rightarrowtail \boxed{3^A}$, which triggers Eve's letting the letter pass.

This causal story, together with our decision about probabilities having to be based exclusively on *causae causantes*, reaffirms that $\mu(\text{A-undec} \rightarrowtail \text{A-sends}) = p^A(\tau_1^A) = 1/6$. For Bob's transition from no information to having received the letter personally, we can start with the following partial answer:

$$\mu(\text{B-noinf} \rightarrowtail \text{B-receives}) = \mu(\{e^A \rightarrowtail \boxed{1^A}, e^E \rightarrowtail \boxed{3^E}\}) = ?$$

The probability of Bob's transition has to be the probability of its two-transition set of *causae causantes*. The question is whether we can use considerations about the causal set-up together with the known probabilities of the basic transitions to narrow down the possible answers, as above, or even to provide a unique answer.

It turns out that in the present case, the causal set-up provides the resources to secure the uniqueness of the causal probability. The adequate representation of this fact is, however, not completely straightforward from the perspective of standard probability theory.

Note, first, that since "B-receives" is a concrete outcome event, represented by an outcome chain or a scattered outcome, the set of *causae causantes* $CC(\text{B-noinf} \rightarrowtail \text{B-receives}) = \{e^A \rightarrowtail \boxed{1^A}, e^E \rightarrowtail \boxed{3^E}\} = \{\tau_1^A, \tau_3^E\}$ is consistent (see Def. 6.2). This does not imply, however, that any other set of basic transitions $\{\tau_i^A, \tau_j^E\}$ is also consistent. In the previous scenario in which we considered two *SLR* initials, the absence of MFB was sufficient to guarantee the consistency of all joint outcomes, and this made it possible to use the familiar probability-theoretic idea of building joint spaces via Cartesian products. In the present case, however, this construction makes no sense. Indeed, by Fact 4.7, only one outcome of the earlier choice point e^A is consistent with the occurrence of the later choice point e^E, namely, the outcome $\Pi_{e^A}\langle e^E \rangle = \boxed{1^A}$. For $i \neq 1$, any set of transitions $\{\tau_i^A, \tau_j^E\}$ ($j \in \{1, \dots, 8\}$) is inconsistent. One might think that, as inconsistent sets of transitions cannot occur in any history, they could simply be assigned probability zero in a standard Cartesian product space, but this move has disastrous consequences. (As we will show in Chapter 7.3, this seemingly innocent move is one root of Humphreys's paradox.)

Before we discuss this problem, we can make a second observation: as above, we can establish a causally motivated marginal probability—only this time, limited to the *consistent* joint outcomes. Given that Alice's die showed 1, Eve's throw of the octahedron has to have one of its eight possible outcomes, and these together constitute the full set of alternatives; the transition from e^E to the disjunctive outcome consisting of all of e^E's basic outcomes, $e^E \rightarrowtail \{\boxed{1^E}, \ldots, \boxed{8^E}\} = e^E \rightarrowtail \breve{I}_{e^E}$, is deterministic. Accordingly, the disjunctive transition from the joint initial, $\{e^A, e^E\} \rightarrowtail \{\boxed{1^E}, \ldots, \boxed{8^E}\}$, only has one originating cause:[17]

$$CC\left(\{e^A, e^E\} \rightarrowtail \{\boxed{1^E}, \ldots, \boxed{8^E}\}\right) = \{e^A \rightarrowtail \Pi_{e^A}\langle e^E \rangle\} = \{e^A \rightarrowtail \boxed{1^A}\}.$$

Given that Alice threw her die and Eve's throw of her octahedron had any of its possible outcomes, it must be that Alice's die showed 1. So, given that causal probabilities depend only on the *causae causantes*, we have another marginal probability result, this time for the $<$-related initials e^A and e^E:

$$\mu\left(\{e^A, e^E\} \rightarrowtail \{\boxed{1^E}, \ldots, \boxed{8^E}\}\right) = \mu(e^A \rightarrowtail \boxed{1^A}) = 1/6.$$

The above result, we claim, is fundamental: given what causal probabilities are, it must be that full "upward" coarse-graining reduces the set of *causae causantes*, and thus makes a consideration of the upper chance event probabilistically superfluous.

We can now venture to propose a related principle of the upward multiplication of causal probabilities. That principle may, however, have some traces of empirical content—it is hard to be sure.[18] So we flag it as a causal-metaphysical Postulate rather than as a result of logico-metaphysical analysis:

Postulate 7.1. *(Finite-case Markov condition) Let $I \rightarrowtail O$ be an indeterministic BST transition to an outcome chain O such that $CC(I \rightarrowtail O) = \{e_0 \rightarrowtail H_0, e_1 \rightarrowtail H_1, \ldots, e_K \rightarrowtail H_K\}$, $e_0 < e_k$ for $k = 1, \ldots, K$, and $\mu(I \rightarrowtail O)$ is defined. Then*

$$\mu(\{e_0 \rightarrowtail H_0, e_1 \rightarrowtail H_1, \ldots, e_K \rightarrowtail H_K\}) =$$
$$\mu(\{e_0 \rightarrowtail H_0\}) \cdot \mu(\{e_1 \rightarrowtail H_1, \ldots, e_K \rightarrowtail H_K\}).$$

Two remarks are in order. Note first that in the precondition of Postulate 7.1, the set $\{e_0 \rightarrowtail H_0, e_1 \rightarrowtail H_1, \ldots, e_K \rightarrowtail H_K\}$ is required to be the full set of

---

[17] In the following, we make use of the interchangeability of proposition-like and event-like transitions, noting that the initial events are point-like.

[18] For some pertinent results, see Brierley et al. (2015) on quantum temporal correlations.

causae causantes for the transition $I \rightarrowtail O$ from initial I to the outcome chain O. That is, if $I \rightarrowtail O$ is to occur, exactly these basic transitions, no more and no less, are needed. Second, as O is an outcome chain, this set of basic transitions is consistent (see Def. 6.4), so in particular there is a history h_E such that $\{e_k \mid 0 < k \leqslant K\} \subseteq h_E$. This fact and the ordering relations $e_0 < e_k$ for $k = 1, \ldots, K$ imply that $H_0 = \Pi_{e_0}\langle h_E \rangle$. It is thus uncontroversial that the transition $I \rightarrowtail O$ is adequately represented as consisting of two consecutive steps. The first step is the basic transition $e_0 \rightarrowtail \Pi_{e_0}\langle h_E \rangle$ that enables the initials $e_1, \ldots, e_K$. The second step consists of K basic transitions from these initials, with the proviso that these transitions might work jointly. The fact that there are two consecutive, separate steps strongly supports the multiplication formula for causal probabilities. However, as the K basic transitions of the second step might work jointly, the second multiplicand need not to factor; that is, it can happen that

$$\mu(\{e_1 \rightarrowtail H_1, \ldots, e_K \rightarrowtail H_K\}) \neq \mu(e_1 \rightarrowtail H_1) \cdot \ldots \cdot \mu(e_K \rightarrowtail H_K).^{19}$$

Given Postulate 7.1, the causal probability of two immediately consecutive basic transitions $\tau_0 = e_0 \rightarrowtail H_0$ and $\tau_1 = e_1 \rightarrowtail H_1$ is given by multiplying the respective individual causal probabilities. Whatever its ultimate metaphysical merit, this result is highly plausible. Given that in order for e_1 to occur, e_0 has to have had outcome $H_0 = \Pi_{e_0}\langle e_1 \rangle$, which is a possible event with causal probability $\mu(e_0 \rightarrowtail H_0)$, and that the causal probability for e_1 to result in outcome H_1 is $\mu(e_1 \rightarrowtail H_1)$, and that no other indeterminism occurs between e_0 and e_1, the causal probability for both transitions to occur consecutively is just given by multiplying the two individual values. Further support for our Postulate comes from considerations of causal probabilities attaching to histories. Earlier, we expressed reservations about that idea as a general approach (see note 3), but it provides an instructive illustration here. Assume that the number of histories under consideration is finite and that all histories are equiprobable. Then there has to be a natural number n of histories containing event e_0, and the size of the bundle of histories $H_0 = \Pi_{e_0}\langle e_1 \rangle$ is some natural number m. Given our assumptions, there are no further indeterministic happenings between e_0 and e_1, so that m is also the number of histories containing event e_1. Let the size of the history bundle H_1 be k. As we have $H_1 \subseteq H_{e_1} = H_0 \subseteq H_{e_0}$, we have $k \leqslant m \leqslant n$. By equiprobability of histories, one can read off the causal probabilities of the BST transitions under consideration as fractions:

---

[19] For a slightly more general discussion in terms of layered spaces, see Müller (2005).

$$\mu(e_0 \rightarrowtail H_0) = \frac{m}{n}; \quad \mu(e_1 \rightarrowtail H_1) = \frac{k}{m}; \quad \mu(e_0 \rightarrowtail H_1) = \frac{k}{n}.$$

It follows that, indeed, we can prove the required instance of Postulate 7.1:

$$\mu(e_0 \rightarrowtail H_1) = \frac{k}{n} = \frac{m}{n} \cdot \frac{k}{m} = \mu(e_0 \rightarrowtail H_0) \cdot \mu(e_1 \rightarrowtail H_1).$$

We can now use Postulate 7.1 to calculate the causal probability of Bob's personally receiving the letter in our "mail gets lost" scenario, as follows:

$$\mu(\text{B-noinf} \rightarrowtail \text{B-receives}) = \mu\left(\{e^A \rightarrowtail \boxed{1^A}, e^E \rightarrowtail \boxed{3^E}\}\right)$$
$$= \mu(e^A \rightarrowtail \boxed{1^A}) \cdot \mu(e^E \rightarrowtail \boxed{3^E})$$
$$= 1/6 \cdot 1/8 = 1/48.$$

Having dealt with the numerical values of the causal probabilities, we can now look into their adequate representation via probability spaces. We said earlier that, as some (in fact, most) combinations of basic outcomes of e^A and of e^E are inconsistent and accordingly cannot have a causal probability, we cannot use the standard Cartesian product construction. The resulting sample space $S^A \times S^E$ contains many inconsistent joint outcomes, such as $\langle \tau_2^A, \tau_5^E \rangle$. We have already warned that it will not do to simply assign these inconsistent outcomes the probability zero and stick to the Cartesian product construction. Especially with a view to our discussion of Humphreys's paradox in Chapter 7.3, it is instructive to see what the consequences would be. Thus, assume that in fact all inconsistent sets of transitions—that is, all pairs $\langle \tau_i^A, \tau_j^E \rangle \in S^A \times S^E$ for which $i \neq 1$—are assigned probability zero. Recall that by the marginal property, we have

$$p^{A \times E}(\{\langle \tau_1^A, \tau_j^E \rangle \mid j = 1, \ldots, 8\}) = p^A(\tau_1^A) = 1/6.$$

Yet, on the other hand, all pairs of transitions in $S^A \times S^E$ whose first element is not equal to τ_1^A have probability zero, and the measure $p^{A \times E}$ has to be normalized. So it has to be that

$$1 = \sum_{i=1}^{6} \sum_{j=1}^{8} p^{A \times E}(\langle \tau_i^A, \tau_j^E \rangle) = \sum_{j=1}^{8} p^{A \times E}(\langle \tau_1^A, \tau_j^E \rangle)$$
$$= p^{A \times E}(\{\langle \tau_1^A, \tau_j^E \rangle \mid j = 1, \ldots, 8\}).$$

This is a contradiction, so something has to give. We argue that, far from showing that the notion of a causal probability for a transition such as $\mu(\text{B-noinf} \rightarrowtail \text{B-receives})$ makes no sense, the contradiction in fact naturally vanishes once we provide a proper analysis of the causal background before which it arises.

The crucial question is: What is an adequate probability space in which the causal probability $\mu(\text{B-noinf} \rightarrowtail \text{B-receives})$ can be represented? Given the causal relations in our scenario, the causal alternatives to $CC(\text{B-noinf} \rightarrowtail \text{B-receives}) = \{e^A \rightarrowtail \boxed{1^A}, e^E \rightarrowtail \boxed{3^E}\}$ are of the following two kinds: (1) Alice's die in fact shows 1, but Eve's octahedron shows an outcome different from 3, or (2) Alice's die shows an outcome different from 1, and Eve never gets to throw her octahedron. That is, instead of the $6 \cdot 8 - 1 = 47$ alternatives according to the Cartesian product construction that was seen to make no sense, there are really only $7 + 5 = 12$ alternatives. Given Postulate 7.1, all of these have well-defined causal probabilities based on the causal probabilities characterizing Alice's and Eve's chance devices: (1) The causal probabilities $\mu(\{e^A \rightarrowtail \boxed{1^A}, e^E \rightarrowtail \boxed{j^E}\})$ can be computed by multiplying the causal probabilities for the respective outcomes of the two chance set-ups. (2) The causal probabilities $\mu(e^A \rightarrowtail \boxed{i^A})$ are all already given via the characterization of Alice's throw of her die. As one can easily check (see Eq. 7.5), on that representation, no problem with marginals ensues.

Building on our causal analysis, we posit the following probability space $\langle S^D, \mathscr{A}^D, p^D \rangle$ for an adequate representation of $\mu(\text{B-noinf} \rightarrowtail \text{B-receives})$: The sample space consists of 13 sets of transitions,

$$S^D =_{\mathrm{df}} \{\{\tau_i^A\} \mid i = 2, \ldots, 6\} \cup \{\{\tau_1^A, \tau_j^E\} \mid j = 1, \ldots, 8\}.$$

The algebra is, as usual, the power-set algebra. For the probability measure, we assign:

$$p^D(\{\tau_i^A\}) = \mu(e^A \rightarrowtail \boxed{i^A}) = 1/6, \quad i = 2, \ldots, 6;$$
$$p^D(\{\tau_1^A, \tau_j^E\}) = \mu(e^A \rightarrowtail \boxed{1^A}) \cdot \mu(e^E \rightarrowtail \boxed{j^E}) = 1/6 \cdot 1/8 = 1/48,$$
$$j = 1, \ldots, 8.$$

We then recover the result of our causal analysis:

$$\mu(\text{B-noinf} \rightarrowtail \text{B-receives}) = p^D(\{\tau_1^A, \tau_3^E\}) = 1/48.$$

As a sanity check, note that our measure is indeed normalized:

$$p^D(1^D) = \sum_{i=2}^{6} p^D(\{\tau_i^A\}) + \sum_{j=1}^{8} p^D(\{\tau_1^A, \tau_j^E\}) = 5 \cdot 1/6 + 8 \cdot 1/48 = 1. \quad (7.5)$$

Variant 3: Alice, Bob, and Eve. Let us quickly put the two previous scenarios together, such that Bob's transition B-noinf $\rightarrowtail$ B-receives now causally depends on Alice's die, on Bob's coin toss, and on Eve's octahedron, which is thrown iff Alice's die showed 3. That is, the *causae causantes* are now

$$CC(\text{B-noinf} \rightarrowtail \text{B-receives}) = \{e^A \rightarrowtail \boxed{1^A}, e^B \rightarrowtail T^B, e^E \rightarrowtail \boxed{3^E}\}.$$

We assume that the spatio-temporal relations are: $e^A \, SLR \, e^B$, $e^A < e^E$, $e^E \, SLR \, e^B$.

By logico-metaphysical analysis, as above, we have the following results concerning marginal probabilities:

$$\mu(\{e^A \rightarrowtail \boxed{1^A}, e^B \rightarrowtail T^B, e^E \rightarrowtail 1^E\}) = \mu(\{e^A \rightarrowtail \boxed{1^A}, e^B \rightarrowtail T^B\});$$

$$\mu(\{e^A \rightarrowtail \boxed{1^A}, e^B \rightarrowtail 1^B, e^E \rightarrowtail \boxed{3^E}\}) = \mu(\{e^A \rightarrowtail \boxed{1^A}, e^E \rightarrowtail \boxed{3^E}\});$$

$$\mu(\{e^A \rightarrowtail \boxed{1^A}, e^B \rightarrowtail 1^B, e^E \rightarrowtail 1^E\}) = \mu(\{e^A \rightarrowtail \boxed{1^A}\}).$$

Note, however, that the expression $\mu(\{e^A \rightarrowtail 1^A, e^B \rightarrowtail T^B, e^E \rightarrowtail \boxed{3^E}\})$, which also looks like a marginal probability, is undefined: as we said when discussing the previous scenario, it makes no sense to consider the causal probability of an inconsistent set of transitions, and that expression invokes, for example the inconsistent set $\{e^A \rightarrowtail \boxed{2^A}, e^B \rightarrowtail T^B, e^E \rightarrowtail \boxed{3^E}\}$.

As there are *SLR* transitions involved whose joint causal probability is only constrained, but not uniquely determined by the underlying individual causal probabilities, our analysis does not yield a unique verdict on the propensity of the transition B-noinf $\rightarrowtail$ B-receives. However, if there are no space-like correlations, we can simply multiply, which yields

$$\mu(\text{B-noinf} \rightarrowtail \text{B-receives}) = \mu(e^A \rightarrowtail \boxed{1^A}) \cdot \mu(e^B \rightarrowtail T^B) \cdot \mu(e^E \rightarrowtail \boxed{3^E})$$
$$= 1/6 \cdot 1/2 \cdot 1/8 = 1/96.$$

The important question is, again, which mathematical probability space is adequate for representing this causal probability. We take our lead from the previous case, in which the recipe was to *consider all consistent causal alternatives* that could be based on the local indeterministic happenings involved and on their local alternatives. That is, with a view toward a general recipe:

- We consider the set $\tilde{T}$ of all basic indeterministic transitions that are either a member of $CC(\text{B-noinf} \rightarrowtail \text{B-receives})$, or a local alternative to such a member.

  Given that the *causae causantes* include transitions from the initials e^A, e^B, and e^E, we have the 16-element set

$$\tilde{T} = \{e^A \rightarrowtail \boxed{i^A} \mid i = 1, \ldots, 6\} \cup \{e^B \rightarrowtail T^B, e^B \rightarrowtail H^B\} \cup \{e^E \rightarrowtail \boxed{i^E} \mid i = 1, \ldots, 8\}.$$

- Many subsets of $\tilde{T}$ are inconsistent and thus cannot have a causal probability assigned.[20] We therefore take the sample space of our probability space to consist of consistent subsets of $\tilde{T}$ only. We already know that some consistent subsets, such as $\{e^A \rightarrowtail \boxed{1^A}\}$, come up via considerations of marginals. These sets should thus not be elements of the sample space, but elements of the algebra. As a general recipe, we take our sample space S to be the set of *maximal consistent subsets of $\tilde{T}$*.
- As this is a finite set, we let the algebra $\mathscr{A}$ be the power set algebra of S.
- The probability measure p on the space defined by S and $\mathscr{A}$ has to represent the corresponding causal probabilities μ. Given the causal structure of our scenario, the measure p is constrained (via marginals) by the measures p^A, p^B, and p^E of the probability spaces representing the individual chance devices at issue, which in turn represent the corresponding single-device causal probabilities. The measure p, however, is not uniquely determined by p^A, p^B, and p^E, due to the possibility of space-like correlations between outcomes of e^A and e^B, and between outcomes of e^E and e^B. Whether such correlations obtain is a fact about the corresponding propensities of sets of transitions.

---

[20] As stated in note 7 we do not categorically rule out extreme cases in which a real possibility may have probability zero. An inconsistent set of transitions, however, does not represent a real possibility at all, and consequently cannot have a probability assigned. Probabilities are, after all, graded possibilities.

- In the (here, highly plausible) case that such space-like correlations are absent, the measure is simply the product measure. But as we are here dealing with a construction that is not based on Cartesian products, we need to write the product measure in a somewhat non-standard way: For $T \in S$ a maximal consistent set of transitions, we have

$$p(T) = \prod_{\tau \in T} p^*(\tau),$$

where p^* is the appropriate single-device probability. Thus, for example, $p(\{\tau_4^A, \tau_2^B\}) = p^A(\tau_4^A) \cdot p^B(\tau_2^B) = 1/6 \cdot 1/2 = 1/12$.

Given this recipe and assuming no *SLR* correlations, we can thus compute

$$p(\{\tau_1^A, \tau_2^B, \tau_3^E\}) = p^A(\tau_1^A) \cdot p^B(\tau_2^B) \cdot p^E(\tau_3^E) = 1/6 \cdot 1/2 \cdot 1/8 = 1/96,$$

in full accordance with our causal analysis for the no correlation case.

7.2.4 General probability spaces in BST

Based on our discussion of a few simple but exemplary cases, we can now extract a general recipe for representing causal probabilities of BST transitions in mathematically well-defined probability spaces. We maintain our two simplifying assumptions: first, we assume that the structures we will be dealing with are all finite: we will always be dealing with transitions $I \rightarrowtail O^*$ for which the set of *causae causantes* is finite, and for any choice point e that is an initial of one of the *causae causantes* (i.e., for any cause-like locus $e \in cll(I \rightarrowtail O^*)$), the number of different local alternatives splitting off at e (the cardinality of Π_e) is finite.[21] Second, we assume that there is no modal funny business of the kind discussed in Chapter 5. Thus, for example, for any

[21] If a set of local alternatives from an initial e, Π_e, is infinite, one can use standard tools of measure theory (e.g., Borel sets) to generalize our approach. In case the set of initials itself is infinite, different approaches appear to be needed for the case in which there is an infinite set of *SLR* initials and for the case in which there is an infinite chain of initials. In the first case, the tools of standard probability theory for infinite product spaces (cylinder sets, zero-one laws) will apply. In the second case, the situation appears to be more challenging, as upward multiplication may trivialize the resulting probabilities. It is metaphysically interesting to investigate these cases, as, for example, in the modal theory of agency (Belnap et al., 2001), 'busy choice sequences' are analyzed whose probabilistic equivalent exactly requires a probability theory for infinite chains of transitions. We will leave this issue to one side and continue working with finite structures.

set of *SLR* initials, *all* combinations of local outcomes can be assumed to be consistent. That is, we assume that there are no *modal* correlations that would restrict the space of possibilities. This assumption, however, leaves open whether or not there are *probabilistic* correlations between distant outcomes, which will be the topic of Chapter 7.2.6.

In our preceding discussions it has been crucial to get the sample space right, and we motivated the choice of the sample space as a set of sets of transitions via a consideration of the causal alternatives of a given BST transition. Following exactly the recipe given above, we now work toward a general definition of a probability space based on a consistent set of basic transitions, such as provided via the *causae causantes* of an outcome chain or of a scattered outcome.

Definition 7.2 (Causal alternatives). Let $\langle W, < \rangle$ be a BST_{92} structure, and let T be a consistent set of basic transitions, the initials of which form the set E. The *causal alternatives for T* are the sets of transitions in S, where

$$S =_{\text{df}} \{T' \subseteq \tilde{T} \mid T' \text{ is maximally consistent}\}, \text{ and where}$$
$$\tilde{T} =_{\text{df}} \{e \rightarrowtail H \mid e \in E \text{ and } H \in \Pi_e\}. \tag{7.6}$$

Before we define probabilistic BST_{92} structures and causal probability spaces, we list again the constraints on the measure μ that have emerged in our examples above, and which we will postulate to hold in general.

Markov condition. First, for the sake of completeness, we repeat the statement of the finite-case Markov condition already introduced above:

Postulate 7.1. *(Finite-case Markov condition) Let $I \rightarrowtail O$ be an indeterministic BST transition to an outcome chain O such that $CC(I \rightarrowtail O) = \{e_0 \rightarrowtail H_0, e_1 \rightarrowtail H_1, \ldots, e_K \rightarrowtail H_K\}$, $e_0 < e_k$ for $k = 1, \ldots, K$, and $\mu(I \rightarrowtail O)$ is defined. Then*

$$\mu(\{e_0 \rightarrowtail H_0, e_1 \rightarrowtail H_1, \ldots, e_K \rightarrowtail H_K\}) =$$
$$\mu(\{e_0 \rightarrowtail H_0\}) \cdot \mu(\{e_1 \rightarrowtail H_1, \ldots, e_K \rightarrowtail H_K\}).$$

Partial function. Fundamentally, we allow μ to be a partial function. We hold, however, that μ should not contain weird gaps in its domain: if μ is defined for some transitions, it should also be defined for parts of their *causae causantes* and for local causal alternatives to them. We can now spell out this constraint in a precise way, as the following causal-metaphysical Postulate:

Postulate 7.2 (Constraints on the domain of μ). *If μ is defined for the* causae causantes *of a BST transition $I \rightarrowtail \mathcal{O}^*$, then $\mu(I \rightarrowtail \mathcal{O}^*) = \mu(CC(I \rightarrowtail \mathcal{O}^*))$, and μ is also defined for the causal alternatives to $I \rightarrowtail \mathcal{O}^*$ and for all subsets of $CC(I \rightarrowtail \mathcal{O}^*)$.*

Unavoidable transitions have causal probability one. The causal probability of a non-trivial unavoidable transition, that is, of a transition to an exhaustive disjunction of alternatives, has to be one. Given that the initial occurs, the world has to continue in some way. All these ways together exhaust the possible alternatives, and therefore, their disjunction is certain to happen. We formulate this constraint as the general logico-metaphysical Postulate of the law of total causal probability:

Postulate 7.3 (Law of total causal probability). *Let $\{T_\gamma \mid \gamma \in \Gamma\}$ be a set of non-empty consistent sets of indeterministic transitions that are pairwise incompatible, i.e., for any $\gamma, \gamma' \in \Gamma$, if $\gamma \neq \gamma'$, then $H(T_\gamma) \cap H(T_{\gamma'}) = \emptyset$. Suppose further that there is an initial event (a consistent subset of W) I such that $H_{[I]} \subseteq \bigcup_{\gamma \in \Gamma} H(T_\gamma)$. Then, as these sets T_γ of transitions partition the possible ways in which I can occur, so that exactly one of them has to occur given that I occurs, if all the $\mu(T_\gamma)$ are defined, we have*

$$\sum_{\gamma \in \Gamma} \mu(T_\gamma) = 1.$$

This postulate applies naturally to transitions to disjunctive outcomes, as follows. Consider such a transition $I \rightarrowtail \breve{\mathbf{O}}$, and let $T_\gamma = CC(I \rightarrowtail \hat{O}_\gamma)$, where $\breve{\mathbf{O}} = \{\hat{O}_\gamma \mid \gamma \in \Gamma\}$. Then the T_γ fulfill the conditions of Postulate 7.3: We have $H(T_\gamma) \cap H(T_{\gamma'}) = \emptyset$ for $\gamma \neq \gamma'$ by the definition of a disjunctive outcome event (see Defs. 4.3(4) and 4.9), and $H_{[I]} = \bigcup_{\gamma \in \Gamma} H(T_\gamma)$ as $I \rightarrowtail \breve{\mathbf{O}}$ is unavoidable. So $\mu(I \rightarrowtail \breve{\mathbf{O}}) = \sum_{\gamma \in \Gamma} \mu(T_\gamma) = 1$.

Note that the premises of Postulate 7.3 are not trivial: there has to be a concrete initial I for which $H_{[I]} \subseteq \bigcup_{i \in \Gamma} H(T_i)$. Such an I may be hard to find. If one tries, for example, to take the set of minimal initials from the union of all the T_is, that set may well be inconsistent, which disqualifies it from being a BST initial. And if one tries to locate I in the common past of these initials (guided by the PCP), the premises may fail because the set $H_{[I]}$ stops being a subset of $\bigcup_{i\in\Gamma} H(T_i)$. Thus, the set $\{T_i \mid i \in \Gamma\}$ has to be of a special kind to admit an initial I satisfying the premises of the Postulate. Reading the premises informally, the T_is have to represent *all* alternative possible continuations that the world can take, provided that I occurs. In that case, the probability that any of these continuations will occur is one, and that is also the sum of the probabilities of each of the continuations taken separately, as these continuations are pairwise incompatible alternatives.[22]

Marginal probabilities. In a similar vein, we formulate a general Postulate for marginal probabilities, written in terms of sets of basic transitions:

Postulate 7.4 (General marginal probabilities). *Let $T = \{e_i \rightarrowtail H_i \mid i \in \Gamma\}$ be a set of basic transitions, and let $E =_{df} \{e_i \mid i \in \Gamma\}$ be the set of initials of transitions from T. Let $e \in W \setminus E$ be a new initial such that for any $e_i \in E$, either $e_i < e$ or $e_i\,SLR\,e$, so that e is maximal in the set $E \cup \{e\}$. Assume that $\mu(T)$ is defined and that also $\mu(T \cup \{e \rightarrowtail H_0\})$ is defined for some $H_0 \in \Pi_e$ (it is then defined for* all *such H, by Postulate 7.2). Then we have*

$$\sum_{H\in\Pi_e} \mu(T\cup\{e \rightarrowtail H\}) = \mu(T).$$

This postulate also results from logico-metaphysical analysis: the summation formula is based on the observation that the basic transitions from a maximal element of $E \cup \{e\}$ are truly alternative; that is, they have pairwise inconsistent outcomes, while not affecting the occurrence or non-occurrence of any of the other transitions (given no MFB). Note that, in contrast to the Cartesian product construction, and in line with our previous discussioin, the summation formula applies only to transitions issuing from a *maximal* element of $E \cup \{e\}$.[23]

---

[22] It is tempting to broaden the postulate to also cover trivial deterministic transitions, such as from a deterministic initial e to its only outcome $\Pi_e = \{H_e\}$, or to the reduced set of *causae causantes* for a deterministic transition to a disjunctive outcome. The problem in these cases is, however, that the set of *causae causantes* is then empty, so that the resulting algebra of subsets would have just one element, $\emptyset$—and there is no way to consistently define a probability measure on an algebra that has only one element, as one can see from Def. 7.1. Metaphysically, a trivial transition that happens anyway surely happens with certainty, but a probability space always has to include at least a probability-zero alternative, which is lacking if the set of *causae causantes* of a transition is empty.

[23] Note that the Cartesian product construction of standard probability theory is adequate precisely if all initials are maximal in the set of all initials, i.e., if they are all pairwise SLR (and there is no MFB). As standard probability theory formalizes neither space nor time (let alone space-time or modal correlations), this precondition of the Cartesian product construction is not—cannot be—made explicit. As we have shown, however, it is indeed crucial, and it fails to apply already in simple, everyday scenarios.

Based on the Postulates summarized above, we now define *probabilistic BST₉₂ structures*:[24]

Definition 7.3 (Probabilistic BST structure). A *probabilistic* BST_{92} structure is a triple $\mathscr{W} = \langle W, <, \mu \rangle$, where $\langle W, < \rangle$ is a BST_{92} structure and the propensity function μ is a partial function defined on sets of indeterministic basic transitions in $\langle W, < \rangle$, $\mu : \mathscr{P}(\mathrm{TR}(W)) \mapsto [0,1]$, that satisfies the Markov Postulate 7.1, Postulate 7.2 concerning the domain of μ, Postulate 7.3 of the law of total probability, and Postulate 7.4 of general marginal probabilities.

We can now give our general definition of causal probability spaces based on an arbitrary BST transition. We will proceed in two steps, first giving a definition for probability spaces based on transitions to outcome chains or scattered outcomes, and then for transitions to disjunctive outcomes. This division results from the fact that causal alternatives look different in these two cases, which implies that the base sets of the probability spaces are constructed differently. We begin with the first case. All the probability spaces discussed in our examples above fulfill this definition, as is easy to check.

Definition 7.4 (Causal probability spaces, $O/\hat{O}$ version). Let $\langle W, <, \mu \rangle$ be a probabilistic BST_{92} structure in which there is no MFB, and let $I \rightarrowtail \mathscr{O}^*$ be an indeterministic transition from an initial I to an outcome chain or a scattered outcome $\mathscr{O}^*$ for which $\mu(CC(I \rightarrowtail \mathscr{O}^*))$ is defined. The *causal probability space based on* $I \rightarrowtail \mathscr{O}^*$, $CPS(I \rightarrowtail \mathscr{O}^*)$, is the probability space $\langle S, \mathscr{A}, p \rangle$, where S is the set of causal alternatives for $CC(I \rightarrowtail \mathscr{O}^*)$ according to Def. 7.2, $\mathscr{A}$ is the power-set algebra over S, and the measure p on $\mathscr{A}$ is induced via the measure assigned to the elements $T \in S$ via $p(T) = \mu(T)$.

We have to establish that the object defined in this way, $\langle S, \mathscr{A}, p \rangle$, is in fact a probability space fulfilling Def. 7.1. This is the subject of the following lemma:

Lemma 7.1. *Let the conditions of Def. 7.4 hold for an indeterministic transition $I \rightarrowtail \mathscr{O}^*$, with $\mathscr{O}^*$ an outcome chain or a scattered outcome, and consider $CPS(I \rightarrowtail \mathscr{O}^*) = \langle S, \mathscr{A}, p \rangle$. That triple is in fact a probability space satisfying Def. 7.1. That is, $CPS(I \rightarrowtail \mathscr{O}^*)$ is well defined and p is a normalized measure on $\mathscr{A}$. Furthermore, we have that*

---

[24] Note the somewhat different status of the Postulates: While Postulates 7.3 and 7.4 result from purely metaphysical considerations, Postulates 7.1 and 7.2 also have causal underpinnings.

$$CC(I \rightarrowtail \mathcal{O}^*) \in S \quad and \quad p(CC(I \rightarrowtail \mathcal{O}^*)) = \mu(I \rightarrowtail \mathcal{O}^*).$$

Proof. Let $E =_{\text{df}} cll(I \rightarrowtail \mathcal{O}^*)$, let $T_0 =_{\text{df}} CC(I \rightarrowtail \mathcal{O}^*)$, let $\tilde{T} =_{\text{df}} \{e \rightarrowtail H \mid e \in E, H \in \Pi_e\}$, and let S be the set of maximally consistent sets of transitions from $\tilde{T}$.

For well definedness, we have to prove that for any $T \in S$, $\mu(T)$ is defined. This follows directly from Postulate 7.2. As a consequence, $p(T)$ is defined via Def. 7.4, inducing the full measure p on $\mathscr{A}$.

To see that $CC(I \rightarrowtail \mathcal{O}^*) = T_0 \in S$, note first that according to Def. 7.2, E is the set of initials of basic transitions from T_0, so that $T_0 \subseteq \tilde{T}$. Furthermore, T_0 is consistent since any set of *causae causantes* of a transition to a scattered outcome or outcome chain is consistent. There is thus some $T_1 \in S$ for which $T_0 \subseteq T_1$. To see that $T_0 = T_1$, assume otherwise, so that there would have to be some $\tau_1 = e_1 \rightarrowtail H_1 \in T_1 \setminus T_0$. By the construction of S, it has to be that $e_1 \in cll(I \rightarrowtail \mathcal{O}^*)$. Note that there is some basic transition $e_1 \rightarrowtail H_0 \in T_0$, $H_0 \in \Pi_{e_1}$, as e_1 is a cause-like locus for $I \rightarrowtail \mathcal{O}^*$. As $\tau_1 \notin T_0$ by assumption, it has to be that $\tau_1 = e_1 \rightarrowtail H_1$ for some $H_1 \in \Pi_{e_1}$, $H_1 \neq H_0$. But then we have $H(T_1) \subseteq H_0 \cap H_1 = \emptyset$, showing that T_1 is inconsistent. So T_0 is in fact maximal consistent, and $T_0 \in S$. The claim that $p(T_0) = \mu(I \rightarrowtail \mathcal{O}^*)$ then follows by definition of p and by our general assumption that $\mu(I \rightarrowtail \mathcal{O}^*) = \mu(CC(I \rightarrowtail \mathcal{O}^*))$, which is part of Postulate 7.2.

It remains for us to show that p is indeed a normalized probability measure. By Def. 7.4, the measure on the algebra is induced in the standard way by the probabilities assigned to the elements of the sample space, so additivity holds by construction. Therefore it suffices to show that the probabilities assigned to the different elements of S sum to one, as then

$$p(1_{\mathscr{A}}) = \sum_{T \in S} p(T) = 1.$$

Our proof uses the law of total probability in the form of Postulate 7.3. We thus have to show that the elements of S partition the set of histories in which the initial I occurs. First we have to show that for two different $T_1, T_2 \in S$, $T_1 \neq T_2$, we have $H(T_1) \cap H(T_2) = \emptyset$. This follows immediately from each T_1 and T_2 being a maximal consistent subset of $\tilde{T}$. We need next to show $H_{[I]} \subseteq \bigcup_{T \in S} H(T)$. Thus, for an arbitrary $h \in H_{[I]}$ we have to find some $T \in S$ for which $h \in H(T)$. There are two cases.

Case 1: $h \in H_{\langle \mathcal{O}^* \rangle}$ for the given $\mathcal{O}^*$ that defined $CPS(I \rightarrowtail \mathcal{O}^*)$. In this case, we have found the element $T_0 = CC(I \rightarrowtail \mathcal{O}^*) \in S$ such that $h \in H(T_0)$, and we are done.

Case 2: $h \notin H_{\langle \mathcal{O}^* \rangle}$. In this case, h is a witness for some $e_0 \in E = cll(I \rightarrowtail \mathcal{O}^*)$, such that $h \perp_{e_0} H_{\langle \mathcal{O}^* \rangle}$. We define the set

$$T' =_{\text{df}} \{e \rightarrowtail \Pi_e \langle h \rangle \mid e \in E \cap h\}.$$

By definition, $h \in H(T')$, so that T' is consistent. Furthermore, its initials belong to E, so we have $T' \subseteq \tilde{T}$. To show that T' is maximally consistent, take some $\tau' = e' \rightarrowtail H' \in \tilde{T} \setminus T'$, for which $e' \in E$ and $H' \in \Pi_{e'}$. We have to show that $T' \cup \{\tau'\}$ is inconsistent. There are again two cases. (1) Either $e' \in h$, whence $H' \neq \Pi_{e'} \langle h \rangle$ by $\tau' \notin T'$, showing that $T' \cup \{\tau'\}$ is (blatantly) inconsistent. (2) The other case is that $e' \notin h$. Let $h' \in H_{e'}$. We have $e' \in h' \setminus h$, and so, by PCP_{92}, there is some $c < e'$ for which $h \perp_c h'$. By Fact 3.8, $h \perp_c H_{e'}$. We claim that $c \in E$. Since $e' \in E = cll(I \rightarrowtail \mathcal{O}^*)$, there is some history $h^* \in H_{[I]}$ for which $h^* \perp_{e'} H_{\langle \mathcal{O}^* \rangle}$. So $H_{\langle \mathcal{O}^* \rangle} \subseteq H_{e'}$, which implies $h \perp_c H_{\langle \mathcal{O}^* \rangle}$. This, however, is the defining expression for $cll(I \rightarrowtail \mathcal{O}^*)$ (note that $h \in H_{[I]}$ by our initial assumption), i.e., we have $c \in E$, and as $c \in h$ as well, we have established that the transition $c \rightarrowtail \Pi_c \langle h \rangle \in T'$. On the other hand, we have $h \perp_c h'$, and as the new initial $c < e'$ and $e' \in h' \setminus h$, we have $H' \subseteq \Pi_c \langle e' \rangle$. But $\Pi_c \langle e' \rangle \neq \Pi_c \langle h \rangle$, so that $T' \cup \{\tau'\}$ is in fact inconsistent. So, in both cases, $T' \in S$ and $h \in H(T')$.

This establishes $H_{[I]} \subseteq \bigcup_{T \in S} H(T)$. With the premises of the law of total causal probability (Postulate 7.3) satisfied, we therefore have

$$\sum_{T \in S} p(T) = \sum_{T \in S} \mu(T) = 1,$$

showing that the measure p is indeed normalized. $\square$

The Definition 7.4 of causal probability spaces requires a small modification to apply to a transition to a disjunctive outcome, $I \rightarrowtail \check{\mathbf{O}}$: For $\check{\mathbf{O}} = \{\hat{O}_\gamma \mid \gamma \in \Gamma\}$, we take S to be the set of maximal consistent subsets of $\tilde{T} =_{\text{df}} \bigcup_{\gamma \in \Gamma} \{e \rightarrowtail H \mid e \in cll(I \rightarrowtail \hat{O}_\gamma) \wedge H \in \Pi_e\}$. The transition to the disjunctive outcome is then represented not via an element of S, but via an element of the algebra $\mathscr{A}$.

Definition 7.5 (Causal probability spaces, $\check{O}$ version). Let $\langle W, <, \mu \rangle$ be a probabilistic BST$_{92}$ structure in which there is no MFB. Let $I \rightarrowtail \check{O}$ be a transition to a disjunctive outcome $\check{O} = \{\hat{O}_\gamma \mid \gamma \in \Gamma\}$, and let $\mu(CC(I \rightarrowtail \hat{O}_\gamma))$ be defined for every $\gamma \in \Gamma$. The *causal probability space based on* $I \rightarrowtail \check{O}$, $CPS(I \rightarrowtail \check{O})$, is the probability space $\langle S, \mathscr{A}, p \rangle$, where S is the set of maximal consistent subsets of $\tilde{T} =_{df} \bigcup_{\gamma \in \Gamma} \{e \rightarrowtail H \mid e \in cll(I \rightarrowtail \hat{O}_\gamma) \wedge H \in \Pi_e\}$, $\mathscr{A}$ is the power-set algebra over S, and the measure p on $\mathscr{A}$ is induced via the measure assigned to the elements $T \in S$ via $p(T) = \mu(T)$.

One might be worried by a difference, both set-theoretical and content-wise, between *causae causantes* for a transition to a disjunctive outcome, $I \rightarrowtail \check{O}$, and the base set S of the causal probability space based on $I \rightarrowtail \check{O}$. To recall, $CC(I \rightarrowtail \check{O})$ is the *set* of $CCr(I \rightarrowtail \hat{O})$, with $\hat{O} \in \check{O}$. This seems right, as each $CCr(I \rightarrowtail \hat{O})$ stands for a separate path leading to the production of disjunctive outcome $\check{O}$. By taking the union of all $CCr(I \rightarrowtail \hat{O})$, we will typically lose the information about these alternative paths. Furthermore, the appeal to reduced sets, $CCr(I \rightarrowtail \hat{O})$, rather than *causae causantes* simpliciter, $CC(I \rightarrowtail \hat{O})$, stems from the fact that, given the disjunctive nature of an outcome in question, not every element of the latter set is needed for the production of that disjunctive outcome. Recall that at the extreme case of a deterministic transition to a disjunctive outcome, $CC(I \rightarrowtail \check{O})$ is (the singleton of) the empty set.

As for set S, which is the set of transitions, it encodes information about alternative paths to a disjunctive outcome, as among members of S there are all sets $CC(I \rightarrowtail \hat{O})$, where $\hat{O} \in \check{O}$. And in constructing S we appeal to $CC(I \rightarrowtail \hat{O})$ rather than to $CCr(I \rightarrowtail \hat{O})$ (literally speaking, to $cll(I \rightarrowtail \hat{O})$ rather than $cllr(I \rightarrowtail \hat{O})$), since even in a case of a deterministic disjunctive outcome, we want to accommodate all the underlying causal information. We take it that determinism should come up at a probabilistic level, by $I \rightarrowtail \check{O}$ getting assigned probability $p(I \rightarrowtail \check{O}) = 1$.

We need to check whether the object thus introduced is a probability space, that is, whether it satisfies Def. 7.1. The following Lemma states that this is indeed the case.

Lemma 7.2. *Let the conditions of Def. 7.5 hold for a transition $I \rightarrowtail \check{O}$ to a disjunctive outcome $\check{O}$, and consider $CPS(I \rightarrowtail \check{O}) = \langle S, \mathscr{A}, p \rangle$. That triple is in fact a probability space satisfying Def. 7.1. That is, $CPS(I \rightarrowtail \check{O})$ is well defined and p is a normalized measure on $\mathscr{A}$. Furthermore, we have that*

$$p(CC(I \rightarrowtail \hat{O}_\gamma)) = \mu(\{T \in S \mid CC(I \rightarrowtail \hat{O}_\gamma) \subseteq T\}) = \sum_{T \in S, \, CC(I \rightarrowtail \hat{O}_\gamma) \subseteq T} \mu(T);$$

$$p(CC(I \rightarrowtail \breve{\mathbf{O}})) = \sum_{\gamma \in \Gamma} p(CC(I \rightarrowtail \hat{O}_\gamma)).$$

Proof. For the proof, one needs to adapt the proof of Lemma 7.1. We leave this task as Exercise 7.1. □

Although officially, Defs. 7.4 and 7.5 require one to base a causal probability space on a concrete BST transition, such a space can be based on any set of consistent basic transitions, as any such set provides a well-defined notion of causal alternatives (see Def. 7.2).[25]

We are now in a position to provide a full answer to the question about the formal structures in which transitions are assigned probabilities.

Full answer to Question 7.2: When $\mu(I \rightarrowtail \mathscr{O}^*)$ is defined, that causal probability is represented in a causal probability space as defined via Defs. 7.4 and 7.5.

7.2.5 Representing transitions in different causal probability spaces

Before we end our introduction of causal probability spaces with a discussion of space-like correlations (which we will call "probabilistic funny business"), there is one topic left to discuss: how are concrete BST transitions represented in different causal probability spaces? This topic is important especially in light of the formal challenge of Humphreys's paradox, which will be treated in Chapter 7.3. In our general approach, BST transitions and their propensities are basic, while causal probability spaces are derivative. We have already stated that, in general, the propensity function μ is not a probability measure, but the connection of propensities to probabilities is, of course, intimate. We stress this connection by discussing the representation

---

[25] Indeed, as one can see from our definitions, it is enough to specify the set E of initials. This approach is investigated in Müller (2005). That paper also provides a discussion of a generalization of the Markov property and an extended discussion of the representation of transitions in different probability spaces. The latter discussion forms the background for the following remarks on representability.

of the propensity of a BST transition in different causal probability spaces.

The formal setting of this investigation is as follows. Let $\langle W, <, \mu \rangle$ be a probabilistic BST structure, and let $I \rightarrowtail \mathcal{O}^*$ be an indeterministic BST transition from initial I to outcome $\mathcal{O}^*$. We assume for simplicity's sake that $\mathcal{O}^*$ is an outcome chain or a scattered outcome. Assume that $\mu(I \rightarrowtail \mathcal{O}^*)$ is defined. Now let $CPS = \langle S, \mathscr{A}, p \rangle$ be some causal probability space definable on the basis of $\langle W, <, \mu \rangle$. We will discuss the following three questions:

- When is the transition $I \rightarrowtail \mathcal{O}^*$ representable in CPS?
- If that transition is representable, how is it represented?
- How can the numerical value of $\mu(I \rightarrowtail \mathcal{O}^*)$ be recovered from p?

The simplest case is, of course, $CPS = CPS(I \rightarrowtail \mathcal{O}^*)$, where the answers are immediate from Lemma 7.1. Almost equally simple is the case in which $T =_{df} CC(I \rightarrowtail \mathcal{O}^*) \in S$, that is, in which the transition in question is a causal alternative to the transition on which the space CPS is based; see again Lemma 7.1.

It is also clear that for $I \rightarrowtail \mathcal{O}^*$ to be representable in CPS, all the transitions in T have to occur in the transition sets that make up the sample space S. Let $\mathscr{S} =_{df} \bigcup S$ be the set of all transitions that belong to some element of the sample space. Then we can say that for representability of $I \rightarrowtail \mathcal{O}^*$ in CPS it is necessary that $T = CC(I \rightarrowtail \mathcal{O}^*) \subseteq \mathscr{S}$. If this condition is fulfilled, there are two cases, depending on how T is situated with respect to the other transitions present in $\mathscr{S}$. The crucial issue is whether or not $\mathscr{S}$ contains any transitions that precede (in the sense of the transition ordering $\prec$) any of the transitions from T.

Case 1: There is no $\tau' \in \mathscr{S}$ for which $\tau' \prec \tau$ for any $\tau \in T$. In this case, consider the element $a \in \mathscr{A}$ defined via

$$a =_{df} \{T' \in S \mid T \subseteq T'\}.$$

That element a is either equal to $\{T\}$ (leading back to the simplest cases above), or it is an extension (we might say, a fine-graining) of T via later or SLR transitions. By the marginal property (Postulate 7.4), perhaps applied a number of times, we therefore have

$$p(a) = \mu(a) = \mu(T) = \mu(CC(I \rightarrowtail \mathcal{O}^*)),$$

thus answering all our questions satisfactorily. For an illustration, consider our discussion of how to represent Alice's throw of her die in the "Variant 2" story above (p. 177).

Case 2: There is some $\tau' \in \mathcal{S}$ and some $\tau \in T$ for which $\tau' \prec \tau$; that is, the transition $I \longmapsto \mathcal{O}^*$ has causal preconditions (e.g., τ') that are made explicit in *CPS*. This case is somewhat more tricky—in fact, it is at the root of the problematic assumption (CI) in the statement of Humphreys's paradox (see Chapter 7.3.3). We can discuss this case in the context of the "Variant 2" story as well. To recall, a letter to Bob is transmitted iff first, Alice's throw of a die (initial e^A) has outcome $\boxed{1^A}$, and then, Eve's throw of an octahedron (initial e^E) has outcome $\boxed{3^E}$. As discussed earlier, the adequate causal probability space $CPS = CPS(\{e^A \longmapsto \boxed{1^A}, e^E \longmapsto \boxed{3^E}\})$ has a sample space consisting of the 13 maximally consistent combinations of basic transitions from e^A and basic transitions from e^E. Now consider the transition $\tau =_{df} e^E \longmapsto \boxed{5^E}$ of Eve's throwing her octahedron with result 5. Clearly, $\tau \in \mathcal{S}$, because the set of transitions $T =_{df} \{e^A \longmapsto \boxed{1^A}, e^E \longmapsto \boxed{5^E}\} \in S$. That set, however, also contains the transition $\tau' =_{df} e^A \longmapsto \boxed{1^A}$, for which $\tau' \prec \tau$.

In this case, the transition τ is sufficient to uniquely identify the set of transitions T: saying that Eve's octahedron has shown 5 amounts to saying that Alice's die has shown 1 and then Eve's octahedron has shown 5, because the former is a causal precondition of the latter. But the two sets of transitions, T and $\{\tau\}$, are different, and their propensities are different as well (unless τ' is inevitable). In fact, by the Markov property (Postulate 7.1), we have

$$\mu(T) = \mu(\tau') \cdot \mu(\tau).$$

It turns out that our questions have no simple answer in this case: there is no element a of the algebra $\mathcal{A}$ that represents τ such that the propensity of that transition (1/8, in our example) could be read off as $p(a)$. It is, however, possible to recover $\mu(\tau)$ as the conditional propensity of T given τ', as we will show in detail in Chapter 7.3.3.

7.2.6 Probabilistic funny business

For our discussion of quantum correlations in Chapter 8, there is one substantial task left: We need to provide a representation of non-local probabilistic correlations in BST. We provide details in the context of a discussion of random variables and their dependence or independence in the framework of causal probabilities.

What we call probabilistic funny business (PFB) consists of the probabilistic correlations between outcomes of space-like related events in a BST_{92} structure in which there is no MFB. Ultimately we will define PFB in terms of dependence of random variables. For ease of presentation, however, we begin with the rudimentary example of a non-local probabilistic correlation involving two basic outcomes $\tau_1 = e_1 \rightarrowtail H_1$ and $\tau_2 = e_2 \rightarrowtail H_2$, with $H_1 \in \Pi_{e_1}$, $H_2 \in \Pi_{e_2}$, and initials $e_1 \, SLR \, e_2$. Given no MFB, the transitions τ_1 and τ_2 are compatible, i.e., $H_1 \cap H_2 \neq \emptyset$. For this simple example, the causal probability space $CPS = \langle S, \mathscr{A}, p \rangle$ is based on the set $\{\tau_1, \tau_2\}$, so $\{\tau_1, \tau_2\} \in S$. We then say that τ_1 and τ_2 exhibit non-local probabilistic correlations iff

$$p(\{\tau_1, \tau_2\}) \neq p(\tau_1) \cdot p(\tau_2),$$

where $p(\tau_1) = p(\bigcup_{T \in S, \tau_1 \in T} T) = \sum_{T \in S, \tau_1 \in T} p(T)$, and analogously for $p(\tau_2)$, are the marginal probabilities for τ_1 and for τ_2. Correspondingly, at the level of the causal probabilities μ represented via p, a non-local probabilistic correlation in this case is of the form

$$\mu(\{\tau_1, \tau_2\}) \neq \mu(\tau_1) \cdot \mu(\tau_2).$$

We can generalize this idea to two generic transitions with SLR initials I and I' and scattered outcomes $\hat{O}$ and $\hat{O}'$, respectively. Further generalizations to more than two transitions, or to transitions to disjunctive outcomes, are natural (see Placek, 2010).

Probabilistic correlations are often defined in terms of the dependence of random variables, and analyses of non-local quantum correlations are typically expressed in the framework of random variables. In Chapter 8 we embark on such an analysis: using probabilistic BST_{92} structures we ask whether, and under what conditions, non-local quantum correlations can be accounted for by hidden local factors. To this end, we need to define random variables and their (in)dependence in the framework of causal probabilities.

As in standard probability theory, in our theory a random variable X is a function defined on the base set S of a causal probability space $\langle S, \mathscr{A}, p \rangle$, where the values of X are real numbers. (One may consider other ranges of random variables, but we fix $\mathbb{R}$ as the range in order to be specific.) Just as in the standard theory, random variables allow for the definition of correlations and (probabilistic) dependence and independence:

Definition 7.6 (Independence and correlations). A family $X_1, \ldots, X_n$ of random variables defined on a causal probability space $CPS = \langle S, \mathcal{A}, p \rangle$ is called *independent* iff for any n-tuple $\langle x_1, \ldots, x_n \rangle$ of respective values of these variables, $p(X_1 = x_1 \wedge \ldots \wedge X_n = x_n) = p(X_1 = x_1) \cdot \ldots \cdot p(X_n = x_n)$. If the family $X_1, \ldots, X_n$ is not independent, it is called *dependent*, or *correlated*.

A causal probability space contains information about causal relations and the locations of events. It is therefore particularly well-suited to capturing the idea of concrete space-like related measurements, each capable of producing a number of alternative possible results. We now discuss how a setup of that sort is represented in a probabilistic BST_{92} structure with NO MFB, and, in particular, which random variables should be selected to describe non-local correlations. Let us thus consider finitely many pairwise-SLR initials $I_1, I_2, \ldots, I_K$. As the initials are thought of as representing a joint measurement, there must be a history in which they all occur. Thus, together the initials form a consistent set $E =_{df} \bigcup_{k=1}^{K} I_k$. Each initial I_k has a family of possible outcomes $\mathbf{1}_k =_{df} \{\hat{O}^k_\gamma \mid \gamma \in \Gamma(k)\}$, so we can represent an individual measurement via a deterministic transition to a disjunctive outcome, $I_k \rightarrowtail \mathbf{1}_k$, where $\bigcup_{\gamma \in \Gamma} H_{\langle \hat{O}^k_\gamma \rangle} = H_{[I_k]}$ and for $\gamma, \gamma' \in \Gamma(k)$, $H_{\hat{O}^k_\gamma} \cap H_{\hat{O}^k_{\gamma'}} = \emptyset$ if $\gamma \neq \gamma'$.[26]

Note now that by NO MFB, every element of $\mathbf{1}_1 \times \mathbf{1}_2 \times \ldots \times \mathbf{1}_K$ is consistent. Thus, by taking the set-theoretical union of an element of $\mathbf{1}_1 \times \mathbf{1}_2 \times \ldots \times \mathbf{1}_K$ (an n-tuple of sets) one obtains a scattered outcome. Since any two different elements of $\mathbf{1}_1 \times \mathbf{1}_2 \times \ldots \times \mathbf{1}_K$ are incompatible, the set of the unions below is thus a disjunctive outcome:

$$\mathbf{1}_E =_{df} \{\textstyle\bigcup Z \mid Z \in \mathbf{1}_1 \times \mathbf{1}_2 \times \ldots \times \mathbf{1}_K\}. \tag{7.7}$$

One can immediately see that $\mathbf{1}_E$ is exhaustive and pairwise exclusive:

$$H_{\mathbf{1}_E} = H_{[E]}, \text{ and for } \hat{O}, \hat{O}' \in \mathbf{1}_E : H_{\langle \hat{O} \rangle} \cap H_{\langle \hat{O}' \rangle} = \emptyset \text{ if } \hat{O} \neq \hat{O}'.$$

By our assumption, E is consistent; furthermore, E is below $\mathbf{1}_E$ in the relevant sense of Def. 4.4. It follows that E together with $\mathbf{1}_E$ constitutes the deterministic transition $E \rightarrowtail \mathbf{1}_E$ to a disjunctive outcome. (The proof of the

---

[26] Compare the discussion following Definition 6.1 on p. 139.

above claims concerning exhaustiveness and the ordering between E and 1_E is left as Exercise 7.2.)

The transition we just constructed, $E \rightarrowtail 1_E$, gives rise to a causal probability space $CPS(E \rightarrowtail 1_E) = \langle S, \mathscr{A}, p \rangle$ that is adequate to capture any PFB induced by the transitions $I_1 \rightarrowtail 1_1, \ldots, I_K \rightarrowtail 1_K$. That space is based on the set S determined by the causal alternatives to $E \rightarrowtail \hat{O}$, for $\hat{O} \in 1_E$ (see Def. 7.5). That is, S is the set of maximal consistent subsets of $\tilde{T} = \bigcup_{\hat{O} \in 1_E} \{ e \rightarrowtail H \mid e \in cll(E \rightarrowtail \hat{O}) \wedge H \in \Pi_e \}$, $\mathscr{A}$ is the powerset algebra over S, and the measure p on $\mathscr{A}$ is induced by the measure assigned to the elements $T \in S$ via $p(T) = \mu(T)$. Now, in our construction of random variables that capture PFB, we need to represent sets like $CC(I_k \rightarrowtail \hat{O}_k)$ in $CPS(E \rightarrowtail 1_E)$; note the different initials, I_k vs. E. Representation is possible due to the following fact:

Fact 7.1. *Let $I_1, \ldots, I_K$ be pairwise SLR initial events in a BST_{92} structure with NO MFB, let $E =_{df} \bigcup_{k=1}^{K} I_k$, and for every $k \in \{1, \ldots, K\}$, let $I_k \rightarrowtail 1_k$ be a deterministic transition from I_k to the disjunctive outcome 1_k. Let S be the base set of $CPS(E \rightarrowtail 1_E)$, where 1_E is defined by Eq. 7.7. Then for every $k = 1, \ldots, K$ and every $\hat{O}_k \in 1_k$, there is $T \in S$ such that $CC(I_k \rightarrowtail \hat{O}_k) \subseteq T$.*

Proof. Let $\tau = (e \rightarrowtail H) \in CC(I_k \rightarrowtail \hat{O}_k)$, so $e \in cll(I_k \rightarrowtail \hat{O}_k)$. There is thus $h \in H_{[I_k]}$ such that $h \perp_e H_{\langle \hat{O}_k \rangle}$. By no-MFB, $e < \hat{O}_k$ (Fact 6.1(4)), so $H = \Pi_e \langle \hat{O}_k \rangle$ and, again by no-MFB (as the I_k are pairwise SLR), there is $h' \in H_{[E]}$ such that $h' \perp_e H_{\langle \hat{O}_k \rangle}$. Hence $e \in cll(E \rightarrowtail \hat{O}_k)$. By no-MFB, and as each 1_k is a disjunctive outcome, $(\hat{O}_1 \cup \ldots \cup \hat{O}_k \cup \ldots \cup \hat{O}_K)$ is consistent. It is also appropriately above E. Hence it is a scattered outcome of E. Also, as it is a scattered outcome, (†) $H_{\langle \hat{O}_1 \cup \ldots \hat{O}_k \cup \ldots \cup \hat{O}_K \rangle} \subseteq H_{\langle \hat{O}_k \rangle}$. Thus, $e \in cll(E \rightarrowtail \hat{O}_1 \cup \ldots \cup \hat{O}_k \cup \ldots \cup \hat{O}_K)$, where each $\hat{O}_i \in 1_i$, so $(\hat{O}_1 \cup \ldots \cup \hat{O}_k \cup \ldots \cup \hat{O}_K) \in 1_E$. It follows that there is $T \in S$ such that

$$(e \rightarrowtail \Pi_e \langle \hat{O}_1 \cup \ldots \cup \hat{O}_k \cup \ldots \cup \hat{O}_K \rangle) \in T.$$

We can now show that the above transition is identical to our given τ: From (†) it follows that $\Pi_e \langle \hat{O}_1 \cup \ldots \cup \hat{O}_k \cup \ldots \cup \hat{O}_K \rangle = \Pi_e \langle \hat{O}_k \rangle = H$, and so we are done. $\square$

The moral of this fact is that a set $CC(I_k \rightarrowtail \hat{O}_k)$, which belongs to the base set S_k of $CPS(I_k \rightarrowtail \hat{O}_k)$, is represented in $CPS(E \rightarrowtail 1_E) = \langle S, \mathscr{A}, p \rangle$ as the following element of the algebra $\mathscr{A}$:

$$\{ T \in S \mid CC(I_k \rightarrowtail \hat{O}_k) \subseteq T \}.$$

Given the construction of our causal probability space and the above observation concerning the definition of our random variables, we are ready to state the general definition of probabilistic funny business, or PFB:

Definition 7.7 (PFB exhibited by a set of transitions and a set of random variables). Let $\mathscr{W} = \langle W, <, \mu \rangle$ be a probabilistic BST$_{92}$ structure (Def. 7.3) with no MFB that contains the pairwise SLR initial events $I_1, \ldots, I_K$, and for every $k \in \{1, \ldots, K\}$, let $I_k \rightarrowtail 1_k$ be a deterministic transition to a disjunctive outcome 1_k. Consider the causal probability space $CPS(E \rightarrowtail 1_E) = \langle S, \mathscr{A}, p \rangle$ determined by the transition $E \rightarrowtail 1_E$ in the sense of Def. 7.5, where $E = \bigcup_{k=1}^{K} I_k$ and $1_E = \{\bigcup Z \mid Z \in 1_1 \times 1_2 \times \ldots \times 1_K\}$. Let $\{X_1, \ldots, X_K\}$ be the set of random variables on $CPS(E \rightarrowtail 1_E)$ defined via

$$X_k : S \mapsto \Gamma(k) \text{ such that for every } T \in S : X_k(T) = \gamma \text{ iff } CC(I_k \rightarrowtail \hat{O}_\gamma) \subseteq T.$$

We say that the set of transitions $\{I_1 \rightarrowtail 1_1, \ldots, I_K \rightarrowtail 1_K\}$ *exhibits PFB* iff the random variables $\{X_1, \ldots, X_K\}$ are correlated in the sense of Def. 7.6.

Observe that each $T \in S$ corresponds to a unique element of $1_1 \times 1_2 \times \ldots \times 1_K$. Hence, such a T corresponds to a unique sequence of indices $\langle \gamma_1, \gamma_2, \ldots, \gamma_K \rangle$, with $\gamma_k \in \{1, \ldots, \Gamma(k)\}$ ($k = 1, \ldots, K$). We may thus view the random variable X_k as the projection function that projects the K-tuple $\langle \gamma_1, \gamma_2, \ldots, \gamma_K \rangle$ on the kth axis, yielding γ_k.

We can illustrate this definition by linking it to non-local quantum correlations which will be investigated in detail in Chapter 8. Assume that there are three space-like related measurement events, which we represent by pairwise SLR initials I_1, I_2, and I_3. Let these initial events have two, three, and four possible outcomes, respectively, i.e., $\Gamma(1) = 2, \Gamma(2) = 3$, and $\Gamma(3) = 4$. A joint outcome of these three measurements is thus represented by a scattered outcome—a union $\hat{O}_{\gamma_1}^1 \cup \hat{O}_{\gamma_2}^2 \cup \hat{O}_{\gamma_3}^3$, where $1 \leqslant \gamma_k \leqslant \Gamma(k)$ and $\hat{O}_{\gamma_k}^k \in 1_k$ ($k = 1, 2, 3$). The base set S of the causal probability space $CPS = \langle S, \mathscr{A}, p \rangle$ thus comprises causal alternatives to each $CC(I_1 \cup I_2 \cup I_3 \rightarrowtail \hat{O}_{\gamma_1}^1 \cup \hat{O}_{\gamma_2}^2 \cup \hat{O}_{\gamma_3}^3)$, for all alternative joint outcomes given by allowable values of γ_1, γ_2, and γ_3. Since we consider all possible outcomes of the initials in our setup, each element of S contains as a subset $CC(I_1 \rightarrowtail \hat{O}_{\gamma_1}^1)$ for some γ_1. Analogously, it contains $CC(I_2 \rightarrowtail \hat{O}_{\gamma_2}^2)$ and $CC(I_3 \rightarrowtail \hat{O}_{\gamma_3}^3)$, for some γ_2 and γ_3. The three random variables, X_1, X_2, and X_3, are so defined that for any $T \in S$, $X_1(T) = \gamma_1$ iff $CC(I_1 \rightarrowtail \hat{O}_{\gamma_1}^1) \subseteq T$, $X_2(T) = \gamma_2$ iff $CC(I_2 \rightarrowtail \hat{O}_{\gamma_2}^2) \subseteq T$, and $X_3(T) = \gamma_3$ iff $CC(I_3 \rightarrowtail \hat{O}_{\gamma_3}^3) \subseteq T$. The setup then exhibits PFB if these random variables are correlated, i.e., if for some triple $\langle \gamma_1, \gamma_2, \gamma_3 \rangle$,

$$p(X_1 = \gamma_1 \wedge X_2 = \gamma_2 \wedge X_3 = \gamma_3) \neq p(X_1 = \gamma_1) \cdot p(X_2 = \gamma_2) \cdot p(X_3 = \gamma_3).$$

This means that for the $T \in S$ such that $T = CC(I_1 \rightarrowtail \hat{O}^1_{\gamma_1}) \cup CC(I_2 \rightarrowtail \hat{O}^2_{\gamma_2}) \cup CC(I_3 \rightarrowtail \hat{O}^3_{\gamma_3})$ we have, via the marginal probabilities:

$$p(T) \neq p(\bigcup\{T' \in S \mid CC(I_1 \rightarrowtail \hat{O}^1_{\gamma_1}) \subseteq T'\}) \cdot$$
$$p(\bigcup\{T' \in S \mid CC(I_2 \rightarrowtail \hat{O}^2_{\gamma_2}) \subseteq T'\}) \cdot$$
$$p(\bigcup\{T' \in S \mid CC(I_3 \rightarrowtail \hat{O}^3_{\gamma_3}) \subseteq T'\}).$$

This formula says that the probability of the joint (triple) outcome of the three SLR events I_1, I_2, and I_3 that corresponds to T does not factor into the probabilities of the three component single outcomes. A concrete system that shows such behavior will be discussed in Chapter 8.4.

We devote the rest of this chapter to a discussion of Humphreys's paradox. We will illustrate how our rigorous theory of causal probability can be employed to handle the problem, thereby also making good on our second condition of adequacy.

7.3 Fending off objections to propensities

As mentioned in Chapter 7.1.1, we consider two conditions of adequacy for our theory of causal probability: the formal condition of fulfilling Kolmogorov's axioms, and the condition of adequately responding to philosophical objections leveled against the notion of causal probabilities, or propensities. In the previous section we dealt with the first condition, showing how to define general causal probability spaces that fulfill the standard axioms of probability theory (starting with Def. 7.4 and Lemma 7.1). In this section we tackle the second task. The challenge we are facing is that our theory belongs to the category of theories of propensities, and propensities do not have a good reputation in philosophy. Propensity theories are often criticized on the grounds of certain "paradoxes" put forward by Humphreys (1985) and others. The aim of this section is to exhibit the reasons why these objections do not apply to our theory of propensities. In order to grasp propensities fully, we begin with a short survey of propensity-related concepts.

7.3.1 Some remarks on propensities

In the philosophical literature, propensities are typically assigned to singular entities. The English language suggests that these entities are objects, processes, or singular events.[27] Propensities can be graded in degrees of more or less, high or low, etc., which makes them at least similar to probabilities in this respect.

Propensities are valenced toward the future and they relate to a time-asymmetric situation. In 1973, Eddy Merckx has a propensity to win the *Tour de France* the next year; but after winning it in 1974, he does not have a propensity for having won the Tour in 1974. Pure probability theory is incapable of making such a distinction. The notion of a BST transition, however, is directly built on possibilities for the future and thereby provides the necessary resources to express the future-directedness of propensities.

As argued in Popper's (1959) influential essay on propensities, propensities are *unashamedly indeterminist*, since they "influence future situations without determining them". There being a propensity for such and such may make it likely, but is not itself a guarantee. Further, propensities are objective (Popper, 1959, p. 32). Both ideas suit causal probabilities of BST.

A further important issue that we want to stress is that *there is always some causal claim involved in an ascription of propensity*. Salmon (1989, p. 86) writes as follows: "Propensities, I would suggest, are best understood as some sort of probabilistic causes". BST theory makes good on this claim by building causal probability spaces directly out of the material provided by the BST analysis of indeterministic causation.

7.3.2 Humphreys's paradox

As propensities come in grades and are assignable to singular entities, a natural move, taken by Popper and others, is to identify propensities with a sub-species of probabilities. Propensities are put forward as providing an objective single-case interpretation of certain probabilities. Of course, everybody agrees that not *all* probabilities could be propensities, as there

---

[27] Phrases suggesting that propensities are ascribed to singular events are harder to find in English. Our usage of ascribing propensities to singular events, however, follows standard philosophical usage.

are, for example, also subjective probabilities, but the claim is that *some* probabilities are indeed propensities, and *all* propensities are probabilities.

The underlying idea that propensities are probabilities was fundamentally questioned by Humphreys's (1985) paper "Why propensities cannot be probabilities", which launched an intense discussion of the relation between these two concepts. Humphreys's claim can either be read as arguing that probability theory in its present form cannot serve as a true theory of propensities, or as arguing that the notion of a propensity as a specific kind of probability makes no sense. Many parties to the debate subscribe to the latter view.

Humphreys always has in mind that propensities are conditional, as in "there is a propensity Pr for an electron in the metal to be emitted, conditional upon the metal being exposed to light above the threshold frequency" (Humphreys, 1985, p. 558), and so his notation is generally of the form $Pr(A \mid B)$.[28] The principal lemma for Humphreys's claim that propensities cannot be probabilities is that *Bayes's theorem fails for propensities*. More generally, his strategy is to take issue with the idea that the inversion principles of probability theory apply to *conditional propensities*. Here "inversion" means that $p(A \mid B)$ and $p(B \mid A)$ are equally grammatical and inter-definable. The conclusion Humphreys draws is that "the theory of probability [is] an inappropriate constraint on any theory of single-case propensities" (Humphreys, 2004, p. 945).

To give a preview of our response to the objection, we agree with Humphreys that the five assumptions of his argument listed below lead to a contradiction. We take issue with them, however, via an analysis of what they could mean. In this analysis, our main goal is not to focus on an exegesis of Humphreys's semi-formal notation, but to attempt to construct a probability space in which all of the assumptions hold. As the assumptions are contradictory, we know that the attempted construction has to fail. By learning how it fails, we will learn which assumptions are untenable, and why. In this context, our causal probability spaces are highly relevant, as they can represent all the causal relations in Humphreys's photon story explicitly. We will argue that his problematic assumptions (iii) and (CI), see below, misrepresent the causal settings of his story. In fact, we already discussed a part of the problem in Chapter 7.2.3. We will thus reject the mentioned

---

[28] We uniformly write $Pr(A \mid B)$ for the propensity in order to heighten the contrast with the standard probability-theory notation for conditional probabilities, $p(A \mid B)$.

assumptions, and we will show how to represent Bayesian inversion in BST causal probability spaces. This means that the inversion principles which Humphreys blames for the contradiction are not problematic for propensities understood as BST causal probabilities. In Chapter 7.3.4 we analyze a more complex Humphreys-inspired case to further highlight the workings of causal probability spaces in the handling of inversion, and of conditional propensities.

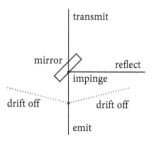

Figure 7.3 A nanosecond in the life of a photon. See the main text for details.

Humphreys's photon argument. Humphreys tells a story in which a Bayesian calculation gives the wrong answer. The story concerns a photon.[29] Figure 7.3 tells of a photon that has just been emitted in a laboratory. The photon either Impinges ($= I$) on a half-silvered mirror, with propensity q, or it drifts off somewhere. If it impinges on the mirror, it has a fixed propensity p to be Transmitted ($= T$) straight ahead through the mirror and onto a detector, and also of course a companion propensity of $r = (1 - p)$ to wind up Reflected ($= R$) off to the right. Figure 7.3 spells this out in a two-dimensional spatial diagram, looking at the apparatus from above.

The list below gives the assumptions of Humphreys's argument. The first four items on the list, namely, (i)–(iii) and (CI), are intended by Humphreys as local assumptions, governing just the emit-impinge-transmit set-up that is depicted in Figure 7.3. The rest, (TP) and (MP), express general principles of probability theory, including conditional-probability theory. The hypothesis under consideration is whether a theory of propensities can serve as an interpretation of conditional-probability theory. On the basis of that

[29] We take a few liberties with the story, such as not explicitly mentioning the background conditions holding at time t_1. The changes we make are irrelevant to the structure of the argument. Our identification of non-transmission ($\overline{T}_{t_3}$) with reflection (R_{t_3}) below is based on Humphreys's (1985) glosses on pp. 561 and 526.

hypothesis, according to Humphreys, those six assumptions should hold good for conditional propensities. Humphreys (1985), however, derives a contradiction and enters this as an argument that classical probability theory does not give a correct account of conditional propensities.

(i) $Pr_{t_1}(T_{t_3} \mid I_{t_2}) = p$, $p > 0$. [*The propensity at t_1 for T to occur at t_3 conditional upon I occurring at t_2 is p.*]

(ii) $Pr_{t_1}(I_{t_2}) = q$, $0 < q < 1$. [*The propensity at t_1 for I to occur at t_2 is q.*]

(iii) $Pr_{t_1}(T_{t_3} \mid \bar{I}_{t_2}) = 0$. [*The propensity at t_1 for T to occur at t_3 conditional upon $\bar{I}$ occurring at t_2 is 0.*]

(CI) $Pr_{t_1}(I_{t_2} \mid T_{t_3}) = Pr_{t_1}(I_{t_2} \mid \overline{T}_{t_3}) = Pr_{t_1}(I_{t_2})$. [*The propensity for a particle to impinge upon the mirror is unaffected by whether the particle is transmitted or not.*]

(TP) $Pr_{t_i}(A_{t_j}) = Pr_{t_i}(A_{t_j} \mid B_{t_k}) \cdot Pr_{t_i}(B_{t_k}) + Pr_{t_i}(A_{t_j} \mid \overline{B}_{t_k}) \cdot Pr_{t_i}(\overline{B}_{t_k})$. [*A version of the principle of total probability is assumed for propensities.*]

(MP) $Pr_{t_i}(A_{t_j}B_{t_k}) = Pr_{t_i}(B_{t_k}A_{t_j}) = Pr_{t_i}(A_{t_j} \mid B_{t_k}) \cdot Pr_{t_i}(B_{t_k})$. [*The standard definition of conditional probabilities is assumed for propensities.*]

The notation $Pr_{t_i}(A_{t_j} \mid B_{t_k})$ is to be read as "the propensity at time t_i for A to occur at time t_j, conditional upon B occurring at time t_k" (cf. Humphreys, 1985, p. 561). Times are as follows: t_0 is the last moment before emission, t_1 is a time just after emission, t_2 is the time of impingement/no impingement, I_{t_2} is the (possible) event of the photon impinging upon the mirror at time t_2, and $\bar{I}_{t_2}$ is the (possible) event of the photon failing to impinge on the mirror at t_2, as it drifts off a moment before. t_3 is the time of transmission/no transmission. T_{t_3} is the (possible) event of the photon being transmitted through the mirror at time t_3. $\overline{T}_{t_3}$ is the (possible) event of the photon failing to be transmitted through the mirror at t_3 after impinging on the mirror, so $\overline{T}_{t_3} = R_{t_3}$, the (possible) event of the photon being reflected at t_3.

The contradiction that Humphreys (1985) derives from the above assumptions is this: We have

$$Pr_{t_1}(I_{t_2} \mid T_{t_3}) = Pr_{t_1}(I_{t_2}) = q < 1$$

by (CI) and (ii). But by (TP) and (MP) (or, equivalently, by Bayes's Theorem) and (i), (ii), and (iii) we get

$$Pr_{t_1}(I_{t_2} \mid T_{t_3}) = \frac{Pr_{t_1}(T_{t_3} \mid I_{t_2}) \cdot Pr_{t_1}(I_{t_2})}{Pr_{t_1}(T_{t_3} \mid I_{t_2}) \cdot Pr_{t_1}(I_{t_2}) + Pr_{t_1}(T_{t_3} \mid \bar{I}_{t_2}) \cdot Pr_{t_1}(\bar{I}_{t_2})}$$
$$= \frac{pq}{pq + 0} = 1.$$

7.3.3 Our diagnosis of Humphreys's paradox

We turn to our evaluation of the preceding assumptions. It has been noted that the notation used has some shortcomings.[30] We will not focus on this line of criticism here, although, for the record, the problematic assumption (iii) that we will discuss is also the target of the mentioned notation-driven criticism. One further notational worry can also be put to rest: the notation $\bar{I}_{t_2}$ seems to involve the negation, or the complement, of a singular event, which may not be well-defined. We can, however, identify $\bar{I}_{t_2}$ with the well-defined event of the photon drifting off just before hitting the mirror, which we take to be the single alternative to I_{t_2}.

The bottom line to our diagnosis is that Humphreys fails to motivate his assumptions by indicating what they could refer to in some probability space. In fact, there is no probability space which satisfies all the assumptions. This can either mean that there is no classical probability space for propensities (which is Humphreys's diagnosis),[31] or that the assumptions are causally flawed. The latter is our diagnosis. Our positive contribution will be to point out which two assumptions are flawed: Assumption (iii), which happens to be numerically salvageable on our analysis, suggests inappropriate causal assumptions, and assumption (CI) fails both causally and numerically. In addition, we can show which correct principles lie behind these problematic assumptions, thus (we hope) removing the whiff of a paradox from them. On top of this diagnosis, we will show how the BST theory of causal probabilities allows for the construction of an adequate probability space.

---

[30] Miller (1994, p. 113) suggests a modified reading of $Pr_{t_i}(A_{t_j} \mid B_{t_k})$, namely "The propensity of the world at time t_i to develop into a world in which A comes to pass at time t_j, given that it (the world at time t_i) develops into a world in which B comes to pass at the time t_k". He also shows that inversion holds after the modification. A detailed discussion of Humphreys's and Miller's notations is provided in Belnap (2007).

[31] Humphreys holds that standard probability theory is "an inappropriate constraint on any theory of single-case propensities" (Humphreys, 2004, p. 945).

We start with assumption (iii), which says: $Pr_{t_1}(T_{t_3} \mid \bar{I}_{t_2}) = 0$. Presumably, $Pr_{t_1}(T_{t_3} \mid \bar{I}_{t_2})$ must be zero because T_{t_3} cannot occur together with $\bar{I}_{t_2}$, because impinging (I_{t_2} rather than $\bar{I}_{t_2}$) is a causal precondition of transmission. Writing $Pr(AB)$ to signify the probability of the joint occurrence of A and B, this means $Pr_{t_1}(T_{t_3}\bar{I}_{t_2}) = 0$.

In the same vein, since reflection (R_{t_3}, assumed in the story to be identical to $\bar{T}_{t_3}$; see Humphreys, 1985, pp. 561 and 563) cannot occur together with $\bar{I}_{t_2}$, we should also have $Pr_{t_1}(R_{t_3} \mid \bar{I}_{t_2}) = 0$, and so $Pr_{t_1}(R_{t_3}\bar{I}_{t_2}) = 0$. But then, as reflection at t_3 means no transmission at t_3, and vice versa, we should have $Pr_{t_1}(\bar{I}_{t_2}) = Pr_{t_1}(T_{t_3}\bar{I}_{t_2}) + Pr_{t_1}(R_{t_3}\bar{I}_{t_2})$, and therefore it should be that $Pr_{t_1}(\bar{I}_{t_2}) = 0$, in contradiction to assumption (ii).

The background of this calculation is exactly the problematic construction that we discussed in our story "Variant 2: Mail gets lost" in Section 7.2.2 (p. 177): a probability space involving causally impossible elements in its sample space. A significant problem with Humphreys's analysis is therefore that the sample space that his notation suggests is inadequate—it fails to properly reflect the causal relations in the photon story. In the spirit of a standard Cartesian product construction, that notation requires there to be the 2×2 fine-grained combinations of impinge/not impinge at t_2 and transmit/reflect at t_3:

$$I_{t_2}T_{t_3}, \quad I_{t_2}R_{t_3}, \quad \bar{I}_{t_2}T_{t_3}, \quad \bar{I}_{t_2}R_{t_3}.$$

These should make up the sample space of the underlying probability space and thus should have probabilities assigned that add up to one. However, given Humphreys's photon story, the last two combinations are causally impossible. Humphreys has to assign them probability zero. This assignment then has numerical consequences that contradict the causal story: the occurrence of event I_{t_2} appears to be inevitable, as the corresponding marginal probability is one.

The problematic consequence of (iii), inevitability of the occurrence of event I_{t_2}, follows from (CI) as well; the problem here is the mirror image of the problem we have just discussed. Given that I_{t_2} is compatible with both T_{t_3} and with R_{t_3}, it is correct to calculate the marginal probability according to principle (TP) as follows:

$$Pr_{t_1}(I_{t_2}) = Pr_{t_1}(I_{t_2} \mid T_{t_3}) \cdot Pr_{t_1}(T_{t_3}) + Pr_{t_1}(I_{t_2} \mid R_{t_3}) \cdot Pr_{t_1}(R_{t_3}),$$

as we will confirm in our analysis below. On that basis, however, assumption (CI) implies

$$Pr_{t_1}(I_{t_2}) = Pr_{t_1}(I_{t_2}) \cdot Pr_{t_1}(T_{t_3}) + Pr_{t_1}(I_{t_2}) \cdot Pr_{t_1}(R_{t_3}),$$

from which it follows that

$$Pr_{t_1}(T_{t_3}) + Pr_{t_1}(R_{t_3}) = 1.$$

As the occurrence of I_{t_2} is a causal precondition of the occurrence of either T_{t_3} or R_{t_3}, it again follows, absurdly, that the occurrence of I_{t_2} is inevitable at t_1.

Given a 2×2 Cartesian product space, both assumption (iii) and assumption (CI) thus allow one to argue that $Pr_{t_1}(I_{t_2}) = 1$, in contradiction to assumption (ii). The Cartesian product space is, however, clearly causally inadequate. Given the causal details of the story there are just three, not four, really possible combinations of the four events in question:

$$I_{t_2}T_{t_3}, \quad I_{t_2}R_{t_3}, \quad \bar{I}_{t_2}.$$

These objects, or rather the corresponding sets of basic transitions, form the base set S of an adequate causal probability space according to Def. 7.4. Notably, the causally impossible combination $\bar{I}_{t_2}T_{t_3}$ is missing from S, so Humphreys's premise (iii) is not even statable without further analysis.

We can construct an adequate causal probability space to substantiate our point. To begin with the impingement I_{t_2}, which is supposed to have a single alternative, $\bar{I}_{t_2}$, we assume that there is a choice event e_1 with exactly two possible outcomes $H_I, H_{\bar{I}} \in \Pi_{e_1}$, to be read as "impingement" and "drifting-off", respectively. These outcomes give rise to two basic transitions, $\tau_1^I = e_1 \rightarrowtail H_I$ and $\tau_1^{\bar{I}} = e_1 \rightarrowtail H_{\bar{I}}$.

In a similar vein, as T_{t_3} and R_{t_3} are the only alternatives at t_3 and each can only occur in the outcome I_{t_2}, we posit a second choice event e_2 that is above e_1 in the I_{t_2}-outcome and that has exactly the two basic outcomes $H_T, H_R \in \Pi_{e_2}$. The resulting basic transitions are $\tau_2^T = e_2 \rightarrowtail H_T$ and $\tau_2^R = e_2 \rightarrowtail H_R$. We now exhibit a causal probability space $CPS = \langle S, \mathscr{A}, p \rangle$ that is adequate to represent the photon story. According to Def. 7.4, given the set of initials $E = \{e_1, e_2\}$, the base set S is

$$S = \{\{\tau_1^{\bar{I}}\}, \{\tau_1^I, \tau_2^T\}, \{\tau_1^I, \tau_2^R\}\}.$$

The algebra $\mathscr{A}$ is the power-set of S, $\mathscr{A} = \mathscr{P}(S)$, and the full measure p on $\mathscr{A}$ is generated from the measure on the elements of S in the standard way, assuming that the three causal alternatives that make up the set S are assigned propensities by μ, so that

$$p(\tau_1^{\bar{I}}) = \mu(\tau_1^{\bar{I}}); \quad p(\{\tau_1^I, \tau_2^T\}) = \mu(\{\tau_1^I, \tau_2^T\}); \quad p(\{\tau_1^I, \tau_2^R\}) = \mu(\{\tau_1^I, \tau_2^R\}).$$

Note that Postulate 7.1 implies

$$\mu(\{\tau_1^I, \tau_2^T\}) = \mu(\tau_1^I) \cdot \mu(\tau_2^T) \quad \text{and} \quad \mu(\{\tau_1^I, \tau_2^R\}) = \mu(\tau_1^I) \cdot \mu(\tau_2^R),$$

and Postulate 7.4 implies

$$\mu(\tau_1^I) = \mu(\{\tau_1^I, \tau_2^T\}) + \mu(\{\tau_1^I, \tau_2^R\}).$$

These relations carry over to the measure p, noting (in accordance with our discussion in Section 7.2.5) that the BST transition τ_1^I is represented in *CPS* as a fine-grained element of $\mathscr{A}$, viz., as $\{\{\tau_1^I, \tau_2^T\}, \{\tau_1^I, \tau_2^R\}\}$.

As Humphreys alleges that propensities do not satisfy the standard inversion principles of probability theory, it will be useful to show that in our causal probability space, inversion is not problematic at all. To illustrate, here is how our theory relates the conditional probabilities of impingement and transmission in both ways.

As we just said, the concrete event τ_1^I of the photon impinging on the mirror at time t_2 is represented in *CPS* not by an element of the sample space, but by the following element of the event algebra $\mathscr{A}$:

$$I_{CPS} = \{\{\tau_1^I, \tau_2^T\}, \{\tau_1^I, \tau_2^R\}\},$$

which consists of the two elements of the sample space that include the "impinge" transition τ_1^I. The concrete event of the photon being transmitted through the mirror at time t_3 is represented by the following element of the event algebra $\mathscr{A}$:

$$T_{CPS} = \{\{\tau_1^I, \tau_2^T\}\},$$

which consists of the single element of the sample space that includes the "transmit" transition τ_2^T. Note that this element of the sample space also includes the "impinge" transition, as this transition is a causal precondition of the "transmit" transition that is represented in our probability space. The "transmit" transition, of course, has lots of other preconditions, such as the installation of the apparatus, but these are not represented in our locally based probability space CPS. In precisely the same way, the "reflect" transition is represented as $R_{CPS} = \{\{\tau_1^I, \tau_2^R\}\}$.

With a view to conditional probabilities, we next need to find out which element of the event algebra corresponds to the conjunctive event "impinge and transmit". As our algebra $\mathscr{A}$ is just the power set algebra of S, this is simple: we have

$$I_{CPS}T_{CPS} = \{\{\tau_1^I, \tau_2^T\}, \{\tau_1^I, \tau_2^R\}\} \cap \{\{\tau_1^I, \tau_2^T\}\} = \{\{\tau_1^I, \tau_2^T\}\} = T_{CPS}.$$

Note that I_{CPS} is a proper superset of T_{CPS}, so assuming that there is a non-zero probability for reflection, probability theory alone suffices to guarantee that $p(T_{CPS}) < p(I_{CPS})$.

We can now calculate conditional probabilities in our space in the standard way. The probability of impingement conditional on transmission turns out to be one, as it should, because impingement is a causal precondition of transmission that is represented in CPS:

$$p(I_{CPS} \mid T_{CPS}) = \frac{p(I_{CPS}T_{CPS})}{p(T_{CPS})} = \frac{p(T_{CPS})}{p(T_{CPS})} = 1.$$

On the other hand, given impingement, transmission is contingent:

$$p(T_{CPS} \mid I_{CPS}) = \frac{p(T_{CPS}I_{CPS})}{p(I_{CPS})} = \frac{p(I_{CPS}T_{CPS})}{p(I_{CPS})} = \frac{p(T_{CPS})}{p(I_{CPS})} < 1.$$

These conditional probabilities clearly fulfill Bayes's theorem. As a sanity check, here is how to calculate $p(I_{CPS} \mid T_{CPS})$ (see below for the expansion of the denominator via the law of total probability):

$$p(I_{CPS} \mid T_{CPS}) = \frac{p(T_{CPS} \mid I_{CPS}) \cdot p(I_{CPS})}{p(T_{CPS})} = \frac{\frac{p(T_{CPS})}{p(I_{CPS})} \cdot p(I_{CPS})}{p(T_{CPS})} = 1.$$

It will be illuminating now to look at Humphreys's assumption (iii), which involves a causally impossible combination of events. In our framework, we have

$$T_{CPS}\bar{I}_{CPS} = \{\{\tau_1^I, \tau_2^T\}\} \cap \{\{\tau_1^{\bar{I}}\}\} = \emptyset.$$

As $\emptyset$ is a valid element of the event algebra $\mathscr{A}$, with probability $p(\emptyset) = 0$ by normalization, this implies $p(T_{CPS}\bar{I}_{CPS}) = 0$, and accordingly,

$$p(T_{CPS} \mid \bar{I}_{CPS}) = \frac{p(T_{CPS}\bar{I}_{CPS})}{p(\bar{I}_{CPS})} = 0,$$

in accord with the numerical claim of Humphreys's assumption (iii). We even recapture the relevant instance of the law of total probability,

$$p(T_{CPS}) = p(T_{CPS} \mid I_{CPS}) \cdot p(I_{CPS}) + p(T_{CPS} \mid \bar{I}_{CPS}) \cdot p(\bar{I}_{CPS}),$$

because the first term evaluates as

$$p(T_{CPS} \mid I_{CPS}) \cdot p(I_{CPS}) = \frac{p(T_{CPS}I_{CPS})}{p(I_{CPS})} \cdot p(I_{CPS}) = p(T_{CPS}I_{CPS}) = p(T_{CPS}),$$

and the second term is zero, as we have just shown.

As we pointed out earlier, however, the causal structure of the set-up is misrepresented by the suggestion that one should work with the Cartesian product of impinge/not impinge and transmit/reflect. Numerically, this misrepresentation shows up as a problem with assumption (CI). To repeat, (CI) says

$$Pr_{t_1}(I_{t_2} \mid T_{t_3}) = Pr_{t_1}(I_{t_2} \mid R_{t_3}) = Pr_{t_1}(I_{t_2}).$$

In our preceding discussion, we have already evaluated the first term involved, and the calculation for the second term is exactly analogous. So, on our analysis,

$$p(I_{CPS} \mid T_{CPS}) = 1; \quad p(I_{CPS} \mid R_{CPS}) = 1; \quad \text{but} \quad p(I_{CPS}) = q < 1.$$

Assumption (CI) is therefore not just causally, but also numerically incorrect on our analysis, and this is how Humphreys's contradiction is avoided. In order to dispel the air of paradox, we should also be able to point out which correct principles lie behind assumption (CI). For starters, it is correct that the propensity at t_1 (in our analysis, at initial event e_1) for I to occur is independent of what happens afterward. As we indicated above, there is a number $q = \mu(\tau_1^I)$ that represents the causal probability of the local indeterministic transition from initial e_1 to the outcome "impinge", and we follow Humphreys's assumption that $0 < q < 1$ (the photon is neither certain to impinge nor certain not to impinge). That number q is the number it is, and it describes a local propensity of a concrete event. In this sense, the gloss of Humphreys's assumption (CI) holds on our analysis as well: the propensity to impinge in unaffected by whether the particle is transmitted or not. But, as we have just shown, the equation (CI) does not hold. One might think that the following (fallacious) reasoning provides good intuitive support for (CI):

> There is a number $q = \mu(\tau_1^I)$, the propensity for the photon to impinge, and given that the photon has impinged, there is a number $p = \mu(\tau_2^T)$ for the photon to be transmitted. The propensity for the photon to impinge and then to be transmitted is $\mu(\tau_1^I, \tau_2^T) = \mu(\tau_1^I) \cdot \mu(\tau_2^T)$, by the Markov condition. So we can calculate the propensity for the photon to impinge, conditional on its being transmitted, as

$$\mu(\tau_1^I \mid \tau_2^T) = \frac{\mu(\tau_1^I, \tau_2^T)}{\mu(\tau_2^T)} = \frac{\mu(\tau_1^I) \cdot \mu(\tau_2^T)}{\mu(\tau_2^T)} = \mu(\tau_1^I),$$

> in support of (CI); and the calculation involving R is analogous.

The problem with this argument is that it fails to take into account how the transitions in question are represented, skipping a crucial step in deriving a mathematically well-defined probability measure p on the algebra $\mathscr{A}$ of a probability space from the propensity function μ. It is correct that μ assigns a value between 0 and 1 to basic transitions and to sets of basic transitions (with the proviso that μ need not be a total function). But it is not correct to assume that μ itself is a probability measure. We already pointed out in Section 7.1.3 that this assumption is untenable. There is some work involved in using the Nature-given causal probabilities (propensities) μ to construct causal probability spaces, and that work depends crucially on how the relevant transitions are represented. Once an adequate probability space has been constructed, all assumptions of probability theory, including conditional probability theory and its inversion principles, hold without any reservations.

At this point we can take up the discussion of case 2 of representing BST transitions in causal probability spaces from Section 7.2.5. The issue with the fallacious reasoning above is exactly that the "transmit" transition τ_2^T is *not* represented in isolation in a causal probability space that includes the "impinge" transition τ_1^I, which is a causal precondition of transmission. There is no element of the event algebra $\mathscr{A}$ of *CPS* that represents the "transmit" transition alone, and therefore, the above calculation involving a conditional propensity is undefined in our *CPS*. It is correct, due to the Markov condition, that

$$\mu(\{\tau_1^I, \tau_2^T\}) = \mu(\tau_1^I) \cdot \mu(\tau_2^T). \tag{7.8}$$

The propensity for the two-step transition from e_1 (before impinging) to transmission, which has the two *causae causantes* τ_1^I and τ_2^T, factors into the propensity for the photon to impinge and the propensity for the photon then to be transmitted. It is, therefore, also correct to calculate

$$\mu(\tau_2^T) = \frac{\mu(\{\tau_1^I, \tau_2^T\})}{\mu(\tau_1^I)} = \frac{p(T_{CPS})}{p(I_{CPS})} = \frac{p(T_{CPS})}{p(T_{CPS}) + p(R_{CPS})}.$$

In this calculation we have used the probability measure p defined on *CPS* wherever possible. But the expression $\mu(\tau_2^T)$, which is well-defined in the probabilistic BST structure in which we are working, has no counterpart in our causal probability space, which represents the two consecutive events of impingement and transmission. In any such space, the transmission event is not represented in isolation, but only together with its causal precondition, impingement. Eq. 7.8 thus connects *two different* causal probability spaces. Conditional probabilities, however, are only defined within *one single* probability space.

Viewed in this light, we can repeat our initial assessment that Humphreys's claim, read charitably, is indeed correct: raw propensities are not probabilities because there is no probability space whose measure they could be. In order to show that, no Bayesian inversion is needed, however: it is enough to note that two incompatible deterministic transitions, each of which has propensity one, would have to give rise to a disjunctive event with propensity 2 (see p. 165). Thus, Humphreys's argument is a red herring. Once propensities have been used to construct causal probability spaces, standard probability theory holds, and there is no problem. Humphreys's paradox is no threat to propensity theory as implemented via BST causal probability theory.

7.3.4 Salmon's corkscrew story: More on conditional propensities and inversion

In this final section we will show how causal probability theory can also handle Bayesian inversion in a more complex case. We will proceed in terms of an analysis of Salmon's (1989) corkscrew story, which provides a standard case illustrating Bayesian inversion and which Salmon had invoked in defending his reluctance to give a probabilistic reading to propensities.[32] Here is how Salmon (1989, p. 88) put the matter.

> Imagine a factory that produces corkscrews. It has two machines, one old and one new, each of which makes a certain number per day. The output of each machine contains a certain percentage of defective corkscrews. Without undue strain, we can speak of the relative propensities of the two machines to produce corkscrews (each produces a certain proportion of the entire output of the factory), and of their propensities to produce defective corkscrews. If numerical values are given, we can calculate the propensity of this factory to produce defective corkscrews. So far, so good. Now, suppose an inspector picks one corkscrew from the day's output and finds it to be defective. Using Bayes's theorem we can calculate the *probability* that the defective corkscrew was produced by the new machine, but it would hardly be reasonable to speak of the *propensity* of that corkscrew to have been produced by the new machine.[33]

In retelling the story, we shall rely on Figure 7.4. Each h_j ($j = 1, \ldots, 5$) in that figure is a history. e_0 is a reference point event before all the action. The ellipses are machines M_i ($i = 1, 2$), with e_i the (idealized) point event of production by M_i. e_3 is the choice for the Inspector to pick one out of the two defective corkscrews. d_i [g_i] is a defective [good] corkscrew produced

---

[32] Salmon told (at least) two "inversion stories", the can opener story, in Salmon (1984), and the corkscrew story, in Salmon (1989). With these stories he subscribed to the position that propensities "make sense as direct probabilities [...], but not as inverse probabilities (because the causal direction is wrong)" (Salmon, 1989, p. 88).

[33] Salmon's language attaches propensities to things; without further comment, we translate his story into language that attaches propensities to transitions.

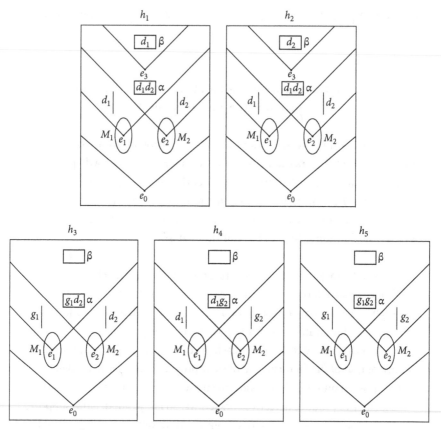

Figure 7.4 The Corkscrew Story. See main text for details.

by machine M_i. In turn, $d_i^*(i = 1, 2)$ stand for the Inspector's picking of a defective corkscrew produced by M_i from Box-α at e_3. To simplify Salmon's story, we assume that each machine makes just one corkscrew per day. The causal analysis of the story as pictured leads us to six basic transitions. First there are four transitions related to the output (good or defective) of the machines:

$$\tau_1 = e_1 \rightarrowtail g_1, \quad \tau_1' = e_1 \rightarrowtail d_1, \quad \tau_2 = e_2 \rightarrowtail g_2, \quad \tau_2' = e_2 \rightarrowtail d_2. \qquad (7.9)$$

The remaining two basic transitions result from possible actions of the Inspector, who ignores any good output in Box α, paying attention only to defective corkscrews. Accordingly, she only has a choice if two defective corkscrews are in Box α, which occurs in histories h_1 and h_2. In these histories the Inspector indeterministically chooses which defective corkscrew to transfer to Box β, and her two possible picks are represented by two alternative basic transitions,

$$\tau_3 = e_3 \rightarrowtail d_1^* \text{ and } \tau_3' = e_3 \rightarrowtail d_2^*. \tag{7.10}$$

The following table gives exemplary values for the propensities of all the basic transitions involved which we will use in our calculations below:

$$\mu(\tau_1) = 0.4 = p_1, \quad \mu(\tau_2) = 0.9 = p_2, \quad \mu(\tau_3) = 0.7 = p_3,$$
$$\mu(\tau_1') = 0.6 = p_1', \quad \mu(\tau_2') = 0.1 = p_2', \quad \mu(\tau_3') = 0.3 = p_3'. \tag{7.11}$$

From this information we can calculate the propensities of relevant non-basic transitions, such as the transition $e_0 \rightarrowtail d_1^*$ from the initial state to the inspector picking the first corkscrew.[34] To this end, one must first locate the sets of *causae causantes* of these transitions:

$$CC(e_0 \rightarrowtail d_1^*) = \{\tau_1', \tau_2', \tau_3\}, \quad CC(e_0 \rightarrowtail d_2^*) = \{\tau_1', \tau_2', \tau_3'\}, \tag{7.12}$$
$$CC(e_0 \rightarrowtail (g_1 \cup d_2)) = \{\tau_1, \tau_2'\}, \quad CC(e_0 \rightarrowtail (g_2 \cup d_1)) = \{\tau_1', \tau_2\}, \tag{7.13}$$
$$CC(e_0 \rightarrowtail (g_1 \cup g_2)) = \{\tau_1, \tau_2\}. \tag{7.14}$$

As an illustration, we will calculate the propensity of $e_0 \rightarrowtail d_1^*$ as represented in an adequate causal probability space $CPS = \langle S, \mathscr{A}, p \rangle$. First, we list the set of alternatives $\tilde{T}$ to elements of $CC(e_0 \rightarrowtail d_1^*)$:

$$\tilde{T} = \{\tau_1, \tau_1', \tau_2, \tau_2', \tau_3, \tau_3'\}. \tag{7.15}$$

Then, given the causal set-up of our story, the base set S, i.e., the set of maximal consistent subsets of $\tilde{T}$, is

$$S = \{\{\tau_1, \tau_2\}, \{\tau_1, \tau_2'\}, \{\tau_1', \tau_2\}, \{\tau_1', \tau_2', \tau_3\}, \{\tau_1', \tau_2', \tau_3'\}\}. \tag{7.16}$$

$\mathscr{A}$ is then the set-theoretic Boolean algebra over S, as it should be. The propensity of $e_0 \rightarrowtail d_1^*$ is thus to be analyzed in the probability space $\langle S, \mathscr{A}, p \rangle$. Given the assignment of propensities μ stated in Eq. 7.11, we need

---

[34] Recall that the g_i, d_i, and d_i^* are scattered outcome events, so if any two of them are consistent, their set-theoretical union is a scattered outcome as well. But if such a pair is inconsistent (like g_1, d_1^*), they do *not* yield a scattered outcome.

to find the numerical value of $p(e_0 \rightarrowtail d_1^*) = \mu(e_0 \rightarrowtail d_1^*) = \mu(CC(e_0 \rightarrowtail d_1^*)) = \mu(\{\tau_1', \tau_2', \tau_3\}) = p(\{\tau_1', \tau_2', \tau_3\})$. To calculate this number, some extra information is needed, over and above the propensities of the basic transitions involved. If each individual *causa causans* works separately and independently of each other, we may multiply probabilities of all ingredient transitions. On the other hand, the SLR *causae causantes* could work jointly. In that case, we have probabilistic funny business, and we cannot just multiply probabilities.

Here we give the numerical values that result from the basic propensities of Eq. 7.11, the Markov condition (Postulate 7.1), and the reasonable assumption of no probabilistic funny business:

$$p(e_0 \rightarrowtail d_1^*) = p_1' \cdot p_2' \cdot p_3 = .6 \cdot .1 \cdot .7 = .042$$
$$p(e_0 \rightarrowtail d_2^*) = p_1' \cdot p_2' \cdot p_3' = .6 \cdot .1 \cdot .3 = .018$$
$$p(e_0 \rightarrowtail (g_1 \cup d_2)) = p_1 \cdot p_2' = .4 \cdot .1 = .04$$
$$p(e_0 \rightarrowtail (g_2 \cup d_1)) = p_1' \cdot p_2 = .6 \cdot .9 = .54$$
$$p(e_0 \rightarrowtail (g_1 \cup g_2)) = p_1 \cdot p_2 = .4 \cdot .9 = .36$$

Since our causal analysis requires that, given the occurrence of e_0, exactly one of the five scattered outcomes must occur, it is hardly a surprise that the sum of these five propensities is 1 (see Postulate 7.3).

For future reference, let us calculate the propensity of the transition from e_0 to "the Inspector picks a defective corkscrew". In our BST structure, this is the transition from e_0 to the disjunctive outcome $\breve{O} = \{\{d_1^*\}, \{d_2^*\}\}$, which in *CPS* is represented as the element $\{\{\tau_1', \tau_2', \tau_3\}, \{\tau_1', \tau_2', \tau_3'\}\} \in \mathscr{A}$. The calculation of the probability is not difficult; just add, using the numerical results above:

$$p(e_0 \rightarrowtail \breve{O}) = .042 + .018 = 0.06. \qquad (7.17)$$

Returning to Salmon's main problem, the crucial question of the story is the following:

Question 7.3. What shall we say about the chosen corkscrew's propensity to have been made by M_1?

Salmon's answer is: "it would hardly be reasonable to speak of the *propensity* of that corkscrew to have been produced by" M_1 (Salmon, 1989, p. 88). Indeed: would anyone want to say that the corkscrew lying quietly in their hand has a certain propensity to—what? A propensity to *have been made* by machine M_1 rather than M_2? The reason why this answer sounds non-sensical has precious little to do with propensities, probabilities, or Bayes: It is a purely causal issue. The corkscrew has either been made by M_1 or by M_2, and even though the Inspector may not know which, that is a settled matter.

Although the appeal to a past-directed propensity does not make sense, we would still like to reformulate Question 7.3 in some propensity-friendly way. We offer the following as an adequate replacement that makes sense.

Question 7.4. What is the propensity for the event e_0 (in Figure 7.4) to give rise to the Inspector's taking a corkscrew made by machine M_1, given that e_0 gives rise to his picking a defective corkscrew?

The two transitions involved are $e_0 \rightarrowtail d_1^*$ and $e_0 \rightarrowtail \check{\mathbf{O}}$. In order to compute the probability of the former conditional on the latter, we also need the element of $\mathscr{A}$ that represents their intersection, which comes out as

$$\{CC(e_0 \rightarrowtail d_1^*)\} \cap CC(e_0 \rightarrowtail \check{\mathbf{O}}) = \{\{\tau_1', \tau_2', \tau_3\}\} = \{CC(e_0 \rightarrowtail d_1^*)\}.$$

Given all this, we can compute the probability of the Inspector taking a corkscrew made by M_1 conditional on the Inspector picking a defective corkscrew:

$$p(\{\{\tau_1', \tau_2', \tau_3\}\} \mid \{\{\tau_1', \tau_2', \tau_3\}, \{\tau_1', \tau_2', \tau_3'\}\}) = \frac{p(e_0 \rightarrowtail d_1^*)}{p(e_0 \rightarrowtail \check{\mathbf{O}})} = \frac{0.042}{0.06} = 0.7.$$

The inversion, although of little interest, makes perfectly good technical sense as well:

$$p(\{\{\tau_1', \tau_2', \tau_3\}, \{\tau_1', \tau_2', \tau_3'\}\} \mid \{\{\tau_1', \tau_2', \tau_3\}\}) = \frac{p(e_0 \rightarrowtail d_1^*)}{p(e_0 \rightarrowtail d_1^*)} = 1.$$

The propensity for e_0 to give rise to "The Inspector taking a defective corkscrew" given that e_0 gives rise to "The Inspector taking a corkscrew made by machine M_1", is a boring 1, since in our story, among the corkscrews made by M_1, the Inspector takes only defective corkscrews. We leave a sanity check of the corresponding instance of Bayesian inversion as Exercise 7.4.

7.4 Conclusions

In this chapter, we have used the analysis of causation in BST in terms of *causae causantes* as a background for a formal theory of objective single-case probabilities, or propensities, defined on BST transitions $I \rightarrowtail \mathscr{O}^*$. Working from a set of simple examples, we established a number of constraints on the (partial) propensity function μ that constitutes an additional ingredient in a probabilistic BST structure (Def. 7.3). We distinguished the propensity function μ from the probability measure p in causal probability spaces according to Defs. 7.4 and 7.5, and we discussed in which way a BST transition $I \rightarrowtail \mathscr{O}^*$ can be represented in different causal probability spaces (Section 7.2.5). With a view especially to later applications (see Chapter 8), we analyzed the notion of space-like probabilistic correlations, or probabilistic funny business (Def. 7.7), in terms of the dependency of random variables.

Our analysis fulfills two crucial conditions of adequacy, which in our view singles it out from among other accounts of propensities. First, the formal structures that we have defined, causal probability spaces, are standard Kolmogorovian probability spaces, showing how our causal probability theory fits in with the mainstream accounts of probabilities. The main distinction to standard approaches is the way in which probability spaces combine. As we have shown, the standard Cartesian product construction is only adequate if causally separated probability spaces with space-like related initials are combined. In the general case, the adequate way to construct the sample space of a causal probability space is to consider a set of causal alternatives (Def. 7.2). Second, we showed that our account of propensities via causal probability spaces is immune to the criticism of Humphreys's paradox. Conditional probabilities and Bayesian inversion constitute no problems for our approach, and we were able to pinpoint exactly which assumption for Humphreys's impossibility result is fallacious. We ended by showing in which way conditional propensities also make sense in more complex scenarios.

To sum up, BST-based probability theory is not an alternative to standard probability theory, but a fine-grained application of it that makes that theory able to treat objective single-case probabilities in a formally perspicuous way.

7.5 Exercises to Chapter 7

Exercise 7.1. Prove Lemma 7.2.

Hint: Consider the finite number of disjuncts separately and note that these disjuncts are mutually exclusive, allowing for the summation of their probabilities. A full proof is given in Appendix B.7.

Exercise 7.2. Let $I_1, I_2, \ldots, I_K$ be a set of pairwise SLR initial events that give rise to deterministic transitions to disjunctive outcomes $I_1 \rightarrowtail 1_1, I_2 \rightarrowtail 1_2, \ldots, I_K \rightarrowtail 1_K$. Let

$$1_E = \{\bigcup Z \mid Z \in 1_1 \times 1_2 \times \ldots \times 1_K\}.$$

Prove that 1_E is a disjunctive outcome and that $E =_{\text{df}} \bigcup_k^K I_k$ and 1_E form a transition (i.e., that E is below 1_E in the relevant sense of Def. 4.4), and that that transition is deterministic.

Exercise 7.3. Exhibit the causal probability space induced by set $CC(e_0 \rightarrowtail g_1)$ in the corkscrew story of Chapter 7.3.4 and then show how this set, $CC(e_0 \rightarrowtail g_1)$, is represented in the probability space *CPS* defined on the basis of Eq. 7.16.

Exercise 7.4. Perform a sanity check of the principle of Bayesian inversion for the corkscrew story of Chapter 7.3.4; that is, verify that, with the appropriate A and B,

$$p(A \mid B) = \frac{p(B \mid A)}{p(B \mid A) \cdot p(A) + p(B \mid \overline{A}) \cdot p(\overline{A})}.$$

PART II

APPLICATIONS

8

Quantum Correlations

In this chapter, we use BST to analyze the phenomenon of quantum mechanical (QM) correlations. Our hope is that our BST analysis, based on rigorously developed notions of transitions, propensities, and funny business, sheds some light on these puzzling phenomena. We conceive of this project as a modest one: we do not aim to produce a general "BST interpretation of QM" or anything similar. We study how the machinery of BST can be applied to the modeling of the strange correlations that QM predicts and which have been so well confirmed experimentally, and what we can learn from attempts at getting rid of the correlations via extended models.

Quantum mechanical correlations seem to be ideally suited for description and analysis in the BST framework. They concern events occurring in space and time, such as the selection of measurement parameters, measurements, or various possible detection events. The spatio-temporal features of the set-up matter, as some of the events involved are space-like related. And the correlations involve modal issues of possibility and impossibility, or of grades of possibility. We have already mentioned purely modal QM correlations when motivating the investigation of what we have called modal funny business (MFB) in BST (Chapter 5.1) via the quantum-mechanical EPR set-up. That set-up involves an entangled particle pair, leading to perfect anticorrelations between space-like separated measurements (see Figure 5.1, p. 107): some joint outcomes are impossible even though the individual outcomes are separately possible. A more complex, but also more interesting case of modal correlations is the so-called Greenberger-Horne-Zeilinger (GHZ) set-up, which involves an entangled three-particle state for which, again, some individually possible measurement outcomes are jointly impossible. Of course, this has more than a whiff of MFB about it. The GHZ set-up will play a prominent role in our discussions in Section 8.3, especially in Section 8.3.3.

Apart from *modal* correlations, and more in the focus of discussions in the philosophy of physics, QM also predicts space-like *probabilistic*

Branching Space-Times: Theory and Applications. Nuel Belnap, Thomas Müller, and Tomasz Placek,
Oxford University Press. © Oxford University Press 2022. DOI: 10.1093/oso/9780190884314.003.0008

correlations, which we have called probabilistic funny business (PFB; see Chapter 7.2.6). In such correlations, joint results of individually possible, space-like related outcomes are possible all together, but the degree to which they are possible—their joint causal probabilities or propensities—cannot be derived from the probabilities of the individual outcomes taken separately. An important set-up in this regard is the so-called Bell-Aspect set-up, which will be a focus for our discussions of PFB in Section 8.4, especially in Section 8.4.4.

Since the famous EPR paper by Einstein et al. (1935), but especially following the ground-breaking works of John Bell in the 1960s (see Bell, 1987a), quantum correlations have been analyzed in terms of the possibility of introducing an additional explanatory structure, that of the so-called hidden variables. The question from the title of the EPR paper, "Can quantum-mechanical description of physical reality be considered complete?", can be answered by studying which extensions, or completions, of the quantum-mechanical surface description of a QM correlation experiment are empirically viable. Over the years, starting with Bell's results, there have been a large number of "no go" theorems concerning the introduction of certain classes of hidden variables for various set-ups. In the BST framework, we can study the introduction of hidden variables as structure extensions that lead from a BST surface structure to an enlarged "hidden" structure that has more desirable features than the surface structure. Ideally, while the surface structure harbors MFB (or PFB), the extended structure represents the same surface phenomena, but is free from MFB (or PFB). The "no go" theorems say that this is not always possible.

In BST, we can reproduce these results in a framework representing both the spatio-temporal and the modal aspects. The BST versions of the "no go" theorems we derive show why it is impossible to explain certain cases of modal or probabilistic funny business by invoking hidden variables, and which modal, causal, propensity-related, and spatio-temporal features of our world these theorems rely on. Another significant contribution of a BST analysis, in our view, lies in the fact that we can spell out in formal detail what it means to analyze a set-up *as an experiment*, and which role this plays in the derivation of the "no go" results. It turns out that it is crucial to keep two types of indeterminism separate: indeterminism due to an experimenter's selection of measurement parameters, and indeterminism due to Nature's choice of a measurement result. BST makes room for a transparent representation of this distinction.

Our analysis in this chapter is framed in terms of QM, but our analysis of structure extensions vis-à-vis modal or probabilistic funny business is of a general nature. It can be applied to any case in which one wonders whether or not funny business is, as it were, empirically inevitable. Quantum mechanics is just the theory that gives us the strongest reasons for thinking that this may be so.

8.1 Introducing quantum correlation experiments

Put in abstract terms, a quantum correlation experiment has the following general form. The experimental set-up contains a source of systems (think of entangled pairs or n-tuples of particles) that are channeled into two or more stations, assigned to experimenters Alice (A), Bob (B), Carol (C), etc. The measurements conducted at each of the stations can be varied. In the fairly simple Bell-Aspect set-up, each experimenter is in control of two alternative settings of her measurement apparatus, 1 and 2 for Alice and 3 and 4 for Bob, and each of the possible measurements has two possible individual outcomes, '+' and '−'. For a given two-particle state, quantum mechanics predicts which joint outcomes are possible for given settings, and what the probabilities for these joint outcomes are. A single run of the experiment can thus be described by listing the selected settings together with the results obtained with these settings. So, in a set-up with just two experimenters, $a_1^+ b_3^-$ describes a run in which Alice selected setting 1, Bob selected setting 3, Alice obtained result '+', and Bob obtained result '−'. The Bell-Aspect set-up is pictured schematically in Figure 8.1.

What we have just described schematically is an experiment rather than a collection of natural happenings. The idea behind doing experiments is to pose questions to Nature by exerting control over the conditions and then observing what happens. Intervention is crucial to an experiment, yet conceptually it cannot be determined or dictated by Nature. There must be a certain independence between what happens anyway and which questions are asked in an experiment. In basic BST terms, we can incorporate the distinction between natural happenings and experimental interventions by distinguishing between two types of local choices: Some of these choices are assumed to be under experimental control, while the rest of the indeterminism in the model is due to Nature alone. This way of describing experiments

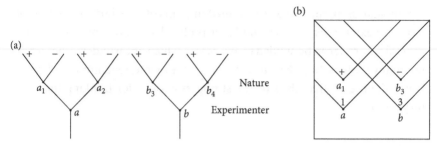

Figure 8.1 Schematic illustration of the Bell-Aspect quantum correlation experiment. Experimenters Alice (*a*) and Bob (*b*) choose settings (1 or 2 for Alice, 3 or 4 for Bob) for their experiments, which have outcomes '+' or '−'. Left: Schematic illustration of the modal splitting involved. Right: Space-time diagram for the history including the joint outcome $a_1^+ b_3^-$.

is in fact deeply ingrained in the whole quantum correlation literature. Here are some relevant quotes. Einstein et al. (1935) speak of the consequences of performing different measurements:

> We see therefore that, as a consequence of two different measurements performed upon the first system, the second system may be left in states with two different wave functions. On the other hand, since at the time of measurement the two systems no longer interact, no real change can take place in the second system in consequence of anything that may be done to the first system. (Einstein et al., 1935, p. 779)

Bell (1987b) speaks of the freedom of experimenters:

> It has been assumed that the settings of instruments are in some sense free variables—say at the whim of experimenters—or in any case not determined in the overlap of the backward light cones. Indeed without such freedom I would not know how to formulate *any* idea of local causality, even the modest human one. (Bell 1987b, p. 61)

Similar remarks can be found in papers in experimental physics. For example, Aspect et al. (1982b) mention "other choices of orientations". In our schematic set-up, we thus contrast the selection of settings, which is under experimental control, and the chancy occurrence of the measurement outcomes, which is due to Nature. Experimental control does not have to mean that the selection of settings is due to a human agent.[1] A distinctive feature of experimental control, which enters as an assumption in the derivation of results about the possibility or impossibility of certain types of hidden variables, such as the derivation of Bell-type theorems (see Section 8.4), is that the selection of settings must be independent of the physical state of the system on which the experiment is performed—including a hypothetical full state that adds hidden variables. Typically, this independence is understood as the existence of truly indeterministic processes of the selection of settings whose outcomes are, in particular, not influenced by the values of any hidden variables.[2] In contrast to the settings, the measurement outcomes may of course depend upon, or even to be dictated by, the full physical state including the values of hidden variables.

8.2 On the BST analysis of quantum correlations

Since Einstein, Podolski and Rosen's (1935) seminal paper, quantum correlations have been viewed as mysterious; they seem spooky, or "spukhaft", as Einstein had it.[3] After all, such correlations coordinate remote space-like separated measurement results, like a_1^+ and b_3^- in our schematic illustration, without the possibility of physical interaction. In some experiments, the spatial distance between the results is enormous.[4] BST has two resources to represent correlations between space-like related measurement results, modal funny business (MFB) and probabilities funny business (PFB). The first approach applies to quantum-mechanical correlations only if some joint outcomes are assigned probability zero. On that modal (MFB) approach, probability zero is interpreted as meaning that a joint outcome is impossible, while a non-zero probability signals that a joint outcome is possible.

---

[1] The experiment of Aspect et al. (1982a), with semi-random selections of settings produced by an optical process, was viewed as incorporating a type of experimental control that constituted a decisive improvement over earlier experiments in which the settings were fixed for whole series of runs of the experiment. Even so, direct human control seems to be the conceptual gold standard for such experiments, despite the fact that the quality of the randomness produced by humans, as assessed via statistical tests, is much inferior to other sources of randomness. For a recent large-scale experiment using direct human control for the selection of parameters in quantum correlation experiments, see Abellán et al. (2018).

[2] For a dissenting view, see Esfeld (2015).

[3] See footnote 2 in Chapter 5.

[4] As of Spring 2019, the longest distance is 144 km on the ground (Scheidl et al., 2010) and 1200 km by satellite communication (Yin et al., 2017).

228 BRANCHING SPACE-TIMES

Measurement events are represented as transitions, and a set-up with modal correlations can be analyzed using the resources developed in Chapter 5. The second approach applies to all cases of quantum-mechanical correlations, including, but not limited to those in which probabilities zero and one occur. On that probabilistic (PFB) approach, probabilities of joint outcomes are interpreted as (single case) propensities, which can be analyzed using the framework developed in Chapter 7.2.6.

According to quantum mechanics, remote correlations can indeed be perfect (at least in theory), meaning that the probability of the joint occurrence of two specific, individually possible results can be zero. This implies that knowing her own result, an experimenter can immediately predict a remote result. This looks as if quantum correlations could beat the limit of the speed of light that is imposed on physical interactions by the special theory of relativity. This leads to the worry of whether quantum correlations indicate a conflict between the theories of relativity and of quantum mechanics, and in particular, whether they permit superluminal signaling. The consensus is that while entanglement can be used to greatly enhance the security of communication (Colbeck and Renner, 2012), it does not provide a resource for superluminal communication. There are space-like correlations involved, but these cannot be exploited by the experimenters to send signals. More generally, the consensus appears to be that there is no logical conflict between relativity and quantum mechanics, and that the two theories live in a kind of peaceful coexistence, as argued first by Shimony (1978). Needless to say, as this very phrase is reminiscent of Cold War talk,[5] the implication is that there are decisive conflicts between the two theories, but these are not lethal, as they do not amount to a head-on contradiction. Thus, the phrase also suggests some hope of resolving the conflict via a future unified theory encompassing large- and small-scale phenomena together.[6]

[5] Shimony attributed the phrase to Chairman Khrushchev; "mirnoye sosushchestvovaniye", as it is called in Russian, was part of Soviet propaganda in the Cold War and afterward.
  A key element in Shimony's (1978) peaceful coexistence strategy is to uphold one premise of Bell's theorem (the independence of measurement outcomes from the choice of remote settings), while rejecting a mathematically similar premise that remote measurement outcomes are independent from each other (for a precise statement of these premises, see Def. 8.24 for Parameter Independence and Def. 8.22 for Outcome Independence, both discussed in Section 8.4). This suffices to block the derivation of Bell's theorem, while prohibiting faster than light signalling. Given the mentioned similarity of the two premises, the move is nevertheless controversial.
[6] Given more recent developments, like the Colbeck-Renner (2011) proof of a Bell-type theorem, or purported progress with Bohmian analyses of EPR correlations (for a survey, see, e.g., Maudlin 2019, ch. 4), the idea of peaceful coexistence now looks more complicated than ever. For a recent reassessment, see Butterfield (2018).

Correlations between space-like separated events raise the question of explanation and especially of constraints on explanations that relate to "locality." If correlated events are space-like separated, a correlation cannot be explained via direct physical interactions—at least this is the consensus view. An explanation would thus have to come from an extended description of the states of the system involved, as already argued by Einstein et al. (1935). The big question is, of course, which extensions are admissible conceptually and empirically, and which kind of explanation they can provide. The initial thrust of the quest for hidden variables is often interpreted as an attempt to eliminate the indeterminism of quantum mechanics, or to reduce quantum-mechanical probabilities to epistemic phenomena.[7] Over the years, however, it has become clear that the real challenge is to explain away any supposed "non-locality" of quantum mechanics, not the theory's indeterminism. Thus, one constraint on the hidden-variable completions of quantum mechanics is that the resulting models be "local", where the informal notion of locality must be made precise in some way. A natural way to explain QM correlations locally is to postulate common causes lying in the past of correlated events, or to postulate instructions that the correlated particles carry along from the source to the measurement. Starting with the work of John Bell in the 1960s (see Bell, 1987a), it has become clear that postulating such hidden variables is not a merely metaphysical maneuver, but can have implications that can be tested empirically. Given the well-established empirical fact that observed correlations vindicate the predictions of QM,[8] the issue has become less one of extending the quantum-mechanical formalism, but of coming up with any viable account of the observed correlations at all. For such an account to count as explanatory, it has to fulfill certain structural constraints that are either physically or philosophically motivated. Crucially, they include the

---

[7] Einstein is often attributed with the slogan "God does not play dice", but the interpretation of this slogan is controversial. In a letter to Max Born dated December 4, 1926, Einstein writes: "Die [Quantenmechanik] liefert viel, aber dem Geheimnis des Alten bringt sie uns kaum näher. Jedenfalls bin ich überzeugt, daß *der* nicht würfelt." See Einstein et al. (1971, p.90): "[Quantum mechanics] says a lot, but does not really bring us any closer to the secret of the 'old one'. I, at any rate, am convinced that *He* is not playing at dice."

[8] A whole literature is devoted to the important question of whether the possible "loopholes" often present in actual experiments can be and have been closed. For example, if detection efficiency is too low, it is impossible to check whether the runs registered constitute a fair sample of all really possible runs. For a discussion of loopholes, see, e.g., Myrvold et al. (2019). Over the years, substantial progress in closing these loopholes has been made. Some recent experiments have been claimed to be "loophole-free" (Giustina et al., 2013; Hensen et al., 2015; Shalm et al., 2015). It appears that at least the scientific community is convinced that in these experiments, the issues of loopholes have been dealt with satisfactorily, so that the quantum-mechanical predictions are really empirically vindicated.

independence of experimenters' choices of measurement parameters from the particles' state.

The basic *local* features of BST can help to analyze the quest for local explanations. In BST, all possible events are well-defined spatio-temporal objects. If such an event is an indeterministic choice point, it has possible outcomes that occur *after* it in the BST pre-causal ordering. This basic locality is supplemented by two restricted and well-defined varieties of non-locality: BST permits modal as well as probabilistic funny business (see Chapter 5 and Chapter 7.2.6, respectively). With these resources, BST is able to resolve issues that are outside the scope of other rigorous, but purely probabilistic frameworks available in the literature.[9] The BST resources will be especially helpful when tackling the independence of experimental interventions and natural indeterminism. This independence has at least two aspects. First, an experimenter, who has a range of possible measurement parameters to choose from, should not be restricted in her choice. This might perhaps sound overly broad: undoubtedly, Nature limits our ways, and it may not be possible to construct an experimental set-up in which one can ask Nature different questions by selecting alternative parameters. The independence that is at stake here, however, is more specific: given that an apparatus is in place that affords the choice of different parameters, there must be no surreptitious or hidden limitation of the experimenters' choices in any given run of the experiment. The other aspect goes in the opposite direction: the experimenter must not be able to restrict or influence Nature's choices at remote (space-like separated) events.

The notion of independence still needs to be made formally precise in a way that accords with the general framework that is used, but here we can already formulate a template for the independence condition, which we will later call C/E independence:

Definition 8.1 (Target notion of C/E independence). Let $\mathscr{W} = \langle W, < \rangle$ be a BST_{92} structure with two disjoint sets of choice points, $C, E \subseteq W$, where C represents the choices of experimenters and E represents indeterminism due to Nature. The structure is said to be C/E *independent* iff for any consistent subsets $C_0 \subseteq C$ and $E_0 \subseteq E$, the outcomes of C_0 and the outcomes of E_0 are independent in the relevant (modal or probabilistic) sense.

---

[9] For such accounts, see, e.g., Hofer-Szabó et al. (1999) and Pitowsky (1989).

We already said that our analysis focuses on the representation of the possible single runs of quantum correlation experiments. Each single run is a complex spatio-temporal happening that involves the particles' emission from the source, the selection of measurement settings, concrete measurement processes and detection events, and perhaps more. A satisfactory account of these possible single runs cannot be a mere list of them, and a satisfactory account of an experiment must be more than a mere chronicle of the mentioned characteristics of the actual runs. The challenge is to come up with a law-like account (i.e., one that is in some sense stable under counterfactual variations). Our BST analysis offers a way to represent arguments along these lines in full formal detail.

The main formal tool that we will use is that of a structure extension. Given a BST surface description of a QM correlation experiment—more or less, a compilation of the really possible runs and their spatio-temporal features—we can study ways of extending the surface structure via hidden variables, or instruction sets. The idea is that such an extended structure could show that the modal or probabilistic funny business present in the surface structure is only apparent, and how the surface correlations can come about on the basis of hidden structure without such troublesome correlations.

We begin our work in the setting of modal correlations in Chapter 8.3. We then move to probabilistic correlations in Chapter 8.4. In our exposition, for simplicity's sake we stick to the BST_{92} framework

8.3   Explaining modal correlations via instruction sets

The motivation for wanting to get rid of modal correlations is, bluntly speaking, that we seem unable to understand how they could occur. If two distant events each have a number of different possible outcomes and these events cannot causally influence one another, then how could some joint outcomes be impossible? But that is exactly the situation that we seem to be facing in certain quantum correlation experiments, and which we have called modal funny business (MFB). The task of getting rid of MFB is, therefore, the following: Given a BST structure $\langle W, < \rangle$ that harbors MFB, we ask whether there is a different BST structure $\langle W', <' \rangle$ that is free from MFB while still representing the same facts. This means that we take the initially given BST structure to be a representation of surface facts only, and that we look for an extension of that structure that consists of copies of histories of the surface structure that are differentiated by hidden factors. The idea is that while each history in the surface structure describes the empirically accessible facts about one possible run of the experiment in question, each history in the extended structure describes such a possible run together with hidden factors that are posited to explain the modal correlations present in the surface structure. In agreement with widespread usage, we call these hidden factors instruction sets, or deterministic hidden variables.[10] Technically, the overall idea is to replace one model of a quantum correlation experiment, the surface model, by an extended model that does not contain modal funny business and in which additional instruction sets prescribe the outcomes of measurements (Nature's choices). Such a prescription has to be given for any possible choice of measurement parameters by the experimenter. In general, therefore, an instruction set has to have a counterfactual character; it has to include instructions, not all of which can be realized together.

Indeed, instruction sets without such a counterfactual character appear fishy. As we said earlier and as we will prove below (Theorem 8.2), the following maneuver is always possible. Assume that all the choice points in a BST structure $\langle W, < \rangle$ that models an experiment lie above some event e^*, forming a set E for which $e^* < E$. Take all the histories h that contain e^*— under our assumptions, this is the full set $\mathrm{Hist}(W)$—and create a unique label $\hat{h}$ for each history. Then build a new structure in which each history h, which we can write as $\{a \mid a \in h\}$, is replaced by a labeled copy $h' =_{df} \{\langle a, \hat{h} \rangle \mid a \in h\}$. Adjust the ordering such that above e^*, the labeled copies are all kept separate, splitting at e^* only. Then the extended structure will have no choice points apart from e^*, so that there can be no modal correlations in the extended structure, but all the histories of the surface structure will still be represented.

Such an extension is formally possible, but it is not considered to be satisfactory as a possible representation of the hidden structure of a quantum experiment. Such extensions are commonly called superdeterministic or conspiratorial, and the reason for this negative verdict is that they

---

[10] As we said, the BST analysis in this section is special because it proceeds in terms of modal correlations, targeting the underlying structure of possibilities. Usual analyses of hidden variables, on the other hand, take an underlying uncorrelated structure of possibilities for granted and introduce a probability distribution over the hidden factors to explain surface correlations. On that probabilistic approach, a difference is made between *stochastic* hidden variables, which fix probabilities for measurement outcomes, and *deterministic* hidden variables, which *dictate* unique measurement results. For obvious reasons, such deterministic hidden variables are also called instruction sets. See, e.g., Fine (1982, p. 291), who writes of "response functions (giving the λ-determined responses to the measurements)", or Mermin (1981, p. 403), who uses the language of "instruction sets".

eliminate the experimenter's freedom to choose measurement parameters. In a superdeterministic extension, the instruction sets that are selected at e^* not only instruct Nature on the outcomes of the measurements, but they also instruct the experimenter on how to choose the measurement parameters. So the instructions encode a conspiracy between the measurement parameters and the measurement outcomes. This goes against the basic assumption that an experimental outcome is Nature's answer to a question that the experimenter has freely chosen to ask.[11] A satisfactory extension of a surface structure, in contrast, has to retain the experimenter's freedom to choose measurement parameters.

Translated into BST terminology, this means that in the description of the surface structure, we have to distinguish between two kinds of choice points: those that represent Nature's choices of the measurement outcomes (E), and those representing the experimenter's independent choices (C) of experimental parameters. Furthermore, an extension of a surface structure will be satisfactory to the extent to which it eliminates the disturbing MFB among outcomes of members of E while retaining the independence of the choices at members of C.

We will discuss the issue of introducing instruction sets in BST structures in two steps. In Section 8.3.1 we first discuss the general idea of extending a BST surface structure, using the notion of generic instruction sets. Such instruction sets are not fit to be used in real applications, but they allow us to define the general formal procedure of extending a surface structure and to prove a number of important general results about such extensions. For example, we can establish Theorem 8.1, which says that extended structures are again BST_{92} structures. The general discussion paves the way for our discussion of the more specific (and actually useful) notions of non-contextual and contextual instruction sets in Section 8.3.2.

Formally, our point of departure is what we will call a BST_{92} surface structure, in which there is indeterminism induced by experimenters' choices (C) as well as indeterminism produced by Nature (E). The motivation for embarking on the project of a structure extension has to come from some instances of MFB present in the surface structure involving the members of E: some experimental outcomes are modally correlated. (If that is not so, there is no incentive to modify the given surface structure, as it already contains a proper causal account of the possible experimental outcomes.)

---

[11] See Section 8.1 for some quotes backing this claim.

While removing the MFB present in E, we want to retain the freedom of the experimenters' choices.

In defining a BST_{92} surface structure, the guiding idea is that we are here considering BST as a formal tool for modeling. This means that idealizing assumptions are warranted, as we are only concerned with the experiment in question. We will assume that there are only finitely many choice points in W and that each of them is only finitely splitting. We will also assume that in the model, all choice points are either due to Nature (providing measurement outcomes) or due to the experimenter (providing choices of measurement parameters). Furthermore, the experimenters' choices of measurement parameters must of course be made before the respective measurement with the chosen parameters occurs. Formally, this means that any choice point under the experimenters' control ($c \in C$) must be below some measurement choice point ($e \in E$).

Definition 8.2. A BST_{92} *surface structure* is a quintuple $\langle W, <, e^*, E, C \rangle$, where $\langle W, < \rangle$ is a BST_{92} structure, $e^* \in W$ is a deterministic point in W, and $E, C \subseteq W$ are two finite sets of finitely splitting choice points fulfilling the conditions that $E \neq \emptyset$, $E \cap C = \emptyset$, and that $E \cup C$ is the set of all choice points in W. With respect to the ordering, we demand $e^* < E$ (i.e., for any $e \in E$, we have $e^* < e$), and for any $c \in C$ there must be some $e \in E$ for which $c < e$. For future use we define

$$\tilde{T}_E =_{df} \{e \longmapsto H \mid e \in E, H \in \Pi_e\}; \quad S_E =_{df} \{T \subseteq \tilde{T}_E \mid T \text{ maximal consistent}\}.$$

Note that $\tilde{T}_E$ and S_E are finite by our assumptions. We will drop the subscript E if it is clear from context.

Here are some simple facts about the structure of the set S of maximal consistent sets of transitions with initials from E.

Fact 8.1. *Let $\langle W, <, e^*, E, C \rangle$ be a BST_{92} surface structure, let $T \in S$, and let $h_1, h_2 \in H(T)$. With respect to the splitting of h_1 and h_2, the following holds: (1) If $h_1 \perp_c h_2$, then $c \in C$. (2) If $h_1 \perp_c h_2$, then there is some $e \in E$ for which $c < e$, but there is no $e \in E$ for which $e \in h_1 \cup h_2$ and for which $c < e$.*

Proof. (1) Let $T_{h_i} =_{df} \{e \longmapsto \Pi_e \langle h_i \rangle \mid e \in E \cap h_i\}$ be the sets of transitions on h_i that have initials in E ($i = 1, 2$). By $h_i \in H(T)$, we have $T \subseteq T_{h_i}$, and as T is a maximal consistent set of transitions with initials in E, in fact $T_{h_1} = T = T_{h_2}$.

So h_1 and h_2 cannot split at a member of E. As $E \cup C$ is the set of choice points in W, it must be that $c \in C$.

(2) By the definition of a BST_{92} surface structure, for any $c \in C$ there must be some $e \in E$ for which $c < e$. By (1), we have $e \in h_1$ iff $e \in h_2$ for members $e \in E$. And if $e \in h_1 \cap h_2$ and $c < e$, then $h_1 \equiv_c h_2$.   □

As we said, the idea of adding instruction sets is to explain away surface MFB via a hidden structure that provides instructions at e^* for what should happen at the choice points in E (which represent Nature's choices). Such instructions have to be counterfactual in the sense of allowing for different choices of the experimenters via the choice points in C. In addition, there should be no MFB between the choices of the measurement settings and the measurement outcomes—this would contradict the basic idea that the choice of measurement settings is completely independent of the outcomes that Nature provides. The following definition captures this idea.

Definition 8.3 (C/E independence). We say that a BST_{92} surface structure $\langle W, <, e^*, E, C \rangle$ *violates* C/E *independence* iff there are two consistent, non-empty sets of transitions, T_C, with initials $E_{T_C} \subseteq C$, and T_E, with initials $E_{T_E} \subseteq E$, for which $T_C \cup T_E$ is combinatorially consistent but inconsistent, so that $T_C \cup T_E$ constitutes a case of combinatorial funny business (see Def 5.6). We say that a BST_{92} surface structure (or its set E) *satisfies* C/E *independence* iff it does not violate it.

In the following we develop the idea of instruction sets in a number of ways. In Section 8.3.1 we exhibit the formal procedure of extending a surface structure with respect to an unconstrained notion of instruction sets, which we call *generic* instruction sets. In Section 8.3.2 we then focus on two actually useful types of instruction sets, non-contextual and contextual ones. In Section 8.3.3 we then turn to the analysis, both in terms of non-contextual and contextual instruction sets, of the GHZ experiment, which is a well-known example of modal correlations due to quantum entanglement. We provide a brief summary in Section 8.3.4.

8.3.1 Extensions of a surface structure by generic instruction sets

Generic instruction sets are defined to be, quite simply, subsets of S, that is, sets of maximal consistent sets of transitions with initials from E, the set of Nature's choice points.

Definition 8.4 (Generic instruction set). Let $\langle W, <, e^*, E, C \rangle$ be a BST_{92} surface structure. A *generic instruction set for* $\langle W, <, e^*, E, C \rangle$ is any nonempty subset λ of S, i.e., $\lambda \subseteq S$ and $\lambda \neq \emptyset$. We write $\mathfrak{I}_g$ for the set of all generic instruction sets.

Note that this definition implies, trivially, that any consistent set of transitions can be expanded to a generic instruction set. The notion of a generic instruction set is too wide for applications. The set S generally contains maximal consistent subsets T_1, T_2 that are openly contradictory, prescribing opposite outcomes to each of their initials, and by Def. 8.4, a generic instruction set may contain both T_1 and T_2. In due course, we will exclude such instruction sets from consideration. We work with the notion of a generic instruction set here in order to show that a substantial part of the theory of instruction sets in BST can be developed without any commitment as to the nature of the instruction sets. The first example of such a general definition concerns the notion of matching:

Definition 8.5 (Matching). Let $\lambda \in \mathfrak{I}_g$ be a generic instruction set and h a history in W. We define the *matching set for h* to be $T_h =_{df} \{e \rightarrowtail \Pi_e \langle h \rangle \mid e \in E \cap h\}$. We say that h *matches* λ iff λ contains the matching set for h, i.e., iff $T_h \subseteq \bigcup \lambda$. As a stylistic variant, we also say that λ *matches h*.

By this definition, if a history h contains no elements of E, then $T_h = \emptyset$, and *any* instruction set $\lambda \in \mathfrak{I}_g$ matches h. For the general case, we can also show that there are always matching pairs of histories and instruction sets.

Fact 8.2. *(1) For any history h in W there is some generic instruction set $\lambda \in \mathfrak{I}_g$ that matches h. (2) Let $\lambda \in \mathfrak{I}_g$ and $T \in \lambda$, and let h be a history. If $h \in H(T)$, then h matches λ. (3) For any generic instruction set $\lambda \in \mathfrak{I}_g$, there is some matching history h.*

Proof. (1) Let $h \in \text{Hist}(W)$ be given. The matching set for h, T_h of Def. 8.5, is consistent, so it can be extended to some maximal consistent set $T \in S$. The set $\lambda =_{df} \{T\}$ is already a generic instruction set according to Def. 8.4. The instruction set λ (and any of its extensions) matches h because $T_h \subseteq \bigcup \lambda$.

(2) Let $\lambda \in \mathfrak{I}_g$ and $T \in \lambda$ be given. The set T is consistent. Assume that $h \in H(T)$, and consider the matching set for h, T_h of Def. 8.5. As h lies in all outcomes of transitions of T, which all have initials in E, it must be that $T \subseteq T_h$. Now T is a maximal consistent subset of $\tilde{T}$, so T_h, which is also consistent, cannot be a proper superset of T, so that $T = T_h$. From $T \in \lambda$ we have $T \subseteq \bigcup \lambda$, so that indeed $T_h \subseteq \bigcup \lambda$, i.e., h matches λ.

(3) As $\lambda \neq \emptyset$, there is some $T \in \lambda$, which is consistent. Take $h \in H(T)$; the claim follows directly from (2). □

If a history h in a BST$_{92}$ surface structure matches an instruction set λ, this means that λ provides a direction (a possible outcome) at each member $e \in E$ that occurs on h, and that direction is "to stay on h" ($\Pi_e \langle h \rangle$). The instruction set λ is, as it were, h-friendly. Given the generality of the definition of a generic instruction set, however, the set of transitions $\bigcup \lambda$ may in such a case also contain a different, incompatible transition with initial e, $e \rightarrowtail H$ with $H \neq \Pi_e \langle h \rangle$. So, on the generic approach, an instruction set can match a history even though the instructions do not *dictate* that the history occur—they only have to *allow* for the history to occur. This feature of the possibility of different instructions for the same initial is what distinguishes non-contextual from contextual instruction sets: for the former, an instruction at one of Nature's choice points $e \in E$ has to be unique, while for the latter, it may depend on the context and thus fail to be unique.

Given the notion of matching, we can work toward our definition of an extended structure (Def. 8.7). In an extended structure, we replace elements $a \in W$ of the surface structure with labeled elements $\langle a, L \rangle$. The labels represent instruction sets, which will be implemented as new elementary outcomes of e^*. So for $a > e^*$, the label L has to be some $\lambda \in \mathfrak{I}_g$. For $a \not> e^*$, on the other hand, an outcome at e^* can have no causal influence on a; in this case we use the label $L = \emptyset$, just in order to preserve a uniform format for the members of the extended structure. Formally, we build the extended structure from lifted histories, which are defined as follows:

Definition 8.6 (Lifted history). Let h be a history in W matching a generic instruction set λ. Then we define the lifted history $\varphi_\lambda(h)$ to be

$$\varphi_\lambda(h) =_{\mathrm{df}} \{\langle a, \emptyset \rangle \mid a \in h, a \not> e^*\} \cup \{\langle a, \lambda \rangle \mid a \in h, a > e^*\}.$$

Using these lifted histories, we define the *extended BST structure* based on a BST$_{92}$ surface structure as follows:

Definition 8.7 (Extended structure). Let $\mathscr{W}_S = \langle W, <, e^*, E, C \rangle$ be a BST$_{92}$ surface structure, and let $\mathfrak{I}_g$ be its set of generic instruction sets. We define the extended structure $\mathscr{W}_E = \langle W', <' \rangle$ corresponding to $\mathscr{W}_S$ to be the union of all lifted histories together with an ordering relation that respects that events can be multiply copied with different instruction sets.

$$W' =_{df} \bigcup_{h \in \text{Hist}(W), \, \lambda \in \mathfrak{I}_g \text{ matching } h} \varphi_\lambda(h);$$

$$\langle e_1, L_1 \rangle <' \langle e_2, L_2 \rangle \text{ iff } e_1 < e_2 \text{ and } (L_1 = \emptyset \text{ or } L_1 = L_2);$$

$$\mathscr{W}_E =_{df} \langle W', <' \rangle.$$

The different copies of elements above e^* are not order related. The following fact about elements of an extended structure shows how elements of W' come from lifted histories:

Fact 8.3. *Let $\mathscr{W}_E = \langle W', <' \rangle$ be the extended structure based on a BST_{92} surface structure $\mathscr{W}_S = \langle W, <, e^*, E, C \rangle$. Then $\langle a, L \rangle \in W'$ iff either $(a \not> e^*$ and $L = \emptyset)$ or $(a > e^*$ and there is some history h matching λ for which $a \in h)$.*

Proof. For $a \not> e^*$, the "$\Rightarrow$" direction follows by Def. 8.7. For the "$\Leftarrow$" direction, note that a belongs to some history $h \in \text{Hist}(W)$, and by Fact 8.2(1), there is some $\lambda \in \mathfrak{I}_g$ matching h.

For $a > e^*$, the claim follows immediately from Def. 8.7. □

We are working toward our first main result about extended structures, Theorem 8.1, which says that these are also BST_{92} structures. As a simple first step, we can show that some basic properties of the ordering $<$ carry over to $<'$ immediately:

Fact 8.4. *Let $\mathscr{W}_E = \langle W', <' \rangle$ be the extended structure based on a BST_{92} surface structure $\mathscr{W}_S = \langle W, <, e^*, E, C \rangle$. Then $\langle W', <' \rangle$ is a non-empty, dense, strict partial ordering.*

Proof. Left as Exercise 8.1 □

We now have to characterize the histories in the extended structure, so that we can prove further BST_{92}-relevant properties. We split the crucial history lemma (Lemma 8.1) into two facts. The one direction is the following.

Fact 8.5. *Let $\langle W, <, e^*, E, C \rangle$ be a BST_{92} surface structure, let $h \in \text{Hist}(W)$, and let $\lambda \in \mathfrak{I}_g$ be a generic instruction set matching h. Then the set $\varphi_\lambda(h)$ is maximal directed, i.e., it is a history in the extended structure $\langle W', <' \rangle$.*

Proof. Let $A' =_{df} \varphi_\lambda(h)$ be the lifted history. We first show that A' is directed: Let $\langle e_1, L_1 \rangle, \langle e_2, L_2 \rangle \in A'$ (where $L_i = \lambda$ or $\emptyset$ depending on whether or not $e_i > e^*$, $i = 1, 2$). These elements of A' were lifted from elements $e_1, e_2 \in h$, and as h is directed, there is some $e_3 \in h$ for which $e_1 \leq e_3$

and $e_2 \leqslant e_3$. The corresponding element $\langle e_3, L_3 \rangle$ of A' is $\leqslant'$-above the two mentioned elements of A', as one can easily verify by the definition of the ordering.

Assume now for reductio that A' is not maximal directed, i.e., that there is a proper superset $A'' \supsetneq A'$, $A'' \subseteq W'$, that is also directed. As a subset of W', A'' has elements $\langle a, L \rangle$, where $L = \emptyset$ iff $a \not> e^*$. We define the set $A =_{\mathrm{df}} \{a \in W \mid$ there is some $\langle a, L \rangle \in A''\}$. We consider two cases.

Case 1: If A' contains some element $\langle a_0, \lambda \rangle$, then any $\langle a, L \rangle \in A''$ for which $a > e^*$ must satisfy $L = \lambda$; this follows by directedness of A'' and by the definition of the ordering. So if $\langle a, L \rangle, \langle a, L' \rangle \in A''$, then $L = L'$; the labels of elements of A'' are unique. We can show that the set $A \subseteq W$ defined above is directed: Pick $e_1, e_2 \in A$, so that the respective $\langle e_1, L_1 \rangle, \langle e_2, L_2 \rangle$ are members of A''. As A'' is directed, these two members have a common upper bound $\langle e_3, L_3 \rangle \in A''$, so that there is $e_3 \in A$ that is a common upper bound of e_1 and e_2 in W. Now there is some $\langle e_0, L_0 \rangle \in A'' \setminus A'$ by our reductio assumption, so that A contains some member $e_0 \notin h$. So A is a proper superset of h that is also directed, which contradicts the definition of histories as maximal directed sets.

Case 2: If A' contains only elements that have the label $\emptyset$, the directed proper superset A'' (whose existence constitutes our reductio assumption) might also contain only elements that have the label $\emptyset$. So the labels of elements of A are unique in A'', and we can reason as in case 1. It might be, however, that A'' contains an element $\langle a_0, \lambda' \rangle$ with $\lambda' \neq \emptyset$. In this case we cannot guarantee that $\lambda' = \lambda$. But it must still be that any $\langle a, L \rangle \in A''$ for which $a > e^*$ must satisfy $L = \lambda'$, again by directedness and by the definition of the ordering. So in this case too, the labels of elements of A are unique, and we can again reason exactly as in case 1. □

For the second direction of the history lemma, we have to use the assumption that the indeterminism in W is finite.[12]

Fact 8.6. *Let $\langle W, <, e^*, E, C \rangle$ be a BST_{92} surface structure. Let $A' \subseteq W'$ be maximal directed. Then there is some $h \in \mathrm{Hist}(W)$ and some λ matching h such that $A' = \varphi_\lambda(h)$.*

---

[12] It is instructive to see how infinite structures can cause trouble here. If, for example, there is an infinite chain of choice points from C that has no maximum in the set called A in the proof below, it may be impossible to find a history containing all of A that matches a given instruction set λ.

Proof. Let $A =_{df} \{a \in W \mid \langle a,L \rangle \in A'\}$. As in the proof of Fact 8.5, using the definition of the ordering $<'$ we can establish that A is directed.

We show that (*) there is some $a^* \in A$ for which any $h^* \in H_{a^*}$ contains all of A. Consider the set of past cause-like loci for members of A,

$$B =_{df} \bigcup_{a \in A} \{c \in W \mid c < a \text{ and there is some } h \text{ for which } h \perp_c H_a\}.$$

Note that B contains only indeterministic events, so by the finiteness of indeterminism in W, B is finite of size N. For each of the finitely many $c_i \in B$, we can pick some $a_i \in A$ for which $c_i < a_i$ ($i = 1,\ldots,N$). At this point we also tend to a subtlety concerning the labels of elements of A', as our sought-after element $a^* \in A$ must admit a label that is appropriate for all of A. So if A' contains some element $\langle a, L \rangle$ with $L \neq \emptyset$, then this $L \in \mathfrak{I}_g$ must be unique in the directed set A' by the definition of the ordering, and so we let $a_0 =_{df} a$. Otherwise, all elements of A' have the label $\emptyset$, and a_0 is not needed; for a uniform construction, we simply set $a_0 =_{df} a_1$. The element a_0 thus keeps track of which labels occur in A'. As A is directed, there is some $a^* \in A$ for which all the finitely many $a_i < a^*$ ($i = 0,\ldots,N$), and thereby $c_i < a^*$ ($i = 1,\ldots,N$). Now pick some $h_A \in H_{[A]}$; such a history exists as A is directed. Consider an arbitrary $h^* \in H_{a^*}$. Note that by the choice of a^*, for any $c_i \in B$ we have $h^* \equiv_{c_i} h_A$ because $a_i \in h^* \cap h_A$ and $c_i < a_i$. We claim that $A \subseteq h^*$. Assume not, then there must be some $a \in A \setminus h^*$, whence $a \in h_A \setminus h^*$. By PCP, there must then be some $c < a$ for which $h^* \perp_c h_A$ and in fact $h^* \perp_c H_a$, so by the definition of B, $c \in B$. This implies that $c = c_i$ for some $i \in \{1,\ldots,N\}$, but we have established $h^* \equiv_{c_i} h_A$, contradicting $h^* \perp_c h_A$. So indeed, any history h^* that contains a^* contains all of A. Now as $a^* \in A$, we have $\langle a^*, L \rangle \in A'$ with $L = \emptyset$ or $L = \lambda$ for some $\lambda \in \mathfrak{I}_g$. Given the way the element a_0 was picked, in the former case all elements of A must have the label $\emptyset$ in A'; pick some $h^* \in H_{a^*}$. By Fact 8.2(1), we can pick some λ that matches h^*. In the latter case, as $\langle a^*, \lambda \rangle \in W'$, by Fact 8.3 there is some $h^* \in H_{a^*}$ that matches λ. In both cases, we have a history $h^* \in H_{a^*}$ that matches λ, and by (*), that h^* contains all of A.

So by Fact 8.5, the set $A^* =_{df} \varphi_\lambda(h^*)$ is a maximal directed subset of W', and as $A \subseteq h^*$, we have that A^* is a superset of A'. Now as A' is maximal directed, it has to be that $A^* = A'$, i.e., $A' = \varphi_\lambda(h^*)$. □

Given the two above Facts, we have established our history lemma:

Lemma 8.1. *Let $\langle W, <, e^*, E, C \rangle$ be a BST_{92} surface structure. The set $A' \subseteq W'$ is a history in W' (maximal directed) if and only if there is some $h \in \text{Hist}(W)$ and some generic instruction set λ matching h such that $A' = \varphi_\lambda(h)$.*

Proof. The two directions have been shown as Facts 8.5 and Facts 8.6. □

Now we can go on to show that extended structures are in fact BST_{92} structures:

Theorem 8.1. *Let $\langle W, <, e^*, E, C \rangle$ be a BST_{92} surface structure, and let $\langle W', <' \rangle$ be the corresponding extended structure. Then $\langle W', <' \rangle$ is a BST_{92} structure.*

Proof. By Fact 8.4, $\langle W', <' \rangle$ is a dense, strict partial ordering. It remains to prove that there are infima for lower bounded chains and history-relative suprema for upper bounded chains, that Weiner's postulate holds, and that the prior choice postulate, PCP_{92}, is satisfied. Given Lemma 8.1, all these properties of $\langle W', <' \rangle$ can be lifted from the respective properties of $\langle W, < \rangle$. We ask the reader to supply details via Exercise 8.2. □

Note that in the whole series of proofs leading up to Theorem 8.1, apart from being a subset of S, the inner structure of the instruction sets has played no role. This means that we are free to work with more restrictive notions of instruction sets in real applications without having to revisit the whole construction.

8.3.1.1 The possibility of superdeterministic extensions

While motivating the distinction between Nature's choices at $e \in E$ and experimenters' choices at $c \in C$, we pointed to the possibility of superdeterministic extensions. We are now in a position to define them in more formal detail.

The basic idea of a superdeterministic extension of a surface structure is that *all* of the choice points in the structure are taken care of via the instruction sets at e^*. A single choice at e^* determines all the outcomes of all the choice points in its future. Formally speaking, we have such a situation in case $C = \emptyset$ and the event e^* is in the common past of all the choice points ($e^* < E$). One might say that this amounts to taking the experimenters to be part of Nature, so that the instruction sets pertain to their actions as well as to the experimental outcomes. As we remarked, this move is generally not considered to be satisfactory.[13] Here we are only concerned with spelling out what the move amounts to in the BST framework.[14]

In case $C = \emptyset$, the set $\tilde{T}$ of transitions from members of E is the total set of indeterministic transitions in W, $\text{TR}(W) = \tilde{T}$. Accordingly, any member $T \in S$ singles out exactly one history from $\text{Hist}(W)$, which we denote h_T; that history's matching set is again T itself:

$$\text{for } T \in S, \quad H(T) = \{h_T\} \quad \text{and} \quad T_{h_T} = T.$$

Accordingly, any $T \in S$ matches exactly one history, viz., h_T. In this case, each appropriate instruction set should single out exactly one history. This amounts to taking superdeterministic instruction sets to be singletons of members of S:

Definition 8.8 (Superdeterministic instruction set). Let $\langle W, <, e^*, E, C \rangle$ be a BST$_{92}$ surface structure in which $C = \emptyset$. A *superdeterministic instruction set for* $\langle W, <, e^*, E, C \rangle$ is a singleton $\lambda = \{T\}$ with $T \in S$. We write $\mathfrak{I}_s$ for the set of all superdeterministic instruction sets.

As we have already remarked, all of our above results about structure extensions stay in place for restrictions of generic instruction sets, including superdeterministic instruction sets. So, for a BST$_{92}$ surface structure $\langle W, <, e^*, E, C \rangle$ with $C = \emptyset$, the *superdeterministic extension* is well-defined:

Definition 8.9. Let $\langle W, <, e^*, E, C \rangle$ be a BST$_{92}$ surface structure in which $C = \emptyset$, and let $\mathfrak{I}_s$ be its set of superdeterministic instruction sets. The corresponding *superdeterministic extension* is the extended structure $\langle W', <' \rangle$ of Def. 8.7, replacing $\mathfrak{I}_g$ by $\mathfrak{I}_s$.

Given these definitions, we can establish our main theorem about superdeterministic extensions:

Theorem 8.2. *Let* $\langle W, <, e^*, E, C \rangle$ *be a BST$_{92}$ surface structure in which* $C = \emptyset$, *and let* $\langle W', <' \rangle$ *be its corresponding superdeterministic extension. Then* $\langle W', <' \rangle$ *is a BST$_{92}$ structure without MFB in which there is exactly one choice point,* $\langle e^*, \emptyset \rangle$. *Furthermore, there is a bijection between the sets of histories* $\text{Hist}(W)$ *and* $\text{Hist}(W')$.

---

[13] For an interesting dissenting voice, see Adlam (2018), who combines a notion of global determinism with a reassessment of quantum correlations from a perspectivalist point of view.

[14] We are convinced that our formal analysis provides useful material for a discussion of the status of experiments in the free will debate, but we leave this matter to one side here.

Proof. As we noted, $\langle W', <' \rangle$ is a BST$_{92}$ structure by Theorem 8.1. Any history $h \in \mathrm{Hist}(W)$ matches exactly one instruction set $\lambda_h =_{\mathrm{df}} \{T_h\}$, so by Lemma 8.1, the mapping

$$h \mapsto \varphi_{\lambda_h}(h)$$

is a bijection between $\mathrm{Hist}(W)$ and $\mathrm{Hist}(W')$. Note that $E \neq \emptyset$, so the set S has at least two members. Accordingly, W and W' have at least two histories. Let $h'_1, h'_2 \in \mathrm{Hist}(W')$ with $h'_1 \neq h'_2$; by the previous observation, we must have $h'_i = \varphi_{\lambda_i}(h_i)$ for $h_i \in \mathrm{Hist}(W)$ and $\lambda_i = \{T_{h_i}\}$ ($i = 1, 2$), $\lambda_1 \neq \lambda_2$. By the definition of the ordering, the point $\langle e^*, \emptyset \rangle$ is maximal in $h'_1 \cap h'_2$; that is, a choice point for the arbitrary histories $h'_1, h'_2 \in \mathrm{Hist}(W')$. As W' contains only one choice point, there can be no MFB in W'. $\qquad\square$

The upshot of this result is simple: if a quantum correlation experiment is modeled in such a way that instruction sets (deterministic hidden variables) are allowed to determine not just measurement outcomes but also all of the experimenters' choices ($C = \emptyset$), then one can replace the surface structure, no matter how many cases of modal funny business it contains, with a very simple structure that represents the same surface facts (has exactly isomorphic histories) while pulling together all indeterminism to a single point in the past of E. This result is not surprising, but it is good as a reality check for our formal approach before we embark on a more general discussion of splittings induced by instruction sets.

8.3.1.2 Splitting in extended structures: The general case

In the general case, $C \neq \emptyset$, and we need to characterize the splitting of histories in an extended structure in relation to the splitting of histories in the initially given surface structure. Our aim was that an extended structure should be instructed to behave in some specific way toward the members of E (Nature's choices should be guided by the instruction set: ideally, Nature will have no choice at all apart from the splitting at e^*), whereas the indeterminsm outside of E should be retained (experimenters should still be free to choose measurement settings at choice points $c \in C$, no matter which instruction set for Nature has been given at e^*). The lemma below shows to what extent these aims can be accomplished for *generic* instruction sets. This will also provide helpful guidance toward the construction of practically useful notions of instruction sets later on.

Lemma 8.2. *We consider a pair of histories in W and a corresponding pair of histories in W', as provided by Lemma 8.1: Let $h_1, h_2 \in \text{Hist}(W)$, let $\lambda_1, \lambda_2 \in \mathfrak{I}_g$ be instruction sets, let λ_i match h_i, and set $h'_i =_{df} \varphi_{\lambda_i}(h_i)$ $(i = 1, 2)$. The splitting of histories in W and in W' can be characterized as follows:*

1. *As to splitting at e^*, we have $h_1 \equiv_{e^*} h_2$, and we have $h_1 \equiv_{\langle e^*, \emptyset \rangle} h_2$ iff $\lambda_1 = \lambda_2$.*
2. *Let $c \not> e^*$ and $c \ne e^*$ (which implies $c \notin E$). We have $h_1 \equiv_c h_2$ iff $h'_1 \equiv_{\langle c, \emptyset \rangle} h'_2$.*
3. *Let $c \notin E$ and $c > e^*$ and assume $\lambda_1 = \lambda_2 = \lambda$. Then $h_1 \equiv_c h_2$ iff $h'_1 \equiv_{\langle c, \lambda \rangle} h'_2$.*
4. *Let $c \in E$ (hence $c > e^*$) and $\lambda_1 \ne \lambda_2$. If $h_1 \perp_c h_2$, then neither $h'_1 \perp_{\langle c, \lambda_1 \rangle} h'_2$ nor $h'_1 \perp_{\langle c, \lambda_2 \rangle} h'_2$, but $h'_1 \perp_{\langle e^*, \emptyset \rangle} h'_2$.*
5. *Let $c \in E$, $\lambda_1 = \lambda_2 = \lambda$, and assume that $\bigcup \lambda$ contains no blatantly inconsistent transitions with initial c. Then it cannot be that $h_1 \perp_c h_2$, nor that $h'_1 \perp_{\langle c, \lambda \rangle} h'_2$.*
6. *Let $c \in E$, $\lambda_1 = \lambda_2 = \lambda$, and assume that $\bigcup \lambda$ contains two blatantly inconsistent transitions $c \rightarrowtail \Pi_c \langle h_1 \rangle \in T_1$, $c \rightarrowtail \Pi_c \langle h_2 \rangle \in T_2$, with $T_1, T_2 \in \lambda$. Then $h_1 \perp_c h_2$, and $h'_1 \perp_{\langle c, \lambda \rangle} h'_2$.*
7. *Let $c \in E$, $\lambda_1 = \lambda_2 = \lambda$, and $h_1 \equiv_c h_2$. Then $h'_1 \equiv_{\langle c, \lambda \rangle} h'_2$.*

Proof. (1) By definition e^* is deterministic, so $h_1 \equiv_{e^*} h_2$. And by the definition of the ordering, if $\lambda_1 \ne \lambda_2$, there is no $e > e^*$ and no label L for which $\langle e, L \rangle \in h'_1 \cap h'_2$. On the other hand, $h_1 \equiv_{e^*} h_2$ implies that there is $e \in h_1 \cap h_2$ with $e^* < e$. So if $\lambda_1 = \lambda_2 = \lambda$, we have $\langle e, \lambda \rangle \in h'_1 \cap h'_2$, and hence $h'_1 \equiv_{\langle e^*, \emptyset \rangle} h'_2$.

(2) Let $c \not> e^*$, so $c \in h_i$ iff $\langle c, \emptyset \rangle \in h'_i$ $(i = 1, 2)$. If $c \notin h_1 \cap h_2$, both undividedness claims are trivially false, so assume $c \in h_1 \cap h_2$, which implies that we have $\langle c, \emptyset \rangle \in h'_1 \cap h'_2$ as well. We have to show that $h_1 \equiv_c h_2$ iff $h'_1 \equiv_{\langle c, \emptyset \rangle} h'_2$.

"$\Rightarrow$": Assume $h_1 \equiv_c h_2$ but $h'_1 \not\equiv_{\langle c, \emptyset \rangle} h'_2$, which by the above implies $h'_1 \perp_{\langle c, \emptyset \rangle} h'_2$. By $h_1 \equiv_c h_2$, there is $a \in h_1 \cap h_2$ above c. If $a \not> e^*$, then $\langle a, \emptyset \rangle \in h'_1 \cap h'_2$ and $\langle c, \emptyset \rangle <' \langle a, \emptyset \rangle$, contradicting $h'_1 \perp_{\langle c, \emptyset \rangle} h'_2$. If $a > e^*$, we will find some $a' > c$ for which $a' \not> e^*$. To this end consider a maximal chain l containing c and a. Then $l' =_{df} \{ x \in l \mid x > e^* \}$ is a non-empty chain (it contains a) that is lower bounded by c, so it has an infimum i. By density, there is some a' for which $c < a' < i$, and $a' \not> e^*$. Thus, $\langle a', \emptyset \rangle \in h'_1 \cap h'_2$ and $\langle c, \emptyset \rangle <' \langle a', \emptyset \rangle$, contradicting $h'_1 \perp_{\langle c, \emptyset \rangle} h'_2$.

"$\Leftarrow$": Assume $h'_1 \equiv_{\langle c,\emptyset \rangle} h'_2$ but $h_1 \not\equiv_c h_2$. The latter implies $h_1 \perp_c h_2$. The former must be witnessed by some $\langle a,L \rangle \in h'_1 \cap h'_2$, where $\langle c,\emptyset \rangle <' \langle a,L \rangle$. Hence, by the definition of the ordering, $a \in h_1 \cap h_2$ and $c < a$, which contradicts $h_1 \perp_c h_2$.

(3) This is shown exactly as (2). We show that if $c \notin h_1 \cap h_2$, then the equivalence holds trivially (each side is false). Next we show that $c \in h_1 \cap h_2$ iff $\langle c,\lambda \rangle \in h'_1 \cap h'_2$. Finally, we run two reductio arguments: (1) assume $h_1 \equiv_c h_2$ but $h'_1 \not\equiv_{\langle c,\emptyset \rangle} h'_2$ and (2) assume $h'_1 \equiv_{\langle c,\emptyset \rangle} h'_2$ but $h_1 \not\equiv_c h_2$. Both assumptions lead to a contradiction, as in (2).

(4) Let $c > e^*$ and $\lambda_1 \neq \lambda_2$. Assume $h_1 \perp_c h_2$. Note that h'_1 and h'_2 contain different copies of c, $\langle c,\lambda_1 \rangle \neq \langle c,\lambda_2 \rangle$, so the presupposition for being undivided ($h'_1 \equiv_{\langle c,\lambda_i \rangle} h'_2$) and for splitting ($h'_1 \perp_{\langle c,\lambda_i \rangle} h'_2$) is violated, so none of these can hold ($i = 1,2$). In this case, by the definition of the ordering, $\langle e^*,\emptyset \rangle$ is maximal in the intersection of h'_1 and h'_2, i.e., $h'_1 \perp_{\langle e^*,\emptyset \rangle} h'_2$.

(5) Assume for reductio that $h_1 \perp_c h_2$. This implies that $\tau_1 =_{\mathrm{df}} (c \rightarrowtail \Pi_c \langle h_1 \rangle) \neq \tau_2 =_{\mathrm{df}} (c \rightarrowtail \Pi_c \langle h_2 \rangle)$. But as h_1 and h_2 both match λ by assumption and $c \in h_1 \cap h_2$, it has to be that $\tau_1 \in \bigcup \lambda$ (by matching h_1) and $\tau_2 \in \bigcup \lambda$ (by matching h_2), which contradicts the assumption that $\bigcup \lambda$ contains no blatantly inconsistent transitions with initial c. The assumption that $h'_1 \perp_{\langle c,\lambda \rangle} h'_2$ can be dealt with via that former case: It cannot be that $h_1 \cap h_2$ contains some $a > c$, for then $\langle a,\lambda \rangle > \langle c,\lambda \rangle$, showing $h'_1 \equiv_{\langle c,\lambda \rangle} h'_2$. So $h_1 \perp_c h_2$, and we continue as above.

(6) As λ matches both h_1 and h_2, the existence of such blatantly inconsistent transitions implies $h_1 \perp_c h_2$. Thereby we have $\langle c,\lambda \rangle \in h'_1 \cap h'_2$. Now it cannot be that $h'_1 \equiv_{\langle c,\lambda \rangle} h'_2$: this would imply that there is $\langle a,\lambda \rangle \in h'_1 \cap h'_2$ for which $\langle a,\lambda \rangle > \langle c,\lambda \rangle$, so that there is $a \in h_1 \cap h_2$ with $a > c$, contradicting $h_1 \perp_c h_2$. Thus, $h'_1 \perp_{\langle c,\lambda \rangle} h'_2$.

(7) As $h_1 \equiv_c h_2$, there is $c_1 > c$ such that $c_1 \in h_1 \cap h_2$. Hence $\langle c,\lambda \rangle < \langle c_1,\lambda \rangle$ and $\langle c_1,\lambda \rangle \in h'_1 \cap h'_2$. Thus, $h'_1 \equiv_{\langle c,\lambda \rangle} h'_2$. $\qquad\square$

Let us summarize these results in plain English. The biggest change that a structure extension brings concerns e^*. The event e^* is deterministic in the original structure by assumption, but its counterpart $\langle e^*,\emptyset \rangle$ is a new seed of Nature's indeterminism in the extended structure, with its elementary outcomes playing the role of instruction sets (clause 1). Next, there are no changes with respect to choices in the region not above e^* (excluding e^* itself), by clause (2). As noted, such choices cannot involve members of E, as $e^* < E$. Together with clause (3), clause (2) implies that choices outside of E are preserved with respect to all appropriate instruction sets λ. Note that counterparts of clauses (2) and (3) also hold for the splitting relations $\perp_c$ and $\perp_{\langle c,L \rangle}$. Clause (4) shows that the structure extension leads to different copies of members of E for different instruction sets, placing the respective splitting at $\langle e^*, \emptyset \rangle$. Clauses (5) and (6) point to an important distinction that we will spell out below, when we specify two useful notions of instruction sets, viz., contextual vs. non-contextual ones. The difference between them lies exactly in whether $\bigcup \lambda$ is allowed to contain blatantly inconsistent pairs of transitions (contextuality) or not (non-contextuality). Clauses (5) and (6) inform about the consequences. They say that in the absence of contextuality (in the absence of blatantly inconsistent pairs of transitions with a given initial c in $\bigcup \lambda$), Nature's choice at $c \in E$ is completely removed and replaced by the new splitting at $\langle e^*, \emptyset \rangle$ (clause (5)), while splittings at a member $c \in E$ are retained in case an instruction set λ does not give a unique verdict for what has to happen at c.

Earlier, we defined the notion of C/E-independence for BST_{92} surface structures (Def. 8.3). The guiding idea was that experimenters' choices (outcomes of choice points $c \in C$) should be independent of the outcomes of Nature's choices at choice points $e \in E$. A BST_{92} surface structure is C/E-independent iff there is no modal funny business involving outcomes of both members of C and members of E. With respect to the splitting in the extended structure, a similar question can be asked: are there modal correlations involving Nature's choice of an instruction set at the new splitting point $\langle e^*, \emptyset \rangle$, or Nature's remaining choices above e^*, and sets of experimenters' choices? The following definition provides the relevant notion of C/Ext-independence.[15]

Definition 8.10 (C/Ext independence). We say that an extended structure $\langle W', < \rangle$ derived from a BST_{92} surface structure $\langle W, <, e^*, E, C \rangle$ violates C/Ext independence iff there is a case of MFB that involves some C-based and some E-based transitions in one of the following two ways:

(1) There is some $\lambda \in \mathfrak{I}_g$ and a transition

$$\tau_{e^*}^\lambda = \langle e^*, \emptyset \rangle \rightarrowtail \Pi_{\langle e^*, \emptyset \rangle} \langle \varphi_\lambda(h) \rangle, \quad \text{with } h \in H_{e^*},$$

and a consistent, non-empty set of transitions T_C' with initials $\langle c, L \rangle, c \in C_0 \subseteq C$ and $L \in \{\emptyset, \lambda\}$, for which $T_C' \cup \{\tau_{e^*}^\lambda\}$ is combinatorially consistent but

---

[15] As with other definitions and results in this general part, we write it out using generic instructions sets for concreteness, but the definition is exactly the same for other types of instruction sets.

inconsistent, thus constituting a case of combinatorial funny business (see Def 5.6).

(2) There is some $\lambda \in \mathfrak{I}_g$, a consistent, non-empty set of transitions T'_E with initials $\langle e, L \rangle$, $e \in E_0 \subseteq E$ and $L \in \{\emptyset, \lambda\}$, and a consistent, non-empty set of transitions T'_C with initials $\langle c, L \rangle$, $c \in C_0 \subseteq C$ and $L \in \{\emptyset, \lambda\}$, for which $T'_C \cup T'_E$ constitutes a case of combinatorial funny business.

We say that the extension *satisfies C/Ext independence* iff it does not violate it.

Here we cannot yet prove any general results about C/Ext-independence, apart from the following triviality:

Fact 8.7. *A superdeterministic extension is C/Ext independent.*

Proof. The only choice point in such an extension is $\langle e^*, \emptyset \rangle$, so there can be no MFB in a superdeterministic extension. □

8.3.2 Non-contextual and contextual instruction sets

So far, we have discussed two extreme cases of structure extensions: generic ones ($\mathfrak{I}_g$), in which there are no constraints on instruction sets at all, and superdeterministic ones ($\mathfrak{I}_s$), in which instructions pertain to *all* choice points in the surface structure, so that an instruction set given out at the new splitting point $\langle e^*, \emptyset \rangle$ amounts to the selection of a single history. We already pointed out that these two extreme cases are not satisfactory from a philosophical point of view. The challenge for a useful notion of instruction sets is to steer between these two extremes and to offer instruction sets that are comprehensive enough to allow for the free choice of experimental parameters while still providing useful guidance for the outcomes of Nature's choice points. In terms of structure extensions, this translates into two demands: (1) to allow for surface structures in which $C \neq \emptyset$ (unlike in superdeterminism) and (2) to provide a determinate outcome at the members of E that occur, given the instruction set λ.

Demand (2) is still a bit vague, and there is a good reason for that, as there appear to be two ways to fulfill it: the already mentioned non-contextual and contextual approaches to instruction sets. In line with the previous notation, we denote the sets of these instruction sets, which will again be sets of subsets of S, as $\mathfrak{I}_n$ and $\mathfrak{I}_c$, respectively. It is easiest to explain non-contextual instruction sets first, as these are more tightly constrained than the properly contextual ones. A non-contextual instruction set λ provides instructions for all or for a part of Nature's choice points $e \in E$, specifying exactly one outcome of e for a transition $\lambda(e) = e \rightarrowtail H$, $H \in \Pi_e$. A contextual instruction set, on the other hand, may specify different outcomes for one and the same choice point $e \in E$, but not arbitrarily: if there is a difference in outcome, there must also be a difference in the measurement context (hence the name).

8.3.2.1 Non-contextual instruction sets

We first define non-contextual instruction sets.

Definition 8.11 (Non-contextual instruction sets). Given a BST_{92} surface structure $\langle W, <, e^*, E, C \rangle$, the set of sets of transitions $\lambda \subseteq S$ is a *non-contextual instruction set for* $\langle W, <, e^*, E, C \rangle$ iff λ is maximal with respect to the conditions that (1) $\bigcup \lambda$ is not blatantly inconsistent and (2) for every consistent set of initials of transitions in $\bigcup \lambda$, the respective set of transitions (which is uniquely determined by (1)) is also consistent. We write $\mathfrak{I}_n$ for the set of all non-contextual instruction sets.

To unpack this definition, by condition (1), any λ can be viewed as a partial function from E into $\tilde{T}$, providing exactly one transition with initial e for every $e \in E_\lambda$, where $E_\lambda \subseteq E$ is the set of initials of transitions in $\bigcup \lambda$. We will write $\lambda(e) = \tau$ to indicate that τ is the unique transition with the initial e that occurs in $\bigcup \lambda$, and we extend this notation to sets of initials, so that for $E_0 \subseteq E_\lambda$, $\lambda(E_0) = T_0$ means that $T_0 \subseteq \bigcup \lambda$ is the unique set of transitions with initials in E_0 that occur in λ. So condition (2) can be written as follows: if $E_0 \subseteq E_\lambda$ is consistent, then $\lambda(E_0) \subseteq \tilde{T}$ is also consistent.

We can characterize condition (1) of Def. 8.11 also in a different way, which paves the way for a generalization to contextual instruction sets. The relevant fact is this:

Fact 8.8. *For $\lambda \subseteq S$, define $H(\lambda) = \bigcup_{T \in \lambda} H(T)$. For a set $\lambda \subseteq S$, $\bigcup \lambda$ is blatantly inconsistent iff there are $h_1, h_2 \in H(\lambda)$ for which $h_1 \perp_e h_2$ for some $e \in E$. Accordingly, $\bigcup \lambda$ is not blatantly inconsistent iff for all $h_1, h_2 \in H(\lambda)$ and for all $c \in W$, if $h_1 \perp_c h_2$, then $c \in C$.*

Proof. "$\Rightarrow$": Assume that $\bigcup \lambda$ is blatantly inconsistent, and let a witnessing pair be $\tau_1 = e \rightarrowtail H_1 \in T_1 \in \lambda$ and $\tau_2 = e \rightarrowtail H_2 \in T_2 \in \lambda$, $H_1 \neq H_2$. Let

$h_1 \in H(T_1)$ and $h_2 \in H(T_2)$; then $e \in h_1 \cap h_2$ and $\Pi_e\langle h_1 \rangle = H_1 \neq H_2 = \Pi_e\langle h_2 \rangle$, i.e., $h_1 \perp_e h_2$.

"$\Leftarrow$": Assume that there are $h_1, h_2 \in H(\lambda)$ for which $h_1 \perp_e h_2$ for some $e \in E$. Then there must be $T_1, T_2 \in \lambda$ for which $(e \rightarrowtail \Pi_e\langle h_i \rangle) \in T_i$ $(i = 1, 2)$, and $\Pi_e\langle h_1 \rangle \neq \Pi_e\langle h_2 \rangle$. As $\bigcup \lambda \supseteq T_1 \cup T_2$, the set $\bigcup \lambda$ contains two different transitions with initial e and is, therefore, blatantly inconsistent. □

This fact says that if a non-contextual instruction set λ is truly counterfactual (that is, if it has at least two different elements), then the difference is due to what happens outside of E: the corresponding histories split at elements of C only. As we will see, contextual instruction sets relax this condition by allowing that the corresponding histories may split at members of C *and* at members of E.

In our discussion of generic instruction sets we remarked that such instruction sets can be built from any consistent set of transitions. A similar, but more specific result also holds for non-contextual instruction sets; it is worth spelling out in detail.

Fact 8.9. *Let $T \subseteq \tilde{T}$ be a consistent set of transitions with initials $E_T \subseteq E$. Then there is some non-contextual instruction set $\lambda \in \mathfrak{I}_n$ for which $E_T \subseteq E_\lambda$ and $\lambda(E_T) = T$.*

Proof. The given T can be extended to a maximal consistent set $T^* \in S$. Note that $\lambda_0 =_{df} \{T^*\}$ fulfills the conditions (1) and (2) of Def. 8.11: (1) holds by construction, and (2) is trivial as T^* is consistent. There is a maximal extension of λ_0, $\lambda \in \mathfrak{I}_n$, which retains conditions (1) and (2). As a superset of λ_0, for λ we have $E_\lambda \supseteq E_{\lambda_0} \supseteq E_T$. And by the choice of λ_0 and by (1), it must be that $\lambda(E_T) = \lambda_0(E_T) = T$. □

Here is another simple fact about non-contextual instruction sets.

Fact 8.10. *Let $\lambda \in \mathfrak{I}_n$. (1) The set of transitions $\bigcup \lambda = \lambda(E_\lambda)$ is downward closed, i.e., if $\tau \in \bigcup \lambda$ and for some $\tau' \in \tilde{T}$ we have $\tau' \prec \tau$, then $\tau' \in \bigcup \lambda$. (2) For $T \in S$, we have $T \in \lambda$ iff $T \subseteq \bigcup \lambda$.*

Proof. (1) This claim follows directly from Def. 8.11, as any $T \in S$ is downward closed by maximality.

(2) "$\Rightarrow$": If $T \in \lambda$, then clearly any element of T is in $\bigcup \lambda$.

"$\Leftarrow$": Let $T \subseteq \bigcup \lambda$, so $\bigcup(\lambda \cup \{T\}) = \bigcup \lambda$. As $\bigcup \lambda$ fulfills the conditions (1) and (2) of Def. 8.11, $\bigcup(\lambda \cup \{T\})$ fulfills these conditions as well, and so $T \in \lambda$ follows by maximality of λ. $\square$

The definition of a non-contextually extended structure follows our template of structure extensions from Section 8.3.1. The definitions of matching and of lifted histories (Defs. 8.5 and 8.6) remain unaltered. We write out the definition of the extended structure for the sake of completeness.

Definition 8.12. Let $\langle W, <, e^*, E, C \rangle$ be a BST_{92} surface structure and let $\mathfrak{I}_n$ be its set of non-contextual instruction sets. The corresponding *non-contextual extension* is the extended structure $\langle W', <' \rangle$ of Def. 8.7, replacing $\mathfrak{I}_g$ by $\mathfrak{I}_n$.

Theorem 8.1 applies and shows that such $\langle W', <' \rangle$ is again a BST_{92} structure. Furthermore, the histories $h \in \mathrm{Hist}(W)$ and $h' \in \mathrm{Hist}(W')$ are related by Lemma 8.1. For the splitting of these histories, clauses (1)–(5) of Lemma 8.2 are relevant; clause (6) is excluded as for $\lambda \in \mathfrak{I}_n$, the set of transitions $\bigcup \lambda$ cannot be blatantly inconsistent by Def. 8.11. This implies that in an extended structure based on non-contextual instruction sets, there can be no MFB between outcomes of members of E any more, independent of such MFB in the surface structure: the respective copies of members of E are no longer choice points. There is, however, a new choice point $\langle e^*, \emptyset \rangle \in W'$, and the crucial question is whether an extended structure will exhibit C/Ext independence. As we will show in our discussion of the GHZ experiment in Section 8.3.3, C/Ext independence can fail even if the surface structure is C/E independent. So, while the extension by non-contextual instruction sets is well-defined for any BST_{92} surface structure, it will not always be satisfactory (see Theorem 8.3).

8.3.2.2   Contextual instruction sets

Contextual instruction sets relax a constraint on the non-contextual ones: the prescribed outcome for some $e \in E$ need not be unique, but may depend on the context provided by outcomes of C. In contrast to generic instruction sets, for which there are no constraints, contextual instruction sets *are*, however, constrained: if they specify different outcomes for some $e \in E$, there must also be a difference in the outcome of some $c \in C$. In the following definition, this is spelled out as the condition of C-splitting.

Definition 8.13 (Contextual instruction sets). Let $\langle W, <, e^*, E, C \rangle$ be a BST_{92} surface structure. We say that $\lambda \subseteq S$ is a *contextual instruction set for* $\langle W, <, e^*, E, C \rangle$ iff λ is maximal with respect to the following condition of *C-splitting*: For any two $T_1, T_2 \in \lambda$, $T_1 \neq T_2$, the T_1- and T_2-histories split at a member of C, i.e.,

$$\forall T_1, T_2 \in \lambda \ [T_1 \neq T_2 \rightarrow \forall h_1 \in H(T_1) \forall h_2 \in H(T_2) \exists c \in C \ [h_1 \perp_c h_2]].$$

We write $\mathfrak{I}_c$ for the set of all contextual instruction sets. We call an instruction set $\lambda \in \mathfrak{I}_c$ *properly contextual* iff it provides inconsistent instructions for at least one $e \in E$, i.e., iff $\bigcup \lambda$ is blatantly inconsistent.

Contextual instruction sets can be constructed starting from any consistent set of transitions. That is, we have the following analogue of Fact 8.9:

Fact 8.11. *Let $T \subseteq \tilde{T}$ be a consistent set of transitions. There is some contextual instruction set $\lambda \in \mathfrak{I}_c$ for which $T \subseteq \bigcup \lambda$.*

Proof. Left as Exercise 8.3. □

It is also clear that contextual instruction sets are downward closed (see Fact 8.10(1)).

Many further remarks on contextual instruction sets parallel our remarks for non-contextual ones. The definition of a contextually extended structure follows our template of structure extensions from Section 8.3.1. The definitions of matching and of lifted histories (Defs. 8.5 and 8.6) remain unaltered. We write out the definition of the extended structure for the sake of completeness.

Definition 8.14. Let $\langle W, <, e^*, E, C \rangle$ be a BST_{92} surface structure and let $\mathfrak{I}_c$ be its set of contextual instruction sets. The corresponding *contextual extension* is the extended structure $\langle W', <' \rangle$ of Def. 8.7, replacing $\mathfrak{I}_g$ by $\mathfrak{I}_c$.

Theorem 8.1 applies and shows that such $\langle W', <' \rangle$ is again a BST_{92} structure. Furthermore, the histories $h \in \text{Hist}(W)$ and $h' \in \text{Hist}(W')$ are related by Lemma 8.1. For the splitting of these histories, all of the clauses (1)–(6) of Lemma 8.2 may be relevant. If there are properly contextual $\lambda \in \mathfrak{I}_c$, then in an extended structure there are choice points, apart from those based on elements of C, of the form $\langle e, \lambda \rangle \in W'$ (besides $\langle e^*, \emptyset \rangle \in W'$), so that the crucial question of C/Ext independence becomes more complex. As we will show in our discussion of the GHZ experiment in Section 8.3.3, it can happen that a surface structure is C/E independent, while the extended structure violates C/Ext independence. So, while the extension by contextual instruction sets is well-defined for any BST_{92} surface structure, it will not always be satisfactory (see Theorem 8.4).

8.3.2.3 On the interrelation of different types of instruction sets

We have provided four different definitions of instruction sets, each singling out a unique set of subsets of S for a given BST_{92} surface structure $\langle W, <, e^*, E, C \rangle$, with the proviso that superdeterministic instruction sets are defined only if $C = \emptyset$. In that case, it turns out that the non-contextual and the contextual instruction sets are singletons of elements of S, thus coinciding with the superdeterministic instruction sets.

Fact 8.12. *Let $\mathscr{W} = \langle W, <, e^*, E, C \rangle$ be a BST_{92} surface structure with $C = \emptyset$, so that E is the set of all choice points in $\mathscr{W}$. Then the superdeterministic, non-contextual, and contextual instruction sets coincide, i.e., $\mathfrak{I}_s = \mathfrak{I}_n = \mathfrak{I}_c$.*

Proof. There can be no contextual instruction set $\lambda \in \mathfrak{I}_c$ that contains more than one member of S, as the condition of external splitting is impossible to fulfill given $C = \emptyset$. Thus, the $\lambda \in \mathfrak{I}_c$ are singletons of elements of S. By Fact 8.8, the same holds for non-contextual instruction sets $\lambda \in \mathfrak{I}_n$. So, any instruction set, superdeterministic, non-contextual, or contextual, is a singleton of an element $T \in S$. $\square$

The corresponding superdeterministic extended structure has already been discussed in Section 8.3.1.1: by Lemma 8.2, it contains just one choice point, $\langle e^*, \emptyset \rangle$, which has as many outcomes as there are histories in the surface structure. Thus, in this superdeterministic extended structure, everything is decided at $\langle e^*, \emptyset \rangle$. By the assumption $C = \emptyset$, there is no agent-induced indeterminism in the surface structure, nor is there such indeterminism in the extended structure.

In the general case, the four notions of instruction sets single out different subsets of S, but they nest in the sense that instruction sets of a more restricted type can be extended to instruction sets of a looser type. The following lemma spells this out in formal detail.

Lemma 8.3 (Nesting of types of instruction sets). *The sets of instruction sets $\mathfrak{I}_s$, $\mathfrak{I}_n$, $\mathfrak{I}_c$, and $\mathfrak{I}_g$ nest in the following way: (1) If $C = \emptyset$, so that $\mathfrak{I}_s$ is defined,* we have $\mathfrak{I}_s = \mathfrak{I}_n = \mathfrak{I}_c \subseteq \mathfrak{I}_g$. (2) For each $\lambda \in \mathfrak{I}_n$, there is some $\lambda' \in \mathfrak{I}_c$ for which $\lambda \subseteq \lambda'$. (3) For each $\lambda \in \mathfrak{I}_c$, there is some $\lambda' \in \mathfrak{I}_g$ for which $\lambda \subseteq \lambda'$.

Proof. (1) The two equalities have been established via Fact 8.12. The inclusion is trivial, as $\mathfrak{I}_g$ is the set of *all* subsets of S.

(2) It is easy to prove that each $\lambda \in \mathfrak{I}_n$ fulfills the condition of C-splitting: As $\bigcup \lambda$ contains no blatantly inconsistent transitions, there can be no T_1, T_2 such that histories $h_1 \in H(T_1)$, $h_2 \in H(T_2)$ split at a member of E at all. (See also Fact 8.8.)

(3) As any $\lambda \in \mathfrak{I}_c$ is a subset of S, λ itself is already a member of $\mathfrak{I}_g$, which comprises *all* subsets of S. □

8.3.3 Instruction sets for GHZ

In this section we put the concept of structure extension to work on a well-known example. We will use it to analyze the three-particle GHZ experiment (Greenberger et al., 1989) in the form presented by Mermin (1990). In this experiment, a source emits triples of particles that fly to remote measurement stations. Each triple is in the quantum state

$$\lambda = 1/\sqrt{2}(|+\rangle|+\rangle|+\rangle - |-\rangle|-\rangle|-\rangle), \tag{8.1}$$

which is a vector in the 8-dimensional Hilbert space $\mathscr{H} = \mathscr{H}_2 \otimes \mathscr{H}_2 \otimes \mathscr{H}_2$. At each station i ($i = 1,2,3$), the experimenter can choose to measure the spin projection on the direction x_i or y_i, where x_i is the direction perpendicular to the plane of flight, whereas y_i is in the plane of flight, perpendicular to the trajectory of the ith particle. For each measurement setting, x_i or y_i, there are two alternative possible outcomes, x_i^+ and x_i^-, or y_i^+ and y_i^-.

Taking the description of the experimenter's choices and alternative measurement outcomes at face value, there are two layers of choice points: the experimenter's choices at the three stations correspond to $C = \{c_1, c_2, c_3\}$, and the measurements at the three stations, with the two possible directions x_i and y_i, correspond to $E = \{x_1, y_1, x_2, y_2, x_3, y_3\}$. The choice of direction is obviously prior to the respective measurements, i.e., $c_i < x_i$ and $c_i < y_i$ ($i = 1,2,3$). We can select a location for e^* that is in the past of all of E, yet *SLR* to all of C, which corresponds to the spatio-temporal arrangement of the particle source. Thus we have specified elements e^*, E, and C for a valid BST_{92} surface structure.

In our BST_{92} analysis we represent the indeterminism in the experiment by transitions, first from an experimenter's indecision c_i at station i to one of the selected settings, x_i, or y_i, and then from the measurement with a given setting, say x_i, to one of its outcomes, x_i^+ or x_i^-. To simplify the notation, we will write c_1^x for the transition $c_1 \rightarrowtail \Pi_{c_1}\langle x_1 \rangle$ and y_2^- for $y_2 \rightarrowtail \Pi_{y_2}\langle y_2^- \rangle$, and so on.[16] So we have the following indeterministic transitions in our structure:

$$c_1^x, c_1^y, \ c_2^x, c_2^y, \ c_3^x, c_3^y \quad \text{(transitions due to agents)};$$
$$x_1^+, x_1^-, y_1^+, y_1^-, \ x_2^+, x_2^-, y_2^+, y_2^-, \ x_3^+, x_3^-, y_3^+, y_3^- \quad \text{(transitions due to Nature)}.$$

Figure 8.2 provides a schematic picture of the two layers of indeterminism.

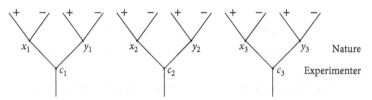

Figure 8.2 Schematic diagram of the GHZ experiment. Choice points c_i are the experimenters' selections of measurement directions x_i or y_i, and x_i (or y_i) are choice points with possible outcomes '+' or '−' determined by Nature ($i = 1, 2, 3$).

To exclude the possibility of physically transmitted causal influences in this experiment, the following *SLR* relations are assumed: (1) between the different experimenter's choices (e.g., $c_1 SLR c_2$), (2) between an experimenter's choice and a remote measurement (e.g., $c_1 SLR x_2$), and (3) between measurements in different stations (e.g., $x_1 SLR y_3$).

So far, there is nothing strange or curious about this experiment. You could think of the source as emitting triples of marbles in the three directions, with the experimenters selecting to measure their size or color, for example. The quantum-mechanical strangeness of the GHZ experiment lies in the fact that only some, but not all triples of *joint* transitions to experimental outcomes such as $\langle x_1^-, y_2^-, y_3^+ \rangle$ are possible, even though all *individual* transitions to the outcomes (such as x_2^-) are possible. Given the

---

[16] That is, in the latter case we use the expression 'y_2^-' both for the transition from initial y_2 to its '−' outcome and for the '−' outcome itself. It will always be clear from context what is meant.

state of Eq. 8.1 and directions x_i, y_i, quantum mechanics assigns non-zero probabilities to 48 out of the 64 possible triples. But for the 16 other triples, quantum mechanics assigns probability zero. In our current modal framework we interpret these probabilistic verdicts as verdicts about possibility and impossibility: triples of the first kind are consistent, while triples of the second kind are inconsistent.[17] Read in this way, quantum mechanics implies that there are modal correlations among the measurement outcomes. These can be described by the following two simple rules:

(yy) A triple with two y's and one x is consistent iff it includes an even number of minuses.

(xxx) A triple with all three x's is consistent iff it includes an odd number of minuses.

The rules imply, among other things, that there is no history that includes the transitions x_1^+, x_2^+, and x_3^+ together. This lacking history indicates combinatorial funny business, as the set of transitions $T =_{df} \{x_1^+, x_2^+, x_3^+\}$ is combinatorially consistent (for the initials x_1, x_2, x_3, we have $x_i SLR x_j$ for each pair $i, j, i \neq j$), but $H(T) = \emptyset$. As we indicated, and as one can check via the rules (yy) and (xxx), MFB in this structure is in fact ubiquitous. In total, 64 combinations of transitions are combinatorially consistent (2^3 combinations of initials by 2^3 outcome combinations), but only 48 of these satisfy both rules, so there are 16 cases of MFB.

Via the above discussion we have defined our BST_{92} surface structure for GHZ:[18]

Definition 8.15. The *surface structure for GHZ* is the BST_{92} surface structure $\langle W, <, e^*, E, C \rangle$ discussed above, in which $C = \{c_1, c_2, c_3\}$, $E = \{x_1, y_1, x_2, y_2, x_3, y_3\}$, and $e^* < E$ and $e^* SLR C$. There are the further SLR relations noted above (e.g., $c_1 SLR c_2$, $c_1 SLR y_2$, $x_1 SLR y_2$). The structure has exactly those 48 histories that satisfy the two given rules (yy) and (xxx).

We can now study structure extensions.

[17] The experiment can also be analyzed probabilistically in BST—see Section 8.4.3 for a BST approach to probabilistic hidden variable models.
[18] In order not to bury the important message of our discussion under a mound of formalism, we do not define the ordering in the structure fully formally, relying instead on the specification of the relevant SLR relations. It is possible to provide a fully formal specification, e.g., as a Minkowskian Branching Structure as discussed in Chapter 9.1.

8.3.3.1 The superdeterministic extension

We mention, only to leave it to the side, the issue of a superdeterministic extension. As $C \neq \emptyset$, our given BST surface structure does not allow for a superdeterministic extension according to Def. 8.8, but it could be modified in the following way: move e^* to the past such that the new $e'^* < C$, and then set $E' =_{df} E \cup C$ and $C' =_{df} \emptyset$. The structure $\langle W, <, e'^*, E', C' \rangle$ is a BST$_{92}$ surface structure that allows for a superdeterministic extension. In that extension, the 48 histories then split at $\langle e'^*, \emptyset \rangle$, and any outcome of $\langle e'^*, \emptyset \rangle$ fixes both the experimenter's choice of settings (x_i or y_i) and the respective measurement outcomes (x_i^+ or x_i^-, for example). This extended structure is philosophically and physically unilluminating: it summarizes which individual runs of the experiment are possible, but it does not describe the experiment as an experiment any more. A useful extension needs to stick with the given surface structure in which the experimenter's choices of measurement directions are formally distinguished from Nature's choices of the measurement outcomes.

8.3.3.2 C/E independence

Before we move to the construction of non-contextual and contextual structure extensions for GHZ, we have to discuss the crucial issue of C/E independence. When specifying the rules (yy) and (xxx), we noted that these rules amount to the introduction of MFB among the measurement outcomes—there are subsets of $\tilde{T}$, such as $\{x_1^+, x_2^+, x_3^+\}$, that are combinatorially consistent but inconsistent. What about MFB involving choice points for the selection of settings c_i and for the selection of measurement outcomes x_j? By the given SLR relations, and as the choice points x_i and y_i are incompatible ($i = 1, 2, 3$), a combinatorially consistent set that includes transitions both from C and from E has to have one of the following schematic forms (where α, β, γ stand for x or y, and m, n stand for $+$ or $-$), leading to a total of 144 sets:

$$c_1^\alpha, c_2^\beta, \gamma_3^m; \quad c_1^\alpha, \beta_2^m, c_3^\gamma; \quad \alpha_1^m, c_2^\beta, c_3^\gamma; \quad \alpha_1^m, \beta_2^n, c_3^\gamma; \quad \alpha_1^m, c_2^\beta, \gamma_3^n; \quad c_1^\alpha, \beta_2^m, \gamma_3^n.$$

That is, such a set either specifies two outcomes for the experimenter's selection of parameters at two of the stations (e.g., c_1^x, c_2^x) and one outcome of a specific measurement at the third station (e.g., x_3^+), or it specifies one outcome of parameter selection (e.g., c_3^x) and two outcomes of specific measurements at the other two stations (e.g., x_1^+,x_2^+). Obviously, by the downward closure of histories, if such a set includes a specific outcome such as x_3^+, this implies the corresponding choice of measurement setting, in this case, c_3^x. By Lemma 5.2, however, such a downward extension preserves both consistency and combinatorially consistency, so we can save the labor of writing out the additional transitions (e.g., we stick to our c_1^x,c_2^x,x_3^+ in place of the longer c_1^x,c_2^x,c_3^x,x_3^+).

Given the structure of the combinatorially consistent sets mixing transitions from C and from E and the structure of the rules (yy) and (xxx), it is easy to see that these mixed sets are all consistent, so that there is no case of MFB among these sets. That is, the GHZ surface model satisfies C/E independence:

Fact 8.13. *The GHZ surface structure defined in Def. 8.15 satisfies the condition of C/E independence of Def. 8.3.*

Proof. We check three exemplary cases. (1) For $T_1 = \{c_1^x,c_2^x,x_3^+\}$, we need to take heed of the (xxx) rule. The set of outcomes x_1^-,x_2^+,x_3^+ satisfies that rule, so there is a corresponding history $h^{x-x+x+} \in H(T_1)$. (2) For $T_2 = \{c_1^x,y_2^y,c_3^y\}$, we need to take heed of the (yy) rule. The set of outcomes x_1^-,y_2^-,y_3^+ satisfies that rule, so there is a corresponding history $h^{x-y-y+} \in H(T_2)$. (3) For $T_3 = \{x_1^+,x_2^-,c_3^x\}$, we again need to take heed of the (xxx) rule. The set of outcomes x_1^+,x_2^-,x_3^+ satisfies that rule, so there is a corresponding history $h^{x+x-x+} \in H(T_3)$. The structural point is that a set that includes transitions both from C and from E always leaves at least one outcome $\pm$ unspecified, and so if one of the rules (yy) or (xxx) applies, that outcome can be chosen accordingly. The upshot is that any combinatorially consistent set of transitions involving both C and E as initials is consistent. □

This fact may raise the hope that an extension of the GHZ surface structure that removes the troublesome cases of MFB could still satisfy the companion notion of C/Ext independence (Def. 8.10). The crucial question is whether this is possible and, if so, for which type of instruction sets.

8.3.3.3 Non-contextual instruction sets for GHZ

The idea of an instruction set is to provide guidance for Nature's choice points $e \in E$ in a way that is independent of the experimenters' choices at $c \in C$. Non-contextual instruction sets are such that an instruction set $\lambda \in \mathfrak{I}_n$ specifies exactly one outcome for each initial in its domain, so that λ can also be read as a partial function from E to $\tilde{T}$. The set-up of GHZ is simple in this respect, because it is layered: the experimenter selects three parameters x_i or y_i ($i = 1, 2, 3$), and then Nature provides one of the possible joint outcomes. The MFB is among the members of $\tilde{T}$ only, as we have just shown (Fact 8.13).

Intuitively, one would expect a non-contextual instruction set to be a function that specifies maximal information, meaning that it specifies one outcome each for each of $x_1, \ldots, y_3$. The sobering fact is that this is not possible in a way that satisfies the GHZ rules.

Fact 8.14. *There is no way to specify one outcome each for the six possible measurements* $x_1, \ldots, y_3$ *in such a way that both rules (yy) and (xxx) are fulfilled.*

Proof. There are eight six-element sets of outcomes that satisfy the (yy) rule: in total, there are $2^6 = 64$ possible combinations, and the (yy) rule cuts this number down by a factor of $2^3 = 8$, as it imposes three parity constraints. We list the eight satisfactory combinations here. The first line gives those for which the outcomes of x_1 and y_1 agree; the second line gives those for which x_1 and y_1 disagree.

$$\{x_1^+ x_2^+ x_3^+ y_1^+ y_2^+ y_3^+\} \ \{x_1^+ x_2^- x_3^- y_1^+ y_2^- y_3^-\} \ \{x_1^- x_2^+ x_3^- y_1^- y_2^+ y_3^-\} \ \{x_1^- x_2^- x_3^+ y_1^- y_2^- y_3^+\}$$
$$\{x_1^+ x_2^- x_3^- y_1^- y_2^+ y_3^+\} \ \{x_1^+ x_2^+ x_3^+ y_1^- y_2^- y_3^-\} \ \{x_1^- x_2^- x_3^+ y_1^+ y_2^+ y_3^-\} \ \{x_1^- x_2^+ x_3^- y_1^+ y_2^- y_3^+\}.$$

As one can see by inspection, none of these eight sets satisfies the (xxx) rule, as the number of minuses on the x-outcomes is always zero or two.   □

Before this background, it is interesting to ask what the non-contextual instruction sets look like for GHZ and which effects it has that these sets are not maximally specific. Our result is that a non-contextual extension for GHZ introduces C/Ext dependence despite the surface structure's C/E independence.

Instruction sets $\lambda \in \mathfrak{I}_n$ are subsets of S maximal with respect to two conditions spelled out in Def. 8.1: (1) such a set must provide a unanimous instruction for each $e \in E$ covered, so that $\bigcup \lambda$ is not blatantly inconsistent, and (2) such a set must actually provide instructions for consistent sets of initials (i.e., if $E_0 \subseteq E_\lambda$ is consistent, then so is the corresponding set of transitions $\lambda(E_0)$).

The set S has 48 members, each corresponding to a history in the surface structure (e.g., $T_0 =_{df} \{x_1^+, x_2^+, x_3^-\} \in S$). It is instructive to see, for example, what happens if one extends T_0 to a non-contextual instruction set λ. We provide one illustration; by Fact 8.14, all alternative attempts at constructing an instruction set will end up in a similar predicament. Consider first, as a warm-up, a construction that is constrained only by condition (1). Basically, the task is to select outcomes for y_1, y_2, and y_3, as the outcomes for the x's are already fixed. So let us try, for example, y_1^+, y_2^-, and y_3^+. The members of S that correspond to this selection are the following:

$$\{x_1^+,x_2^+,x_3^-\}, \{x_1^+,x_2^+,y_3^+\}, \{x_1^+,y_2^-,x_3^-\}, \{y_1^+,x_2^+,x_3^-\},$$
$$\{y_1^+,x_2^+,y_3^+\}, \{y_1^+,y_2^-,x_3^-\}, \{y_1^+,y_2^-,y_3^+\}.$$

What is missing from this list is a set that provides guidance as to the selection of settings x_1, y_2, y_3—the set that would have to be included, as forced by no blatant inconsistency, is $\{x_1^+,y_2^-,y_3^+\}$, but this set violates the (yy) rule, and so there is no history in the GHZ surface structure that includes it.

For the real instruction sets, we have to consider both conditions (1) and (2). Obviously, the resulting λ coming from T_0 is a subset of the set displayed above. The result is

$$\lambda = \{\{x_1^+,x_2^+,x_3^-\}, \{x_1^+,x_2^+,y_3^+\}, \{y_1^+,x_2^+,x_3^-\}, \{y_1^+,x_2^+,y_3^+\}\}. \quad (8.2)$$

For example, the set $\{x_1^+,y_2^-,x_3^-\}$, which is the third one displayed above, cannot be included because given the two previous sets $\{x_1^+,x_2^+,x_3^-\}$ and $\{x_1^+,x_2^+,y_3^+\}$, adding it violates condition (2): the set of initials $\{x_1,y_2,y_3\}$ would otherwise be a consistent subset of the set of initials of these three sets of transitions, and the corresponding set of transitions, $\{x_1^+,y_2^-,y_3^+\}$, violates the (yy) rule. The reader is invited to check that the given λ cannot be extended while still satisfying both conditions (1) and (2); see Exercise 8.5.

Note that when viewed as a partial function, λ is defined on five out of the six members of E, viz., x_1, x_2, x_3, y_1, y_3. By Fact 8.14, that is the maximum domain for a consistent partial function. Lacking an outcome instruction for y_2, our displayed instruction set λ does not provide instructions for four out of the eight possible selections of settings. Thus, the extended structure violates C/Ext independence: if Nature selects our instruction set λ at $\langle e^*,\emptyset \rangle$, then the experimenter must be prohibited, among other things, from choosing the parameters y_1, y_2, and y_3. At any rate, the choice of y_2

must be prevented somehow. Formally, to show how our definitions apply, we can spell out this result in the form of the following "no go" Theorem:

Theorem 8.3. *There are no non-contextual hidden variables for the GHZ experiment.*

More precisely: For the GHZ surface structure $\langle W, <, e^*, E, C \rangle$ defined in Def. 8.15, which is C/E-independent, the extension by non-contextual instruction sets $\mathfrak{I}_n$ results in an extended structure $\langle W', <' \rangle$ that violates C/Ext independence (Def. 8.10).

Proof. The C/E-independence of the surface structure has been shown via Fact 8.13. To prove C/Ext dependence formally, consider the non-contextual instruction set of Eq. (8.2) discussed above,

$$\lambda = \{\{x_1^+, x_2^+, x_3^-\}, \{x_1^+, x_2^+, y_3^+\}, \{y_1^+, x_2^+, x_3^-\}, \{y_1^+, x_2^+, y_3^+\}\}.$$

The set of transitions in W' (where we abbreviate the outcomes in a mnemonic way)

$$T = \{\langle e^*, \emptyset \rangle \rightarrowtail H^\lambda, \langle c_1, \emptyset \rangle \rightarrowtail H^y, \langle c_2, \emptyset \rangle \rightarrowtail H^y, \langle c_3, \emptyset \rangle \rightarrowtail H^x\}$$

is combinatorially consistent, as all initials are pairwise *SLR*, but it is inconsistent, as no history h matching λ includes the transitions c_1^y, c_2^y, and c_3^x. □

That is, the extended structure is conspiratorial, and the process of extending the surface structure turns out to be a cure worse than the disease: from an initially hard-to-understand case of modal funny business without any Nature-experimenter conspiracies, the process of structure extension has created a structure in which the modal funny business between Nature's outcomes is removed at the expense of introducing modal funny business between Nature's single choice and the experimenter's choice of measurement parameters. This is certainly not a satisfactory kind of structure extension.

8.3.3.4 Contextual instruction sets for GHZ

The failure of the non-contextual approach to GHZ motivates the use of contextual instruction sets. These are subsets of S that are less strictly constrained than non-contextual sets, as the instructions given for Nature's choice points may depend on the measurement context. For the GHZ set-up, such a measurement context consists of exactly one choice of direction x or y at each station $i = 1, 2, 3$. As with the previous case of non-contextual instruction sets, it is instructive to check in formal detail what the structure extension turns out to be like.

By Def. 8.13, a contextual instruction set is a subset of S that is maximal with respect to what is there called C-splitting. It is easy to check that the sets of transitions $T_1, T_2 \in S$ are C-splitting exactly if they differ in at least one measurement parameter. Thus, for example, the sets $\{x_1^+, x_2^+, x_3^-\}$ and $\{x_1^+, x_2^+, y_3^+\}$ are C-splitting, but the sets $\{x_1^+, x_2^+, x_3^-\}$ and $\{x_1^-, x_2^+, x_3^+\}$ are not. Any selection of parameters allows for a consistent choice by Nature, as one can see by inspecting the (yy) and (xxx) rules, and there are no further constraints on contextual instruction sets. Thus, a contextual instruction set for GHZ has exactly eight elements, one for each possible combination of parameter choices, $x_1 x_2 x_3, x_1 x_2 y_3, \ldots, y_1 y_2 y_3$. A fortiori, such an instruction set provides information pertaining to all six possible measurement choice points $E = \{x_1, \ldots, y_3\}$. By Fact 8.14, it is impossible to specify one outcome each for all six possible measurements in such a way that the (yy) and (xxx) rules are satisfied. Therefore, any $\lambda \in \mathfrak{I}_c$ has to be properly contextual, i.e., $\bigcup \lambda$ must be blatantly inconsistent. Here is one exemplary $\lambda \in \mathfrak{I}_c$:

$$\lambda = \{\{x_1^-, x_2^-, x_3^-\}, \{x_1^-, x_2^-, y_3^-\}, \{x_1^-, y_2^-, x_3^-\}, \{x_1^- y_2^- y_3^+\},$$
$$\{y_1^+, x_2^-, x_3^-\}, \{y_1^+, x_2^-, y_3^-\}, \{y_1^+, y_2^-, x_3^-\}, \{y_1^-, y_2^-, y_3^+\}\}. \tag{8.3}$$

To see that λ is properly contextual, note that

$$\bigcup \lambda = \{x_1^-, x_2^-, x_3^-, y_1^-, y_2^-, y_3^+, y_3^-\},$$

so that there are two different outcomes prescribed for $y_3 \in E$, depending on the context (compare the second and the fourth elements displayed, $\{x_1^-, x_2^-, y_3^-\}$ and $\{x_1^-, y_2^-, y_3^+\}$). By Lemma 8.2(6), this means that the element $\langle y_3, \lambda \rangle \in W'$ in the contextual extension $\langle W', < \rangle$ remains a choice point.

Now given our λ and looking at the instance of blatant inconsistency just described, one sees that the outcome of y_3 depends on which measurement parameter is chosen at station 2. (Our λ also includes a case of sensitivity for the choice of station 1.) Such a dependence looks as if the structure extension has introduced novel cases of MFB between experimenters' choices and some of Nature's choices, thus showing that the attempt to provide instruction sets for GHZ in the contextual way is also futile. The following "no go" Theorem spells out in formal detail that this is indeed the case.

Theorem 8.4. *There are no contextual hidden variables for the GHZ experiment.*

More precisely: For the GHZ surface structure $\langle W, <, e^*, E, C \rangle$ defined in Def. 8.15, which is C/E-independent, the extension by contextual instruction sets $\mathfrak{I}_c$ results in an extended structure $\langle W', <' \rangle$ that violates C/Ext independence (Def. 8.10).

Proof. The C/E-independence of the surface structure has been shown via Fact 8.13. To prove C/Ext dependence formally, consider the contextual instruction set λ of Eq. (8.3) discussed above and the following set of transitions in W' (where we abbreviate the outcomes in an obvious way):

$$T = \{ \langle c_1, \emptyset \rangle \rightarrowtail H^x, \langle c_2, \emptyset \rangle \rightarrowtail H^x, \langle y_3, \lambda \rangle \rightarrowtail H^+ \}.$$

That set is combinatorially consistent, as all initials are pairwise *SLR*, but it is inconsistent: a history h' including these transitions would have to include $\langle y_3, \lambda \rangle$, so by Fact 8.3, there would have to be some $h \in \text{Hist}(W)$ matching λ for which $h' = \varphi_\lambda(h)$. Furthermore, $\{x_1, x_2, y_3\} \subseteq h$, by the members of T. Now the only history in W that matches λ and which contains these three measurement initials is h^{x-x-y-}, for which the initial y_3 has outcome '$-$'; but by the third element of T, y_3 would have to have outcome '$+$'. $\quad\square$

That is, similarly to the case of non-contextual instruction sets, the structure extended by contextual instruction sets is conspiratorial, and the process of extending the surface structure turns out to be a cure worse than the disease: from an initially hard-to-understand case of modal funny business without any Nature-experimenter conspiracies, the process of structure extension has created a structure in which the modal funny business between Nature's outcomes is removed at the expense of introducing modal funny business between one of Nature's remaining choice points and the experimenter's choice of measurement parameters. Again, this is certainly not a satisfactory kind of structure extension.

8.3.4 Summary of the BST approach to modal structure extensions

In this part of our investigation of quantum correlations and their modeling, we chose an approach that is specific to BST, as it uses the formally well-defined notion of modal funny business to motivate structure extensions, to define them, and to gauge their success.

We provided a template for the general case of structure extensions, showing how these create BST_{92} structures from BST_{92} surface structures. Our framework allowed us to specify exactly what a superdeterministic (or conspiratorial) extension is, and it allowed us to discuss the issue of dependence and independence between experimenters' and Nature's choices in formal detail. We spelled out general conditions of adequacy for structure extensions, which require that such an extension should not introduce novel cases of modal funny business involving choices of the experimenters and outcomes provided by Nature: the objective is to remove surface MFB by (1) modifying choices given by Nature, while (2) leaving experimenters' choices intact.

We defined the well-known approaches of non-contextual and contextual structure extensions in the BST framework. In order to put our definitions to the test, we tackled the GHZ experiment, which is known not to admit sensible hidden variable extensions. We could reproduce these findings in the form of two formally well-specified no go-theorems.

The upshot is that BST allows for structure extensions, both contextual and non-contextual, that are always well-defined. Whether they are satisfactory or not, however, depends on the set-up in question. There are quantum set-ups that exhibit modal correlations that cannot be satisfactorily explained via structure extensions, neither non-contextual nor contextual ones.

8.4 Probabilistic correlations

Our task in the following sections is to investigate whether it is possible to explain *probabilistic* funny business (PFB) by invoking *probabilistic* hidden variables. This task sounds technical, but it concerns a non-technical philosophical problem. We will define the notion of extending BST_{92} structures with probabilistic hidden variables for the explanation of PFB. This notion is required to fulfill a number of intuitive desiderata concerning explanatoriness as well as the modal and spatio-temporal features of the phenomena in question. The technical task of defining structure extensions is motivated by the philosophical question of whether our world fulfils the mentioned desiderata. The literature on Bell-type theorems suggests an answer in the negative, and our analysis will support that verdict. Yet, the negative answer merely marks the point where our real investigation starts. We want to find out exactly which features of our world enforce this negative answer. Given BST, we have at our disposal a mathematically rigorous modal and spatio-temporal framework with gradable possibilities (propensities). Our hope is that this rich framework can deliver a precise answer as to why certain cases of PFB do not allow for explanatory extensions by probabilistic hidden variables. Before we embark on our analysis, we provide a brief discussion of the notion of probabilistic hidden variables.

8.4.1 Probabilistic hidden variables

In our analysis of modal correlations (Section 8.3), surface structures were extended by instruction sets, or deterministic hidden variables. Such instruction sets determine the measurement results, given the measurement settings. In contrast, probabilistic hidden variables work indeterministically: they do not fully determine a result, but instead they have a propensity to bring about a result. In the extreme, there can of course also be the "deterministically looking" propensities zero or one.

If the explanation of a case of PFB by means of probabilistic hidden variables is feasible, one can argue that the PFB is merely epistemic, as it will then be absent at the deepest level at which the hidden variables operate, and arise only at a shallower surface level, at which the distribution of hidden variables is averaged over. In studying this problem, we will use a format parallel to the one used in Section 8.3. We will assume that there is a surface structure, in this case a *probabilistic* BST_{92} surface structure harboring PFB, and we will ask whether this structure can be appropriately extended, so that the original PFB is removed in the resulting extended structure. As we appeal to probabilistic hidden variables, we do not aim to remove chanciness: instead, the aim is rather to remove the probabilistic surface correlations via an explanatory extension.

In contrast to our analysis in Section 8.3, which only made sense for structures in which MFB is present, in what follows we assume no MFB. This is in line with our analysis of probabilistic BST structures in Chapter 7. In particular, we assume NO MFB in the initial BST_{92} structure that represents phenomena with probabilistic funny business. When constructing an initial structure to represent phenomena that appear to lack some joint outcomes, like in the GHZ case discussed in Section 8.3.3, we will nevertheless represent such joint outcomes as possible in our initial BST_{92} structure, but we will assign them propensity zero to represent their absence. This move is very much in line with quantum mechanics, whose actual empirical predictions are always probabilistic.

The obvious aim of an explanatory probabilistic structure extension is to remove PFB. The extended structure, which must not harbor PFB, must of course adequately represent the probabilistic data encoded in the surface structure, including the probabilistic correlations. If an extension is successful, the surface probabilities will be recovered by collecting together the non-problematic probabilities in the extended structure. There are several constraints that limit the range of possible extensions. An important feature of our approach is the distinction between indeterminism resulting from experimental control and Nature's indeterminism, formally represented by sets of choice points C and E in the surface structure (see, e.g., Def. 8.2). Quantum mechanics predicts, and experiments confirm, probabilistic correlations between measurement outcomes, so that a structure extension would mainly operate on elements of E. However, this may also have some side-effects for the behavior of the choice points from C in the extended structure, and not all such side effects will be tolerable. An obvious desideratum is to require that the removal of PFB should not compromise the freedom of the experimenters. Another desideratum of this type is that in the extended structure, the experimental results should be independent from the remote choices of the experimenters—the motivation being that "remote" is identified with space-like relatedness and that faster-than-light signalling is prohibited. In the literature, the first desideratum is spelled out via a condition known as No Conspiracy, whereas the other desideratum goes by the name of Parameter Independence. (There is also a third condition that is frequently used in the literature, Outcome Independence: see Def. 8.22.) The independence that is invoked in these postulates can be interpreted modally or probabilistically, and thus the exact mathematical formulation is far from straightforward. The formal rigor of BST helps to state the conditions precisely and to analyze the proofs of theorems in which they are assumed. See Defs. 8.22 and 8.24 for our formulations.

We will analyze set-ups with differing degrees of complexity. The concrete way to analyze cases of PFB may depend on the complexity (or on our decision to acknowledge the complexity) of the set-up in question. The simplest case of PFB corresponds to an experiment with two fixed measurements performed in space-like separated regions. Such a set-up is represented in BST_{92} by a surface structure with a single case of PFB; that is, with a family of transitions with pairwise SLR initials (see our schematic representation of the EPR set-up, Figure 5.1, p. 107). As no choices of agents are present (there is no selection of alternative measurement settings), the set C of the experimenter-controlled choice points is empty, $C = \emptyset$. The set E, on the other hand, consists of the two SLR choice points corresponding to the initials of the transitions from the measurements to their possible results. It will not be surprising to learn that in such a case, a successful extension of the surface structure by probabilistic hidden variables is always possible.

More complex set-ups involve incompatible measurements, that is, measurements that cannot all occur together. Such set-ups can include the selection of the measurement settings by an experimenter at a choice point $c \in C$. The spatio-temporal ordering brings in yet another dimension of complexity, as there can be several experimenters, and their selections of parameters may be SLR. We may thus have several instances of PFB involving the results of pairwise incompatible experiments. BST represents such a set-up by several families of transitions, with the initials of the transitions from each family being SLR, and there always being some choice points responsible for the incompatibility of the initials of transitions from different families. These choice points can correspond to experimenter-induced selections of measurement settings. Similarly to our analysis in Section 8.3, we then face the task of spelling out, and of checking, the requirement of the independence of experimenters' and Nature's choices, thus making precise the general notion of C/E independence of Def. 8.1.

8.4.2 Extension of a probabilistic surface structure

Generally speaking, to explain probabilistic correlations between remote outcomes in BST, we first need to represent the phenomenon in question in a probabilistic BST surface structure, and then extend this structure in such a way that the extended structure harbors no correlations. In this section we discuss how an initial probabilistic BST_{92} structure, possibly containing probabilistic funny business, can be extended to a structure with a probabilistic hidden variable. As the journey to our target is rather lengthy, we offer a preview.

Starting with an initial BST_{92} structure with a case of PFB (but No MFB), our first task is order-theoretical: we introduce a multiplied structure that corresponds to the initial one, and we show it to be a BST_{92} structure. Apart from having some parts of the initial structure multiplied, this extended structure contains an extra choice point. The outcomes of this choice point together with propensities for these outcomes will serve as values of the hidden variable.[19] Apart from this extra choice point, the extended structure will be shown to be conservative with respect to the relations of branching and undividedness present in the initial structure.

Our guiding idea is that there should be no correlations in the probability space and among the random variables in the extended structure, but these objects have to *correspond* to the PFB-infested probability space and random variables in the surface structure. We discuss the correspondences between various objects in the surface structure and in the extended structure, defining corresponding transitions, corresponding probability spaces, and corresponding random variables.

The notion of correspondence is needed to formulate a restriction on the function μ' that generates the causal probability spaces in the extended structure. We explain what it means that the propensity assignment μ' in the extended structure adequately represents the propensity assignment μ in the surface structure. Finally, we introduce the BST_{92} notion of a structure with a probabilistic hidden variable for PFB that incorporates two desiderata: (1) the adequate representation of probabilities from the surface structure and (2) no correlations among the hidden variables corresponding to the random variables exhibiting PFB in the surface structure.

To give a roadmap of our formal construction, Def. 8.16 introduces probabilistic BST_{92} surface structures, which form a subclass of probabilistic BST_{92} structures specified in Def. 7.3. Definition 8.18 then explains the notion of an N-multiplied structure corresponding to a probabilistic BST_{92} surface structure. At that stage it is not yet decided whether the latter structure fulfils the axioms of BST_{92}. This issue is decided by Theorem 8.5, which says that an N-multiplied structure corresponding to a probabilistic

---

[19] Although there is a certain ambiguity in the literature whether to use the plural, "hidden variables", or the singular, we say "values of a (single) hidden variable" since the outcomes of a choice point are representable by a single random variable.

BST_{92} surface structure *is* a BST_{92} structure. The question whether it is a probabilistic BST_{92} structure is still left open. To handle the latter question, we need the notion of an adequate propensity assignment, which is the subject of Def. 8.20. This definition concerns a probabilistic BST_{92} surface structure (with its propensity function μ) and a corresponding N-multiplied BST_{92} structure. The definition spells out what it takes for μ', defined on the transitions of the latter structure, to be *adequate* for the surface structure and N, the size of multiplication. With the notion of adequacy to hand, our next Lemma 8.5 says that an N-multiplied BST_{92} structure corresponding to a surface structure, taken together with a propensity function adequate for the surface structure and for the size N, is a probabilistic BST_{92} structure. At this stage we finally have established the notion of an N-multiplied probabilistic BST_{92} structure corresponding to a probabilistic BST_{92} surface structure. Our construction ends with Def. 8.22, which singles out, from among the N-multiplied probabilistic BST_{92} structures corresponding to a given surface structure with a case of PFB, the *structures with a probabilistic hidden variable for the given case of PFB*.

We will now develop the above-mentioned formal machinery. First, we define the notion of a probabilistic BST_{92} surface structure:[20]

Definition 8.16. A *probabilistic BST_{92} surface structure* is a sextuple $\langle W, <, \mu, e^*, E, C \rangle$, where $\langle W, <, \mu \rangle$ is a probabilistic BST_{92} structure as in Def. 7.3, $e^* \in W$ is a deterministic point in W, and $E, C \subseteq W$ are disjoint sets of choice points that jointly comprise all choice points in $\mathscr{W}$, $e^* < E$ (i.e., for any $e \in E$, we have $e^* < e$), and in W there are only finitely many choice points, and each one is only finitely splitting.

Exactly as in the modal case discussed in Section 8.3, E represents Nature-given indeterminism, whereas C represents indeterminism related to the experimenters' choices of parameters.

We next introduce the auxiliary notion of a lifted history, which we then use to define an N-multiplied structure.

Definition 8.17 (Lifted history). Let $\mathscr{W} = \langle W, <, \mu, e^*, E, C \rangle$ be a probabilistic BST_{92} surface structure and let $h \in \text{Hist}(W)$. The lifted history $\varphi^n(h)$

---

[20] In contrast to the non-probabilistic surface structures of Def. 8.2, we do not require here that every point in C be below some point in E, as this condition does not play any role in the arguments that follow. The assumption that e^* is deterministic is not essential, but it simplifies the calculations considerably.

for $n \in \mathbb{N}$ is defined as:

$$\varphi^n(h) =_{\mathrm{df}} \{\langle x,0\rangle \mid x \in h \wedge x \not> e^*\} \cup \{\langle x,n\rangle \mid x \in h \wedge x > e^*\}.$$

Note that if $e^* \notin h$, then $\varphi^n(h) = \{\langle x,0\rangle \mid x \in h\} = \varphi^m(h)$ for any $m, n \in \mathbb{N}$.

Definition 8.18 (*N-multiplied structure*). Let $\mathscr{W} = \langle W, <, \mu, e^*, E, C\rangle$ be a probabilistic BST$_{92}$ surface structure. The *N-multiplied structure corresponding* to $\mathscr{W}$ is $\mathscr{W}' = \langle W', <'\rangle$, where

$$W' =_{\mathrm{df}} \bigcup_{h \in \mathrm{Hist}(W), n \in \{1,\ldots,N\}} \varphi^n(h),$$

and the ordering $<'$ is given by

$$\langle x_1, n\rangle <' \langle x_2, m\rangle \Leftrightarrow_{\mathrm{df}} x_1 < x_2 \text{ and } (n = m \text{ or } n = 0),$$

where $n, m \in \{0, 1, \ldots, N\}$.

We need to show next that an N-multiplied structure is a BST$_{92}$ structure. The first step toward this objective is the following lemma:

Lemma 8.4. *Let* $\mathscr{W} = \langle W, <, \mu, e^*, E, C\rangle$ *be a probabilistic BST$_{92}$ surface structure and* $\mathscr{W}'$—*the N-multiplied structure corresponding to* $\mathscr{W}$. *Then*

(1) *For every history* $h \in \mathrm{Hist}(W)$ *the set* $\varphi^n(h)$ *is a maximal directed subset of* W', *i.e., a history in* W'.
(2) *For every maximal directed subset* $A' \subseteq W'$ *there is a history* $h \in \mathrm{Hist}(W)$ *and* $n \in \{1, \ldots, N\}$ *for which* $A' = \varphi^n(h)$.

Proof. See Appendix A.4. □

Our next fact tells us about choice points in the N-multiplied structure corresponding to a probabilistic BST$_{92}$ surface structure.

Fact 8.15. *Let* $\mathscr{W} = \langle W, <, \mu, e^*, E, C\rangle$ *be a probabilistic BST$_{92}$ surface structure, and let* $\mathscr{W}'$ *be the N-multiplied structure corresponding to* $\mathscr{W}$. *Then*

(1) *for every* $n \in \{1, \ldots, N\}$ *and every* $h_1, h_2 \in H_{e^*}$: $\varphi^n(h_1) \equiv_{\langle e^*, 0\rangle} \varphi^n(h_2)$.
(2) *for every* $n, m \in \{1, \ldots, N\}$ *such that* $n \neq m$ *and every* $h \in H_{e^*}$: $\varphi^n(h) \perp_{\langle e^*, 0\rangle} \varphi^m(h)$.

(3) $\langle e^*, 0 \rangle$ is a choice point with N outcomes $\Pi_{\langle e^*, 0 \rangle} \langle \varphi^n(h) \rangle$, where h is an arbitrary history from H_{e^*};

(4) for every $n \in \{1, \ldots, N\}$, every $e \in W$, and every $h_1, h_2 \in \text{Hist}(W)$: $h_1 \perp_e h_2$ iff $\varphi^n(h_1) \perp_{\langle e,l \rangle} \varphi^n(h_2)$, where $l = n$ iff $e^* < e$, and $l = 0$ otherwise;

(5) for every $m, n, l \in \{1, \ldots, N\}$ with $m \neq n$, every $e \in W$ such that $e^* < e$, and every $h_1, h_2 \in \text{Hist}(W)$: neither $\varphi^m(h_1) \equiv_{\langle e,l \rangle} \varphi^n(h_2)$, nor $\varphi^m(h_1) \perp_{\langle e,l \rangle} \varphi^n(h_2)$;

(6) for every $m, n \in \{1, \ldots, N\}$ with $m \neq n$, every $e \not> e^*$, and every $h \in H_e$: $\varphi^m(h) \equiv_{\langle e,0 \rangle} \varphi^n(h)$.

Proof. See Appendix A.4. □

The N-multiplied structure adds to the surface structure a choice point $\langle e^*, 0 \rangle$ with N outcomes $\Pi_{\langle e^*, 0 \rangle} \langle \varphi^n(h) \rangle$, where h is any history from H_{e^*}. Once the propensities are added, these outcomes will play the role of values of a hidden variable. Note the difference between clauses (4) and (5)—the former concerns identical superscripts, and the latter is stated for different superscripts. This reflects the fact that any e above e^* is copied into N "new" events $\langle e, n \rangle$, each occurring in a separate history $\varphi^n(h)$, so that it cannot serve as a choice point (nor a point of undividedness) for histories with different superscripts n, m.

To sum up Fact 8.15, as far as choice points are concerned, the N-multiplied structure adds just one single choice point with N outcomes; whatever is above the designated event e^* in the surface structure is multiplied N times: an l-fold choice point is multiplied into N different l-fold choice points, and a deterministic point is multiplied into N deterministic points. In contrast, events that are not above e^* are just copied, not multiplied.

With these results to hand, in full analogy to Theorem 8.1, we can prove that an N-multiplied structure corresponding to a probabilistic BST$_{92}$ surface structure is again a BST$_{92}$ structure.

Theorem 8.5. Let $\mathscr{W} = \langle W, <, \mu, e^*, E, C \rangle$ be a probabilistic BST$_{92}$ surface structure, and let $\mathscr{W}'$ be the N-multiplied structure corresponding to $\mathscr{W}$. Then $\mathscr{W}'$ is a BST$_{92}$ structure.

Proof. See Exercise 8.6. □

Note that we have not yet established that an N-multiplied structure corresponding to $\mathscr{W}$ is a *probabilistic* BST$_{92}$ structure. This is essential for analyzing *probabilistic* hidden variables. Before we prove the relevant result, we need to introduce a number of auxiliary notions.

The first notion is that of correspondence. We will advance claims that an N-multiplied structure has no PFB in the region *corresponding* to a PFB-infested region of a surface structure. At the end of the day we will claim that in the N-multiplied structure, transitions and random variables that *correspond*, respectively, to troublesome transitions and troublesome random variables in the surface structure, do not exhibit PFB. Toward this end we need to explain correspondences between various kinds of objects in a probabilistic BST$_{92}$ surface structure and in the extended, N-multiplied probabilistic structure. We begin with transitions:

Definition 8.19 (Corresponding transitions and sets thereof). Let $\mathscr{W} = \langle W, <, \mu, e^*, E, C \rangle$ be a probabilistic BST$_{92}$ surface structure, and let $\mathscr{W}'$ be the N-multiplied structure corresponding to $\mathscr{W}$.

Let $\tau = e \rightarrowtail \Pi_e \langle h \rangle \in \mathrm{TR}(W)$, $T \subseteq \mathrm{TR}(W)$, and $S \subseteq \mathscr{P}(\mathrm{TR}(W))$ be, respectively, a basic transition, a set of basic transitions, and a set of sets of basic transitions in $\mathscr{W}$. The corresponding objects, respectively, τ^n, T^n and S^n in $\mathscr{W}'$ are defined as follows, for $n \in \{1, \dots, N\}$:

(1) $\tau^n = \langle e, l \rangle \rightarrowtail \Pi_{\langle e, l \rangle} \langle \varphi^n(h) \rangle$, where $l = n$ iff $e^* < e$, and $l = 0$ otherwise;

(2) $T^n = \{ \tau^n \mid \tau \in T \}$ and

(3) $S^n = \{ T^n \mid T \in S \}$.

As to clause (1), since $\tau = e \rightarrowtail \Pi_e \langle h \rangle$, we have $e \in h$, and hence, with the mentioned caveat about the location of e, $\langle e, l \rangle \in \varphi^n(h)$. Given Fact 8.15(4) it is straightforward to see that τ^n is indeed a basic transition in $\mathscr{W}'$.

Observe that the corresponding basic transitions (and hence sets thereof) are different, depending on the location of their initials. The difference is explained by the following Fact:

Fact 8.16. *Let* $\mathscr{W} = \langle W, <, \mu, e^*, E, C \rangle$ *be a probabilistic BST$_{92}$ surface structure, and let* $\mathscr{W}'$ *be the N-multiplied structure corresponding to* $\mathscr{W}$. *For* $e \in W$, $\tau = e \rightarrowtail \Pi_e \langle h \rangle \in \mathrm{TR}(W)$, *and* $m, n \in \{1, \dots, N\}$ *with* $m \neq n$:

(1) *If* $e^* < e$, *then* $\tau^n = \langle e, n \rangle \rightarrowtail \Pi_{\langle e, n \rangle} \langle \varphi^n(h) \rangle \neq \langle e, m \rangle \rightarrowtail \Pi_{\langle e, m \rangle} \langle \varphi^m(h) \rangle = \tau^m$.

(2) *If* $e^* \not< e$, *then* $\tau^n = \langle e, 0 \rangle \rightarrowtail \Pi_{\langle e, 0 \rangle} \langle \varphi^n(h) \rangle = \langle e, 0 \rangle \rightarrowtail \Pi_{\langle e, 0 \rangle} \langle \varphi^m(h) \rangle = \tau^m$.

Proof. (1) If $e^* < e$, then e can be only associated with $1, 2, \ldots, N$, and $\langle e, m \rangle \neq \langle e, n \rangle$ for $m \neq n$, so $\tau^m \neq \tau^n$. (2) If $e^* \not< e$, e can be only associated with 0; the claim then follows by Fact 8.15(6). □

Thus, a basic transition $(e \rightarrowtail H) \in \mathrm{TR}(W)$ either has N copies or just one copy in $\mathscr{W}'$, depending on whether $e^* < e$ or not. Note that the clauses exclude the case $e = e^*$, for which a deterministic transition $e^* \rightarrowtail H_{e^*}$ gives rise to N basic transitions $\langle e^*, 0 \rangle \rightarrowtail H^n$, where $H^n \in \Pi_{\langle e^*, 0 \rangle}$.

We work toward assigning propensities to transitions in an N-multiplied structure with a view to turning that structure into a *probabilistic* BST$_{92}$ structure. With the concept of correspondence to hand, we define adequate propensity assignments as follows:

Definition 8.20 (Adequate propensity assignment). Let $\mathscr{W} = \langle W, <, \mu, e^*, E, C \rangle$ be a probabilistic BST$_{92}$ surface structure, and let $\mathscr{W}' = \langle W', <' \rangle$ be the N-multiplied BST$_{92}$ structure corresponding to $\mathscr{W}$. Let

$$E_1 = \{e \in W \mid \mu(\{e \rightarrowtail H\}) \text{ is defined for some } H \in \Pi_e\};$$

by Postulate 7.2, μ is defined on any $Y \subseteq \tilde{T}_{E_1}$, and hence on any element of S_{E_1}. We say that $\mu' : \mathscr{P}(\mathrm{TR}(W')) \mapsto [0, 1]$ is *adequate for $\mathscr{W}$ and N* iff

(1) μ' is defined for every basic transition $\langle e^*, 0 \rangle \rightarrowtail H^n$, where $H^n \in \Pi_{\langle e^*, 0 \rangle}$;

(2) μ' is defined on each T^n corresponding to some $T \in S_{E_1}$, and
   - if there is an initial $e \in E_T$ such that $e^* < e$, then
   $\mu(T) = \sum_{n=1}^{N} \mu'(\{\langle e^*, 0 \rangle \rightarrowtail H^n\}) \cdot \mu'(T^n)$;
   - if there is no initial $e \in E_T$ such that $e^* < e$, then $\mu(T) = \mu'(T^n)$;

(3) μ' is defined on every consistent set $\{\langle e^*, 0 \rangle \rightarrowtail H^m\} \cup T^n$ (where $m, n \in \{1, \ldots, N\}$), as follows:

$$\mu'(\{\langle e^*, 0 \rangle \rightarrowtail H^n\} \cup T^m) = \mu'(\{\langle e^*, 0 \rangle \rightarrowtail H^n\}) \cdot \mu'(T^m).$$

The conditions on μ' can be glossed as follows. The first clause requires one to assign values only to the really new transitions (i.e., to transitions having no corresponding transitions in the surface structure). The second clause concerns counterparts of elements of S_{E_1}. Its two parts reflect the difference between corresponding transitions established in Fact 8.16. The first part concerns multiplied transitions, whereas the second concerns non-multiplied transitions (note that in the second part, $T^n = T^m$ for every $m, n < N$). The last clause concerns the mixed case, that is, a set consisting of a new transition and a counterpart of an "old" set (an element of S_{E_1}). In the mixed case, the rule prescribes to multiply the probabilities, as, after all, the indeterministic event $\langle e^*, 0 \rangle$ is posited to explain cases of PFB. Allowing for probabilistic dependence between $\langle e^*, 0 \rangle$ and a counterpart of the "old" set of transitions would contravene this project. The clause is nevertheless controversial. As we will see, it might encode some independence conditions between C and E, and as we will argue in our analysis of the Bell-Aspect experiment in Section 8.4.4, the condition might actually fail, showing that the deep-structure explanation of Bell-Aspect that we are after is not achievable.

Note that, although the conditions of Def. 8.20 concern elements of S_{E_1}, they induce a propensity assignment to other sets corresponding to subsets of $\tilde{T}_{E_1}$ and their combinations with new transitions as well. That is, for every $Y^n \in \mathrm{TR}(W')$ that corresponds to some $Y \subseteq \tilde{T}_{E_1}$:

- $\mu'(Y^n) = \sum_{T \in S_Y} \mu'(T^n)$, where $S_Y =_{\mathrm{df}} \{T \in S_{E_1} \mid Y \subseteq T\}$, and
- if $\{\langle e^*, 0 \rangle \rightarrowtail H^m\} \cup Y^n$ is consistent, then

$$\mu'(\{\langle e^*, 0 \rangle \rightarrowtail H^m\} \cup Y^n) = \mu'(\{\langle e^*, 0 \rangle \rightarrowtail H^m\}) \cdot \mu'(Y^n).$$

Moreover, a μ' that is adequate for $\mathscr{W}$ and N delivers the surface propensities for any $Y \subseteq \tilde{T}_{E_1}$, not just for elements of S_{E_1}. This can be established as follows:

$$\sum_{n \leqslant N} \mu'(Y^n) \mu'(\{\langle e^*, 0 \rangle \rightarrowtail H^n\}) = \sum_{n \leqslant N} \sum_{T \in S_Y} \mu'(T^n) \cdot \mu'(\{\langle e^*, 0 \rangle \rightarrowtail H^n\}) =$$

$$\sum_{T \in S_Y} \sum_{n \leqslant N} \mu'(T^n) \cdot \mu'(\{\langle e^*, 0 \rangle \rightarrowtail H^n\}) = \sum_{T \in S_Y} \mu(T) = \mu(Y).$$

$$(8.4)$$

Observe that adequate μ' is sensitive to details of the surface structure: the values of μ, on which sets of transitions μ is defined, and the number N that gives the size of multiplication. All other features of an N-multiplied structure do not bring anything new, as they are copied from the surface structure. This justifies our terminology of μ' being adequate to $\mathscr{W}$ and N.

As expected, the notion of μ' being adequate for $\mathscr{W} = \langle W, <, \mu, E, C \rangle$ and N provides for the following: if one supplements $\mathscr{W}'$ with μ', creating $\mathscr{W}' = \langle W', <', \mu' \rangle$, then the result will be a *probabilistic* BST$_{92}$ structure. This is what our next lemma establishes: a BST$_{92}$ structure corresponding to a probabilistic BST$_{92}$ surface structure is a *probabilistic* BST$_{92}$ structure in the sense of Def. 7.3.

Lemma 8.5. *Let $\mathscr{W} = \langle W, <, \mu, e^*, E, C \rangle$ be a probabilistic BST$_{92}$ surface structure, and let $\mathscr{W}' = \langle W', <' \rangle$ be the N-multiplied BST$_{92}$ structure corresponding to $\mathscr{W}$. Let the partial function $\mu' : \mathscr{P}(\mathrm{TR}(W')) \mapsto [0, 1]$ be adequate for $\mathscr{W}$ and N. Then $\mathscr{W}'' =_{\mathrm{df}} \langle W', <', \mu' \rangle$ is a probabilistic BST$_{92}$ structure.*

We call $\mathscr{W}''$ an N-multiplied probabilistic BST$_{92}$ structure corresponding to $\mathscr{W}$.

Proof. See Exercise 8.7. □

The requirement of an adequate propensity assignment is part of the notion of an N-multiplied probabilistic BST$_{92}$ structure corresponding to $\mathscr{W}$. Since the requirement is fulfilled by many (partial) functions μ' that are adequate for a given surface probabilistic structure $\mathscr{W}$ and N, there are many N-multiplied probabilistic BST$_{92}$ structures corresponding to $\mathscr{W}$. However, they all recapture the propensities of the surface structure, due to clause (2) of Def. 8.20.

At this stage we have defined probabilistic BST$_{92}$ surface structures and the N-multiplied probabilistic BST$_{92}$ structures corresponding to them. We now define the remaining correspondences we need, first between probability spaces and then between random variables.

Definition 8.21 (Corresponding probability spaces and corresponding random variables). Let $\mathscr{W} = \langle W, <, \mu, e^*, E, C \rangle$ be a probabilistic BST$_{92}$ surface structure, and let $\mathscr{W}' = \langle W', <', \mu' \rangle$ be an N-multiplied probabilistic BST$_{92}$ structure corresponding to $\mathscr{W}$. Let $CPS = \langle S, \mathscr{A}, p \rangle$ be a causal probability space, with $S \subseteq \mathrm{TR}(W)$.

Then a triple $CPS^n = \langle S^n, \mathscr{A}^n, p^n \rangle$ with $n \leqslant N$ is *the n-th causal probability space corresponding to CPS* iff S^n and $\mathscr{A}^n$ correspond to S and $\mathscr{A}$, respectively (in the sense of Def. 8.19), and p^n is induced by the propensity assignment μ' ($n \leqslant N$).

Further, for random variables X and X^n defined within the corresponding probability spaces CPS and CPS^n, respectively, we say that *X and X^n correspond* iff X and X^n have the same range and for every $T \in S$: $X(T) = x$ whenever $X^n(T^n) = x$,

where $T^n \in S^n$ is the set corresponding to $T \in S$.

Note that, typically, corresponding random variables have different probabilities, i.e., $p(X(T) = x) \neq p^n(X^n(T^n) = x)$, because typically μ assigns to T a different propensity than μ' assigns to T^n.

Having all the machinery in place, we can now apply it to a surface structure with a case of PFB. To this end let us first recall our description of PFB, as offered by Def. 7.7:

Given a probabilistic BST_{92} structure, a locus of PFB is provided by pairwise SLR initial events $I_1, \ldots, I_K$, each associated with a transition to an unavoidable disjunctive outcome $\mathbf{1}_k$ with cardinality $\Gamma(k)$. The set of transitions $\{I_1 \rightarrowtail \mathbf{1}_1, \ldots, I_K \rightarrowtail \mathbf{1}_K\}$ is said to exhibit PFB iff the random variables $\{X_1, \ldots, X_K\}$ are correlated. These random variables are defined on the causal probability space $CPS(E \rightarrowtail \mathbf{1}_E) = \langle S, \mathscr{A}, p \rangle$ determined by the transition $E \rightarrowtail \mathbf{1}_E$, where $E = \bigcup_{k=1}^{K} I_k$ and $\mathbf{1}_E = \{\bigcup Z \mid Z \in \mathbf{1}_1 \times \mathbf{1}_2 \times \ldots \times \mathbf{1}_K\}$, by the formula:

$$X_k : S \mapsto \Gamma(k) \text{ where for every } T \in S : X_k(T) = \gamma \text{ iff } CC(I_k \rightarrowtail \hat{O}_\gamma) \subseteq T. \tag{8.5}$$

Having recalled the BST_{92} analysis of PFB, we end our construction by singling out, from among the N-multiplied probabilistic BST_{92} structures corresponding to a given surface structure that includes a case of PFB, a *structure with a probabilistic hidden variable for the given case of PFB.*

Definition 8.22 (Structure with a probabilistic hidden variable for PFB). Let $\mathscr{W} = \langle W, <, \mu, e^*, E, C \rangle$ be a probabilistic BST_{92} surface structure with transitions $\{I_1 \rightarrowtail \mathbf{1}_1, \ldots, I_K \rightarrowtail \mathbf{1}_K\}$, where $\mathbf{1}_k = \{\hat{O}_{k,\gamma(k)} \mid \gamma(k) \in \Gamma(k)\}$, and random variables $X_1, \ldots, X_K$ of Def. 8.5 that exhibit PFB. Let also $cll(I_k \rightarrowtail \hat{O}_{k,\gamma(k)}) \subseteq E$, for every $k \leqslant K$ and every $\gamma(k) \in \Gamma(k)$.

A *structure with a probabilistic hidden variable for the given PFB in $\mathscr{W}$* is an N-multiplied probabilistic BST_{92} structure $\mathscr{W}' = \langle W', <', \mu' \rangle$ (for some $N \in \mathbb{N}$) that corresponds to $\mathscr{W}$ and which satisfies the following condition:

for every $n \leqslant N$, the random variables $X_1^n, \ldots, X_K^n$ corresponding to $X_1, \ldots, X_K$ are independent. [Outcome Independence]

To see that our condition captures the condition of Outcome Independence as it is known from the literature, observe that each random variable $X_k : S \mapsto \Gamma(k)$ represents the results of a measurement I_k; a corresponding random variable X_k^n represents the counterparts of these results in the N-multiplied structure for the n-th value of the hidden variable. The initials $I_1, \ldots, I_K$ are pairwise *SLR*, and so are their images. The statement that the latter random variables are independent means that the probabilities of the images of results in the N-multiplied structure (for each $n \leqslant N$) are independent (i.e., they multiply to yield the probability of the joint result). Observe that the defined notion incorporates two conditions, Outcome Independence and Adequate Propensity Assignment.

Note that the general procedure of N-multiplication can introduce a case of PFB that has no counterpart in a surface structure. This is connected with the fact that the factorization of propensities in the surface structure does not guarantee the factorization of propensities in an N-multiplied structure. That is, $\mu(T_1 \cup T_2) = \mu(T_1) \cdot \mu(T_2)$ does not imply $\mu'(T_1^n \cup T_2^n) = \mu'(T_1^n) \cdot \mu'(T_2^n)$, where T_i and T_i^n are a set of transitions in the surface structure and a set corresponding to it in an N-multiplied structure, respectively. The adequacy condition only requires that the μ-propensity results as the weighed average of the μ' propensities. Whether the notion of N-multiplied BST$_{92}$ probabilistic structure corresponding to a surface structure should prohibit such new cases of PFB, depends on one's attitude as to what the notion is to achieve. If its aim is to *remove* the PFB, the prohibition is very much in place. But if its aim is to *explain* the PFB in the surface structure, the prohibition should not necessarily be imposed, as there might be no way to account for the surface PFB apart from also acknowledging PFB on a deeper level. In what follows, we do not impose the mentioned restriction.

Having explained our framework for the construction of a probabilistic hidden variable for PFB, we finally turn to interesting real questions: Do various set-ups with cases of PFB, of varying complexity, admit structures with probabilistic hidden variables for their cases of PFB?

8.4.3 Single and multiple cases of PFB, and super-independence

Before we investigate BST$_{92}$ structures with PFB, which come with specified sets of experimenter-controlled choice points C and Nature's choice points E, we need to add more substance to the notion of C/E independence.

We introduced C/E independence as a target notion in Def. 8.1, and we already gave it a rigorous formulation in the context of deterministic hidden variables (see Def. 8.3 for surface structures and Def. 8.10 for extended structures). We now have to flesh out the notion in the context of probabilistic hidden variables, by relating it to propensities. For this approach to go through, however, we need to assign propensities to certain sets of transitions that include both an agent-based transition and a Nature-given transition. This move may be controversial, so we try to keep our commitments to propensities of agent-based transitions minimal. In our view, an unproblematic case is a transition $I \rightarrowtail \hat{O}$ from before an experimenter's making a measurement decision to the occurrence a measurement result. The set $CC(I \rightarrowtail \hat{O})$ contains both the experimenter-based transition and a Nature-given transitions. In such a case, when a set T with "mixed" transitions is identical to the set of *causae causantes* of some transition, it may (we think) have a propensity assigned. Once the propensity is assigned to such a set, the notion C/E propensity independence is easily formalized, by equating it with the factorization of propensities. (Note that by Postulate 7.2, if a propensity is defined on a set, it is defined on any of its subsets.) These ideas are reflected by the following definition:

Definition 8.23 (C/E propensity independence). Let $\mathscr{W} = \langle W, <, \mu, E, C \rangle$ be a probabilistic BST_{92} structure with two designated disjoint sets E, C of choice sets such that $E \cup C$ is the set of all choice points in $\mathscr{W}$. We say that $\mathscr{W}$ *violates C/E propensity independence* iff there is a transition Tr in W and $T = CC(Tr)$ with $cll(T) \cap C = C_0 \neq \emptyset$ and $cll(T) \cap E = E_0 \neq \emptyset$ such that $\mu(T) \neq \mu(T_{C_0}) \cdot \mu(T_{E_0})$, where T_{C_0} is the subset of T with initials in C_0, and T_{E_0} is the subset of T with initials in E_0, i.e., $T_{C_0} =_{df} \{(e \rightarrowtail H) \in T \mid e \in C_0\}$ and $T_{E_0} =_{df} \{(e \rightarrowtail H) \in T \mid e \in E_0\}$.

We say that a probabilistic BST_{92} structure (or its set E) *satisfies C/E independence* iff it does not violate it.

In what follows we apply this definition to both probabilistic surface structures and to N-multiplied probabilistic structures; for this reason we have not included e^* in the specification of $\mathscr{W}$.

8.4.3.1 A structure with a single case of PFB

We first investigate a simple set-up that exhibits a single case of PFB and which does not involve any relevant choices of experimenters. We produce a probabilistic BST_{92} surface structure for this set-up and ask if it can be given an N-multiplied BST_{92} probabilistic structure with a probabilistic hidden variable for this case of PFB. The absence of relevant choices of experimenters means that $C = \emptyset$ in the description of the surface structure for this set-up. The following lemma proves that the answer to our question is "yes", given that some minor conditions on the surface structure are satisfied.

Lemma 8.6. *Let* $\mathscr{W} = \langle W, <, \mu, e^*, E, C \rangle$ *be a probabilistic BST_{92} surface structure in which transitions* $\{I_1 \rightarrowtail \mathbf{1}_1, \ldots, I_K \rightarrowtail \mathbf{1}_K\}$ *with random variables* $X_1, \ldots, X_K$ *exhibit PFB, where* $\mathbf{1}_k = \{\hat{O}_{k,\gamma(k)} \mid \gamma(k) \in \Gamma(k)\}$, $\Gamma(k)$ *are index sets and* $1 \leqslant k \leqslant K$. *Let* $cll(I_k \rightarrowtail \hat{O}_{k,\gamma(k)}) \subseteq E$ *for every* $k \leqslant K$ *and every* $\gamma(k) \in \Gamma(k)$ *and* $C = \emptyset$. *Then there exists a structure with a probabilistic hidden variable for this case of PFB. Moreover, the structure satisfies C/E propensity independence.*

Proof. See Appendix A.4.   □

This lemma says that if we focus on a single case of PFB, if only there is a deterministic event below all cause-like loci involved in this case of PFB and our common finitistic assumptions are satisfied, then there is a probabilistic BST_{92} structure that explains away the PFB in question. The result holds no matter what the background of the case of PFB consists of. In the background there might be some other transitions, possibly forming other cases of PFB. And, on some sets of such neglected transitions, μ might be undefined in the surface structure. Still, we can ignore all such complexities, the lemma says, and construct a probabilistic BST_{92} structure explaining away a single case of PFB. The fact that μ might fail to be defined on a set of transitions has an impact, however, on the feasibility of a more general result concerning multiple cases of PFB.[21] Accordingly, in our next result, we assume that μ is defined on all consistent subsets of the set of all basic transitions.

8.4.3.2 A structure with multiple cases of PFB — super-independence

Given the success of the construction of a BST_{92} structure with a probabilistic hidden variable for a set-up with a single instance of PFB and no choices produced by agents, it is natural to ask whether the construction can be successfully applied to set-ups with two or more instances of PFB. This category includes set-ups with and without agent-based selections of parameters. We can prove that all set-ups with finitely many cases of PFB and

---

[21] For starters, attempt to analyze two cases of PFB, one "above" the other, with some transitions "between" the two, that do not form a set to which μ is assigned.

no agent-based choices admit an N-multiplied BST$_{92}$ probabilistic structure in which the instances of PFB are removed. The sought-after N-multiplied structure can be produced in a way that is quite similar to construction of the quasi-deterministic structure investigated in the proof of Lemma 8.6. The difference is that the present lemma concerns *all* basic transitions in the surface structure, in contrast to just those on which μ is defined. The relevant lemma is as follows.

Lemma 8.7. *Let* $\mathscr{W} = \langle W,<,\mu,e^*,E,C \rangle$ *be a probabilistic BST$_{92}$ surface structure harboring multiple cases of PFB for which $C = \emptyset$. Let μ be defined on every subset of* TR(W). *Then there is an N-multiplied probabilistic BST$_{92}$ structure corresponding to $\mathscr{W}$ that provides a hidden variable for every case of PFB in $\mathscr{W}$. Moreover, that extended structure satisfies C/E propensity independence.*

Proof. See Appendix A.4. □

We will call the option of removing *all* cases of PFB, without paying attention to other constraints, "super-independence", in analogy to "super-determinism". Superdeterminism, as investigated in Section 8.3.1.1, is an option for removing all the (finitely many) cases of MFB in a BST$_{92}$ surface structure in which $C = \emptyset$. Theorem 8.2 shows that if our single aim is to remove cases of surface MFB by positing instruction sets, we can always achieve this by assuming that the experimenters' choices are in fact due to Nature ($C = \emptyset$). Similarly, if our single aim is to explain away multiple cases of PFB, we can always achieve this—we just need to assume that no indeterminism comes from experimenters' choices, so that $C = \emptyset$. In this case, any constraints related to members of C are satisfied vacuously. Lemma 8.7 guarantees that such super-independent extensions always exist.

As we said repeatedly, the challenge in constructing hidden variable extensions of surface structures consists in analyzing the set-ups in question *as experiments*; that is, in a way that upholds a separation between Nature's and experimenters' choices ($C \neq \emptyset$). We now turn to the Bell-Aspect experiment,[22] in which experimenters' choices *are* present. It is generally assumed that a good explanation of this experiment should accommodate such choices, and superdeterministic or super-independent accounts are dismissed as conspiratorial. We will show that the existence of an extended

[22] The set-up was proposed in Bell (1964), and a breakthrough experiment with the set-up was carried out by Aspect et al. (1982b).

structure with a hidden variable, given the cases of PFB present in this experiment, implies the so-called Bell-CH inequality. As the Bell-CH inequality is violated by quantum mechanical predictions and in many experiments, we arrive at a "no go" result concerning a propensity-based account of non-local correlations. Our formal derivation of the Bell-CH inequality is standard,[23] but our construction, which focuses on both the modal (propensity) and spatio-temporal aspects, is novel. The thrust of our analysis is an attempt to understand, having support of all the resources of BST, what the premisses for the derivation of the Bell-CH inequality amount to. In this way we hope to contribute to the answering of an ultimately metaphysical question: What must our world be like for the Bell-CH inequality to fail?

8.4.4 The Bell-Aspect experiment

In the last section we developed the technique of structure multiplication. A multiplied surface structure is intended to explain away instances of PFB present in the surface structure. If successful, the construction permits one to interpret PFB epistemically, by claiming that on a deeper level, as described by the N-multiplied structure, there is no PFB. PFB can be seen only on a less than fully fine-grained description of the phenomena; it comes from averaging over the deep level probabilities. We have seen some successes with this program: We proved that any single case of PFB can be explained away by N-multiplication. Also, we showed that, if we ignore agent-based choices as a separate category (subsuming them under the category of Nature-given indeterminism as well), any finite number of cases of PFB can be explained away by the same method. Given the unproblematic development of the technique and its initial successes, the rules of suspense suggest that one should now expect a plot twist. The literature on mysterious quantum correlations additionally enhances the feeling of an imminent catastrophe: an inexplicable set-up with PFB is around the corner.

These feelings are fully justified. In this section we analyze an experiment, the famous Bell-Aspect experiment, for which our N-multiplication technique fails. We will prove, following similar arguments in the literature, that there is no BST_{92} structure with a probabilistic hidden variable for the

---

[23] Our derivation is a BST rendition of J. S. Bell's reasoning sketched in the introduction to his book (Bell, 1987a).

cases of PFB present in this experiment. Acknowledging that our argument is standard may raise doubts as to why to run it again. The argument has a certain set of premises, so presenting it anew may show that some new premises are needed, or, to the contrary, that some standard premises are superfluous. We will, however, not present any such discoveries about premises of this argument. We rerun the argument to see what it takes, in terms of the modal, causal, propensity-like and spatio-temporal features of our world, for the derivation to go through—or, to put it more technically, to see what it takes for the argument's premises to hold. BST is a suitable framework to address these questions, as it offers all the resources needed in an integrated framework and, importantly, it is mathematically rigorous. Thus, it seem well-suited to address the question that Jeremy Butterfield (1992) raised almost three decades ago: "Bell's theorem: what does it take?"

Our plan is thus to run in the BST framework a standard argument for a "no go" theorem for the existence of local hidden variables for the Bell-Aspect set-up, relying on the Bell-CH inequality. We will identify the premises in terms of BST notions, and, using BST resources, we will investigate what these premises amount to. The aim is to shed some light on what the theorem intimates about our world. What must our locally indeterministic and relativistic world be like for the Bell-CH inequality to fail?

8.4.4.1 The set-up of the Bell-Aspect experiment

The set-up of the Bell-Aspect experiment is outlined in Figure 8.1 (p. 226). Here are some more details. A source emits pairs of particles with spin $\frac{1}{2}$, with each pair being in the singlet spin state already mentioned in Chapter 5, written in the basis $|\pm, \pm\rangle = |\pm\rangle_1 \otimes |\pm\rangle_2$ as

$$|\psi\rangle = \frac{1}{\sqrt{2}}(|+, -\rangle - |-, +\rangle).$$

The members of each pair fly in opposite directions toward remote measurement stations (wo)manned by Alice on the left and Bob on the right. For each emission, in the left station there is Alice's selection a of one of the two settings, a_1 or a_2, of her measuring apparatus, and in the right station there is Bob's selection b of one of the two settings, b_3 or b_4, of his measuring apparatus. The measurement with the setting a_i has two possible results, $a_{i,+}$ or $a_{i,-}$; the measurement with the setting b_j has also two possible results, $b_{j,+}$ or $b_{j,-}$. To write quantified formulas, we henceforth assume that the range of i, i' is $\{1,2\}$, the range of j, j' is $\{3,4\}$ and the range of m, m' is $\{+,-\}$. The relevant *SLR* relations are as follows. In each round of the experiment, each selection event is *SLR* to the measurement event in the remote station (i.e., a *SLR* b_j and b *SLR* a_i). Next, outcomes in different stations are *SLR* (i.e., $a_{i,m}$ *SLR* $b_{j,n}$). Also, a measurement event in one station is *SLR* to an outcome of a measurement event in a remote station (i.e., a_i *SLR* $b_{j,m'}$ and b_j *SLR* $a_{i,m}$). Finally, there are non-local correlations: for each pair (i, j), the remote results $a_{i,m}$ and $b_{j,m'}$ are probabilistically correlated. Such non-local correlations, if understood as reflecting underlying propensities, provide evidence for PFB. The Bell-Aspect experiment involves two kinds of chanciness (like the GHZ experiment): experimenter-induced indeterminism in transitions $a \rightarrowtail a_i$ and $b \rightarrowtail b_j$, and Nature-given chanciness to be seen in transitions $a_i \rightarrowtail a_{i,m}$ and $b_j \rightarrowtail b_{j,m'}$.

8.4.4.2 The surface structure for the Bell-Aspect experiment

We turn now to the construction of a probabilistic BST_{92} surface structure representing the Bell-Aspect experiment, $\mathscr{W}_{BA} = \langle W, <, \mu, e^*, E, C \rangle$. W must contain at least 15 events, which we assume to be all point-like. These are: $a, b, a_i, b_j, a_{i,m}, b_{j,m'}, e^* \in W$. The ordering relations, including *SLR* relations and compatibility, are specified as follows: $a < a_i, b < b_j, a_i < a_{i,m}, b_j < b_{j,m'}$, then $\Pi_a\langle a_1 \rangle \neq \Pi_a\langle a_2 \rangle, \Pi_b\langle b_3 \rangle \neq \Pi_b\langle b_4 \rangle, \Pi_{a_i}\langle a_{i,+} \rangle \neq \Pi_{a_i}\langle a_{i,-} \rangle$, $\Pi_{b_j}\langle b_{j,+} \rangle \neq \Pi_{b_j}\langle b_{j,-} \rangle$. Next, a *SLR* b_j, b *SLR* a_i, $a_{i,m}$ *SLR* $b_{j,m'}$, and a_i *SLR* $b_{j,m'}$ and b_j *SLR* $a_{i,m}$. And e^* is *SLR* or below each of a, b but $e^* < a_i$ and $e^* < b_j$. It is easy to calculate that $\mathscr{W}_{BA}$ has 16 histories that can be identified via outcomes $a_{i,m} \cup b_{j,m'}$. The a and b-based transitions are agent-induced, whereas the a_i and b_j-based transitions are thought of as Nature-given. Accordingly, $E = \{a_1, a_2, b_3, b_4\}$ and $C = \{a, b\}$. We may further assume that the structure satisfies C/E propensity independence.[24] The next element, the propensity function μ, is assigned only to those objects for which quantum mechanics offers numerical predictions. In this vein, $\mu(a_i \cup b_j \rightarrowtail a_{i,m} \cup b_{j,m'})$ is identified with the QM probability for the joint

---

[24] For this assumption we need to define propensity on transitions based on subsets of C, however.

outcome $(a_{i,m}, b_{j,m'})$, and $\mu(a_i \rightarrowtail a_{i,m})$ is the QM probability for a single outcome $a_{i,m}$, and analogously for transitions based on b_j.[25]

The ordering relations and the splitting of histories allow us to speak of a number of transitions: $a \rightarrowtail a_i$, $b \rightarrowtail b_j$, $a_i \rightarrowtail a_{i,m}$ and $b_j \rightarrowtail b_{j,m'}$. Although this notation clearly suggests event-like transitions, we use it freely to refer to the corresponding proposition-like transition, canonically written as $a \rightarrowtail \Pi_a\langle a_i \rangle$, $b \rightarrowtail \Pi_b\langle b_j \rangle$, etc. The assumed idealizations imply that all these transitions are basic, so for each transition its set of *causae causantes* consists only of the set in question (i.e., we have identities like $CC(a \rightarrowtail a_i) = \{a \rightarrowtail a_i\}$).

Note that by constructing a surface probabilistic structure in accordance with Def. 8.16, we decide to have just one point event e^* to take care of four cases of PFB present in the set-up. An alternative idea is to postulate four such points, one for each case of PFB. We argue in App. A.4.1 that the option with four events either reduces to our option with one event, or it gives up on explaining PFB.

8.4.4.3 Probabilistic funny business

Our surface structure contains four cases of PFB, and we have to decide whether to follow a "big space" approach, or, alternatively, a "small spaces" approach (see Butterfield 1992). A "big space" approach takes the selection events a and b to construct one "big" causal probability space based on transitions to disjunctive outcomes, $a \cup b \rightarrowtail \{a_i \cup b_j \mid i \in \{1,2\}, j \in \{3,4\}\}$. This approach has a certain mathematical elegance, but disturbingly it assigns propensities to sets like $a \rightarrowtail a_1$. Ideologically, one might oppose to assigning numerical propensities to agent's choices, but even if one does not oppose to such an assignment on ideological grounds, it is unclear where such numbers could come from. We are thus after a "small spaces" approach, which constructs four separate causal probability spaces, each based on transitions $S_{ij} = \{a_i \cup b_j \rightarrowtail \{a_{i,m} \cup b_{j,m'} \mid m, m' \in \{-,+\}\}\}$ (for some fixed i and j). The algebra $\mathscr{A}_{ij}$ of subsets of S_{ij} is induced automatically, and the probabilities

---

[25] As the settings stand for directions of spin projection, the quantum mechanical probability for a joint measurement on a pair in the singlet state is

$$p_{qm}(a_{i,+}, b_{j,+}) = \frac{1}{2}\cos^2\left(\frac{\angle(i,j)}{2}\right),$$

which means that there are no correlations only if the angle $\angle(i,j)$ between polarization directions equals $\pi/2$. Angles in a typical Bell-type experiment are $\angle(1,3) = \frac{2}{3}\pi = \angle(2,4)$, $\angle(1,4) = \frac{4}{3}\pi$ and $\angle(2,3) = 0$. Thus, for each of these angles there is PFB.

p_{ij} are induced by μ, the values of which are in turn dictated by quantum mechanics. Importantly, for small probability spaces, we need no other values of probabilities than those that are ascribed by quantum mechanics to joint and single measurement results. In this "small spaces" approach, experimenters' choices are out of the picture, as transitions like $a \rightarrowtail a_i$ are not in the algebra of subsets of S_{ij}. (Such transitions will nevertheless return in our discussion, when we attempt to justify the assumptions of Bell's theorem or discuss C/E independence.) The result of this construction is that each case of PFB present in the Bell-Aspect set-up is analyzed in a different probability space.

Given a causal probability space $CPS_{ij} = \langle S_{ij}, \mathscr{A}_{ij}, p_{ij} \rangle$, a case of PFB is specified by transitions $a_i \rightarrowtail 1_{a_i}$ and $b_j \rightarrowtail 1_{b_j}$, where $1_{a_i} = \{a_{i,+}, a_{i,-}\}$ and $1_{b_j} = \{b_{j,+}, b_{j,-}\}$. As for the associated random variables, there is a certain subtlety in our notation, since we need to indicate the probability space on which they are defined so as to distinguish X_i as defined on S_{ij} and X_i as defined of $S_{ij'}$. So we use extended subscripts, writing $X_{i,ij}$, to indicate that the random variable in question is defined on the causal probability space CPS_{ij}. With this little notational complication, we assume that every associated random variable has the same range $\Gamma = \{+, -\}$, and we define it in accordance with Def. 7.7: For any $T \in S_{ij}$,

$$X_{i,ij}(T) = m \text{ iff } a_i \rightarrowtail a_{i,m} \in T \text{ and } X_{j,ij}(T) = m' \text{ iff } b_j \rightarrowtail b_{j,m'} \in T. \quad (8.6)$$

By the quantum mechanical probabilities for joint outcomes in the Bell-Aspect experiment, the random variables $X_{i,ij}$ and $X_{j,ij}$ are dependent. Thus, $a_i \rightarrowtail 1_{a_i}$ and $b_j \rightarrowtail 1_{b_j}$, together with the associated random variables $X_{i,ij}$ and $X_{j,ij}$, exhibit PFB.

8.4.4.4 Derivation of the Bell-CH inequality

We now turn to the crucial question: is there a BST_{92} structure $\mathscr{W}'_{BA}$ with a probabilistic hidden variable for all four cases of PFB present in the surface structure $\mathscr{W}_{BA}$? We will run a standard derivation of the Bell-CH inequality[26] within the BST framework, intended to show that the answer is in the negative. To recall the logic of the argument, it attempts to show that if the mentioned structure exists, then the Bell-CH inequality has to hold. However, quantum mechanical predictions, supported by overwhelming

---

[26] See Clauser and Horne (1974); Myrvold et al. (2019).

experimental evidence, show that the inequality is violated. Hence the sought-after structure with a hidden variable for the four cases of PFB does not exist. As our main task is the analysis of the premises of the derivation from the perspective offered by BST, we present the derivation without much ado, to focus later on the justification of its premises.

As to the premises, apart from the conditions of Adequate Propensity Assignment and Outcome Independence that are part and parcel of the definition of a structure with a probabilistic hidden variable for PFB, we assume a constraint known as Parameter Independence, which we define as follows:

Definition 8.24 (Parameter Independence). Let CPS_{ij}^n, $CPS_{ij'}^n$, and $CPS_{i'j}^n$ be causal probability spaces corresponding to spaces CPS_{ij}, $CPS_{ij'}$, and $CPS_{i'j}$, respectively, and $X_{i,\,ij}^n$, $X_{i',\,i'j}^n$, and let $X_{j',\,ij'}^n$ be random variables corresponding to random variables defined by Eq. (8.6). Then we say that $X_{i,\,ij}^n$ and $X_{i,\,ij'}^n$ satisfy *Parameter Independence* iff for every m: $p_{ij}^n(X_{i,\,ij}^n = m) = p_{ij'}^n(X_{i,\,ij'}^n = m)$. Analogously, $X_{j,\,ij}^n$ and $X_{j,\,i'j}^n$ satisfy *Parameter Independence* iff for every m: $p_{ij}^n(X_{j,\,ij}^n = m) = p_{i'j}^n(X_{j,\,i'j}^n = m)$.

We now assume that there is a BST$_{92}$ structure $\mathcal{W}_{BA}'$ with a probabilistic hidden variable for all four cases of PFB present in the surface structure $\mathcal{W}_{BA}$ and satisfying Parameter Independence. In more detail, $\mathcal{W}_{BA}'$ is an N-multiplied probabilistic BST$_{92}$ structure that corresponds to the probabilistic BST$_{92}$ surface structure $\mathcal{W}_{BA}$. Our derivation starts from the arithmetical fact that for any real numbers u, u', v, and v' from the unit interval $[0, 1]$,

$$-1 \leqslant uv + uv' + u'v' - u'v - u - v' \leqslant 0. \tag{8.7}$$

As values of probabilities fall into the unit interval, we next make these substitutions:

$$u = p_{13}^n(X_{1,13}^n = +) \quad u' = p_{23}^n(X_{2,23}^n = +) \tag{8.8}$$
$$v = p_{13}^n(X_{3,13}^n = +) \quad v' = p_{14}^n(X_{4,14}^n = +). \tag{8.9}$$

Then, using Outcome Independence and Parameter Independence a few times, we arrive at the following:

$$uv = p_{13}^n((X_{1,13}^n = +) \wedge (X_{3,13}^n = +)) \tag{8.10}$$

$$uv' = p_{13}^n(X_{1,13}^n = +)p_{14}^n(X_{4,14}^n = +) = p_{14}^n(X_{1,14}^n = +)p_{14}^n(X_{4,14}^n = +) =$$
$$p_{14}^n((X_{1,14}^n = +) \wedge (X_{4,14}^n = +))$$
$$(8.11)$$

$$u'v' = p_{23}^n(X_{2,23}^n = +)p_{14}^n(X_{4,14}^n = +) = p_{24}^n(X_{2,24}^n = +)p_{24}^n(X_{4,24}^n = +) =$$
$$p_{24}^n((X_{2,24}^n = +) \wedge (X_{4,24}^n = +))$$
$$(8.12)$$

$$u'v = p_{23}^n((X_{2,23}^n = +) \wedge (X_{3,23}^n = +)).$$
$$(8.13)$$

Finally, after making the substitutions (8.8)–(8.9) and then processing the transformations (8.10)–(8.13) in inequality (8.7), we multiply the resulting formula side-ways by $\mu'(\langle e^*, 0 \rangle \rightarrowtail H^n)$, and then sum over $n \in \{1, \ldots, N\}$. Since μ' is a propensity function, $\sum_{n=1}^{N} \mu'(\langle e^*, 0 \rangle \rightarrowtail H^n) = 1$. By the condition of Adequate Propensity Assignment (Def. 8.20), we have these identities:

$$\sum_{n \leqslant N} \mu'(\langle e^*, 0 \rangle \rightarrowtail H^n) \cdot p_{ij}^n((X_{i,ij}^n = +) \wedge (X_{j,ij}^n = +))$$
$$= p_{ij}((X_{i,ij} = +) \wedge (X_{j,ij} = +)).$$
$$(8.14)$$

Putting these observations together, we arrive at the Bell-CH inequality:

$$-1 \leqslant p_{13}((X_{1,13} = +) \wedge (X_{3,13} = +)) + p_{14}((X_{1,14} = +) \wedge (X_{4,14} = +)) +$$
$$p_{24}((X_{2,24} = +) \wedge (X_{4,24} = +)) - p_{23}((X_{2,23} = +) \wedge (X_{3,23} = +)) -$$
$$p_{13}(X_{1,13} = +) - p_{14}(X_{4,14} = +) \leqslant 0.$$
$$(8.15)$$

A reader familiar with Bell's theorems might ask where we used the assumption that the values of a hidden variable and the measurement settings are independent (i.e., the so-called No Conspiracy assumption). This assumption is at work in Eq. (8.14), which says that there is a settings-independent probability distribution on outcomes of $\langle e^*, 0 \rangle$. (Note that settings-dependence of μ' would block the derivation.) No Conspiracy, in one formulation, says that "[t]he probability distribution μ' of [a hidden variable] should not be allowed to depend on (a_i, b_j); this is the mathematical meaning of the assumption... that the control parameters a_i, b_j are "randomly and freely chosen by the experimenters" " (Goldstein et al.,

2011).[27] The assumptions of No Conspiracy and Adequate Propensity Assignment point to problem involving a causal discrepancy in the analysis, which we are about to uncover. We proceed with the premises of the derivation, asking the question of what the world must be like in order to justify the premises and the subsequent steps in the above derivation.

8.4.4.5 Analysis of the derivation

As a way toward a BST_{92} analysis of the derivation, recall first that, by the definition of surface probabilistic structures, e^* is below every a_i and b_j. Consider then the causal situation, first in the surface structure $\mathcal{W}_{BA}$, and focus on $CC(e^* \rightarrowtail a_{i,m} \cup b_{j,m'})$. Given the location of e^*, this set must contain $a_i \rightarrowtail a_{i,m}$ and $b_j \rightarrowtail b_{j,m'}$. But does $CC(e^* \rightarrowtail a_{i,m} \cup b_{j,m'})$ contain other transitions as well? Clearly, e^* is not, and cannot be, above a; otherwise, since a is a choice point for measurement outcomes, e^* would prohibit the occurrence of some outcome of a. For an analogous reason, e^* cannot be above b. It follows that e^* is SLR or below each a and b. Hence $a \rightarrowtail a_i$ and $b \rightarrowtail b_j$ must belong to $CC(e^* \rightarrowtail a_{i,m} \cup b_{j,m'})$, i.e.,

$$CC(e^* \rightarrowtail a_{i,m} \cup b_{j,m'}) = \{a_i \rightarrowtail a_{i,m}, b_j \rightarrowtail b_{j,m'}, a \rightarrowtail a_i, b \rightarrowtail b_j\}. \quad (8.16)$$

We turn next to the transition $\langle e^*, 0 \rangle \rightarrowtail (a_{i,m} \cup b_{j,m'})^n$ in $\mathcal{W}'_{BA}$, where $(a_{i,m} \cup b_{j,m'})^n$ is the n-th counterpart in $\mathcal{W}'_{BA}$ of the event $a_{i,m} \cup b_{j,m'}$. Its set of *causae causantes* almost mirrors the above set of transitions, the difference being the inclusion of $\langle e^*, 0 \rangle \rightarrowtail H^n$, where H^n is the n-th elementary outcome of $\langle e^*, 0 \rangle$. To write down the set:

$$CC(\langle e^*, 0 \rangle \rightarrowtail (a_{i,m} \cup b_{j,m})^n) =$$
$$\{\langle e^*, 0 \rangle \rightarrowtail H^n, (a_i \rightarrowtail a_{i,m})^n, (b_j \rightarrowtail b_{j,m'})^n, (a \rightarrowtail a_i)^n, (b \rightarrowtail b_j)^n\}. \quad (8.17)$$

Clearly, this is different from the set

$$CC^* = \{\langle e^*, 0 \rangle \rightarrowtail H^n, (a_i \rightarrowtail a_{i,m})^n, (b_j \rightarrowtail b_{j,m'})^n\}. \quad (8.18)$$

---

[27] We adapted the symbolism in this quote to match the present text.

The difference lies precisely in the C-based (agent-induced) transitions $(a \rightarrowtail a_i)^n$ and $(b \rightarrowtail b_j)^n$, which are members of the first set but not of the second one.

Now, returning to the condition that the N-multiplied structure must reflect the propensities of the surface structure, the underlying idea is that a (non-disjunctive) event $\hat{O}$ in the surface structure is "replaced" by a disjunctive event $\check{O} = \{\hat{O}^n \mid n \leqslant N\}$ in the N-multiplied structure; the propensity of $\mu(e^* \rightarrowtail \hat{O})$ (in the surface structure) is identified with the propensity $\mu'(\langle e^*, 0\rangle \rightarrowtail \check{O})$ (in the N-multiplied structure), the latter being equal to $\sum_{n \leqslant N} \mu'(\langle e^*, 0\rangle \rightarrowtail \hat{O}^n)$, by the properties of propensity functions. In accordance with the central tenet of our theory of propensities, these latter propensities should fully supervene on the propensities of appropriate sets of *causae causantes*. So, we should have: $(*)$ $\mu(e^* \rightarrowtail \hat{O}) = \sum_n \mu'(CC(\langle e^*, 0\rangle \rightarrowtail \hat{O})^n)$.

In the context of the Bell-Aspect experiment, the formula $(*)$ above requires one to calculate $\mu'(\langle e^*, 0\rangle \rightarrowtail (a_{i,m} \cup b_{j,m})^n)$ by using the larger set given by Eq. (8.17), whereas in our derivation of the Bell-CH inequality we used the smaller set specified by Eq. (8.18). That is, in the structure with a probabilistic hidden variable for the PFB in Bell-Aspect set-up, the contribution of the agent-based transitions, $a \rightarrowtail a_i$ and $b \rightarrowtail b_j^n$, is ignored. The discrepancy between the two sets of *causae causantes* shows a gap in our derivation: the condition of Adequate Propensity Assignment does not apply correctly in the context of the Bell-Aspect set-up, as it distorts the causal situation in question. Can one salvage the derivation despite this discrepancy? Although we only reluctantly assign propensities to agent-based transitions, we have to assign propensity to the large set of Eq. (8.17), which includes agent-based transitions, in order to understand what this discrepancy involves.

Let us thus assign a propensity μ' to the mentioned set. By Postulate 7.2, μ' is then defined for all sets obtained by varying i, j, m, m', and n in the mentioned formula. Let us focus on the last step of our derivation, "multiply by μ' and sum", and on Eq. (8.14). To justify this step, μ' needs to factor in a three-fold way:

$$p_{ij}^n((X_{i,ij}^n = m) \wedge (X_{j,ij}^n = m')) =$$
$$\mu'(\langle e^*, 0\rangle \rightarrowtail H^n, (a_i \rightarrowtail a_{i,m})^n, (b_j \rightarrowtail b_{j,m'})^n, (a \rightarrowtail a_i)^n, (b \rightarrowtail b_j)^n) =$$
$$\mu'(\langle e^*, 0\rangle \rightarrowtail H^n) \cdot \mu'((a_i \rightarrowtail a_{i,m})^n, (b_j \rightarrowtail b_{j,m'})^n) \cdot \mu'((a \rightarrowtail a_i)^n, (b \rightarrowtail b_j)^n),$$
$$(8.19)$$

and the following propensities should be constant:

$$
\mu'((a \rightarrowtail a_i)^n, (b \rightarrowtail b_j)^n) = \mu'((a \rightarrowtail a_{i'})^n, (b \rightarrowtail b_{j'})^n) =
$$
$$
= K(n), \text{ with } \sum_{n \leqslant N} K(n) = 1. \tag{8.20}
$$

Significantly, note that the choice points in $\mathscr{W}_{BA}'$ are divided into C and E in such a way that the counterparts of each a and b belong in C, whereas $\langle e^*, 0 \rangle$ and counterparts of each a_i and of each b_j are in E. The former serve as initials of agents-based transitions, whereas the latter are initials of Nature-given transitions. Given this division, C/E independence (see Def. 8.23) implies the following factorization, for every allowable i, j, m, m' and n:

$$
\mu'(\langle e^*, 0 \rangle \rightarrowtail H^n, (a_i \rightarrowtail a_{i,m})^n, (b_j \rightarrowtail b_{j,m'})^n, (a \rightarrowtail a_i)^n, (b \rightarrowtail b_j)^n) =
$$
$$
\mu'((a \rightarrowtail a_i)^n, (b \rightarrowtail b_j)^n) \cdot \mu'(\langle e^*, 0 \rangle \rightarrowtail H^n, (a_i \rightarrowtail a_{i,m})^n, (b_j \rightarrowtail b_{j,m'})^n). \tag{8.21}
$$

We next note that our finitistic version of the Markov Principle permits a further factorization: The premises of the Markov Principle are satisfied as e^* is below each a_i and each b_j in $\mathscr{W}_{BA}$, these ordering relations carry over to $\mathscr{W}_{BA}'$, and the relevant finitistic assumptions hold as well.

$$
\mu'(\langle e^*, 0 \rangle \rightarrowtail H^n, (a_i \rightarrowtail a_{i,m})^n, (b_j \rightarrowtail b_{j,m'})^n) =
$$
$$
\mu'(\langle e^*, 0 \rangle \rightarrowtail H^n) \cdot \mu'((a_i \rightarrowtail a_{i,m})^n, (b_j \rightarrowtail b_{j,m'})^n). \tag{8.22}
$$

Putting the two factorizations together, we see that C/E independence together with the Markov Principle justifies the three-fold factorization of Eq. (8.19), which in turn, taken together with the assumption stated in Eq. (8.20), justifies the last step of the derivation of the Bell-CH inequality (i.e., to multiply by μ' and to sum over). It is important to stress that we needed C/E independence to derive this inequality. Finally, now observe that the constancy of Eq. (8.20) implies Parameter Independence.[28]

---

[28] Given the causal discrepancy described above, the "big space" approach looks more attractive, since it assigns μ-propensity to sets like in Eq. (8.16), and consequently the condition of Adequate Propensity Assignment has a proper causal underpinning. Nevertheless, the problem with obtaining the three-fold factorization (Eq. 8.19) needed for the derivation of the Bell-CH inequality persists.

8.4.4.6 Consequences from our analysis

Let us reflect on the meaning of the above findings. The main message concerns the causal situation of the Bell-Aspect set-up. The set-up includes agent-based transitions, and this has consequences both for the sets of *causae causantes* and for the propensities. Given these agent-based transitions, there is a mismatch between propensities used in the analysis and the sets of *causae causantes* operating in the set-up. The mismatch can nevertheless be bridged by accepting some intuitive-looking postulates. In this context, the idea of C/E independence comes to the fore, which, given that propensities are assigned to agent-based transitions, receives the precise formulation of Def. 8.23. Given the Markov Principle (in our innocuous finitistic formulation), C/E independence justifies the use of settings-independent propensities $\mu'(\langle e^*, 0 \rangle \rightarrowtail H^n)$ in the derivation of the Bell-CH inequality. This settings-independent measure captures the condition of No Conspiracy. There were two more assumptions, Outcome Independence assumed in the notion of a structure with a hidden variable for PFB, and the constancy of propensities of Eq. (8.20), which is tantamount to Parameter Independence. Our postulates, which govern propensities and often have a causal motivation, thus entail the standard premises of Bell's theorem (i.e., Outcome Independence, Parameter Independence, and No Conspiracy). To have a full list of our postulates, we used Outcome Independence, Eq. (8.20) (Parameter Independence), C/E propensity independence, and the Markov Principle. This set arguably gives us a better insight into what is implicated by the failure of the Bell-CH inequality than the standard set of premises, as there are some differences in status between our postulates. On the one hand, we have the high-level claims of C/E independence and the (finitistic) Markov Principle. They have a high-level status as no experiment-based argument for them looks feasible; on the other hand, it is hard, if not impossible, to conceive of a world without them. The high-level status of these two postulates suggests that they should be retained. In contrast, Eq. (8.20) looks more like a testable statement, as it concerns agents' propensities to choose alternative settings in the experiment. Agents can be trained to make unbiased choices, which arguably informs about propensities of the relevant agent-based transitions: they could be made numerically the same. We concede, however, that this observation is not very persuasive, as words like 'testing' and 'training' have limited sense in the deep realm of hidden variables.[29] In any case,

---

[29] A popular argument for accepting Parameter Independence (see Jarrett, 1984; Shimony, 1984), which relates to superluminal communication, can also be rehearsed in the BST framework. Its gist

given the empirical overtones of Eq. (8.20), and assuming that our everyday observations about agents carry over to their working in the realm of hidden variables, our recommendation is to accept it as well. The option that remains is to reject Outcome Independence, which *prima facie* agrees with the majority view. The novelty that BST brings is that a failure of Outcome Independence now concerns propensities. In non-technical language, that failure means that on a deep level, the Nature-given propensities fail to factor. That is, the degree of possibility of a complex happening does not supervene on the degrees of possibility of its components. In contrast, there is independence of agent-based transitions and Nature-given transitions, as expressed via C/E independence. This is the lesson about modalities that Bell's theorem brings, if it is analyzed from a BST perspective. We hope that with this lesson we are somewhat closer to achieving the great task that J. S. Bell (1997, p. 93) once posed.

> I think you must find a picture in which perfect correlations are natural, without implying determinism, because that leads you back to nonlocality. And also ... as far as our individual experiences goes, our independence of the rest of the world is also natural. So the connections have to be very subtle ...

8.5 Exercises to Chapter 8

Exercise 8.1. Prove Fact 8.4 (i.e., show that the generic-extended structure of a given BST_{92} surface structure is a non-empty, dense, strict partial).

Hint: Use Fact 8.3 and the definition of the ordering.

Exercise 8.2. Prove Theorem 8.1 (i.e., show that the generic-extended structure of a given BST_{92} surface structure is a BST_{92} structure).

Hint: Use Lemma 8.1 and the definition of the ordering.

Exercise 8.3. Prove Fact 8.11.

Hint: See Appendix B.8 for a proof based on the finiteness of S.

---

is that, if an observer knows how Eq. (8.20) fails, and knows "in which" hidden variable she is, she can learn in a faster than light way what settings her partner was choosing.

Exercise 8.4. Construct a BST_{92} surface structure $\langle W, <, e^*, E, C \rangle$, a non-contextual instruction set $\lambda \in \mathfrak{I}_n$, and a directed set $A \subseteq W$ with the following property: For every $a \in A$, there is $h_a \in H_a \subseteq \text{Hist}(W)$ such that h_a matches λ, but there is no $h^* \in \text{Hist}(W)$ such that $A \subseteq h$ and h^* matches λ.

Hint: Assume that A contains an infinite chain $\{a_i\}_{i \in \mathbb{N}}$ of binary choice points such that the chain occurs provided each a_i has outcome $+$. Let $E = \{e\}$, with e a binary choice point that is SLR to the chain, and $T = e \rightarrowtail +$. The CFB results from the assumption that any initial finite segment of pluses on the chain is consistent with T, but all the pluses on the chain are not consistent with T. There might be more choice points needed to secure that the construction is a BST structure.

Exercise 8.5. Show that the non-contextual instruction set λ of Eq. (8.2) on p. 259 is in fact maximal.

Exercise 8.6. Prove Theorem 8.5 (i.e., prove that the N-multiplied structure corresponding to a probabilistic BST_{92} surface structure $\mathscr{W}$ with a designated event e^*, as defined by Def. 8.18, is a BST_{92} structure as well).

Hint: Use Def. 8.18 of the ordering $<'$ and the form of histories in $\mathscr{W}'$ established via Lemma 8.4.

Exercise 8.7. Prove Lemma 8.5 (i.e., prove that a BST_{92} structure corresponding to a probabilistic BST_{92} surface structure is a *probabilistic* BST_{92} structure in the sense of Def. 7.3).

Hint: Note that any set $Y \subseteq \text{TR}(W')$ on which μ' is defined either corresponds to a subset of $\text{TR}(W)$ or has the form $\{\langle e^*, 0 \rangle \rightarrowtail H^m\} \cup T^n$, where T^n corresponds to some $T \subseteq \text{TR}(W)$. Assume that μ satisfies Postulates 7.1, 7.2, 7.3, and 7.4. Then use the conditions of an adequate propensity assignment of Def. 8.20 together with Def. 8.18 of the ordering $<'$ to show that μ' satisfies these Postulates as well. ·

9

Branching in Relativistic Space-Times

We advertised Branching Space-Times theory as a theory of local indeterminism, playing out in our spatio-temporal world, where the space-time is
to be at least rudimentarily relativistic. The theory thus promises to describe
how to combine local indeterminism and relativistic space-times. In this
chapter we will make good on this promise, as we will introduce here
particular BST structures in which histories are similar to the space-times
of relativistic physics. We can approach these tasks differently, with various
degrees of modesty, and our first construction, the so-called Minkowskian
Branching Structures (MBSs), is intended to be a modest one: we focus on
how alternative histories, all happening on Minkowski space-time, can be
seen as developing from some shared past. Of course, a history, realistically
speaking, cannot be a bare Minkowski space-time, as it should involve some
properties that are not spatio-temporal (e.g., matter fields). To have more
full blooded objects, we first make a simplifying assumption regarding how
physical properties are associated to space-time points, and then leave the
matter to physics in order to check whether some of its theories support this
construction. The simplifying assumption is that whatever content a history
has, it comes from the ascription of some quantities (values, or strengths of
a field) to *points* in the space-time. This is of course pointilisme pure and
simple, and the resulting concept of history is a case of a Humean mosaic.
The next idea is that whilst a physics theory is pointilistic as suggested earlier,
perhaps it acknowledges alternative property ascriptions. How can that be
the case? A natural response is what is known as the ill-posedness of the
initial value problem, which means that the theory's equations of evolution
allow that for some initial values, there are multiple global solutions to
these equations, with the solutions representing evolutions occurring in
Minkowski space-time. Whether there are such theories of physics, and
whether their solutions satisfy some further constraints necessary for the
construction to go through, we leave it for physics to judge. And if the
judgement is in the negative, that would mean that, as far as physics goes,

Branching Space-Times: Theory and Applications. Nuel Belnap, Thomas Müller, and Tomasz Placek,
Oxford University Press. © Oxford University Press 2022. DOI: 10.1093/oso/9780190884314.003.0009

there is no local indeterminism of the sort described by BST, and occurring in Minkowski space-time. In the second part (Section 9.3) we turn toward linking BST to general relativity, which is both an ideologically different and more demanding project.

9.1 Minkowskian Branching Structures

We now turn to constructing a special class of BST structures, called Minkowskian Branching Structures (MBSs), in which each history is isomorphic to Minkowski space-time. Our aim here is to show that our MBSs satisfy the axioms of BST_{NF} structures (Def. 3.15). An alternative construction, in which an MBS comes out as a BST_{92} structure, was investigated in our earlier work.[1] In this book, we focus on MBSs that are BST_{NF} structures because we want the structures to admit locally Euclidean topologies; this provides a better integration with General Relativity (GR). Section 9.3 investigates the relations between BST and GR.

9.1.1 Basic notions

Our construction proceeds in terms of the assignments of physical properties (which we think of as values of physical fields) to space-time points. As we will see, MBSs seamlessly permit the introduction of space-time locations in the sense of Def. 2.9. For the construction to succeed, however, some additional physical conditions must be satisfied. As before, by Minkowski space-time we will understand the set $\mathbb{R}^n$ with the Minkowskian ordering, $<_M$, defined in the usual way (see Eq. (2.1)) as:[2]

$$ x <_M y \quad \text{iff} \quad -(x^0 - y^0)^2 + \sum_{i=1}^{n-1} (x^i - y^i)^2 \leqslant 0 \text{ and } x^0 < y^0. \qquad (9.1)$$

Usually we assume that $n = 4$. The ordering $<$ (or $\leqslant$) on the right-hand side refers to the natural strict (or non-strict) ordering of the reals. As usual, we

---

[1] See the papers by Müller (2002), Wroński and Placek (2009), and Placek and Belnap (2012).
[2] For the record, in physics it is common to take Minkowski space-time to be a set of points together with a metric, which allows one to derive the causal ordering. The structures we are working with thus contain less information, but are also simpler to handle, than those of the physicists.

define: $x \leqslant_M y$ iff ($x <_M y$ or $x = y$), and furthermore, $x \, SLR_M \, y$ iff neither $x \leqslant_M y$ nor $y \leqslant_M x$.

Some physical theories ascribe physical properties, typically the strengths of physical fields, to points of a space-time, or can be viewed as involving such an ascription. If the underlying space-time is a Minkowski space-time, the properties are ascribed to points of Minkowski space-time (i.e., elements of $\mathbb{R}^4$). A necessary condition for a theory to exhibit indeterminism is that it allows for many "scenarios" ascribing alternative possible properties to points of Minkowski space-time. In other words, one point of $\mathbb{R}^4$ may have alternative properties assigned, depending on the scenario.

Ultimately, we will define an MBS as a triple $\mathfrak{M} = \langle \Sigma, F, P \rangle$ (see Def. 9.5). We will begin by partly characterizing Σ, F, and P. To help capture the informal concept of possible "scenarios" abstractly, we assume a non-empty set Σ of labels, understood as labels for "scenarios." We let σ, η, γ range over Σ. We want to think of a scenario as Minkowski space-time filled with some "content," where the content of a scenario should be representable by an attribution of properties to each Minkowski space-time point. That is, the content of a single scenario, σ, may be represented by a function in the set $\mathbb{R}^4 \to \mathscr{P}(P)$, where P is a nonempty set of properties attributable to points of $\mathbb{R}^4$. Our purposes do not require putting any structure on P. A system of such contents can then be represented by a global attribution of properties $F : \Sigma \times \mathbb{R}^4 \to \mathscr{P}(P)$. We will call such an F a "property attribution on Σ and P," noting that it is in effect a modal notion because it refers to alternative possible properties for the same space-time point. Writing $\langle \sigma x \rangle$ for a pair from $\Sigma \times \mathbb{R}^4$, we may read "$F(\langle \sigma x \rangle)$" as "the set of properties instantiated at space-time point x in scenario σ". We informally think of the set in question as only containing compatible properties. The function F evidently dictates for each space-time point, x, whether two scenarios, σ, η, are qualitatively the same there ($F(\langle \sigma x \rangle) = F(\langle \eta x \rangle)$) or not ($F(\langle \sigma x \rangle) \neq F(\langle \eta x \rangle)$).—To avoid clutter, from now on we simply write "$F(\sigma x)$".

Clearly, many property attributions yield a pattern of scenarios without any similarity to what one might call indeterminism. Indeed, there is a consensus that indeterminism involves many scenarios that agree over some region (typically, an initial region) and then disagree over some (typically, later) region. In what follows, we will single out those special property attributions that we will call *proper* property attributions. We will find that proper property attributions lead to a pattern of indeterminism that is describable by Branching Space-Times. This means that we will derive from a triple $\mathfrak{M} = \langle \Sigma, F, P \rangle$ a BST-like pair $\langle B, <_R \rangle$, and BST-like notions of history and choice set, and ultimately show that the BST_{NF} axioms are satisfied in the defined model.

We turn to the task of defining a proper property attribution, F. A part of this task is to single out a set of particular points of $\mathbb{R}^4$, to be interpreted, loosely speaking, as locations of chanciness, as where the scenarios diverge. We will call such points "splitting points". We first assume that any two scenarios differ somewhere, i.e.,

$$\forall \sigma, \eta \in \Sigma \, (\sigma \neq \eta \to \exists x \in \mathbb{R}^4 \, F(\sigma x) \neq F(\eta x)), \qquad \text{(SDiff)}$$

where F is a property attribution on Σ and P.

Condition (SDiff) assures us that any two scenarios are qualitatively different. We further require that the pattern of differences for two scenarios be rather special: We postulate that for every two scenarios there is (at least) one point $s \in \mathbb{R}^4$ such that the scenarios disagree at s, but agree everywhere in the past of s. A point satisfying these two conditions will be defined as a splitting point for the two scenarios.

Definition 9.1 (Splitting points). Given $\mathfrak{M} = \langle \Sigma, F, P \rangle$, where F is a property attribution on Σ and P and $\sigma, \eta \in \Sigma$ and $s \in \mathbb{R}^4$, s is a *splitting point* between scenarios σ, η iff s satisfies the condition

$$F(\sigma s) \neq F(\eta s) \wedge \forall y \in \mathbb{R}^4 \, [y <_M s \to F(\sigma y) = F(\eta y)]. \qquad \text{(PastsAgree)}$$

$S_{\sigma\eta} \subseteq \mathbb{R}^4$ is defined as the set of all splitting points between scenarios $\sigma, \eta \in \Sigma$.

Splitting points for two scenarios allow us to define a region of $\mathbb{R}^4$ that we will soon prove to be the region in which the two scenarios are qualitatively the same (see Fact 9.1(4)). But note that the two scenarios are qualitatively different somewhere else as well. We allow for split scenarios to largely reconverge qualitatively. Regions of overlap were first introduced by Müller (2002).

Definition 9.2 (Region of overlap). Given $\mathfrak{M} = \langle \Sigma, F, P \rangle$, for $\sigma, \eta \in \Sigma$, $R_{\sigma\eta} := \{x \in \mathbb{R}^4 \mid \neg \exists s \, (s \leqslant_M x \wedge s \in S_{\sigma\eta})\}$.

Splitting is a qualitative notion that is derived from the differences of properties in scenarios, in contrast to the cause-like notion of choice sets of BST_{NF}. Note also that no $s \in S_{\sigma\gamma}$ can belong to $R_{\sigma\gamma}$. For a pattern of qualitative differences between scenarios to deliver a BST_{NF} structure, it must be somewhat restricted. The restrictions are incorporated in what we call "proper property attribution". Yet, before we state it we need to define the auxiliary notion of the set $\Sigma_\eta(x)$ of those labels for a given $x \in \mathbb{R}^4$ that specify one and the same event in an MBS.

Definition 9.3. Given $\mathfrak{M} = \langle \Sigma, F, P \rangle$, where F is a property attribution on Σ and P and $\eta \in \Sigma$:

$$\Sigma_\eta(x) =_{\mathrm{df}} \{\sigma \in \Sigma \mid x \in R_{\sigma\eta}\}.$$

Note that $\Sigma_\eta(x)$ is not empty for any x, since $x \in R_{\eta\eta}$. This notion is needed in clause (2) of the definition below.

Definition 9.4 (Proper property attribution). Given $\mathfrak{M} = \langle \Sigma, F, P \rangle$, F is a *proper property attribution* on a set of scenarios Σ and a set of properties P iff $F : \Sigma \times \mathbb{R}^4 \to \mathscr{P}(P)$, and F satisfies the condition SDiff that every two scenarios differ, and for all $\sigma, \eta \in \Sigma$,

1. $\forall x \in \mathbb{R}^4 \, [F(\sigma x) \neq F(\eta x) \to \exists s \in S_{\sigma\eta} \, [s \leqslant_M x]]$;
2. for every chain l in $\langle \mathbb{R}^4, <_M \rangle$, if for each $x \in l$ there is a unique $\gamma_x \in \Sigma$ such that for any finite $Z \subseteq l$ we have $\bigcap_{x \in Z} \Sigma_{\gamma_x}(x) \neq \emptyset$, then $\bigcap_{x \in l} \Sigma_{\gamma_x}(x) \neq \emptyset$;
3. for every lower bounded chain l in $\langle \mathbb{R}^4, <_M \rangle$,

$$(\forall x \in l \, \exists s \in S_{\sigma\eta} \, [s \leqslant_M x]) \to (\exists s_0 \in S_{\sigma\eta} \, [s_0 \leqslant_M \inf l]).$$

The first clause links points of qualitative difference to splitting points: either a point of qualitative difference is a splitting point, or there is a splitting point below it. The second clause is to curb the wealth of splittings, and its significance comes out only later (in Fact 9.4), when we consider the form of histories in a BST_{NF} structure that implements an MBS.[3] The clause is automatically satisfied by any finite chain l. The last clause is to guarantee the satisfaction of the Prior Choice Principle, PCP_{NF}.

We can show that for any $\sigma, \eta \in \Sigma$, the set $S_{\sigma\eta}$ of splitting points induced by a proper property attribution on Σ and P, as well as the corresponding region of overlap $R_{\sigma\eta}$, have the following natural properties:

---

[3] The clause is modeled after the Chain Condition 7 of Wroński and Placek (2009).

Fact 9.1. *Assume that* Σ *and* P *are non-empty and that* F *is a proper property attribution on* Σ *and* P. *Then for any* $\sigma, \eta, \gamma \in \Sigma$ *and for any* $x \in \mathbb{R}^4$:

1. $\sigma = \eta$ *iff* $S_{\sigma\eta} = \emptyset$;
2. $S_{\sigma\eta} = S_{\eta\sigma}$;
3. $\forall s, s' \in S_{\sigma\eta}\ (s \neq s' \rightarrow s\,SLR_M\,s')$;
4. $x \in R_{\sigma\eta} \rightarrow F(\sigma x) = F(\eta x)$, *and*
5. $R_{\sigma\eta} \cap R_{\eta\gamma} \subseteq R_{\sigma\gamma}$.

Proof. (1) Immediate from Definition 9.4.

(2) Immediate from Definition 9.1.

(3) By (PastsAgree), neither $s <_M s'$ nor $s' <_M s$. So if $s \neq s'$, then $s\,SLR_M\,s'$;

(4) This is essentially the contrapositive of Definition 9.4 clause (1).

(5) For reductio, assume $x \in R_{\sigma\eta} \cap R_{\eta\gamma}$ but $x \notin R_{\sigma\gamma}$. The latter implies that there is an $s \in S_{\sigma\gamma}$ such that $s \leqslant_M x$ so (†) $F(\sigma s) \neq F(\gamma s)$. Since regions of overlap are evidently closed downward, we get $s \in R_{\sigma\eta} \cap R_{\eta\gamma}$. By item (4) of this Fact, $F(\sigma s) = F(\eta s)$ and $F(\eta s) = F(\gamma s)$, and hence $F(\sigma s) = F(\gamma s)$. Contradiction with (†). $\qquad\square$

Observe that in the proof above we used clause (1), but neither clause (2) nor clause (3) of Def. 9.4.

9.1.2 Defining MBSs

After this preliminary work, we turn now toward defining MBSs and showing that they generate structures satisfying the postulates of BST_{NF}. We thus first officially define Minkowskian Branching Structures, then prove desired facts about the form of B-histories, and finally show that the postulates of BST_{NF} are satisfied.

Definition 9.5 (MBS). A triple $\mathfrak{M} = \langle \Sigma, F, P \rangle$ is a *Minkowskian Branching Structure* (an *MBS*) iff Σ is a non-empty set of scenarios, P is a nonempty set of properties, and F is a proper property attribution on Σ and P.

In order to link MBSs to BST, our first task is to find a correlate for the BST notion of *Our World* (i.e., a base set), and for the BST ordering. Given $\mathfrak{M} = \langle \Sigma, F, P \rangle$, we take the elements of the base set to be equivalence classes of a certain relation $\equiv_R$ on $\Sigma \times \mathbb{R}^4$, where the "R" in the subscript indicates that the relation crucially depends on an assumed region of overlap R—the idea is to "identify" points in regions of overlap. The relation $\equiv_R$ is defined as below.

Definition 9.6 (MBS equivalence relation). Given an MBS $\mathfrak{M} = \langle \Sigma, F, P \rangle$, the relation $\equiv_R$ on $\Sigma \times \mathbb{R}^4$ is defined as:

$$\langle \sigma, x \rangle \equiv_R \langle \eta, y \rangle \quad \text{iff} \quad x = y \text{ and } x \in R_{\sigma\eta}. \tag{9.2}$$

It is easy to check that $\equiv_R$ is an equivalence relation on $\Sigma \times \mathbb{R}^4$: Fact 9.1(1) implies $R_{\sigma\sigma} = \mathbb{R}^4$ from which reflexivity follows, symmetry follows by Fact 9.1(2), and transitivity follows by Fact 9.1(5). With this relation to hand, we introduce our candidate for a BST$_{\text{NF}}$ structure via the equivalence classes under $\equiv_R$:

Definition 9.7 (MBS base set and MBS ordering). Let $\mathfrak{M} = \langle \Sigma, F, P \rangle$ be an MBS. We define the *MBS base set for $\mathfrak{M}$*, B, to be

$$B =_{df} \{ [\sigma x] \mid \sigma \in \Sigma, x \in \mathbb{R}^4 \}, \quad \text{where} \quad [\sigma x] =_{df} \{ \langle \eta, x \rangle \mid \langle \sigma, x \rangle \equiv_R \langle \eta, x \rangle \}.$$

The *MBS ordering* $<_R$ on B is defined by

$$[\sigma x] <_R [\eta y] \quad \Leftrightarrow_{df} \quad x <_M y \wedge \langle \sigma, x \rangle \equiv_R \langle \eta, x \rangle.$$

The pair $\langle B, <_R \rangle$ will be called *the structure generated by the MBS $\mathfrak{M}$*. As usual, we will write $\leqslant_R$ for the weak counterpart of $<_R$.

It is again easy to check that $<_R$ is a strict partial ordering on B. It is antireflexive because $<_M$ is antireflexive. Transitivity follows from Fact 9.1(5); see Exercise 9.1. Similarly, it is straightforward to prove density of $<_R$ (see Exercise 9.2). Note that if x is not in the region $R_{\sigma\eta}$ of overlap of σ and η, then $[\sigma x] \not<_R [\eta y]$ (for all y), but if x is in $R_{\sigma\eta}$, for $[\sigma x] <_R [\eta y]$ we only need to check the Minkowski ordering, $x <_M y$.

Given $\mathfrak{M} = \langle \Sigma, F, P \rangle$, a natural definition for the course of events corresponding to scenario σ is the set $\{ [\sigma x] \mid x \in \mathbb{R}^4 \}$ of equivalence classes. Knowing the set, that is, knowing each equivalence class from it, gives us all there is to be known about this course of events, that is, a property assignment for σ and every $x \in \mathbb{R}^4$. This motivates defining $\{ [\sigma x] \mid x \in \mathbb{R}^4 \}$ as a "B-history."

Definition 9.8 (*B*-histories). Given $\mathfrak{M} = \langle \Sigma, F, P \rangle$, the *B-history correspond-ing to* $\sigma \in \Sigma$ is defined to be $h_\sigma =_{df} \{ [\sigma x] \mid x \in \mathbb{R}^4 \}$. *B*-Hist is the set of all *B*-histories.

Given $\mathfrak{M} = \langle \Sigma, F, P \rangle$, our plan is to show that the pair $\langle B, <_R \rangle$ is a BST_{NF} structure. The next three facts about MBSs concern the form of histories:

Fact 9.2. *Suppose that* $\langle B, <_R \rangle$ *is a structure determined by MBS* $\mathfrak{M} = \langle \Sigma, F, P \rangle$. *Then every B-history is a maximal directed subset of B.*

Proof. Consider a B-history h_σ. Since for any $x_1, x_2 \in \mathbb{R}^4$ there is $x_3 \in \mathbb{R}^4$ such that $x_1 \leqslant_M x_3$ and $x_2 \leqslant_M x_3$, h_σ is directed. To argue that it is maximal directed, suppose for reductio that there is a directed set $g \subseteq B$ such that $h_\sigma \subsetneq g$. There is thus some $[\eta x] \in g \setminus h_\sigma$. Hence (†) $[\eta x] \neq [\sigma x]$. Since $[\sigma x] \in h_\sigma \subsetneq g$ and g is directed, there is $[\alpha y] \in g$ such that $[\sigma x] \leqslant_R [\alpha y]$ and $[\eta x] \leqslant_R [\alpha y]$. It follows that $[\sigma x] = [\alpha x] = [\eta x]$, which contradicts (†). □

Fact 9.3. *Let* $\langle B, <_R \rangle$ *be the structure determined by the MBS* $\mathfrak{M} = \langle \Sigma, F, P \rangle$. *Then*

(1) *if* $[\sigma_1 x_1], [\sigma_2 x_2] \in B$ *and* $[\sigma_1 x_1] \leqslant_R [\sigma_2 x_2]$, *then* $\Sigma_{\sigma_2}(x_2) \subseteq \Sigma_{\sigma_1}(x_1)$;
(2) *for every directed set* $h \subseteq B$ *and every finite subset* $X \subseteq h$, $\bigcap_x \{ \Sigma_\gamma(x) \mid [\gamma x] \in X \} \neq \emptyset$.

Proof. (1) Let $[\sigma_1 x_1] \leqslant_R [\sigma_2 x_2]$, and $\eta \in \Sigma_{\sigma_2}(x_2)$, so $[\sigma_2 x_2] = [\eta x_2]$. Then $[\sigma_1 x_1] = [\eta x_1]$. Hence $\eta \in \Sigma_{\sigma_1}(x_1)$. Thus, $\Sigma_{\sigma_2}(x_2) \subseteq \Sigma_{\sigma_1}(x_1)$.

(2) By directedness of h and finiteness of X, there is an upper bound of X in h, say $[\sigma y]$. Thus, every element $[\eta x] \in X$ can be written as $[\sigma x]$, so $\sigma \in \Sigma_\eta(x)$ for every $[\eta x] \in X$. □

Fact 9.4. *Suppose that* $\mathfrak{M} = \langle \Sigma, F, P \rangle$ *is an MBS.*

1. *Every maximal directed subset of B is a B-history;*
2. *To every B-history there corresponds a unique* $\sigma \in \Sigma$, *i.e., for every pair* h_σ, h_η *of B-histories,* $h_\sigma = h_\eta$ *iff* $\sigma = \eta$;

Proof. (1) Let h be a maximal directed subset of B. We first show that $h = \{ [\sigma x] \mid x \in Y \}$, for some $\sigma \in \Sigma$ and $Y \subseteq \mathbb{R}^4$. For reductio, let us suppose that this is not true, i.e., for any $\sigma \in \Sigma$ there is some $[\eta x] \in h$ such that $[\eta x] \neq [\sigma x]$. We will construct a chain in $\langle \mathbb{R}^4, <_M \rangle$ that contradicts clause 2 of Def. 9.4.

We pick some $[\eta y] \in h$, well-order $\Sigma_\eta (y)$ in some way, and define function $\Theta : \Sigma_\eta (y) \mapsto \mathscr{P}(\mathbb{R}^4)$ such that $\Theta(\sigma) = \{x \in \mathbb{R}^4 \mid [\sigma x] \notin h\}$. $\Theta(\sigma)$ thus comprises all "bad for σ" points from $\mathbb{R}^4$; that is, those points that, taken with σ, produce no elements of h. By the reductio assumption, the value of Θ is never the empty set. To pick a representative from $\Theta(\sigma)$, we use a selection function $T : \mathscr{P}(\mathbb{R}^4) \mapsto \mathbb{R}^4$ such that $T(\Theta(\sigma)) \in \Theta(\sigma)$ for any $\sigma \in \Sigma_\eta (y)$. We label the values of the composition $T \circ \Theta$ so that we have $T \circ \Theta(\sigma_l) = x_l$ (note that l is a label, resulting from $\Sigma_\eta (y)$ being well-ordered). Accordingly, the ordering of $\Sigma_\eta (y)$ is carried over to the image $X =_{df} (T \circ \Theta)[\Sigma_\eta (y)]$. For each x_l, we pick $\sigma_{x_l} \in \Sigma_y(\eta)$ such that $[\sigma_{x_l} x_l] \in h$. Observe that by this construction, (†) for any $\sigma_l \in \Sigma_\eta (y)$: $\sigma_l \notin \Sigma_{\sigma_{x_l}} (x_l)$. Thus, $\bigcap_{x_l \in X} \Sigma_{\sigma_{x_l}} (x_l) = \emptyset$.

We now use the set $X \subseteq \mathbb{R}^4$ to produce a chain in $\langle \mathbb{R}^4, <_M \rangle$ above y. For simplicity's sake, in this construction we work directly in terms of coordinates of events like $y \in \mathbb{R}^4$. This amounts to choosing an arbitrary reference frame. The elements of this chain have different values of the first coordinate, whereas the remaining three coordinates are fixed. To construct the chain, we use a function $up : \mathbb{R}^4 \times \mathbb{R}^4 \mapsto \mathbb{R}^4$ that for any pair of points z_1, z_2 in $\mathbb{R}^4$ yields the minimal element of the set of their upper bounds that lie on a vertical line passing through z_1. Our chain is defined by $z_0 = y$, $z_1 = up(z_0, x_1), \ldots, z_{n+1} = up(z_n, x_{n+1})$. (For the details of the chain construction, in particular, for a limit step, see Exercise 9.3.) The result is a time-like chain $E \subseteq \mathbb{R}^4$ starting with y. Its cardinality is determined by the cardinality of $\Sigma_\eta (y)$ and the location of points of X.

In the next step, we observe that each $y, x_1, x_2, \ldots$, if associated with a proper label, is an element of h. Since h is directed, each $z_0, z_1, z_2, \ldots$, if associated with a proper label, is an element of h. For each z_l, we thus pick a label σ_{z_l} such that $[\sigma_{z_l} z_l] \in h$. Since h is directed, by Fact 9.3(2), for any finite subset $Z \subseteq E$, we have $\bigcap_{z_l \in Z} \Sigma_{\sigma_{z_l}} (z_l) \neq \emptyset$. Accordingly, the premise of clause (2) of Def. 9.4 is satisfied, so by this clause $\bigcap_{z_l \in E} \Sigma_{\sigma_{z_l}} (z_l) \neq \emptyset$. Since $[\sigma_{x_l} x_l] \leqslant_R [\sigma_{x_l} z_l]$, for every $x_l \in X$, $z_l \in E$, by Fact 9.3(1) $\Sigma_{\sigma_{z_l}} (z_l) \subseteq \Sigma_{\sigma_{x_l}} (x_l)$, and hence $\bigcap_{x_l \in X} \Sigma_{\sigma_{x_l}} (x_l) \neq \emptyset$. We thus have arrived at a contradiction with (†).

We thus showed that $h = \{[\sigma x] \mid x \in Y\}$ for some $\sigma \in \Sigma$ and $Y \subseteq \mathbb{R}^4$. By Fact 9.2, if $Y \subsetneq R^4$, then h is not maximal directed. Hence $Y = \mathbb{R}^4$.

(2) Clearly, if $\sigma = \eta$, then $\{[\sigma x] \mid x \in \mathbb{R}^4\} = \{[\eta x] \mid x \in \mathbb{R}^4\}$, so $h_\sigma = h_\eta$. In the other direction, if $\sigma \neq \eta$, then by Fact 9.1(1) $S_{\sigma\eta} \neq \emptyset$, so there are $x \in \mathbb{R}^4$ and $s \in S_{\sigma\eta}$ such that $s <_M x$. Hence $[\sigma x] \neq [\eta x]$. Hence $h_\sigma \neq h_\eta$.   □

There is thus a perfect match between scenarios $\sigma \in \Sigma$, B-histories h_σ, and maximal directed subsets of B. As we now know what the histories in $\langle B, <_R \rangle$ look like, we next address the form of choice sets. Recall that a choice set is a particular subset of the base set B, whereas a splitting point $s \in S_{\sigma\eta}$ for two scenarios σ and η is just a point $s \in \mathbb{R}^4$. We will prove that the two notions correspond nicely. But since we need history-relative suprema in this proof, we first prove a fact concerning suprema and infima of bounded chains in $\langle B, <_R \rangle$, showing that the structure satisfies the two respective postulates of BST_{NF}.

Fact 9.5. *Let $\langle B, <_R \rangle$ be a structure determined by the MBS $\langle \Sigma, F, P \rangle$. Then*
  (1) $\langle B, <_R \rangle$ contains infima for all lower bounded chains,
  (2) $\langle B, <_R \rangle$ contains history-relative suprema for all upper bounded chains.

Proof. Since every chain extends to a history, by the form of histories, every chain l in $\langle B, <_R \rangle$ can be written as $l = \{[\sigma x] \mid x \in l_x\}$ for some $\sigma \in \Sigma$ and chain l_x in $\langle \mathbb{R}^4, <_M \rangle$. Observe that if a chain $l \subseteq B$ is lower (upper) bounded, then the corresponding chain l_x is lower (upper) bounded. By the properties of $<_M$, l_x has then an infimum i_x (supremum s_x). Thus, for any history h_γ such that $l \subseteq h_\gamma$, $[\gamma i_x]$ is a history-relative (with respect to h_γ) infimum of l, and $[\gamma s_x]$ is a history-relative (with respect to h_γ) supremum of l. The second conjunct already proves part (2) of the fact. As for part (1), since histories are downward closed and $[\sigma i_x] \leqslant_R l$, $[\sigma i_x]$ belongs to every B-history that contains l. Accordingly, $[\sigma i_x]$ is an infimum, and not merely a history-relative infimum, of l. $\square$

Having introduced history-relative suprema to MBSs, we are ready to return to the form of choice sets; we will prove the following correspondence:

Fact 9.6. *Let $\sigma, \eta \in \Sigma$ and h_σ, h_η be B-histories in structure $\langle B, <_R \rangle$ that is determined by MBS $\langle \Sigma, F, P \rangle$. Then*
$s \in S_{\sigma\eta}$ *iff* $[\sigma s], [\eta s]$ *belong to a choice set $[\ddot{\sigma} s]$ at which h_σ and h_η branch.*

Proof. $\Rightarrow$ By Defs. 9.1 and 9.6, $[\sigma s] \neq [\eta s]$. As there are no minimal elements in $\mathbb{R}^4$, $[\sigma s]$ and $[\eta s]$ are not minimal elements in B-histories, so the top clause of Def. 3.11 does not apply—we need to only check the bottom clause. Consider thus an arbitrary chain $l \in \mathscr{C}_{[\sigma s]}$. It means that $l \subseteq h_\sigma$, $\sup_{h_\sigma} l = [\sigma s]$, and $[\sigma s] \notin l$. It follows that $l_x = \{x \in \mathbb{R}^4 \mid [\sigma x] \in l\}$ has a supremum s in $\langle \mathbb{R}^4, <_M \rangle$. Further, since every element of l_x is below s

(with respect to $<_M$), it cannot be above some $s' \in S_{\sigma\eta}$, since then $s' <_M s$, which contradicts Fact 9.1 (3). Accordingly, $l_x \subseteq R_{\sigma\eta}$, so $l \subseteq h_\eta$. Thus, l has a supremum in h_η, $\sup_{h_\eta} l = [\eta s]$. Since l is arbitrary, $[\sigma s], [\eta s] \in \bigcap_{l \in \mathscr{C}_{[\sigma s]}} \mathscr{S}(l)$, and hence $[\ddot{\sigma} s]$ is a choice set, with $[\sigma s], [\eta s] \in [\ddot{\sigma} s]$. Further, by Def. 9.8 of B-histories, $h_\sigma \cap [\ddot{\sigma} s] = \{[\sigma s]\}$ and $h_\eta \cap [\ddot{\sigma} s] = \{[\eta s]\}$, and $[\sigma s] \neq [\eta s]$. It thus follows by Def. 3.13 that h_σ and h_η branch at $[\ddot{\sigma} s]$.

$\Leftarrow$ Let $[\sigma s], [\eta s], [\ddot{\sigma} s], h_\sigma$, and h_η be as the RHS of the Fact says. By Def. 3.13, $[\sigma s] \neq [\eta s]$, so there is $s' \in S_{\sigma\eta}$ such that $s' \leqslant_M s$. Let us next suppose that $s' <_M s$. Clearly, $[\sigma s'] \neq [\eta s']$, and for any $x \in \mathbb{R}^4$, if $[\sigma s'] \leqslant_R [\sigma x]$, then $[\sigma x] \notin h_\eta$. Pick then any chain $l \in \mathscr{C}_{[\sigma s]}$ that contains $[\sigma s']$. Since $[\sigma s'] <_R [\sigma s]$, by density there is non-empty upper segment l' of l such that $l' \cap h_\eta = \emptyset$. Thus, $[\eta s]$ is not a h_η-relative supremum of l, which entails $[\eta s] \notin [\ddot{\sigma} s]$, contradicting the Fact's premise. Thus, $s = s'$, with $s' \in S_{\sigma\eta}$. $\square$

We next turn our attention to two more interesting postulates of BST$_{NF}$: PCP$_{NF}$ and Weiner's postulate. We show that each is satisfied in structure $\langle B, <_R \rangle$ determined by an MBS.

Fact 9.7. *The structure $\langle B, <_R \rangle$ determined by an MBS satisfies PCP$_{NF}$, as defined by Def. 3.14.*

Proof. Let h_σ, h_η be B-histories in $\langle B, <_R \rangle$, and let $l = \{[\sigma x] \mid x \in l_x\} \subseteq h_\sigma$ be a lower bounded chain (so l_x is a lower bounded chain in $\langle \mathbb{R}^4, <_M \rangle$) such that $l \cap h_\eta = \emptyset$. Thus, for every $z \in l_x$: $[\sigma z] \neq [\eta z]$, and hence for every $z \in l_x$ there is $s \in S_{\sigma\eta}$ such that $s \leqslant_M z$. By the infima postulate, l has an infimum $[\sigma i_x]$, and hence l_x has an infimum i_x. Accordingly, l_x satisfies the premise of clause (3) of Def. 9.4, so by this very clause there is $s_0 \in S_{\sigma\eta}$ such that $s_0 \leqslant_M inf(l_x)$. It follows that $[\sigma s_0] \leqslant_R l$ and by Fact 9.6, $[\sigma s_0]$ gives rise to choice set $[\ddot{\sigma} s_0]$, at which h_σ and h_η branch, i.e., $h_\sigma \perp_{[\ddot{\sigma} s_0]} h_\eta$. $\square$

Fact 9.8. *The structure $\langle B, <_R \rangle$ determined by an MBS satisfies Weiner's postulate 2.6.*

Proof. To check Weiner's postulate, let $l_1, l_2 \subseteq h_\sigma \cap h_\eta$ be upper bounded chains in B-histories h_σ and h_η. Let s_1, s_2 be relative to h_σ suprema of l_1 and l_2, respectively. If $s_1 = s_2$, then $s_1 = s_2 = [\sigma x]$ for some $x \in \mathbb{R}^4$, and hence there are $c_1 = c_2 = [\eta x] \in h_\eta$ that are h_η-relative suprema of l_1 and l_2, respectively. Analogously, if $s_1 <_R s_2$, then, as $s_1 = [\sigma x]$ and $s_2 = [\sigma y]$ for some $x, y \in \mathbb{R}^4$, it follows that $x <_M y$. Clearly, $c_1 = [\eta x]$ and $c_2 = [\eta y]$ are h_η–relative suprema of l_1, l_2, and $c_1 <_R c_2$. □

The last few facts testify that $\langle B <_R \rangle$ determined by an MBS satisfies the postulates of BST$_{NF}$. We are thus ready to state our main theorem about MBSs:

Theorem 9.1. *The structure $\langle B, <_R \rangle$ generated by an MBS (see Definition 9.7) is a BST$_{NF}$ structure.*

Proof. We need to check if $\langle B, <_R \rangle$ satisfies all the BST$_{NF}$ postulates, as stated in Def. 3.15. Given the definition of MBS, Σ is not empty, so B is not empty, either. As we noted (Def 9.7), $<_R$ is a dense strict partial order. Fact 9.5 states that the infima postulate and the suprema postulate are satisfied. Weiner's postulate is satisfied by Fact 9.8. PCP$_{NF}$ holds by Fact 9.7 and it implies Historical Connection; see Exercise 3.5. Thus, all the postulates of BST$_{NF}$ are satisfied by $\langle B, <_R \rangle$. □

Apart from being a BST$_{NF}$ structure, an MBS has some additional welcome features. The fact below focuses on order-related similarities; that is, the existence of relevant order-isomorphisms.

Fact 9.9. *Let $\langle B, <_R \rangle$ be a structure determined by MBS $\langle \Sigma, F, P \rangle$. Then:*

1. *Every B-history in $\langle B, <_R \rangle$ is order-isomorphic to Minkowski space-time;*
2. *$\langle B, <_R \rangle$ permits the introduction of common space-time locations $\langle S, <_S \rangle$ in the sense of Def. 2.9;*
3. *$\langle S, <_S \rangle$ is order-isomorphic to Minkowski space-time.*

Proof. (1) A required isomorphism is $\varphi_j : \mathbb{R}^4 \mapsto h_\sigma$ such that $\varphi(x) = [\sigma x]$. It is clear to see that for $x, y \in \mathbb{R}^4$, $x \leqslant_M y$ iff $[\sigma x] \leqslant_R [\sigma y]$.

(2) By Def. 2.9, the set S of space-time locations should be a specific partition of B. For our $\langle B, <_R \rangle$ determined by MBS $\langle \Sigma, F, P \rangle$ we define: $S = \{\{[\sigma x] \mid \sigma \in \Sigma\} \mid x \in \mathbb{R}^4\}$. The ordering $<_S$ on S is given by: for $s, s' \in S$ such that $s = \{[\sigma x] \mid \sigma \in \Sigma\}$ and $s' = \{[\sigma y] \mid \sigma \in \Sigma\}$, $s <_S s'$ iff $x <_M y$. Its weak companion $\leqslant_S$ is defined via $s \leqslant_S s'$ iff $x \leqslant_M y$. Clearly, the intersection of any element $s = \{[\sigma x] \mid \sigma \in \Sigma\} \in S$ with any B-history h_η contains exactly one element, $[\eta x]$. And $\leqslant_S$ respects the ordering $\leqslant_R$, i.e., for $s, s' \in S$ and h_σ, h_η, if $s \cap h_\sigma = s' \cap h_\sigma$, then $s \cap h_\eta = s' \cap h_\eta$, and analogously for $<_S$ and $<_R$. (3)

The required isomorphism is $\varphi : S \mapsto \mathbb{R}^4$ with $\varphi(\{[\sigma x] \mid \sigma \in \Sigma\}) =_{df} x$; by definition of $<_S$, φ is an order-isomorphism indeed. □

The similarity between a B-history and Minkowski space-time is even more intimate than an order-isomorphism. After all, as far as the structure goes, a B-history can be viewed as simply Minkowski space-time with the label $\sigma \in \Sigma$. The label σ is important, however, as it determines the physical content of the history, by the proper property attribution F and the set P of properties.

MBSs have interesting topological properties as well, to which we turn in Section 9.2.2, after introducing the required notions from the theory of differential manifolds. For the completeness of exposition, here we put down a few informal observations, while postponing rigorous arguments to the mentioned section. The natural topology on $\mathbb{R}^4$ is given by the base of open balls in $\mathbb{R}^4$. Since every B-history h_σ is, structurally speaking, just $\mathbb{R}^4 \times \{\sigma\}$, it admits a topology given by slightly fancier open balls, viz., standard balls with σ attached, which we call b-balls. We can show that, whenever o is an open ball on $\mathbb{R}^n$, $o_\sigma =_{df} \{[\sigma x] \mid x \in o\}$ is indeed an *open* b-ball in the natural topology on h_σ. This means that the intersection of any two b-balls is the union of (possibly infinitely many) b-balls. Open b-balls generate a topology on h_σ. The open-ball topology on $\mathbb{R}^4$ and the open-b-ball topology on a B-history are thus topologically the same; they are homeomorphic. Further, the open ball topology on $\mathbb{R}^n$ and the open b-ball topology on h_σ share the same separation properties, including the Hausdorff property (see Def. 4.15). That is, if $[\sigma x]$ and $[\sigma y]$ are distinct elements of h_σ, they can be made centers of sufficiently small non-overlapping open b-balls in h_σ. One consequence of this is that the open b-ball topology on a B-history is both locally Euclidean (see Def. 4.16) and Hausdorff, so that a B-history is a topological manifold (see Def. 9.9 and footnote 7).

The first part of this observation carries over, perhaps somewhat surprisingly, to the base set B of a BST_{NF} structure $\langle B, <_R \rangle$ derived from an MBS $\langle \Sigma, F, P \rangle$, even if the structure comprises multiple B-histories. This topology is indeed Euclidean. For this construction to work, it is essential to have PCP_{NF} rather than PCP_{92}, as the former guarantees that the intersection of any two b-balls is open. This construction would not work if $\langle B, <_R \rangle$ were a BST_{92} structure with multiple histories—see Exercise 9.4.

The second part of the observation, the one concerning Hausdorffness, does not carry over from a B-history to $\langle B, <_R \rangle$ if the structure comprises multiple histories. In this case, B contains at least one choice set with distinct elements, say, $[\sigma x]$ and $[\eta x]$. By Lemma 3.1, these two elements have exactly the same proper past. Accordingly, no matter how small the b-balls centered at these elements are that one chooses, they will overlap. Thus, the Hausdorff property fails in the b-ball topology on B.

To put these observations together, the b-ball topology on a B-history is both locally Euclidean and Hausdorff. In the terminology introduced in Chapter 9.2, a B-history is a topological manifold. In contrast, the b-ball topology on the base set B of a BST_{NF} structure derived from an MBS is locally Euclidean but typically non-Hausdorff. Such an object is called a generalized topological manifold. One may then ask if it is possible to extend a B-history by adding some elements from B, while preserving the local Euclidicity and Hausdorffness of the b-ball topology on the extended set. The answer turns out to be no (provided one assumes a further intuitive condition, connectedness). B-histories are the largest subsets of B on which b-balls deliver a locally Euclidean, Hausdorff, and connected topology. In short, B-histories are maximal connected topological sub-manifolds of the generalized topological manifold B.

Having seen these topological developments in Minkowskian Branching Structures and witnessing the debate in General Relativity about the status of the Hausdorff property in this theory's concept of space-time, it is tempting to try to provide topological foundations for BST (i.e., to define histories by a topological condition rather than by our order-theoretical one). More specifically, one might require that the base set W of *Our World* $\mathscr{W}$ admit a generalized (i.e., possibly non-Hausdorff) topological manifold structure, and then define a history in $\mathscr{W}$ to be a maximal subset of W that admits a connected Hausdorff topological manifold structure. In a strengthened version, one might further require that each history in $\mathscr{W}$ admit a differential manifold structure. We discuss this topic in Chapters 9.2 and 9.3, by first introducing the required topological notions, and then relating Branching Space-Times to General Relativity.

Before we turn to these tasks, let us take stock of MBSs.

9.1.3 Taking stock

Starting with three modest building blocks for Minkowskian Branching Structures, a set of labels for scenarios, a set of properties, and a function attributing properties to point-scenario pairs (where points are points of $\mathbb{R}^4$), we arrived at a BST_{NF} structure. The construction relied heavily on the constraints imposed on property attribution.[4]

An important result of this construction is that histories generated by an MBS are isomorphic to Minkowski space-time. Such histories are (and must be) different by having different physical contents (furnished by a property attribution), but they all share the same spatio-temporal structure. We do not claim that our construction is fully "physics-friendly". Even if the spatio-temporal structure of our universe were adequately represented by Minkowski space-time (which it is not), a few conditions must be satisfied on the part of physics for our construction of MBSs to be admitted. First, the physical description must come in the form of a property attribution to spatio-temporal points. Second, the property attribution must be quite specific: in Def. 9.4 we required it to be "proper". These notions are very much needed to guarantee that histories have the desired form, and that the prior choice principle, PCP_{NF}, is satisfied.[5]

Finally, each history in a BST_{NF} structure derived from an MBS can be viewed as a topological manifold (i.e., locally Euclidean and Hausdorff), whereas the structure itself can be seen as a generalized topological manifold (i.e., locally Euclidean but not necessarily Hausdorff). This establishes a

---

[4] Some remarks on the history of MBSs are in order. The notion of a Minkowskian Branching Structure was first introduced informally in Belnap (1992), denoting a BST structure in which every history is a Minkowski space-time. Placek (2000) first attempted to produce a BST model out of (copies of) Minkowski space-time, but failed. The first correct construction of MBSs (but with finitistic assumptions) is in Müller (2002). The construction presented here is a BST_{NF} version of the construction given by Placek and Belnap (2012), which in turn diverged from earlier constructions of Müller (2002) and Wroński and Placek (2009), as it aimed to be more physics-oriented. The latter authors begin their work with specifying a set Σ of labels for scenarios and a collection $\{S_{\sigma\eta}\}_{\sigma,\eta\in\Sigma}$ of sets of splitting points, where each $S_{\sigma\eta}$ possesses properties listed in Fact 9.1(i)-(iv). Given the two primitive notions, that is, labels for scenarios and sets of splitting points, they define MBSs and show, on the assumption of certain additional conditions, that MBSs satisfy BST_{92} postulates. The authors diverge over these additional conditions: Müller assumes finitistic requirements whereas Wroński and Placek accepts a "topological" postulate that is equivalent to the chain condition. This difference notwithstanding, an MBS model is, in their sense, a pair $\langle\Sigma,\{S_{\sigma,\eta}\}_{\sigma,\eta\in\Sigma}\rangle$. In contrast, our point of departure is a property attribution to points in scenarios. Accordingly, an MBS model is, in our sense, a triple $\langle\Sigma, F, P\rangle$—cf. Def. 9.5. Splitting points are then a derived notion—see Def. 9.1 and, as Fact 9.1(i)-(iv) shows, they satisfy the conditions assumed by Müller and Wroński and Placek. Accordingly, given that $\mathfrak{M} = \langle\Sigma, F, P\rangle$ is an MBS in the sense of Definition 9.5, $\langle\Sigma,\{S_{\sigma,\eta}\}_{\sigma,\eta\in\Sigma}\rangle$ with $\{S_{\sigma,\eta}\}_{\sigma,\eta\in\Sigma}$ defined by Def. 9.1 is an MBS in the sense of Wroński and Placek (2009). If finitistic constraints concerning sets $S_{\sigma\eta}$ are assumed, $\langle\Sigma,\{S_{\sigma,\eta}\}_{\sigma,\eta\in\Sigma}\rangle$ is an MBS in the sense of Müller (2002). A discussion of the topological properties of the different ways of pasting together Minkowski space-times is given in Müller (2013).

[5] This is not to say that some other condition on property attributions and a matching definition of splitting points would not do the job. The point is that this notion must be quite regimented to be of use in producing BST models.

link between BST and General Relativity, and suggests adding topological notions to the foundations of BST theory. These are two topics to which we turn in the next two sections.

9.2 Differential manifolds and BST_{NF}

In this section we first present some basic notions from the theory of differential manifolds. We will then use these notions to prove some facts about MBSs that we have already alluded to. We will also investigate BST_{NF} structures generally, asking if they admit differential manifold structure. We will need these notions later on, when we turn our attention to the space-times of General Relativity.

9.2.1 Differential manifolds

We introduce the already mentioned notions of topological manifold and generalized topological manifold. In General Relativity we need more complex structures, differential manifolds, which admit differential structure. We begin with the latter, defining the former as a special case.

Definition 9.9 (Chart, atlas, manifold: generalized, Hausdorff, non-Hausdorff). Let M be a non-empty set and Γ an index set. A collection of pairs $\{\langle u_\gamma, \varphi_\gamma \rangle \mid \gamma \in \Gamma\}$, where each $u_\gamma \subseteq M$, is a C^r n-atlas on M iff $\bigcup_{\gamma \in \Gamma} u_\gamma = M$, each φ_γ is a bijection between u_γ and an open subset of $\mathbb{R}^n$, and for any two $\langle u_\gamma, \varphi_\gamma \rangle$ and $\langle u_\tau, \varphi_\tau \rangle$, if $u_\gamma \cap u_\tau \neq \emptyset$, then $\varphi_\gamma[u_\gamma \cap u_\tau]$ and $\varphi_\tau[u_\gamma \cap u_\tau]$ are open subsets of $\mathbb{R}^n$ and the composite functions $\varphi_\gamma \circ \varphi_\tau^{-1}$ and $\varphi_\tau \circ \varphi_\gamma^{-1}$ are C^r on their domains. Each $\langle u_\gamma, \varphi_\gamma \rangle$ is called a *chart* of the atlas.

A pair $\langle M, A \rangle$, where M is a non-empty set and A is a maximal C^r n-atlas on M, is a C^r *n-dimensional generalized differential manifold*.

If a C^r n-dimensional generalized differential manifold $\langle M, A \rangle$ satisfies the condition that for any distinct $p, q \in M$ there are $\langle u_\gamma, \varphi_\gamma \rangle, \langle u_\tau, \varphi_\tau \rangle \in A$ such that $p \in u_\gamma$, $q \in u_\tau$ and $u_\gamma \cap u_\tau = \emptyset$, then it is called a C^r n-dimensional Hausdorff differential manifold.[6]

If a C^r n-dimensional generalized differential manifold does not satisfy the

---

[6] An equivalent way of defining Hausdorff differential manifolds is to say that the induced topology is Hausdorff.

above condition, it is called a C^r *n-dimensional non-Hausdorff differential manifold.*

Next, in the degenerate case $r = 0$, we speak of a C^0 *n*-dimensional generalized (Hausdorff / non-Hausdorff) *topological* manifold.[7]

Finally, given a C^r *n*-dimensional generalized differential manifold $\langle M, A \rangle$, we say that the atlas A induces a topology $\mathscr{T}$ on the set M, which is given by the condition: $O \in \mathscr{T}$ iff for all $x \in O$ there is $\langle u_\gamma, \varphi_\gamma \rangle \in A$ such that $x \in u_\gamma$.

In what follows, if confusion is unlikely, we omit the qualifications "C^r, *n*-dimensional, differential", and just write "generalized" (or Hausdorff or non-Hausdorff) d-manifold, with "d" for "differential". By the above definitions, a generalized (or Hausdorff or non-Hausdorff) d-manifold counts as a generalized (or Hausdorff or non-Hausdorff) *topological* manifold as well.

9.2.2 Differential manifolds and MBSs

We are now going to make good on our informal observations about the topological features of MBSs, by proving some more general facts. The first fact says that any B-history admits a Hausdorff differential manifold structure, and hence, a Hausdorff topological manifold structure as well.

Fact 9.10. *Let h_σ be a B-history in the BST$_{NF}$ structure$\langle B, <_R \rangle$ determined by an MBS. Then h_σ admits a C^∞ 4-dimensional Hausdorff d-manifold structure.*

Proof. Define b-balls as $o_\sigma =_{df} \{[\sigma x] \mid x \in o\}$, where o is an open ball in $\mathbb{R}^4$. For a b-ball o_σ, define $\varphi_\sigma : h_\sigma \mapsto \mathbb{R}^4$ such that $\varphi_\sigma([\sigma x]) =_{df} x$. Note next that the composition $\varphi_\sigma \circ \varphi_\sigma^{-1}$ is the identity function on $\mathbb{R}^4$, which is differentiable to an arbitrarily large degree. Thus, it follows immediately that $A =_{df} \{\langle o_\sigma, \varphi_\sigma \rangle \mid o$ an open ball in $\mathbb{R}^4\}$ is a C^∞ 4-dimensional atlas on h_σ. Thus, $\langle h_\sigma, A \rangle$ is a C^∞ 4-dimensional d-manifold, which is moreover Hausdorff: For distinct $[\sigma x], [\sigma y] \in h_\sigma$ there obviously are non-overlapping open balls $o^x, o^y \subseteq \mathbb{R}^4$ centered at x and y, respectively. These open balls determine non-overlapping b-balls centered at $[\sigma x]$ and $[\sigma y]$, o_σ^x and o_σ^y, respectively. □

---

[7] This is equivalent to the more typical definition of a topological manifold (Hausdorff or non-Hausdorff) as a locally Euclidean topological space (satisfying or not satisfying the Hausdorff condition).

Our next Fact concerns the base set of a BST_{NF} structure derived from an MBS. It says that the base set admits a 4-dimensional generalized differential manifold structure. Further, the manifold has to be non-Hausdorff if the structure comprises more than one B-history.

Fact 9.11. *Let* $\langle B, <_R \rangle$ *be the* BST_{NF} *structure determined by an MBS* $\langle \Sigma, F, P \rangle$. *Then* B *admits a* C^∞ *4-dimensional generalized (possibly non-Hausdorff) d-manifold structure.*

Proof. Define a b-ball as before: $o_\sigma = \{[\sigma x] \mid x \in o\}$, where o is an open ball in $\mathbb{R}^4$ and $\sigma \in \Sigma$. For each open ball o and each $\sigma \in \Sigma$ define the function φ_σ : $o_\sigma \mapsto \mathbb{R}^4$ such that $\varphi_\sigma([\sigma x]) = x$. Next, consider intersections of the form $o_\sigma \cap o'_\alpha$, which are equal to $\{[\sigma x] \mid x \in o \cap o' \wedge \neg \exists s \in S_{\sigma\alpha}\, s \leqslant_M x\}$. Since the defining condition of this set picks an open subset of $\mathbb{R}^4$, clearly $\varphi_\sigma(o_\sigma \cap o'_\alpha)$ and $\varphi_\alpha(o_\sigma \cap o'_\alpha)$ are open (this holds even if the intersection is empty). Note, as above, that the composition of φ_σ and φ_α is the identity function on an appropriate domain in $\mathbb{R}^4$ (which may be empty), so it is differentiable to an arbitrarily large degree. Thus, the set $\{\langle o_\sigma, \varphi_\sigma \rangle \mid \sigma \in \Sigma, o \in \mathscr{B}^4\}$, with $\mathscr{B}^4$ the set of open balls in $\mathbb{R}^4$, induces a maximal C^∞ 4-dim atlas A on B. Thus, $\langle B, A \rangle$ is a C^∞ 4-dim generalized d-manifold.

If B comprises two B-histories, say h_σ and h_α ($\sigma, \alpha \in \Sigma$), then $\langle B, A \rangle$ is not Hausdorff, however. There is then a splitting point $s \in S_{\sigma\alpha}$, so that $[\sigma s] \neq [\alpha s]$. Then any b-ball centered at $[\sigma s]$ and any b-ball centered at $[\alpha s]$ overlap non-emptily, because $[\sigma s]$ and $[\alpha s]$ share the same proper past.   □

We next show, for $\langle B, <_R \rangle$ derived from an MBS, that each B-history in this structure is a maximal subset of B that admits a connected Hausdorff d-submanifold structure; this manifold is a sub-manifold of the generalized d-manifold admitted by B.

Fact 9.12. *Let* $\langle B, <_R \rangle$ *be the* BST_{NF} *structure determined by an MBS* $\langle \Sigma, F, P \rangle$. *Then for every B-history* h_σ *in* B, $\langle h_\sigma, A^\sigma \rangle$ *is a maximal connected Hausdorff d-submanifold of the generalized d-manifold* $\langle B, A \rangle$, *where* $\mathscr{B}^4$ *is the set of open balls in* $\mathbb{R}^4$, *the maximal atlas* A *is induced by* $\{\langle o_\gamma, \varphi_\gamma \rangle \mid \gamma \in \Sigma, o \in \mathscr{B}^4\}$, *and the maximal atlas* A^σ *is induced by* $\{\langle o_\sigma, \varphi_\sigma \rangle \mid o \in \mathscr{B}^4\}$.

Proof. Observe first that $A^\sigma \subseteq A$, and that the inclusion is strict if Σ has more than one element; otherwise the two d-manifolds are identical. It is also immediate to see that h_σ is open and connected in the manifold topology induced on B by A (see Exercise 9.5). Suppose thus that $A^\sigma \subsetneq A$, and assume for reductio that $\langle h_\sigma, A^\sigma \rangle$ is not maximal; that is, there is an set $g \subseteq B$ endowed with topology induced by atlas A, so that g is an embedded sub-manifold in the manifold on A. Assume further that the topology on g is Hausdorff and connected, and that $h_\sigma \subsetneq g$. By definition (Lee, 2012, p. 99) g has no boundary in the sub-manifold topology. One immediately notes that in that topology h_σ is open. Thus, since the topology on g is connected, the boundary of $d =_{\mathrm{df}} g \setminus h_\sigma$, ∂d, is non-empty. Consider now some $e \in \partial d$. Then for any b-ball $o_\beta \subseteq g$, if $e \in o_\beta$, then $o_\beta \cap h_\sigma \neq \emptyset$. Since $e \notin h_\sigma$, it must be that $e = [\alpha x]$ for some $\alpha \in \Sigma$, $\alpha \neq \sigma$. Hence there must be some $s \in S_{\sigma\alpha}$ such that $s \leqslant x$. Moreover, it is impossible that $s < x$, because then the distance between s and x would be non-zero, so there would be a b-ball o_α such that $[\alpha x] \in o_\alpha$ and $o_\alpha \cap h_\sigma = \emptyset$. Thus, $x = s$. Observe next that $[\sigma s], [\alpha s] \in g$. We then argue, like at the end of the proof of Fact 9.11, that $[\sigma s]$ and $[\alpha s]$ witness a failure of the Hausdorff property in the topology on g that is induced by atlas A on B. Thus, g is not a connected and Hausdorff submanifold of $\langle B, A \rangle$, which contradicts our reductio hypothesis. Combined with Fact 9.10, this implies that any B-history is a maximal subset of the base set B, the topology on which is connected and Hausdorff; the history with the endowed differential structure is thus a maximal Hausdorff sub-manifold embedded in a generalized d-manifold on set B. □

Our next Fact concerns a relation between the diamond topology (see Def. 4.14) and the topology induced by a differential manifold. In Section 4.4.1 we advertised the diamond topology as a natural topology for BST, and wrote that one argument for naturalness is that this topology, if appropriately restricted, coincides with the standard open-ball topology on $\mathbb{R}^n$. The fact below says that this is indeed so: the manifold topology $\mathscr{T}^A$ induced by atlas A on history h_σ and the diamond topology $\mathscr{T}_{h_\sigma}$ on h_σ are identical. It follows that the diamond topology on a B-history is homeomorphic to the open ball topology on $\mathbb{R}^4$.

Fact 9.13. *Let h_σ be a B-history in $\langle B, <_R \rangle$ that is the BST$_{NF}$ structure determined by an MBS $\langle \Sigma, F, P \rangle$ and $\langle h_\sigma, A \rangle$ be a Hausdorff d-manifold on h_σ. Then the manifold topology $\mathscr{T}^A$ induced by atlas A on h_σ and the diamond topology $\mathscr{T}_{h_\sigma}$ on h_σ are identical.*

Proof. Clearly, the two topologies $\mathscr{T}^A$ and $\mathscr{T}_{h_\sigma}$ have the same base set, h_σ. We need to see whether $Z \in \mathscr{T}^A$ iff $Z \in \mathscr{T}_{h_\sigma}$. To argue in the left to right direction, let $[\sigma x] \in Z \in \mathscr{T}^A$, so for some b-ball u_o: $[\sigma x] \in u_o \subseteq Z$, where o is an open ball in $\mathbb{R}^4$. Consider then the set $MC_{\langle \mathbb{R}^4, <_M \rangle}(x)$ of maximal chains in $\langle \mathbb{R}^4, <_M \rangle$ that contain x. By properties of real numbers, for every $t \in MC_{\langle \mathbb{R}^4, <_M \rangle}(x)$ there are $x_1, x_2 \in t \cap o$ such that $x_1 < x < x_2$, so the diamond $\tilde{D}_{x_1, x_2} \subseteq o$. As every diamond $\tilde{D}_{x_1, x_2}$ in $\langle \mathbb{R}^4, <_M \rangle$ determines a unique diamond $D_{[\sigma x_1], [\sigma x_2]}$ in $\langle h_\sigma, <_R \rangle$, we get that for every $t \in MC_{\langle h_\sigma, <_R \rangle}([\sigma x])$ there are $[\sigma x_1], [\sigma x_2] \in t$ such that $[\sigma x_1] < [\sigma x] < [\sigma x_2]$ and the diamond $D_{[\sigma x_1], [\sigma x_2]} \subseteq u_o \subseteq Z$, which proves $Z \in \mathscr{T}_{h_\sigma}$. In the opposite direction, $Z \in \mathscr{T}_{h_\sigma}$ means that for every $t \in MC_{\langle h_\sigma, <_R \rangle}([\sigma x])$ there are $[\sigma x_1], [\sigma x_2] \in t$ such that $[\sigma x_1] < [\sigma x] < [\sigma x_2]$ and the diamond $D_{[\sigma x_1], [\sigma x]} \subseteq Z$. This implies in particular that for the time-like chain $t^* \in MC_{\langle h_\sigma, <_R \rangle}$ there are $[\sigma x_1^*], [\sigma x_2^*] \in t^*$ such that $[\sigma x_1^*] < [\sigma x] < [\sigma x_2^*]$ and the diamond $D_{[\sigma x_1^*], [\sigma x^*]} \subseteq Z$. Diamond $D_{[\sigma x_1^*], [\sigma x_2^*]}$ determines diamond $\tilde{D}_{x_1^*, x_2^*}$ in $\langle \mathbb{R}^4, <_M \rangle$; thanks to time-likeness of t^*, there is an open ball $o \subseteq \tilde{D}_{x_1^*, x_2^*}$, with $x \in o$. Accordingly there is an associated b-ball u_o such that $u_o \subseteq D_{[\sigma x_1^*], [\sigma x_2^*]} \subseteq Z$, with $[\sigma x] \in u_o$. This proves $Z \in \mathscr{T}^A$.   □

The very welcome message of this section is that the base set of a structure derived from an MBS admits a generalized (typically non-Hausdorff) d-manifold structure, whereas each B-history in this structure comes out as an embedded sub-manifold of the above manifold that is maximal with respect to having a connected and Hausdorff topology. This result ties in neatly with the situation in General Relativity, in which space-times are standardly identified with Hausdorff d-manifolds, but larger (generalized) d-manifolds are constructible as well. Before we explore this affinity, we need to take a look at BST$_{NF}$ structures generally: Do they admit a d-manifold structure?

9.2.3 Differential manifolds and BST$_{NF}$, generally

Do the above results concerning MBSs carry over to BST$_{NF}$ structures generally? Given the frugality of the BST$_{NF}$ postulates, one should not expect this to be the case. Indeed, it is already problematic whether the base set of a BST$_{NF}$ structure admits a d-manifold structure. One might try to find constraints on BST$_{NF}$ structures that would ensure the admittance of a d-manifold structure on their base sets. The general case is, however, again unwieldy. A more promising line of enquiry is to see whether "nice" topological properties of histories in a BST$_{NF}$ structure carry over to the structure itself. To some extent this is satisfied, as Theorem 4.1 testifies. It says that, for any BST$_{NF}$ structure $\mathscr{W} = \langle W, < \rangle$, if there is an $n \in \mathbb{N}$ such that every history $h \in \text{Hist}(\mathscr{W})$ admits a generalized topological manifold structure $\langle h, A_h \rangle$ with dimension n, then W admits a generalized topological manifold structure $\langle W, A_W \rangle$ with dimension n as well. In short, we have "local Euclidicity in, local Euclidicity out" for the topological manifold structure admitted by BST_{NF} histories. But General Relativity needs differential manifolds to model space-times, rather than more frugal topological manifolds, which do not come with a differential structure. So a natural question is whether there is a generalization of Theorem 4.1 to C^r generalized d-manifolds with $r > 0$. Unfortunately, by simply putting together d-manifold structures admitted by individual BST_{NF} histories, we will not produce a (generalized) d-manifold that has as its base set the resulting BST_{NF} structure. This is a consequence of the fact that charts from atlases belonging to *different* d-manifolds need not properly combine in the way required by Def. 9.9. To see this, consider the BST_{NF} structure outlined at the bottom of Figure 3.1, with two histories $h_i = \{[\langle x, i \rangle] \mid x \in \mathbb{R}\}$, where $i = 1, 2$, with topologies given by the open intervals $u_{x_1,x_2} = \{[\langle x, 1 \rangle] \mid x_1 < x < x_2\}$ and $v_{x_1,x_2} = \{[\langle x, 2 \rangle] \mid x_1 < x < x_2\}$ $(x_1, x_2 \in \mathbb{R}, x_1 < x_2)$. Let the atlas of the manifold based on h_1 be induced by $\{\langle u_{x_1,x_2}, \varphi_{x_1,x_2} \rangle \mid x_1, x_2 \in \mathbb{R}, x_1 < x_2\}$, where $\varphi_{x_1,x_2} : u_{x_1,x_2} \mapsto \mathbb{R}$ such that $\varphi([\langle x, 1 \rangle]) = x$. Similarly, let the atlas of the manifold based on h_2 be induced by $\{\langle v_{x_1,x_2}, \psi_{x_1,x_2} \rangle \mid x_1, x_2 \in \mathbb{R}, x_1 < x_2\}$, where $\psi_{x_1,x_2} : v_{x_1,x_2} \mapsto \mathbb{R}$ such that $\varphi([\langle x, 2 \rangle]) = x^{1/3}$. Now, the charts from the atlas of the topology on h_1 properly combine, as do the charts from the atlas of the topology on h_1. In each case, the composite functions, $\varphi_{y_1,y_2} \circ \varphi_{x_1,x_2}^{-1}$ and $\psi_{y_1,y_2} \circ \psi_{x_1,x_2}^{-1}$ (if defined), are the identity functions on their domains, which are differentiable to an arbitrary high order. Thus, each d-manifold, on h_1 and on h_2, is C^∞. However, two charts, each from the atlas of a different topology, might fail to properly combine. Pick any two charts u_{x_1,x_2} and v_{x_1,x_2} such that $x_1 < 0 < x_2$. Then $\psi_{x_1,x_2} \circ \varphi_{x_1,x_2}^{-1} : (x_1, x_2) \mapsto \mathbb{R}$ is given by $(\psi_{x_1,x_2} \circ \varphi_{x_1,x_2}^{-1})(x) = x^{1/3}$, which is C^0 but no C^1 on (x_1, x_2). And such charts cannot be removed from the atlases, as they are needed for the distinct elements, $[\langle 0, 1 \rangle]$ and $[\langle 0, 2 \rangle]$, which form a choice set in the structure considered. Thus, although histories admit C^∞ d-manifold, the atlases of these manifolds do not produce a C^r atlas on the base set of the whole structure for any $r > 0$.

This illustration suggests a different approach, however, as one might choose "nicer" functions on the domains of charts in the topology on h_2. Generally, one might try to take advantage of Theorem 4.1. The idea is to begin with a collection of BST_{NF} histories, each of which admits a

C^r d-manifold of the same dimension. Since a C^r d-manifold counts as a topological manifold as well, by the theorem the base set of the whole BST_{NF} structure admits a *topological* manifold. To recall, the charts in the atlas of a topological manifold are only required to be C^0. One might thus hope that it is possible to smoothen these charts somewhat such as to make them C^r, for $r > 0$.[8] The question is thus whether a differential structure can be assigned to any topological manifold. The answer is that this is doable for many topological manifolds, but not for all. There are topological manifolds that do not admit a C^r atlas for any $r > 0$, as was proved by Kervaire (1960). Thus, to sum up, we do not know any (and there might be no) general method of obtaining a d-manifold admissible by a BST_{NF} structure from d-manifolds admissible by histories of that structure, but the smoothing approach can work in many cases.

The fact that the elegant results concerning MBSs and d-manifolds do not carry over to all BST_{NF} structures should not be seen as disconcerting, however. This is to be expected given the frugality of the BST_{NF} postulates. The important question is whether those d-manifolds and generalized (non-Hausdorff) manifolds that occur in the GR literature could be read as d-manifolds and generalized d-manifolds admissible by BST_{NF} histories and BST_{NF} structures, respectively. The results concerning MBSs and d-manifolds suggest a picture in which each space-time is a maximal connected Hausdorff d-submanifold within a generalized d-manifold, representing all space-times, such that any two of them share some initial segment.

Interestingly, this picture is similar to some situations in GR in which there is a failure of the initial value problem, to be discussed at length in Section 9.3.1. In these cases one has maximal Hausdorff manifolds, as well as larger non-Hausdorff manifolds. In the physics literature, the former are standardly interpreted as GR space-times, but the latter are problematic; i.e., their physical interpretation is unclear. Our suggestion is to read these non-Hausdorff d-manifolds very much like non-Hausdorff d-manifolds in MBS contexts, that is, as the representations of multiple alternative spatio-temporal histories, with each such history given by a maximal Hausdorff d-submanifold. Before we turn to this topic, however, we need to recall some notions of General Relativity.

---

[8] More precisely, it is the compositions of functions from different charts, like $\psi \circ \varphi^{-1}$, that are required to be C^r, if their domains are non-empty. Note also that the crucial step is from topological manifolds to C^1 d-manifolds, as a C^1 manifold can always be transformed into a C^r manifold, for any $r > 1$; see Hirsch (1976, Theorem 2.9).

9.2.4 Differential manifolds in GR

We now review some concepts and terminology needed for GR.[9] Standardly, an n-dimensional GR space-time is identified with a pair $\langle M, g_{ab} \rangle$, where M is a connected n-dimensional C^∞ Hausdorff d-manifold (without boundary) and g is a smooth, non-degenerate, pseudo-Riemannian metric of Lorentz signature $(-, +, \ldots, +)$ defined on M (aka a Lorentzian metric). To be explicit about the atlas, A_W, we sometimes write a GR space-time as $\langle W, A_W, g \rangle$, where $M = \langle W, A_W \rangle$ is a connected n-dimensional C^∞ Hausdorff d-manifold (without boundary). Two space-times $\langle M, g \rangle$ and $\langle M', g' \rangle$ are defined to be *isometric* if there is a diffeomorphism (smooth bijection) $\varphi : M \to M'$ such that the induced pull-back function φ^* satisfies $\varphi^*(g') = g$. A space-time $\langle M', g' \rangle$ is an *extension* of the space-time $\langle M, g \rangle$ if there exists an embedding $\Lambda : M \mapsto M'$ (i.e., Λ is a diffeomorphism onto its image) and $\Lambda^*(g' \mid_{\Lambda(M)}) = g$ and $\Lambda(M) \neq M'$. A space-time is *maximal* iff it has no extension.

With each point $p \in M$ there is associated a vector space, called the *tangent space*, M_p, on which g induces a cone structure, so that each vector $\xi^a \in M_p$ is either *timelike*, or *null*, or *spacelike*, depending on whether $\xi^a \xi^b g_{ab}$ is positive, zero, or negative, respectively. Here the superscripts a and b are abstract indexes, indicating that the object is a covariant vector. Time-orientable space-times permit a distinction between future and past lobes of light-cones; technically, a time-orientable space-time has a continuous timelike vector field on "its" manifold.

A continuous curve $\gamma : I \to M$ (where I is an interval of $\mathbb{R}$) is timelike (resp., spacelike, or null) iff its tangent vector ξ^a at each point in $\gamma[I]$ is timelike (resp., spacelike, or null). A curve is *causal* iff its tangent vector at each point is either null or timelike. A curve is *inextendible* iff it has no endpoints. A *geodesic* in a space-time $\langle M, g_{ab} \rangle$ is a curve $\gamma : I \to M$ that satisfies, for every vector $\xi^a \in M_p$ ($p \in \gamma[I]$) tangent to the curve, the geodesics equation: $\xi^a \nabla_a \xi^b = 0$, where ∇_a is the (unique) derivative operator compatible with g_{ab}. For any set $S \subseteq M$, the *domain of dependence* of S, written $D(S)$, is the set of points $p \in M$ such that every inextendible causal curve through p intersects S. S is an achronal subset of M iff no two points in S can be joined by a timelike curve. A Cauchy surface in $\langle M, g_{ab} \rangle$ is a smooth and achronal spacelike hypersurface such that $D(S) = M$.

---

[9] Explanations of the mathematics of GR can be found in mathematically oriented books on GR, such as Malament (2012, Chs. 1–2).

9.3 GR space-times

General Relativity is currently the best theory of space-time and matter. There are a variety of GR models, aka GR space-times, making the BST analysis of GR space-times rather complex. Yet, a fairly large class of GR-space-times, viz., time-orientable space-times that do not end abruptly and that do not contain closed causal curves, is easily amenable to a BST analysis (although, to be fair, some metric information provided by a GR space-time is not present in a bare BST structure). We put down this observation as a fact:

Fact 9.14. *Let $\langle M, g \rangle$ be a GR space-time that is (1) time-orientable, (2) without closed causal curves, and (3) subject to the condition that for any $x, y \in M$ there is $z \in M$ that is reachable from x and from y by future-directed causal curves. Then $\langle M, g \rangle$ induces a one-history BST_{NF} structure $\mathscr{W} = \langle M, < \rangle$, where for $e_1, e_2 \in M$, $e_1 \neq e_2$, we set $e_1 < e_2$ iff there is a continuous future-directed causal curve from e_1 to e_2. Being a one-history structure, that structure is also a BST_{92} structure.*

Proof. (Sketch) Asymmetry of $<$ comes from time-orientability (1) and the absence of closed causal curves (2). Transitivity of $\prec$ results from the composition of causal curves (a method of getting a continuous causal curve from piece-meal continuous causal curve is needed (see, e.g., Chruściel 2011, sec. 2). M is directed by (3), so it is a single BST history. The postulate of history-relative suprema thus simplifies to the condition of suprema simpliciter for upper bounded chains in $\langle M, < \rangle$. Both the infima postulate and the suprema postulate follow from the definition of causal curves. PCP_{92}, PCP_{NF}, and Weiner's postulate are vacuously satisfied since the structure has only one history. $\square$

This result, of course, only constitutes a first step, as it does not touch upon the interesting question of whether BST can be used to model indeterminism occurring in a general relativistic world. Some GR models have indeterministic features. One indication of indeterminism, which is interesting from a BST perspective, arises in the context of the initial value problem (IVP). A failure of the IVP means, generally, that a theory's evolution equations allow for multiple global solutions that coincide over some region. If the region of coincidence is "nice", one may read a failure of IVP as the existence of multiple evolutions of a given system that develop from that common region. The picture of branching histories then naturally springs to mind. Accordingly, in our attempt to relate BST to GR, we begin with an overview of the initial value problem in GR: if there is a motivation for a non-trivial BST structure consisting of GR space-times, this motivation should come from a failure of the IVP in General Relativity.

In the next stage we will be concerned with two interconnected problems in our attempts to analyze GR space-times from a BST perspective. The first is that of GR space-times with closed causal curves. Given BST's reliance on an asymmetric ordering, such GR space-times cannot be modeled as BST structures, as long as the BST ordering is defined via causal curves. We will indicate how to resolve this problem, arguing that a BST-style theory with a *locally* asymmetric ordering (but not necessarily a globally asymmetric ordering on the whole base set) is available and is in line with the topological features of GR space-times—see Section 9.3.6. Conceptually more demanding is the second problem, which concerns the kind of indeterminism that might arise from failures of the IVP in GR. Recall that BST captures local indeterminism, the key idea being that a local entity, like a point-like event, has alternative possible futures. It might happen (the issue is not fully clear) that a failure of the IVP does not deliver local indeterminism: it could produce alternative developments of a region of GR space-time without there being an indeterministic trajectory of any entity (i.e., a trajectory that splits), with each continuation leading into a different possible development. In other words, it might be that each failure of the IVP produces a case of global indeterminism without there being local indeterminism. The big question is, of course, whether that odd combination reflects some incompleteness on the part GR, like its failure to accommodate quantum phenomena, or, alternatively, if it discloses a feature of our physical world.

The two problems, the existence of closed causal curves and the odd kind of indeterminism coming from the violation of the IVP, are related: the known cases of GR space-times that harbor indeterminism contain such curves or other causal anomalies. Furthermore, Theorem 2 of Clarke (1976) implies that a non-Hausdorff manifold with a Lorentzian metric has bifurcating curves (of the second kind – see below) or violates strong causality.[10] We would welcome bifurcating causal curves as they are very much needed

---

[10] The violation of strong causality means that, although a causal curve does not intersect itself, it comes arbitrarily close to intersecting itself.

in order to read a non-Hausdorff manifold as a BST structure with multiple histories (see Section 9.3.4). However, the data that we review below suggests that there might be no bifurcating causal curves in non-Hausdorff manifolds which are naturally constructible in GR (see Section 9.3.2). By Clarke's theorem then, such manifolds harbor violations of strong causality. Thus, in the context of the violation of IVP, causal anomalies might be inevitable. Our approach gives reasons for being optimistic with respect to accommodating closed causal curves in BST, and a violation of strong causality without closed causal curves is not problematic from a BST perspective. Therefore, in what follows we focus on what we believe to be more problematic for BST: the global variety of indeterminism in GR coming from a failure of the IVP. Accordingly, we now provide a short overview of the IVP in General Relativity.

9.3.1 The initial value problem in GR

The question of determinism presupposes the notion of a system evolving in time. This latter notion is not always well-defined in general relativity, as a GR space-time need not come with a distinguished time coordinate. Yet, a somewhat similar issue can be considered in GR: suppose we are given a 3-dimensional space Σ with possibly some data on it. (Technically, this should be a manifold with a metric "appropriate for space"; that is, i.e., a Riemannian metric.) The question now is: Can this space be uniquely extended to a 4-dimensional space-time–that is, to a manifold with a Lorentzian metric that satisfies the properties required from a manifold representing a GR space-time (listed in Chapter 9.2.4)–and in which the Einstein field equations (EFE) hold? The answer depends crucially on the kind of data assumed on the space and on the properties the sought-for space-time is supposed to have. As for the latter, one relevant factor is the existence of a matter field, and (if it is assumed to exist) the kind of model for the matter field; another issue is the value of the cosmological constant in the EFE. However, given the extension problem we will consider, a simple case is enough for our purposes. We focus on space-times with a vanishing Ricci tensor, the so-called vacuum space-times, and we consider the EFE without the cosmological constant. A satisfactory data set for this case consists of a Riemannian metric $\tilde{g}$ and a symmetric covariant 2-tensor $\tilde{k}$ that represents incremental changes of the metric in the direction normal to Σ. In this case the initial value problem amounts to constructing a 4-dimensional manifold M with a Lorentzian metric g and an embedding $i : \Sigma \to M$ such that if k is the second fundamental form on $i(\Sigma) \subseteq M$, then $i^*(g) = \tilde{g}$ and $i^*(k) = \tilde{k}$, where i^* is the pull-back function induced by the embedding i. Further, there is a set of equations relating $\tilde{g}$ and $\tilde{k}$, known as (vacuum) constraint equations, which guarantee the satisfaction of the EFE in the sought-for space-time. A space Σ with tensors $\tilde{g}$ and $\tilde{k}$ that satisfy the (vacuum) constraint equations is said to form a (vacuum) initial data set $\langle \Sigma, \tilde{g}, \tilde{k} \rangle$.

A result that is highly relevant to the initial value problem in the vacuum case was obtained by Choquet-Bruhat and Geroch (1969) in the context of globally hyperbolic space-times. Such space-times have particularly nice causal properties. To recall the definition, $\langle M, g_{ab} \rangle$ is said to be *globally hyperbolic* iff there is an achronal subset $S \subseteq M$ whose domain of dependence is the whole space-time (see Wald, 1984, Ch. 8). One consequence of this definition is that a globally hyperbolic space-time can be foliated by Cauchy surfaces (although the foliation is non-unique). Choquet-Bruhat and Geroch restrict their attention to globally hyperbolic space-times $\langle M, g \rangle$ that (1) are vacuum solutions to the EFE and which can be developed from a given vacuum initial data set such that (2) the image of the space Σ from that data set under the development embedding is a Cauchy surface in $\langle M, g \rangle$. A space-time satisfying these conditions is called a "vacuum Cauchy development" (VCD) of the initial data set. Note that condition (2) implies that a VCD is a globally hyperbolic space-time. The theorem proved by Choquet-Bruhat and Geroch says:

Theorem 9.2. *Let $\langle \Sigma, \tilde{g}, \tilde{k} \rangle$ be an initial vacuum data set. Then there is a unique, up to isometry, maximal VCD $\langle M, g \rangle$ of $\langle \Sigma, \tilde{g}, \tilde{k} \rangle$.*

The phrase "unique, up to isometry, maximal VCD" means that if there is another maximal VCD $\langle M', g' \rangle$ of the same initial data set, then there is a time-orientation preserving isometry $\varphi : M \to M'$. Thus, taking isometry to amount to the physical identity of vacuum space-times of GR (which is a typical move), the result ensures the uniqueness of maximal *globally hyperbolic* space-times compatible with vacuum initial data sets.

It is important to note that the theorem concerns globally hyperbolic developments only: it puts no restrictions on *other* developments of an initial vacuum data set. This raises the question of whether a maximal globally hyperbolic development of an initial data set can be further extended (where, of course, the resulting extension cannot be globally hyperbolic). Here, controversial questions of the physicality of such extensions become important. Some parties to the debate exclude non-globally hyperbolic space-times, on the basis that they involve causal anomalies, like closed causal curves, and these might be unphysical.[11] In recent research on the initial value problem in GR, some authors hold that non-globally hyperbolic developments of initial data sets are *rare*, in some measure-theoretical sense, with respect to a measure defined on the space of relevant solutions to the EFE. A view that is gaining ground is that "for generic initial data to Einstein's equations, the maximal globally hyperbolic development has no extension" (Ringström, 2009, p. 188). Without entering into the voluminous debate here, we nevertheless investigate here non-globally hyperbolic solutions, even if they are non-generic or rare.

To sum up, the Choquet-Bruhat and Geroch theorem has the consequence that evidence for the indeterminism of GR (if there is such evidence) in the vacuum case must consist of multiple non-isometric *extensions* of a maximal globally hyperbolic vacuum space-time. The question is, therefore, whether some maximal globally hyperbolic space-times (satisfying the EFE) have multiple non-isometric extensions (satisfying the EFE). The next section provides a positive answer to this question.

9.3.2 An example of the failure of the IVP: Non-isometric extensions of Taub space-time

We will describe below the construction of multiple non-isometric extensions of Taub space-time. Our discussion is based on a paper by Chruściel and Isenberg (1993), which also investigates a more realistic class of space-times, polarized Gowdy space-times, that also have multiple non-isometric extensions. But since these are mathematically more demanding, we describe here the simpler case of extensions of Taub space-time.

Taub space-time is a vacuum solution to the EFE. The manifold is $M = (t_-, t_+) \times S^3$, and the metric g is given by

$$ds^2 = -U^{-1}dt^2 + (2l)^2 U(d\psi + \cos\Theta d\varphi)^2 + (t^2 + l^2)(d\Theta^2 + \sin^2\Theta d\varphi^2),$$

---

[11] The considerable controversy over how to interpret Theorem 9.2 is related to the idea of cosmic censorship due to Penrose (1969), and especially to its later formulation in terms of the so-called Strong Cosmic Censorship Conjecture. That conjecture says (very roughly) that space-times that are not globally hyperbolic are unphysical.

where m and l are real positive constants, Θ, φ and ψ are Euler coordinates on the 3-sphere S^3, and

$$U(t) = \frac{(t_- - t)(t - t_+)}{l^2 + t^2}, \quad \text{where} \quad t_\pm = m \pm (m^2 + l^2)^{1/2}.$$

Note that $U(t_\pm) = 0$, and hence the metric is not defined at $t_\pm$. Taub space-time is globally hyperbolic, and maximally so, the Cauchy surfaces being identified by the condition $t = const$ for $t \in (t_-, t_+)$. As Newman, Tamburino and Unti (1963) showed, by using appropriate coordinate transformations $\langle M, g \rangle$ can be extended above t_+, the result being two non-hyperbolic space-times $\langle M^{\uparrow+}, g^{\uparrow+} \rangle$ and $\langle M^{\uparrow-}, g^{\uparrow-} \rangle$, known as Taub-NUT space-times. In a similar vein, Taub space-time can be extended below t_- into two non-hyperbolic space-times $\langle M^{\downarrow+}, g^{\downarrow+} \rangle$ and $\langle M^{\downarrow-}, g^{\downarrow-} \rangle$. Each of $\langle M^{\uparrow+}, g^{\uparrow+} \rangle$, $\langle M^{\uparrow-}, g^{\uparrow-} \rangle$, $\langle M^{\downarrow+}, g^{\downarrow+} \rangle$, and $\langle M^{\downarrow-}, g^{\downarrow-} \rangle$ satisfies the EFE and contains closed causal curves in the region new with respect to M.[12] As shown by Chruściel and Isenberg (1993), the pair $M^{\uparrow+}, M^{\uparrow-}$ and the pair $M^{\downarrow+}, M^{\downarrow-}$ are isometric.

To produce non-isometric extensions of Taub space-time, we need to glue together an upward extension together with a downward extension of Taub space-time. "Gluing" means, in mathematical parlance, finding an equivalence relation $\equiv$ on the union of two manifolds, say $M^{\downarrow+} \cup M^{\uparrow-}$, and then taking the set of equivalence classes with respect to this equivalence relation. The result is the quotient structure $(M^{\downarrow+} \cup M^{\uparrow-})/\equiv$.

Consider now four results of the gluing (for the equivalence relation used, consult Chruściel and Isenberg, 1993, p. 1619):

$$M^{xy} = (M^{\downarrow x} \cup M^{\uparrow y})/\equiv, \text{ where } x, y \in \{-, +\},$$

each result being associated to the metric g^{xy}, defined in terms of $g^{\uparrow x}$ and $g^{\downarrow y}$. Each $\langle M^{xy}, g^{xy} \rangle$ is a non-hyperbolic extension of Taub space-time $\langle M, g \rangle$ and satisfies the EFE. As for isometries, there are the following results (Chruściel and Isenberg, 1993, Theorem 3.1):

1. $\langle M^{+-}, g^{+-} \rangle$ is isometric to $\langle M^{-+}, g^{-+} \rangle$.
2. $\langle M^{++}, g^{++} \rangle$ is isometric to $\langle M^{--}, g^{--} \rangle$;

---

[12] See Misner and Taub (1969). The existence of closed causal curves raises the worry of the applicability of BST to such space-times; we address this concern in Section 9.3.6.

3. yet, $\langle M^{--}, g^{--} \rangle$ is not isometric to $\langle M^{-+}, g^{-+} \rangle$, and

4. $\langle M^{++}, g^{++} \rangle$ is not isometric to $\langle M^{+-}, g^{+-} \rangle$.

Each pair of the non-isometric extensions of Taub space-time above provide evidence for indeterminism in the sense of Butterfield's (1989) definition of determinism, which is tailored to applications to GR. It says:

Definition 9.10 (Butterfield's definition of determinism). A theory with models $\langle M, O_i \rangle$ is **S**-deterministic, where **S** is a kind of region that occurs in manifolds of the kind occurring in the models, iff:
given any two models $\langle M, O_i \rangle$ and $\langle M', O'_i \rangle$ containing regions S, S' of kind **S** respectively, and any diffeomorphism α from S onto S':
if $\alpha^*(O_i) = O'_i$ on $\alpha(S) = S'$, then there is an isomorphism β from M onto M' that sends S to S', i.e., $\beta^*(O_i) = O'_i$ throughout M' and $\beta(S) = S'$.

Here the O_i stand for geometric object fields that are either definable in terms of a space-time's metric, or which characterize the matter field of the space-time. In our (vacuum) case the definition simplifies considerably, since in the absence of objects not definable in terms of the metric, the notion of isomorphy coincides with that of isometry, thus β can be an isometry and the condition on α^* above concerns only objects defined in terms of the metric.

To check that the definition yields the verdict of the indeterminism of GR, note that the space-time $\langle M^{++}, g^{++} \rangle$ contains the region $S^{++} = \Lambda^{++}[M]$, and the space-time $\langle M^{+-}, g^{+-} \rangle$ contains the region $S^{+-} = \Lambda^{+-}[M]$, where $\Lambda^{xy} : M \rightarrow M^{xy}$ is an embedding, which ensures that $\langle M^{xy}, g^{xy} \rangle$ is an extension of $\langle M, g \rangle$. For the diffeomorphism α we take $\alpha = \Lambda^{+-} \circ (\Lambda^{++})^{-1} :$ $S^{++} \rightarrow S^{+-}$. Then the push-forward α^* induced by α satisfies $\alpha^*_{|S^{++}}(g^{++}) = g^{+-}$ (by the condition on embedding), and hence $\alpha^*(O_i) = O'_i$ for any object field defined in terms of the metric. On the other hand, however, $\langle M^{++}, g^{++} \rangle$ and $\langle M^{+-}, g^{+-} \rangle$ are not isometric, according to Chruściel and Isenberg's result quoted earlier. It follows that GR is indeterministic in the sense of Def. 9.10.

Physicists and philosophers alike have argued that non-isometric extensions of a globally hyperbolic space-time provide evidence for the indeterminism of GR.[13] Both from the perspective of Butterfield's definition

---

[13] To quote a philosopher's diagnosis, Belot (2011, p. 2876) says: "Instances in which globally hyperbolic solutions admit non-isometric extensions are instances of genuine indeterminism [...]". For a similar assessment by a physicist, see Ringström (2009) or Costa et al. (2015).

of determinism and from the perspective of identifying determinism with unique solutions to a theory's evolution equations (EFE in GR), such cases *are* indeterministic. The crucial question for us is whether this kind of indeterminism can be captured by BST. Before we tackle this question head on, in the next section we discuss a "topological" characterization of indeterminism in GR. The gist of this characterization is to interpret a non-Hausdorff generalized d-manifold as a representation of a family of alternative space-times that extend a given spatio-temporal region.

9.3.3 Can non-Hausdorff manifolds in GR be interpreted modally?

Non-isometric extensions of a maximal globally hyperbolic space-time, like the extensions of Taub space-time just reviewed, are constructed to be Hausdorff d-manifolds, as they have to satisfy the conditions for a GR space-time. Can non-Hausdorff d-manifolds arise in the context of a failure of the IVP in General Relativity? The answer is yes.

Hawking and Ellis (1973, p. 173) exhibit extensions of so-called Misner space-time, which are Hausdorff d-manifolds that can be further extended to a d-manifold that is not Hausdorff. More precisely, there is a generalized (non-Hausdorff) d-manifold such that its maximal Hausdorff submanifolds are the mentioned extensions of Misner space-time. The same authors (p. 177) also discuss the above described non-isometric extensions of Taub space-time and show that they can be viewed as maximal Hausdorff d-submanifolds of some generalized (non-Hausdorff) d-manifold. More generally, Luc and Placek (2020) develop a pasting technique by means of which they prove some results concerning the relations between Hausdorff and non-Hausdorff d-manifolds (equipped with Lorentzian metrics). First, they show that non-isometric space-times (Hausdorff d-manifolds) that have some isometric regions can almost always be used to produce a generalized (non-Hausdorff) d-manifold with a Lorentzian metric. More precisely, in the collection of Hausdorff manifolds that is to be pasted any pair has to have isometric regions; and the "almost always" qualification comes from the fact that the technique yields topological non-Hausdorff manifolds rather than C^r non-Hausdorff manifolds. Thus, the final step in obtaining the sought-after d-manifold depends on whether the resulting topological manifolds admit a differentiable structure. As we already mentioned in

Chapter 9.2.3, there are (relatively rare) so-called Kervaire cases that do not admit such a structure. Second, Luc and Placek (2020) describe in detail the gluing together of non-isometric extensions of Taub space-time to form a generalized non-Hausdorff d-manifold. Finally, a result slightly generalizing Hájíček's (1971b) construction says that any non-Hausdorff d-manifold with a metric can be constructed by the gluing technique from a collection of Hausdorff d-manifolds with metrics (GR space-times). Thus, Kervaire cases aside, there is a general technique that permits one to take the non-isometric (Hausdorff) extensions of some globally hyperbolic space-time and paste them together into a generalized (non-Hausdorff) manifold with a metric. Maximal Hausdorff sub-manifolds of this generalized manifold will then be identifiable with the non-isometric extensions one started with.

This opens the door to a modal interpretation of non-Hausdorff d-manifolds, at least of the variety constructible in the context of a failure of the initial value problem by the mentioned technique. This interpretation sticks to the standard GR identification of space-times with Hausdorff d-manifolds (with some more constraints), and reads a non-Hausdorff manifold resulting from gluing together non-isometric extensions as representing alternative space-times, all of which develop from a certain common region, which is given by the maximal globally hyperbolic d-manifold. This interpretation gives physical meaning to a significant variety of non-Hausdorff d-manifolds occurring in GR. By the construction, a non-Hausdorff manifold of this variety witnesses indeterminism in Butterfield's sense, because it comprises non-isometric extensions of a space-time. Finally, the interpretation dissolves most objections leveled against non-Hausdorff manifolds in the GR literature; see Luc (2020). For more details on the modal interpretation of non-Hausdorff d-manifolds in GR, see Luc and Placek (2020).

The big question now is whether a generalized d-manifold constructed from non-isometric extensions of some space-time can be viewed as a BST structure. The next section suggests a negative answer to this question. This negative answer reflects a philosophically interesting point about indeterminism, which we explore in Section 9.3.5.

9.3.4 On bifurcating curves in GR

BST is a theory of local indeterminism. For such indeterminism, there is a small and well-defined locus at which one of a set of alternative possibilities

is realized, whereas all the other alternatives stop being possible. The locus
is idealized to a set of particular point events. Depending on whether we
assume PCP_{92} or PCP_{NF}, there is either a maximal event in the intersection
of any two histories, or there is a minimal event in the difference of any two
histories. We can express the same point in terms of modal forks, by which
we mean a pair of maximal chains that (1) share an initial segment, (2) the
shared segment is fully within the overlap of some histories, $h_1 \cap h_2$, and (3)
the separate segment of each chain is in a different difference of histories,
$h_1 \setminus h_2$, or $h_2 \setminus h_1$. A modal fork is thus a pair of maximal chains that belong
to two histories in a particular way. Since chains are obviously defined by
the pre-causal ordering, read as "something can happen after something
else", a maximal chain is readily interpreted as a potential trajectory of a
point-like object. Thus, a modal fork can naturally be seen as representing
initially coinciding alternative possible evolutions for some point-like object.
Now, some modal forks have maximal elements in the shared segments—we
call them *modal forks of the first kind*. Other modal forks have no maximal
elements in the shared segments (they rather have minimal elements in each
separate arm)—we call them *modal forks of the second kind*. Clearly, a BST_{92}
structure with more than one history contains at least one modal fork of the
first kind, whereas in BST_{NF} structures, all modal forks are of the second
kind (cf. Fact 3.21).

To recall our topological discussion of Section 9.2.3, BST_{92} structures with
multiples histories are non-starters for yielding generalized d-manifolds,
whereas BST_{NF} structures stand a good chance. Indeed, we have already
seen cases in which the base sets of BST_{NF} structures admit a generalized
d-manifold structure. Thus, as BST_{92} is out of the game, and as BST_{NF} allows
for modal forks of the second kind only, the question is whether generalized
d-manifolds in GR admit bifurcating trajectories of point-like entities that
are topologically like modal forks of the second kind. The required notion,
introduced in the GR literature by Hájíček (1971a), is that of bifurcating
curves of the second kind, defined as follows:

Definition 9.11. A *bifurcating curve of the second kind* on a C^r generalized
d-manifold M is a pair $\langle C_1, C_2 \rangle$ of C^r-continuous curves $C_1 : [0, 1] \to M$,
$C_2 : [0, 1] \to M$ such that for some $k \in (0, 1]$: $\forall x \in I \, [x < k \Leftrightarrow C_1(x) = C_2(x)]$.

Thus, in a bifurcating curve of the second kind there is no maximal element
in I at which C_1 and C_2 agree. A bifurcating curve of the first kind is
defined analogously (by the condition $\forall x \in I \, [x \leqslant k \Leftrightarrow C_1(x) = C_2(x)]$), and

topologically it is like a modal fork of the first kind. Note that, in contrast to modal forks, bifurcating curves of GR as defined by Hájíček (1971a) do not have any overt modal aspect.

We know that there are non-Hausdorff d-manifolds with bifurcating curves of the second kind; for a simple example, pick the bifurcating real line at the bottom of Figure 3.1 (p. 44). Relying on the visual intuition from that example, one might think that given a failure of Hausdorffness, such bifurcating curves should be easily available. After all, there must be a pair of points witnessing non-Hausdorffness, and one might think that one could simply pick two curves that overlap everywhere below the pair, but then pass through different elements of the pair. However, concrete cases from GR argue against this intuition: There are no bifurcating geodesics (a special class of causal curves) of the second kind in the generalized (non-Hausdorff) d-manifold representing non-isometric extensions of Misner space-time. Neither do the generalized (non-Hausdorff) d-manifolds resulting from pasting together non-isometric extensions of Taub space-time admit bifurcating geodesics of the second kind (Hawking and Ellis, 1973).[14]

Thus, the question of under what conditions non-Hausdorff d-manifolds admit bifurcating curves of the second kind is non-trivial. The question is answered, by providing necessary and sufficient conditions, by Hájíček's (1971a) theorem. To state it, we need one auxiliary notion, which is illustrated in Figure 9.1.

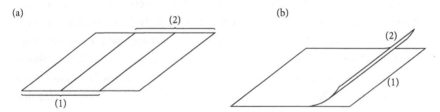

(a)

(2)

(1)

(b)

(2)

(1)

Figure 9.1 Gluing of two surfaces in two dimensions. (a) Not continuously extendible; (b) continuously extendible.

Definition 9.12 (Continuously extendible gluing). Let $\mathscr{W}_1 = \langle W_1, A_{W_1}, g_1 \rangle$ and $\mathscr{W}_2 = \langle W_2, A_{W_2}, g_2 \rangle$ be GR space-times. Then $\varphi : U_1 \mapsto U_2$, where $U_1 \subseteq W_1, U_2 \subseteq W_2$, is a *gluing function* if (1) U_1 is open and (2) φ is an isometry.

---

[14] To see why the visual intuition is wrong, note that these manifolds are constructed in a particular way, viz., by pasting via an equivalence relation. Margalef-Bentabol and Villaseñor (2014) show how non-Hausdorffness combines with the absence of bifurcating curves in Misner space-time.

Moreover, φ is said to be *continuously extendible* iff there exist U_1', U_2', φ' such that $U_1 \subsetneq U_1' \subseteq W_1, U_2 \subseteq U_2' \subseteq W_2, \varphi' : U_1' \mapsto U_2', \varphi'$ is continuous and $\varphi'|_{U_1} = \varphi$.

Clearly, the definition implies that U_2 is open as well, and that $U_2 \subsetneq U_2'$. Significantly, however, φ' above need not be an isometry between U_1' and U_2', thus it need not be a gluing function.

Hájíček's theorem says:

Theorem 9.3. *The necessary and sufficient condition for a d-manifold constructed by gluing together Hausdorff d-manifolds to admit bifurcating curves of the second kind is that the gluing be continuously extendible. (Hájíček, 1971a)*

According to the already mentioned results of Luc and Placek (2020), every non-Hausdorff d-manifold is constructible via the gluing of Hausdorff d-manifolds. Therefore, Hájíček's theorem produces a universal method to determine whether, for any non-Hausdorff d-manifold, it admits bifurcating curves of the second kind or not.

In plain English, Theorem 9.3 says that a non-Hausdorff d-manifold admits the sought-for bifurcating curves of the second kind exactly if among the component space-times that give rise to it, there are two particularly related space-times. These space-times are pasted together by a gluing function φ, with isometric domain and counter-domain U_1 and U_2, respectively. That gluing function φ can be extended to a function φ' on a larger domain, where φ' is to be continuous (but not necessarily an isometry). This does not mean that one can improve on the gluing φ; it merely says that φ, U_1, and U_2 are particularly related.

We have already encountered non-Hausdorff d-manifolds that result from gluing together Hausdorff d-manifolds. A case in point are non-Hausdorff d-manifolds on base sets of MBSs with multiple histories: see Fact 9.11. Another example is provided by the one-dimensional non-Hausdorff topological manifold M_b that results from gluing together two copies of the real line, sketched at the bottom of Figure 3.1 (p. 44).[15] Significantly, in each of the mentioned cases, it can be proved (see Exercise 9.7) that the gluing that is used is continuously extendible. Via Theorem 9.3, each of

---

[15] For an argument that this non-Hausdorff topological manifold gives rise to a C^∞ d-manifold, see Exercise 9.6. The Lorentzian metric g is trivially derived from the standard metric on the real line. The individual lines are Hausdorff d-manifolds, and in fact sub-manifolds of the bifurcating lines.

these non-Hausdorff d-manifold thus harbors bifurcating curves of the second kind. Do these MBS examples provide evidence that there are in GR non-Hausdorff d-manifolds, constructible from non-isometric GR space-times that arise from a failure of IVP for GR *and* that contain bifurcating curves? We do not think so. In the case of GR, one glues together non-isometric space-times. In our BST constructions, in contrast, the Hausdorff d-manifolds that form the basis of the resulting non-Hausdorff d-manifold are isometric. Even more importantly, in the case of a failure of the IVP in GR, the non-isometric solutions to the EFE are not globally hyperbolic—each extends a maximal globally hyperbolic space-time (guaranteed to exist by the Choquet-Bruhat and Geroch theorem). In contrast, the manifold on the real line, as well as Minkowski space-time which forms the base set of the mentioned MBS constructions, *are* globally hyperbolic. The fact that there are bifurcating curves of the second kind in MBS constructions does not imply that analogously bifurcating non-Hausdorff d-manifolds can be constructed from the multiples solutions to the EFE that are available from a failure of the IVP. And, to repeat, the non-Hausdorff d-manifolds resulting from pasting together extensions of Misner space-time[16] or non-isometric extensions of Taub space-time do not contain bifurcating geodesics of the second kind. It is a general and difficult open problem to prove whether there are non-Hausdorff d-manifolds with a Lorentzian metric that are constructible from non-isometric extensions of a maximal hyperbolic space-time by continuously extendible gluing.[17] Only such manifolds harbor bifurcating curves of the second kind. If they exist, one may hope to obtain a GR-based BST_{NF} structure with multiple histories that are individually identifiable with maximal Hausdorff d-submanifolds of the structure.

Yet, in the cases that we know, gluing together non-isometric extensions of a maximal hyperbolic space-time produces a non-Hausdorff d-manifold *without* bifurcating geodesics. Gluing the extensions of Taub space-time is a case in point. Interestingly, such non-isometric extensions witness indeterminism according to the definition that represents the received view in the philosophy of science (see Def. 9.10). We would like to capture this kind of indeterminism in BST as well. After all, BST theory promises to analyze *local* indeterminism, as occurring in spatio-temporal contexts. Yet BST cannot capture these GR cases, since it requires bifurcating curves.

---

[16] Extensions of Misner space-time are isometric, though, see Chruściel and Isenberg (1993).

[17] A qualification is needed to avoid, e.g., cunningly gluing two Taub space-times on regions that are a bit smaller than isometric regions and then presenting standard gluing as an extension.

It appears thus that the philosophy of science notion of determinism and indeterminism is different from that assumed by BST: the former is global, whereas the latter is local.[18] We turn to this issue in the next section.

9.3.5 Global and local determinism and indeterminism

There are two traditions of thinking about determinism, one centered on individual objects and the other centered on the entire universe. The former tradition focuses on relatively small objects or processes (that is, small if compared with the universe) and asks if these objects or processes could evolve differently than they actually did. The cloak story that Aristotle tells in *De Interpretatione* 19a clearly exemplifies this way of thinking: the issue is that the cloak might wear out, but that it could also be cut up first. If we deliberate over whether the cloak case argues in favor of indeterminism or not, the data we look at are limited in space and time. It is of course the cloak that matters, but some of its surroundings are relevant as well. However large these surroundings are compared to the cloak, we typically do not extend them to the entire universe. That is, we limit the data for the determinism question to a relatively small region of our spatio-temporal universe. This observation also applies to another great example we owe to Aristotle, that of tomorrow's sea battle. Although armadas of military vessels, together with sailors, their commanders, weather conditions, etc., occupy a relatively large area of the sea, this area is just a tiny spatio-temporal region of the entire universe. In some examples used in this tradition, namely those involving human agents and their decisions, the data is even further restricted—to a particular person, or more precisely, to some particular period in that person's life.

This local approach to determinism and future contingents is a characteristic feature of theories of agency. Putting philosophical disputes aside, it is this approach that is used in everyday contexts, including in science labs. A chemistry student investigating a catalytic reaction may wonder, seeing different outputs of seemingly identical processes in subsequent runs of her experiment: is the varying output due to the indeterministic nature of the process, or to some tiny differences in the reaction's initial conditions in

---

[18] A distinction between global and local notions of determinism has been argued for, e.g., by Belot (1995), Melia (1999), and Sattig (2015).

subsequent runs of the experiment? In an attempt to clarify the issue, she focuses on local matters of fact: is the catalyst, as well as other chemical substances used, sufficiently similar in subsequent runs of the reaction? Are the temperature, pressure, concentrations, and other relevant characteristics the same in all these runs? For these questions, the universe as a whole and its possible global evolutions play no role.

The second tradition centers on global notions like that of the universe, the world or its history, or a theory's models. That tradition is invariably linked to Laplace's vision of determinism. In Laplace's well known metaphor, a super intelligence is capable of "seeing" the entire past and future of the universe, thanks to its grasp of the instantaneous state of the universe and its knowledge of all the forces acting in the universe. After the removal of its epistemic overtones, signalled by words like "knowledge" or "seeing", the vision forms the backbone of the current received analysis of the determinism of *theories*. The basic intuition of this approach is that of "once similar, always similar", that is, a theory is deterministic iff whenever two models of the theory agree on initial segments, they agree as wholes. A theory's model, like a possible world, is a global notion. In GR, models are of course GR space-times, and similarity is identified with isometry. A general notion of determinism and indeterminism in the global style that is applicable to GR has been given via Def. 9.10.

The local and the global ways of thinking about determinism and indeterminism are in conflict. The combination of global determinism and local indeterminism is known from the literature (Belot, 1995; Sattig, 2015).[19] Non-isometric extensions of a maximal hyperbolic space-time point in the opposite direction, as these cases seem to combine global indeterminism with local determinism. We find this combination paradoxical, wondering how to conceive of a world that faces alternative possible evolutions, whereas each object in this world has a deterministically fixed evolution.

How then shall we define the *local* determinism and indeterminism of theories? Our point of departure is a given theory, together with some interpretation. An appeal to interpretation is needed, since in the next step we ask what individuals (that is, local objects that persist over time) are admissible,

---

[19] The context of their examples is Lewis's (1983) definition of determinism, applied to an ideally (axially) symmetric column with a critical weight on it, known as the 'buckling column'. Common sense and elasticity theory say that the direction of the column's buckling is not determined, whereas Lewis's analysis delivers the verdict "determinism". Lewis's analysis is global, whereas the common sense analysis is local.

and the response may vary from interpretation to interpretation of a given theory. We focus on systems admitted by a given theory, together with its interpretation. In GR, systems are space-times. We then ask if a given system of the theory is locally deterministic or locally indeterministic. To address this question, we need to consider smaller objects, the constituents of the system, that are admitted by the theory and its interpretation. There might be theories with systems without constituents—we leave such theories and systems aside. Considering a given system, we need to investigate whether it has constituents that evolve indeterministically. If a system contains at least one indeterministically evolving admissible object, we call the system *locally indeterministic*. If a theory has at least one locally indeterministic system, the theory is also called locally indeterministic accordingly.[20]

How can we learn that a small object admitted by a theory (plus its interpretation) evolves indeterministically? Here the theory's evolution equations come to the fore. Finding that the equations allow for multiple global solutions for a small object is a clear signal that the object might evolve indeterministically. Clearly, solutions are not to be outright identified with possible evolutions, as their difference might come from the mathematical features alone, that is, they might represent the same physical reality, with their difference residing in mathematical surplus structure only. In the parlance of physics, one says that such solutions exhibit gauge freedom. Further, since solutions to evolution equations describe states posited by a given theory, a further question is how these posited states are related to reality, as conceived by the theory (with its interpretation). Clearly, going local does not absolve one from the usual toil involved in attempts to infer determinism or indeterminism for particular interpreted theories.

Turning to GR, in order to get to grips with local determinism vs. indeterminism, we need the concept of the trajectory of a particle in that theory. In GR, the (potential) worldlines of free test point particles are standardly assumed to be geodesics (i.e., curves that satisfy the geodesic equation; see Section 9.2.4). Here "test" means that the particles do not alter the geometry of the space-time they move in. By the Einstein Field Equations, a particle's motion is governed by the space-time metric, but the metric is generally influenced by the particle's motion as well. In the case of *test* particles, however, their influence on the space-time metric is assumed to

---

[20] For a formal rendering of similar ideas, but assuming the simpler framework of Branching Time, see Müller and Placek (2018). For a different analysis of local indeterminism, see Sattig (2015).

be negligible; the test particle is thought of as moving before the background of an independent metric, following a geodesic.

Earlier, we distinguished bifurcating curves of the first vs. second kind; that distinction carries over to bifurcating geodesics. Now, quite generally, there are no bifurcating geodesics of the first kind in GR space-times, even if the usual requirement of Hausdorffness is dropped—this follows from the local uniqueness result for geodesics (see Chruściel, 2011, p. 6).[21] Thus, only bifurcating geodesics of the second kind are left on the stage. If the Hausdorff property is assumed, it is possible to glue together locally unique solutions to obtain a globally unique solution, and then there is no room for bifurcating geodesics of any kind. In summary, only bifurcating geodesics of the second kind stand a chance at all in GR, and they require non-Hausdorff d-manifolds.

The quest for bifurcating geodesics to support local indeterminism in GR might seem paradoxical, since they have been considered a bad thing by the physics community. To give some examples, Earman (2008, p. 200) asks: "how would such a particle [moving along a bifurcating geodesic] know which branch of a bifurcating geodesic to follow?" In a similar vein, Hawking and Ellis (1973, p. 174) opine that "a [bifurcating] behavior of an observer's world-line would be very uncomfortable", with "one branch going into one region and another branch going into another region". Hájíček (1971b, p. 79) observes that a system cannot have two solutions unless these solutions form a bifurcating curve, and concludes: "Therefore, in view of the classical causality conception coinciding with determinism it is sensible to rule out the bifurcate curves".

The underlying assumption of these objections is determinism. We agree that a bifurcating *actual* trajectory is barely understandable, echoing Hawking and Ellis's uneasiness of there being an observer present simultaneously in two regions. But this is not what a modal interpretation offers, as it takes separate branches of a bifurcating geodesic to be alternative *possible* trajectories of a test particle. Note also that given indeterminism, there are no answers to questions like "why did a particle go along a particular trajectory, which is but one of many alternative possible trajectories?" Taking indeterminism seriously means acknowledging that, sometimes, there are no such contrastive explanations for what happens.

---

[21] For the particular subtleties related to the uniqueness result, see Chruściel (1991, Appendix F).

We return to the case of non-isometric extensions of a maximal hyperbolic space-time, exemplified for instance by the extensions of maximal hyperbolic Taub space-time. As we saw above, the Taub example satisfies Def. 9.10 of global indeterminism (see Chapter 9.3.2). Is the example *locally* indeterministic as well? Suppose that there are some objects in Taub space-time. Could then at least one of these object face indeterministic evolutions, with each possible evolution going to a different non-isometric extension?[22]

To be more specific, what are the objects in Taub space-time and its non-isometric extensions? We focused on geodesics in the last section, which are standardly interpreted as trajectories of unaccelerated test particles. One might wonder what the trajectories of "real" particles in GR are. A dominant tradition, going back to Einstein and Grossmann (1913), assumes that particles of sufficiently small mass and size move along geodesics as well. That tradition is supported by topological theorems to the effect that, given certain idealizations are assumed, the particle moves along a geodesic. A theorem to this effect is proved by Ehlers and Geroch (2004).[23]

Consider, therefore, a Taub space-time inhabited only by photons which satisfy the required idealizations. In this case, all individual objects can be safely assumed to move along geodesics. Now consider a photon that moves along a lightlike geodesic in Taub space-time (such geodesics are called "null"). This space-time has two non-isometric extensions, $\langle M^{++}, g^{++} \rangle$ and $\langle M^{+-}, g^{+-} \rangle$. What happens to the photon as it leaves the initial region? That is, what does the photon's geodesic look like as the photon leaves Taub space-time and proceeds to a new region in one of the two extensions? By the discussions of Chapter 9.3.2, we know that there are no bifurcating geodesics. Thus, there are two classes of null geodesics in Taub space-time. Geodesics of the first class are completed in one extension, and geodesics of the other class are completed in the second extension (cf. Hawking and Ellis, 1973, pp. 170–178, and Chruściel and Isenberg, 1993, Lemma 3.2). The photon's evolution appears predestined: depending on which class the photon's geodesic belongs to, it will continue to one extension or to the other. Thus, no object living in Taub space-time faces an indeterministic evolution. The moral is that non-isometric extensions of Taub space-time inhabited by photons satisfying the mentioned idealizations are locally deterministic,

---

[22] More mathematical sophistication is needed to formulate this question precisely; see Chruściel and Isenberg (1993).

[23] Of course, the direct way to study a particle's trajectory is to find an exact solution to a (relevant) problem of motion of GR, yet there are only very few exact solutions of this kind.

yet they satisfy the definition of global indeterminism, Definition 9.10. We thus have a disturbing combination of global indeterminism and local determinism.

This result confronts us with a dilemma concerning how to further develop branching style-theories. The problem we face is metaphysical: How can we capture the notion of a possible history, which intuitively is a maximal possible course of events? Belnap (1992) opts for an order-theoretical criterion that is based on the "later witness" intuition (see Chapter 2.2): a history is a maximal directed subset of the base set W of a BST_{92} (or BST_{NF}) structure $\mathscr{W}$. That definition captures local indeterminism, but not the combination of global indeterminism plus local determinism that we discussed in this section. A remedy, suggested by the topological results concerning MBSs and discussed at the end of Chapter 9.1.2, might be to resort to topological foundations as an alternative to defining BST structures in terms of an ordering. The idea would be to identify a BST structure with a generalized manifold, and to define histories to be maximal Hausdorff submanifolds. This idea is further supported by the gluing technique applied to non-isometric extensions of maximal hyperbolic space-times of GR. With this remedy, a new BST-style theory could accommodate both varieties if indeterminism, local and global. Whether this remedy is attractive and worth pursuing depends on one's stance on the combination of global indeterminism and local determinism that we described: Does it reflect a feature of our world, or is it merely a mathematical gimmick coming from the theory of differential manifolds? We remain skeptical.

9.3.6 A note on closed causal curves and BST

As we have signaled, there is another relevant issue at the interface of GR and BST: some GR space-times admit closed causal curves, which means that the (strict) ordering determined by these curves is not asymmetric. This contradicts a basic a postulate of BST, viz., that $\mathscr{W} = \langle W, < \rangle$ is a (strict) partial order. Thus, in general, causal curves in GR allow one to define just an irreflexive and transitive relation $\prec$, called a strict pre-order, rather than the strict partial ordering that BST calls for.[24] In the following section we show that this ordering problem can be resolved by slightly generalizing

---

[24] A strict pre-order has a reflexive companion, $\preccurlyeq$, called a pre-order.

BST. For simplicity's sake, we limit our attention to structures without modal funny business as discussed in Chapter 5. We will construct modal structures (i.e., possibly with multiple histories) in which the ordering can be non-asymmetric. The construction of generalized BST, call it genBST, is motivated by the following theorem of GR.[25]

Theorem 9.4. *For every event p in an arbitrary GR space-time there exists an open set U with $p \in U$ such that for every $q, r \in U$ there is a unique geodesic connecting q and r, and staying entirely in U.*

Since geodesics fall into three classes, namely time-like, space-like, and null-like geodesics, the uniqueness of connectability means that in a time-orientable GR space-time the geodesics can be used to define a strict partial ordering $\prec$ on any U of the kind that the theorem above guarantees to exist: $q \prec r$ iff q and r are different events and q is connectible to r by a future directed time-like or null-like geodesic. On each U we can thus construct a BST structure (being topologically prudent, one might prefer BST_{NF}, preparing for cases with bifurcating geodesics). Since the U's (the "patches") cover the entire space-time, the genBST structure needs to somehow combine together all these little BST structures.

For our definitions, we recall the terminology introduced in Chapter 4.4.1: Let $\langle W, \prec \rangle$ be non-empty strict pre-order. Then

1. *MC* is the set of maximal chains in $\langle W, \prec \rangle$, and $MC(e) =_{df} \{ t \in MC \mid e \in t \}$;
2. $t^{\prec x} = \{ z \in t \mid z \prec x \}$, where $t \in MC(x)$ and $x \in W$ ($t^{\preceq x}$, $t^{\succ x}$, and $t^{\succeq x}$ are similarly defined).

Note that elements of *MC*, of *MC(e)*, as well as chains $t^{\prec x}$, $t^{\preceq x}$, $t^{\succ x}$, and $t^{\succeq x}$ can contain loops, i.e., there can be $y, z \in t$ for which both $y \prec z$ and $z \prec y$.

Definition 9.13 (genBST structure). Let $W \neq \emptyset$, $\prec$ be a strict dense pre-order on W, and $\mathcal{O} \subseteq \mathscr{P}(W)$. A triple $\mathscr{W} = \langle W, \prec, \mathcal{O} \rangle$ is a *genBST structure*, iff for every $e \in W$ there is a patch $O_e \in \mathcal{O}$ around e such that:

1. $e \in O_e$;
2. $\langle O_e, \prec_{|O_e} \rangle$ is a nonempty dense strict partial order satisfying the following:

---

[25] See Wald (1984, Theorem 8.1.2).

(a) $\forall e' \in O_e \; \forall t \in MC(e') \; \exists x,y \in t \cap O_e \; [x \prec_{|O_e} e' \prec_{|O_e} y \wedge t^{\succ x} \cap t^{\prec y} \subseteq O_e]$;
(b) every chain in $\langle O_e, \prec_{|O_e} \rangle$ with a lower bound in O_e has an infimum in O_e;
(c) if a chain C in $\langle O_e, \prec_{|O_e} \rangle$ is upper bounded by $b \in O_e$, then $B_b =_{\mathrm{df}} \{x \in O_e \mid C \precsim_{|O_e} x \wedge x \precsim_{|O_e} b\}$ has a unique minimum,
(d) if $x,y \in O_e$ and $x \prec z \prec y$, then $z \in O_e$.

In a genBST structure $\langle W, \prec, \mathscr{O} \rangle$, W and $\prec$ form a non-empty dense strict pre-order, and $\mathscr{O}$ contains local patches, at least one for each $e \in W$. One may think of W and $\prec$ as a non-Hausdorff d-manifold with a Lorentzian metric and the ordering relation determined by geodesics on this manifold. A patch for e must satisfy some conditions: First, O_e contains e (it is a patch for e, after all) and the pre-order $\prec$ restricted to O_e is a dense strict *partial* order (asymmetric, thus containing no loops). Any maximal chain passing through e extends in O_e below and above e. The conditions (b) and (c) emulate the infima postulate and the suprema postulate of common BST structures. Note that (c) makes room for multiple history-relative suprema of a chain, provided that there are multiple upper bounds of the right kind. The next condition forbids O_e from having holes. A genBST structure may contain causal loops, but for any e there is a patch $O_e \in \mathscr{O}$, within which the order is partial, so the patch does not contain any causal loops.

In the usual way, for each O_e we define choice sets in O_e (see Def. 3.11). And we say that a subset $E \subseteq W$ is a choice set in $\mathscr{W}$ if it is a choice set in some $O_e \in \mathscr{O}$ for some $e \in W$. The existence of choice sets hinges on how the condition (2c) of Def. 9.13 is satisfied. If for every upper bounded chain in O_e there is just one minimum for all sets B_b defined in this condition, then there are no choice sets in O_e. This is exactly what happens in a generalized d-manifold (with a Lorentzian metric) with no bifurcating geodesics.

One may wonder how the global pre-order $\prec$ meshes with choice pairs. Somewhat worryingly, our postulates so far allow for distinct elements of a choice set to have an upper bound. Recall that we identified a choice set with something at which alternative possibilities become modally separated: while before the choice set all the relevant alternative possibilities are available, at each element of the choice set only one alternative possibility is available. Allowing for a common bound of distinct elements of a choice set thus sounds like permitting previously excluded alternative possibilities to be possible again. Before certain events, both alternative possibilities are open, then no matter how the world develops, only one of them is open and the other is excluded, but then again, we have both the alternative possibilities available. As this return of once excluded possibilities contradicts the basic intuition of no backward branching, we prohibit it by accepting the following postulate:

Postulate 9.1 (Separation). *For every choice set $\ddot{c} \subseteq W$, and any $x, x' \in \ddot{c}$: if $x \neq x'$, then there is no $z \in W$ such that $x \prec z \wedge x' \prec z$.*

This postulate restricts Def. 9.13, as it implies that not every genBST structure is metaphysically sound. Significantly, it prohibits one kind of loop: those that pass through distinct elements of a choice set.[26] Note the interplay between local and global notions: if x and x' are separated by elements of a choice set $\ddot{c}$ in some patch O_e in the sense that $c \leqslant x$ and $c' \leqslant x'$ for $c, c' \in \ddot{c}$ with $c \neq c'$, then x and x' have no common upper bound, no matter how far we go along $\prec$, possibly outside O_e.

We next define consistency in order to anchor the notion of a history. Note that our definition excludes the possibility of modal funny business as discussed in Chapter 5.[27]

Definition 9.14 (Compatibility and consistency). $e, e' \in W$ are compatible iff there is no choice set $E \subseteq W$ with distinct $x, x' \in E$ such that $x \precsim e$ and $x' \precsim e'$.

$A \subseteq W$ is consistent iff $\forall e, e' \in A : e$ and e' are compatible.

Provably, there are maximal consistent subsets of W in a genBST structure.[28] We identify them with histories:

Definition 9.15. Let $\mathscr{W} = \langle W, \prec, \mathscr{O} \rangle$ be a genBST structure. Histories in $\mathscr{W}$ are maximal consistent subsets of W.

It can be proved that histories in genBST are downward closed, and that genBST structures satisfy PCP$_{\mathrm{NF}}$. However, histories in genBST are not necessarily directed, as they need not satisfy one direction of the later witness intuition: there might be e, e' in some history that have no upper bound in that history. However, genBST structures satisfy the following weaker condition:[29]

---

[26] Note that a bifurcating and reconvening geodesic that involves a choice set is different from a closed causal curve in a single GR space-time, as the latter does not involve a choice set.

[27] To accommodate MFB, one needs to keep track of which sets of elements of choice sets are consistent, and which are inconsistent. We do not discuss this topic, as the aim of this section is only to serve as an illustration of how to handle a non-asymmetric ordering in a branching approach.

[28] This is the proof of Lemma 9.1 of Placek (2014).

[29] See Fact 9.14 of Placek (2014). See that paper for a proof.

Fact 9.15. *If $e, e', e^* \in W$ and $e \preccurlyeq e^*$ and $e' \preccurlyeq e^*$, then there is a history h such that $e, e', e^* \in h$.*

In this sense, histories in genBST generalize the properties of histories in BST_{NF} structures.

Histories in genbBST structures and the structures themselves have welcome topological properties, as discussed in Placek (2014). We do not develop genBST here further, as it does not fully resolve the main obstacle to modeling the indeterminism of GR arising out of a failure of the IVP—the lack of bifurcating geodesics. The structures of genBST do, however, show that causal loops are not a fatal problem for BST.

9.3.7 Summary on General Relativity

In our discussion of indeterminism in GR and its modeling in BST, we focused our attention on indeterminism arising from a failure of the IVP for GR, restricted to vacuum solutions. The salient feature of the known cases of this sort is that a manifold representing the involved space-times is non-Hausdorff, but does not contain bifurcating geodesics of the second kind. To analyze such cases in BST, local Euclidicity (definitionally assumed in manifolds) compels the use of BST_{NF}. Then, in order to represent indeterminism and satisfy PCP_{NF}, one needs modal forks of the second kind (see Sec. 9.3.4). The GR analogue of a modal fork of the second kind is a bifurcating causal curve of the second kind. Thus, a bare minimum for modelling indeterminism in GR in the BST_{NF} framework is the existence of bifurcating causal curves of the second kind in realistic GR manifolds.[30] However, the lack of bifurcating geodesics in the known GR constructions suggests that there might be no bifurcating causal curves of the second kind in GR.

Such GR manifolds, which are non-Hausdorff but do not contain bifurcating causal curves of the second kind, are curious. Our diagnosis is that such cases present a surprising combination of global indeterminism and local determinism. That is the reason why they cannot be modeled by BST, which focuses on local notions like choice sets or choice points. To contrast this with Minkowskian Branching Structures, which *are* BST structures: these are

---

[30] The word "realistic" is meant to put aside some artificial constructions like the one mentioned in footnote 17.

constructed by gluing together copies of Minkowski space-time (a GR space-time indeed), but *without* motivation provided by GR's dynamical laws. That is, the gluing does not come from the fact that these copies of Minkowski space-time, with different fields ascribed, are multiple solutions of GR's laws of evolution.

We do not know whether the absence of natural cases of bifurcating geodesics is just limited to known cases, or whether it is general (i.e., concerning the results of gluing together non-isometric extensions of *any* maximal hyperbolic space-time of GR). We thus do not know whether there are cases of local indeterminism resulting from a failure of IVP in GR. The known non-trivial cases of indeterminism in GR are global, not local.

Does this mean that GR has refuted the metaphysical version of local indeterminism underlying BST, showing that that it is not what the world is like? Undoubtedly, there is a conflict between BST and GR but, as to informing us about the world, GR is the theory of the large. It does not easily integrate with our best theory of the small, quantum mechanics. And, by the very nature of the BST project, it is the behavior of the small that is decisive for the success or failure of local indeterminism. Thus, the physics data for local indeterminism, if it is ever to emerge, will be from quantum gravity, a theory that would unify GR and QM, and not from GR alone.

9.4 Conclusions

In this chapter we made good on our promise of exhibiting BST_{NF} structures in which histories are isomorphic to Minkowski space-time, with their content being given by the attribution of physical properties to points. Our construction of Minkowskian Branching Structures relies on a number of conditions on property attributions. A significant result of this construction is that each history in a BST_{NF} structure derived from an MBS can be viewed as a Hausdorff d-manifold, whereas the whole structure itself can be seen as a generalized d-manifold, which is non-Hausdorff iff it contains multiples histories. These results suggest a more prominent role for topological notions in constructing BST structures.

We next discussed whether differential manifolds can generally be built on BST_{NF} structures. The BST postulates are too frugal to always allow for a topological, let alone a differential manifold structure. However, the slogan "nice input in, nice output out" is vindicated to a large extent. That is, if one begins with BST_{NF} histories that all admit Hausdorff C^r d-manifold for the same r, the entire structure will be a topological manifold; furthermore, if this manifold is not a Kervaire-like case, it admits a C^r generalized d-manifold structure.

Having reviewed some of the basic notions of GR, we attempted a BST-based analysis of indeterminism in GR. We described in detail one case of this sort, namely non-isometric extensions of Taub space-time. Although this case has the desired topological description, with the set of extensions interpretable as a generalized d-manifold, and each extension a maximal Hausdorff d-submanifold, it cannot be given a BST reading, because it does not contain bifurcating geodesics. Bifurcating causal curves are, however, needed to introduce alternative possibilities in the BST framework. Furthermore, bifurcating geodesics are not present in other non-Hausdorff manifolds naturally constructible from multiple solutions to the Einstein Field Equations, at least as far as we know. Whether this signals a general non-existence theorem on bifurcating geodesics in GR is an open problem.

Our diagnosis of this situation is that non-isometric extensions of Taub space-time (and similar systems) oddly combine global indeterminism with local determinism. Whether this combination is a universal feature of non-Hausdorff manifolds constructible from multiple space-times is not clear. Nevertheless, we take it that this opposition of local vs. global varieties of indeterminism is the major obstacle to modeling GR indeterminism in the order-theoretic framework of BST. The presence of closed causal curves in some space-times of GR, however, is not a devastating obstacle for a BST analysis, as a pertinent generalization of BST_{NF}, genBST, is available.

9.5   Exercises to Chapter 9

Exercise 9.1. Prove the transitivity of $<_R$ of Def. 9.7.

Exercise 9.2. Prove the density of $<_R$ of Def. 9.7.

Exercise 9.3. Furnish the detail of the chain construction in the proof of Fact 9.2(2).

Hint: This construction is given in the proof of Lemma 8 of Wroński and Placek (2009). We copy it in Appendix B.9.

Exercise 9.4. "Old" MBSs, as defined by Müller (2002), Wroński and Placek (2009), or Placek and Belnap (2012), yield BST_{92} (not BST_{NF}) structures of the form $\langle B, <_R \rangle$. Show that the open-ball topology on $\mathbb{R}^4$ does not yield the b-ball topology on B, if the structure comprises multiple B-histories.

Hint: By the premise, B has a choice point for some histories h_σ and h_η. This choice point is then in b-balls with label σ and in b-balls with label η. Observe that the intersection of such two b-balls with different labels is not open (i.e., cannot be constructed as an arbitrary union of b-balls). Thus, b-balls are not open and hence they fail to deliver a topology on B.

Exercise 9.5. Let $\langle B, <_R \rangle$ be the BST_{NF} structure determined by an MBS $\langle \Sigma, F, P \rangle$ and $\langle B, A \rangle$ be a generalized d-manifold on B. Show that for any B-history h_σ in B, h_σ is open and connected in the manifold topology induced by atlas A on B.

Exercise 9.6. Show that the non-Hausdorff topological manifold depicted by Figure 3.1(b) and described below it can be equipped with a C^∞ atlas A and a Lorentz metric g, the result being a C^∞ generalized non-Hausdorff d-manifold.

Hint: Consider an open ball $b = \{[\langle x, i \rangle] \mid x \in (x_1, x_2)\}$, where (x_1, x_2) is an open interval in the reals and $i = 1$ or 2, and the mapping is given by $\varphi([\langle x, i \rangle]) = x$, restricted to the ball. Take the atlas A generated by such maps. Assume that g has signature -1, so the metric is defined via $r_1 r_2 = -1 \mid r_1 \parallel r_2 \mid$, where r_i are co-vectors in this manifold. Argue finally that local Euclidicity is satisfied, but the Hausdorff condition is not.

Exercise 9.7. Show that our construction of MBSs involves continuously extendible gluing.

Hint: Write down the function that glues together two copies of Minkowski space-times into two MBS histories. Argue that their shared region is open.

10

A Branching Space-Times Perspective on Presentism

10.1 Introduction

Our commonsense metaphysics of time is, arguably, best expressed by *presentism*, which holds that "the present simply *is* the real considered in relation to two particular species of unreality, namely the past and the future" (Prior, 1970, p. 245). However, it is often claimed that presentism is in conflict with the theory of special relativity, which holds that the simultaneity of distant events is frame-relative. Is there really a conflict? The issue is complicated, to say the least. One might doubt whether relativity theory can have an impact on metaphysics or on everyday notions at all. And even if it can, what precisely is the notion of the present whose independent reality is threatened by relativity theory, and how can that threat be spelled out in a formally precise way?

Our aim in this chapter is to flesh out a notion of the present that can serve the metaphysical role that presentism requires while being relativity-friendly. To this end, we will distinguish two different notions of the present, one based on simultaneity and one based on co-presentness. Simultaneity invokes a static role of the present in singling out something like a temporal location of an event (a time coordinate). Co-presentness, on the other hand, invokes a dynamic role of the present in separating a fixed past from an open future and thereby anchoring a notion of coexistence. We hold that it is the latter role that is important for presentism as a doctrine in the metaphysics of time, and we will show that a relativity-proof notion of the present in its dynamical role can be defended by exploiting the idea that dynamic change must be based on the indeterministic realization of possibilities for the future. In working out the formal details of this idea, we will make use of the fact that BST offers a rich notion of modal correlations (see Chapter 5), based on which we will be able to extend the notion of a fixed past such that it contains more than just an event's past light cone.

Branching Space-Times: Theory and Applications. Nuel Belnap, Thomas Müller, and Tomasz Placek,
Oxford University Press. © Oxford University Press 2022. DOI: 10.1093/oso/9780190884314.003.0010

The chapter is structured as follows. We begin with Chapter 10.2 by reflecting on the supposed conflict between presentism and special relativity and by charting the options available for avoiding that conflict. Next, in Chapter 10.3, we describe the main idea of the chapter, which is to focus on a notion of dynamic time based on real indeterministic change in contrast to static coordinate time. Chapter 10.4 summarizes the formal desiderata for a notion of dynamic time and briefly motivates two different approaches to defining dynamic time. In Chapter 10.5 we describe in detail the first of these approaches, in which dynamic time is analyzed in terms of *causae causantes*. In the second approach, dynamic time is analyzed in terms of an open future; that approach is described in Chapter 10.6. In Chapter 10.7 we draw a unifying conclusion from both approaches, viz., that the fulfillment of all the desiderata on dynamic time in a BST structure can be secured by a certain strong kind of modal correlation, which we call "sticky modal funny business". In Chapter 10.8, we illustrate our results in the framework of Minkowskian Branching Structures, one which will be familiar to the reader from Chapter 9.1. As usual, we end with Conclusions and Exercises.

10.2 The problem of defining the present in special relativity

There appears to be a conflict between our manifest, intuitive notions of the past, present, and future, and what special relativity says about time. Here are seven important features of manifest time:[1] (i) it assumes a mind-independent tripartite division of worldly events into past, present, and future. (ii) These three partitions are supposed to continuously change as future events turn into present events and then into past events. (iii) There is a further difference with respect to openness vs. settledness: the future is viewed as open, in contrast to the past, which is viewed as settled, or closed. (Whether the present is settled is a subject of a small controversy, with the majority view opting for its being settled, like the past). Concerning the present, the manifest view of time suggests that (iv) it is global (so any object existing before a given present and living sufficiently long hits upon it), (v) it cannot be repeated, (vi) it does not extend in time, and (vii) no two presents overlap.

---

[1] For a recent characterization of manifest time, see Callender (2017, Ch. 1).

In the Minkowski space-time of special relativity, each so-called event (i.e., each element of the space-time) can be uniquely identified via its space-time coordinates, a set of four real numbers. There is, however, no unique way to divide up these coordinates into a three-dimensional spatial and a one-dimensional temporal part. Such a division is always relative to an inertial reference frame, and none of those frames is preferred—the principle of relativity states that all frames have to be treated as being on a par. Some important relations among events are frame-invariant. For example, whether one event can causally influence another one is independent of the choice of reference frame: the causal order on Minkowski space-time is frame-invariant. Accordingly, the notion of two events being space-like related is also frame-invariant. Now, one might believe that only frame-invariant properties and relations have independent, objective reality, whereas other properties and relations cannot be taken metaphysically seriously. Problematically, the simultaneity of space-like related events is frame-relative: it depends on which frame one considers whether two space-like related events occur at the same time or not.

These basic truths about the structure of Minkowski space-time can be translated into a formal claim about the definability of a notion of *simultaneity*. There is widespread agreement that such a notion of simultaneity has to be transitive, reflexive, and symmetric (i.e., it has to be an equivalence relation).[2] It follows that the simultaneity relation cannot be the relation of space-like relatedness, as that relation is not transitive. And there are no other sensible options either, as shown by Van Benthem's theorem:[3] If a relation R is definable on the basis of Minkowski space-time alone, it has to be invariant under that structure's automorphisms, which include the Poincaré group and contractions. But once there are x, y for which $x \neq y$ and xRy, one can employ suitable automorphisms to show that xRz for *any* event z. Thus, there are only two equivalence relations that can be defined on Minkowski space-time, identity and the universal relation. None of these provides a sensible notion of simultaneity: on the first option, as each event is identical only to itself, each event would be simultaneous only with itself, and on the second option, simultaneity would not discriminate among events at all. Therefore, no non-trivial equivalence relation can be defined on the

---

[2] See, e.g., Van Benthem (1983); Stein (1991); Clifton and Hogarth (1995); Rakić (1997b).
[3] See Van Benthem (1983, pp. 25f.).

basis of the Minkowski space-time of special relativity alone, and there is no frame-invariant notion of simultaneity.

Now, the technical notion of invariance under a structure's automorphism is meant to single out those notions that are fully objective. It seems, therefore, that the notion of simultaneity, which is not frame-independent, cannot be an objective relation. This, in turn, might mean that the present is just a subjective notion, or even an illusion, which would completely undermine presentism. This challenge concerns the tenability of an objective notion of simultaneity as a necessary, not as a sufficient condition of the tenability of the doctrine of presentism as a whole. The challenge, therefore, arises prior to and independently of the additional question of how, assuming that such an objective notion is available, one should model the phenomenon of the passage of time. In this chapter we do not discuss the latter question since there is fairly widespread agreement in the literature that an indexical treatment of the passage of time is appropriate.[4]

The metaphysical consequences of the mentioned formal result—no frame-invariant notion of objective simultaneity is definable in special relativity theory—are debatable. There appear to be four main ways of reacting:

1. *Rejection of any metaphysical status of special relativity.* It is not implausible to just shrug off any suggested metaphysical import of special relativity, pointing out that that theory is only valid within its range of applicability, which is far from universal.

   Many well established empirical facts, from the details of the orbit of the planet Mercury to gravitational effects on satellites or, recently, to gravitational waves cannot be modeled on the basis of special relativity theory alone. In this sense, special relativity is empirically refuted, and therefore it is implausible to expect to get any metaphysical mileage out of it. If we are looking for a space-time theory to provide metaphysical guidance, we need to look at the general theory of relativity, or even at a successor to that theory describing some form of quantum gravity. It may well be that such a theory will provide additional resources for defining a notion of simultaneity. For example, some cosmological models of general relativity allow for the definition of a class of fundamental observers that can anchor an absolute notion of cosmic time.

---

[4] See, e.g., Belnap et al. (2001, Ch. 6), Beer (1994), or Reichenbach (1952, p. 277).

Given these resources, one can then define two events to be absolutely simultaneous iff they happen at the same cosmic time.[5] So the whole discussion involving special relativity might be a non-starter.

2. *Acceptance and revision of temporal notions.* If one accepts the apparent indefinability of simultaneity as proof that the notion of the present makes no objective sense, one can try to live without it.

While this attitude had already been recommended (for different reasons) by Spinoza,[6] it appears practically impossible: "now" is an essential indexical which has both theoretical and practical import for us.[7]

3. *Acceptance of relativization of temporal notions.* Each concrete act of communication employing temporal determinations comes from the perspective of a corporeal being. Reflecting on this fact, one can relativize temporal determinations to the rest frame of that corporeal being,[8] and one can additionally point out that relativistic effects can be neglected for most practical purposes.[9] An absolute notion of simultaneity is not needed to account for our communication practices— even in hypothetical cases in which relativistic effects become important. If I say that events e and f are simultaneous, and you, speeding by in your space-ship, deny this, then we can understand that we are not in fact disagreeing, but saying different things: I say that e and f are *simultaneous for me*, and you say that they are *not simultaneous for you*.

Such relativizations are in fact common: if I say, "It is raining", and you say, in a different place, "It is not raining", then we are not in fact disagreeing, and we can make the compatibility of our assertions explicit by mentioning our respective locations. We can also live with

---

[5] See Smeenk (2013) for a discussion of results and for some pertinent qualifications.

[6] See his *Ethics*, Book IV, Proposition 62: "Insofar as the mind conceives of things by the dictate of reason, it is equally affected whether the idea is of something in the future or in the past or in the present" (Spinoza, 1677).

[7] See, e.g., Perry (1979).

[8] See Balashov (2010) for a discussion of some subtle qualifications that pertain to the definition of a relativistic object's center of mass. The resulting imprecision is negligible for our purposes. Additionally, it is enough that a speaker may provide a frame of reference in *some* way. The easiest way would certainly be via her body, but there are other possibilities. Compare the similarly imprecise "here".

[9] See Butterfield (1984) for a succinct, quantitative assessment of the practical lack of impact of relativity theory for everyday communication. It should be added that the situation has changed somewhat since 1984, at least if relativistic effects grounding everyday technology are considered as well. Most of us nowadays carry around GPS receivers whose underlying satellite infrastructure relies heavily on (special and general) relativistic effects. This technology, however, has no direct impact on our use of temporal determinations in communication.

relativization when it comes to relativistic frames of reference. In fact, employing the Lorentz transformation between our frames, we will be able to make precise sense of the apparent disagreement and come to agree on the underlying objective facts about space-time.

4. *Addition of structure.* It is possible to add some structure to plain Minkowski space-time that will allow the objective anchoring of a non-trivial equivalence relation to be read, for example, as absolute simultaneity.

In fact, nothing about the results mentioned above rules out such additions, and Rakić (1997a) has shown precisely in which way an equivalence relation of simultaneity can be added as a conservative extension to the structure of a single Minkowski space-time.

Which of these options should a defender of presentism choose? While option (2) seems unavailable, given the importance of the notion of simultaneity, option (1) can easily be invoked. Dialectically, however, that option is not fully satisfactory: the defense of the present either becomes hostage to specific empirical facts about the actual general-relativistic space-time we inhabit, or, going beyond general relativity, the issue is deferred to a future theory of quantum gravity about which there is no consensus yet. It would be better to provide a different response, and that is what we will try in this chapter. In fact, we will provide *two* different responses, one based on option (3) and one based on option (4), which are geared toward two different questions about the present that are mostly run together, but which need to be kept apart.

As already stated in Section 10.1, the notion of the present plays a double role, one static (concerning a time coordinate) and one dynamic (concerning existence). Terminologically, we will distinguish the two relations that characterize these two different roles as simultaneity vs. co-presentness. We hold that both of these relations have to be equivalence relations,[10] but they need not be the same, and simultaneity can be relativized to a frame.

Simultaneity characterizes the present as the time of now, indicating a temporal location. Present events in this static sense are those that are

---

[10] We therefore do not discuss the strategy of denying that the relevant notions of simultaneity or co-presentness have to be equivalence relations. This strategy is followed by many proponents of an extended present, such as Hestevold (2008) or Baron (2012), who allow for overlapping but distinct nows, which leads to a failure of transitivity. Dialectically, denying the requirement of an equivalence relation comes with an additional burden of justification, and so it will be good if we can avoid it.

simultaneous with now, having the same temporal coordinate. This static role of the present has no immediate metaphysical or ontological import, and it should therefore not be the target of our modeling efforts in defense of presentism. In our view, the present in the sense of the time coordinate of now can be fully accounted for by the relativizing strategy (3), making it a matter of perspective. The dependence on a concrete being's rest frame is not problematic, as full agreement in communication can be ensured. As relativity theory poses no obstacle to defining an observer-relative notion of simultaneity anchoring the static present, we will not comment further on the notion of simultaneity here.[11]

Co-presentness, on the other hand, characterizes the present as that which is currently (now) real, indicating an objective, dynamic boundary between the fixed past and the open future of possibilities. These modal notions have ontological import and must not be relativized to an observer or an agent.[12] Considering the above list of options, it is clear, therefore, that we need to invoke option (4): Additional formal structure over and above that provided by a single Minkowski space-time is needed to define a dynamic relation of co-presentness among events.

Rakić's strategy of adding an equivalence relation to the basic structure of a single space-time is one route that might be used to anchor a dynamic relation of co-presentness. Following that recipe, one arrives at a relation that can in fact fulfill both the static and the dynamic requirements of a notion of the present: Rakić's (1997a) result allows for a foliation of Minkowski space-time into space-like hypersurfaces to be added conservatively, and events on the same hypersurface can then be taken to be both simultaneous *and* co-present. While this may be an advantage, one might also be critical of the combination, as there is a price to be paid: first, there can be no empirical test of the chosen equivalence relation, and second, one undercuts the independently motivated strategy (3) of accounting for the static (coordinate) notion of simultaneity via the relativization to a speaker's rest frame.[13]

In what follows we will work toward a different objective notion of dynamic co-presentness that is fully anchored in the modal notions of fixed

---

[11] See Müller (2006, §2) for formal details of how to work out the mentioned relativization.

[12] See, e.g., Gödel (1949, p. 258n), who says that "existence by its nature is something absolute", or Prior (1996, p. 50), who insists that "you can't have a thing existing from one point of view but not from another".

[13] In fact, such an attempt would then involve an error theory: speakers who posit the present of their rest frame as the objective present would almost certainly fail to identify the true objective notion of simultaneity, but would have no empirical means to find out about this.

past vs. open future. This relation will generally not work as a static relation of simultaneity, as the region of co-presentness will normally be extended both spatially *and* (coordinate-)temporally. The formal resources are provided by the BST notions of modal funny business (MFB), discussed in Chapter 5, of *causae causantes* as sets of transitions, analyzed in Chapter 6, and of Minkowskian Branching Structures, described in Chapter 9.1. But first we have to argue that the notion of an extended dynamic present makes good sense.

10.3 Making room for an extended dynamic present

The dynamic role of time is to account for the possibility of dynamic change, both with respect to which things exist and what their properties are. Change in that sense needs to be contrasted with so-called Cambridge change, which is just a thing's having different properties at (or with respect to) different temporal locations. Dynamic change must be more than that if it really requires a dynamic notion of time, because the static notion of temporal location is sufficient to account for Cambridge change. It is, however, notoriously difficult to spell out what dynamic change amounts to.

As announced earlier, we will explore a radical view of dynamic change: change as the indeterministic realization of one option from among a set of alternatives. Such indeterministic happenings clearly amount to change: if Alice orders fries in a Pittsburgh restaurant, or if a radium atom decays, or if a cat jumps to catch a bird, these are indeterministic events that did not have to happen, and their occurrence makes a difference to what the world is like, realizing one possibility for the future in contrast to all the others. In a nutshell, a dynamic change is a transition from open possibilities to settled facts.

Given this indeterministic notion of dynamic change, we need a corresponding notion of dynamic time to anchor the indeterministic realization of possibilities. In a second radical move, we will explore the view that, just as dynamic time is necessary for real change, so real change is necessary for real, dynamic time: No change without time, but also no time without change. In this way, we strongly dissociate the static notion of coordinate time (temporal location) from the dynamic notion of real time. This makes room for yet another move that may be perceived as radical: we will allow a moment of dynamic time to be extended not just spatially, but also coordinate-temporally.

Whether this move is really radical is debatable. A conflict between our semantic or metaphysical intuitions and an indeterminism-based notion of dynamic time could only arise if indeterminism was scarce, so that the dynamic present of an event (e.g., of an utterance) extended for so long that it included events that we would speak of as future. In such a case, the dynamic past, present, and future of our analysis might conflict with the grammatical tenses in English. Whether there is such a conflict thus depends on broadly empirical matters. We will not enter into a lengthy discussion here, but state just one observation that we take to be relevant. Consider a radioactive particle in a lab that has not decayed yet. On a standard understanding of radioactivity, at each moment in the past since the particle was brought into the lab, the particle *could* have decayed. Thus, there were many chancy events in the small spatio-temporal region of our lab—if we stick with the idealization of events as point-like, there could even have been uncountably many. In BST, these chancy events should be modeled as transitions involving choice points or elements of choice sets. Clearly, once we relax the idealization of point-like events, or stop individuating events by non-extended instants of time, we will end up with a smaller number of chancy events. But in any case, given radioactivity as standardly understood, there is really no scarcity of chancy events, and so the extension of dynamic time along the coordinate-time dimension should not pose a problem.

Our view of dynamic time needs to be distinguished sharply from other theories of an "extended present" that are neither based on indeterminism, nor on a distinction between static (coordinate) and dynamic (indeterministic) time. Taking into account indeterministic change, we have at our disposal a richer background before which to define dynamic time. This allows us to hold on to (dynamic) co-presentness as an equivalence relation, in contradistinction to theories that posit overlapping present moments (Hestevold, 2008; Baron, 2012). Before we show how, we first comment on the consequences of the assumption that there is no dynamic time without indeterministic change.

One consequence of the doctrine of dynamic time is that determinism implies no real change, and further, no dynamic time. That is, if real time and real change presuppose indeterminism, it follows that there is no real change, and no real time, in a deterministic world. This may seem outrageous. Take a simple deterministic world, modeled via a single Newtonian space-time, in which a number of point particles move about on continuous trajectories. If initial conditions are properly chosen so that there are no collisions or other problematic configurations, the motion of the particles in such a world may indeed be without physically possible alternatives, thus exhibiting determinism. According to our approach, we have to say that in such a world, there is no real, indeterministic change. There is never a non-trivial range of options from among which only one is realized; there is always and everywhere just one single option to begin with. But the particles in that world move around, changing their absolute locations, as well as their relative ones. Surely that amounts to change in that world?

Given the distinctions we are making, we can agree that such a world harbors Cambridge change: the particles have different locations at different times. But from a dynamic perspective, nothing is really happening. The temporal coordinate is just like another spatial coordinate, along which there can of course be some variation in the configuration of the particles. But it is all just one four-dimensional block without any real dynamics. Everything is accounted for by four-dimensional geometry. From the point of view of dynamic time, every event in the whole deterministic four-dimensional space-time is co-present with every other event (and, of course, with itself). The dynamic present of the deterministic world is maximally extended to the whole space-time block. Matters are only different if one introduces indeterminism.

Linking time to indeterminism has had a few well-known proponents. William James offers a strong image of determinism which deprives the world of dynamics: "The whole is in each and every part, and welds it with the rest into an absolute unity, an iron block, in which there can be no equivocation or shadow of turning" (James, 1884, p. 150). A similar position is advocated by Whitrow (1961, pp. 295f.):

Strict causality would mean that the consequences pre-exist in the premises. But, if the future history of the universe pre-exists logically in the present, why it is not already in the present? If, for the strict determinist, the future is merely "the hidden present", whence comes the illusion of temporal succession? The fact of transition and 'becoming' compels us to recognize the existence of an element of indeterminism and irreducible contingency in the universe. The future is hidden from us—not in the present, but in the future. Time is the mediator between the possible and the actual.

Similar elaborations of this view can be found in Eddington (1949, 1953).[14] More recently, this position underlies Ellis's models of an evolving block universe:

> Things could have been different, but second by second, one specific evolutionary history out of all the possibilities is chosen, takes place, and gets cast in stone. (Ellis, 2006, pp. 1812f)

The doctrine that real time requires modal indeterminism has been vigorously opposed.[15] But neither friends nor foes of the doctrine have expressed the underlying association between time and indeterminism with enough rigor to enable a formal treatment of the doctrine. It is precisely this task to which the rest of this chapter is devoted.

10.4   The dynamic present, past, and future: Two approaches

We have agreed to link dynamic time to indeterminism; we still need to decide how this link is to be defined. Apart from the dynamic present, we need to consider the accompanying notions of the dynamic past and the dynamic future. This triad constitutes the flow of (dynamic) time, and so the past, the present, and the future should be explicated together as equally dynamic. Our distinction between the static present (defined in terms of simultaneity) and the dynamic present (defined in terms of co-presentness and, ultimately, in terms of indeterminism) carries over to the past and the future as well. We will thus contrast the static past and future, explained in terms of relations between coordinates, with the dynamic past and future, explained in terms of indeterminism. As we remarked, the dynamic present can be (coordinate-)temporally extended. The dynamic past and future can therefore be different from their static counterparts as well.

As we stated in Section 10.2, the main challenge for presentism posed by relativity theory is that the space-time of special relativity does not contain

---

[14] The view also bears some affinity to Reichenbach (1952, p. 276), who argues that "[t]he distinction between the indeterminism of the future and the determinism of the past has found, in the end, an expression in the laws of physics". Reichenbach refers to quantum mechanics, and claims that "[t]he consequences for the time of our experience [...] are evident" (Reichenbach, 1952, p. 276). He also advocates an indexical (or, as he says, "token-reflexive", p. 277) treatment of the passage of time.

[15] See, e.g., Gale's (1963) attempt to rebut Whitrow's and Eddington's arguments.

enough structure to define a frame-invariant notion of simultaneity. With respect to the Minkowski space-time of special relativity, our discussion of Minkowskian Branching Structures in Chapter 9.1 has shown that BST offers additional resources, viz., the formal representation of indeterminism via transitions, while the BST ordering in such structures is directly taken from the Minkowskian ordering of cause-like relatedness defined in special relativity. So in this sense, BST is relativity-friendly, and we can consider constructions that base a notion of dynamic time on the primitive notions of BST to be relativity-friendly as well. As we discussed at length in Chapter 9, it is difficult to say whether BST is relativity-friendly in the more general sense of fully capturing possible cases of indeterminism in space-times of General Relativity. But at least there are encouraging positive results also with respect to this question, such as the fact that we can represent single General Relativistic space-times as BST orderings. And the philosophical discussion about the definability of simultaneity focuses on special, not on General Relativity—not the least because some models of General Relativity arguably admit an objective notion of simultaneity (see footnote 5). With respect to our topic here, we thus feel justified to proceed from the assumption that a construction in terms of the primitive notions of BST will be relativity-friendly.

How can we characterize the notion of dynamic time precisely? Here we state the set of desiderata that the sought-for relativity-proof notions of dynamic past, present, and future should ideally satisfy. These desiderata were motivated by our preceding discussion.

Definition 10.1 (Desiderata for dynamic past, present, and future). For any event e,

- D1. $Past(e)$, $Present(e)$, and $Future(e)$ are defined in a relativity-friendly way;
- D2. no two of $Past(e)$, $Present(e)$, and $Future(e)$ overlap;
- D3. $Past(e)$ and $Future(e)$ are symmetric in the sense that e' is in $Past(e)$ iff e is in $Future(e')$;
- D4. $Past(e)$ is settled in the modal sense that if a possible scenario includes e, it also includes $Past(e)$;
- D5. any possible scenario to which e belongs is fully partitioned by $Past(e)$, $Present(e)$, and $Future(e)$;
- D6. in any possible scenario, the co-presentness relation (e is co-present with e' iff e is in $Present(e')$) is reflexive, symmetric, and transitive.

As we argued above, desideratum (D1) is satisfied once we define the notions of dynamic time in terms of the primitive notions of BST. In BST, a (maximal) possible scenario is a history, so that items (D4) and (D5) should be read in terms of histories. As we will show in what follows, there are non-trivial BST structures that satisfy all the desiderata (D1)–(D6).[16] Furthermore, we can spell out exact necessary and sufficient conditions for BST structures that fulfill these desiderata. Perhaps surprisingly, these conditions involve a strong form of modal funny business. We postpone the formal statement of these general results to Section 10.7. Before that, we provide some more motivation by characterizing two different routes to explicating co-presentness via the resources of BST.

The first route focuses on what delineates co-present events from below (i.e., on the events' causal past). It is reasonable to require that dynamically co-present events should share the same *indeterministic causal factors*. In BST, causally significant factors for a given event e are captured by the notion of the *causae causantes* for e, which are indeterministic transitions. Furthermore, the initial of a *causa causans* for e can lie in the past light-cone of e, but it can also be *SLR* to e if modal funny business is present (see Fact 6.4(3)). Therefore, a focus on causal factors permits us, in the end, to remove the restriction involved in focusing on factors in the causal past only. It is natural to define co-present events as those events that share the same set of *causae causantes*. With co-presentness so defined, it is straightforward to define the notions of the dynamic present, the dynamic past, and the dynamic future of an event e. The resulting notion of dynamic time is relativity-proof. As we will show, this explication of dynamic time does not guarantee that all of the desiderata of Def. 10.1 are fulfilled— the problematic item is (D5), the full partitioning of any history into past, present, and future. All the desiderata are, however, fulfilled in structures in which there is a quite specific form of modal funny business, so that the co-presentness relation reaches all across a history. The *causae causantes*-based approach is described in detail in Section 10.5. In spelling it out, we will write co-presentness of e and e' as "$CP_C(e, e')$", with subscript "C" for "*causae causantes*", and similarly for "$\text{Present}_C(e)$".

The second route, described in Section 10.6, takes a language-oriented turn that reflects how we link our talk about the future with indeterminism.

---

[16] There are also trivial structures that satisfy the desiderata, e.g., deterministic structures that contain only a single history structures, or structures that are basically linear. See Section 10.7 for details.

As not every future event is undetermined (contingent), some work is needed to spell out what the open future of a given event is. The resulting notion is used to explain the relation "e' belongs to the dynamic future of e". This relation naturally permits one to define a sister notion of "e belongs to the dynamic past of e'". Hence we arrive at the concepts of the dynamic past and dynamic future of a given event. Finally, the dynamic present of e from a given history h is identified with whatever remains from h once the dynamic past and the dynamic future of e are removed. This approach also delivers relativity-proof notions of the dynamic present, past, and future, but again, not all of the desiderata of Def. 10.1 are fulfilled automatically. Here, item (D6) is problematic: the notion of co-presentness may fail to be transitive, even when restricted to the events in one single history. Again, a strong form of modal funny business provides a sufficient condition for fulfilling all the desiderata. On this semantics-based approach, we will write co-presentness of e and e' as "$CP_S(e, e')$" and the present of e as "$Present_S(e)$", with subscript "S" for "semantic".

Both approaches suggest that our intuitive ideas about the natural features of dynamic time, as far as they can be made exact, place significant demands on the indeterministic structure of *Our World* $\mathscr{W}$: for all of our intuitive desiderata of Def. 10.1 to hold without restrictions, a strong form of MFB is required. In Section 10.7 we spell out this dependence in the form of a number of precise formal results.

As our discussion concerns the space-times of physics, it is advisable to use a BST framework in which the local Euclidicity condition can be satisfied. For this reason we use BST_{NF}. For the purposes of illustrating our results in Section 10.8, we also use the version of Minkowskian Branching Structures that builds upon BST_{NF}, as described in Chapter 9.1.[17]

10.5 Dynamic time via *causae causantes*

Given the formal background of BST_{NF}, we have at our disposal a precise candidate definition of dynamic time: time passes at exactly those events that belong to choice sets. At other events, there is no indeterminism, no dropping off of histories, no realization of one possibility in contrast to

---

[17] In this book we do not attempt to model any additional indeterministic *dynamics* of the passage of time. As we said in Section 10.2, we assume that an indexical treatment of the "dropping off of histories" is appropriate for our communication purposes.

others, no real change, and therefore no passing of dynamic time. Of course, as stressed above, static (coordinate) time also passes at other events—but our target here is exclusively dynamic time.

To provide the necessary background, we recall the definitions of cause-like loci and *causae causantes* in BST_{NF}, in the version that allows for MFB. Based on Def. 5.12 of cause-like loci (*cll*) for outcome chains and on Def. 6.7 of the set of *causae causantes* of a transition in an MFB context (compare also Def. 6.3), we arrive at the following:

Definition 10.2 (Cause-like loci and *causae causantes* for e).

$$cll(e) =_{df} \{\ddot{c} \subseteq W \mid \exists h \in \text{Hist } h \perp_{\ddot{c}} H_e\}.$$

$$CC(e) =_{df} \{\ddot{c} \rightarrowtail \Pi_{\ddot{c}}\langle H_e \rangle \mid \ddot{c} \in cll(e)\}, \text{ where}$$

$$\Pi_{\ddot{c}}\langle H_e \rangle = \bigcup\{H \in \Pi_{\ddot{c}} \mid H \cap H_e \neq \emptyset\} = \bigcup\{H_c \subseteq \text{Hist} \mid c \in \ddot{c} \wedge H_c \cap H_e \neq \emptyset\}.$$

That is, the second element of a *causa causans* for e is the union of the elementary outcomes of $\ddot{c}$ that permit the occurrence of e. And, to recall, if there is no MFB, every *causa causans* of e is a basic proposition-like transition with an initial in $cll(e)$, i.e., $\Pi_{\ddot{c}}\langle H_e \rangle \in \Pi_{\ddot{c}}$ (see the discussion of Fact 5.7).

A definition of co-presentness in terms of an event's *causae causantes* is not hard to come by (note the subscript "C" for "*causae causantes*"):

Definition 10.3 (Co-presentness based on *causae causantes*). Events e_1, $e_2 \in W$ are *co-present*, written $CP_C(e_1, e_2)$, iff $CC(e_1) = CC(e_2)$.

Being based on an identity, the relation $CP_C(\cdot, \cdot)$ is clearly an equivalence relation on W. And it is well-defined no matter whether there are modal correlations or not, by the generality of Def. 10.2.

There is an alternative way of characterizing $CP_C(\cdot, \cdot)$, viz., in terms of the sameness of histories, as shown by the following fact:

Fact 10.1. *We have* $CP_C(e_1, e_2)$ *iff* $H_{e_1} = H_{e_2}$.

Proof. "$\Leftarrow$": From the Definition 10.2 of *causae causantes*, if $H_{e_1} = H_{e_2}$, then $CC(e_1) = CC(e_2)$, and hence $CP_C(e_1, e_2)$.

"$\Rightarrow$": For this direction, we show that the set of histories H_e in which an event e occurs is determined by its *causae causantes* $CC(e)$. We show that for any $e \in W$,

$$H_e = \bigcap \{\Pi_{\ddot{c}}\langle H_e \rangle \mid (\ddot{c} \rightarrowtail \Pi_{\ddot{c}}\langle H_e \rangle) \in CC(e)\}. \qquad (*)$$

Given $(*)$, from $\mathrm{CP}_C(e_1, e_2)$, i.e., $CC(e_1) = CC(e_2)$, we immediately have $H_{e_1} = H_{e_2}$.

For the "$\subseteq$" direction, by Definition 10.2, if $h \in H_e$, then for every $\ddot{c} \in cll(e)$ there is $c \in \ddot{c}$ such that $h \in H_c$, so $h \in \Pi_{\ddot{c}}\langle H_e \rangle$, by the same definition.

For the "$\supseteq$" direction, let us assume for reductio that there is $(\dagger)$ $h \in \bigcap \{\Pi_{\ddot{c}}\langle H_e \rangle \mid (\ddot{c} \rightarrowtail \Pi_{\ddot{c}}\langle H_e \rangle) \in CC(e)\}$, but $h \notin H_e$. Take some $h' \in H_e$. As $e \in h' \setminus h$, by $\mathrm{PCP}_{\mathrm{NF}}$ there is a choice set $\ddot{c}$ at which $h \perp_{\ddot{c}} h'$, and $c \in \ddot{c}$ such that $c \leqslant e$. This implies $h \perp_{\ddot{c}} H_e$, so that $\ddot{c} \in cll(e)$, and $(\ddot{c} \rightarrowtail \Pi_{\ddot{c}}\langle H_e \rangle) \in CC(e)$. Since $h \perp_{\ddot{c}} H_e$, we get $h \notin \Pi_{\ddot{c}}\langle H_e \rangle$, and hence $h \notin \bigcap \{\Pi_{\ddot{c}}\langle H_e \rangle \mid (\ddot{c} \rightarrowtail \Pi_{\ddot{c}}\langle H_e \rangle) \in CC(e)\}$, which contradicts $(\dagger)$. $\square$

This is a welcome result: even though BST allows for modal correlations, there are still two differently motivated definitions of co-presentness that characterize the same relation.[18] Note that the identity $(*)$ still obtains if we restrict $\ddot{c}$ to the *past* cause-like loci of e—this is the subject of Exercise 10.2.

A typical shape of a region of co-presentness in the absence of modal correlations is shown in Figure 10.1. Modal correlations allow for more extended regions of co-presentness. A pertinent example is shown in Figure 10.2. The generalization to larger sets of correlated choice sets is suggestive: if many space-like related, modally correlated choice sets exist, a region of co-present events can spatially extend arbitrarily far, up to universal Sticky MFB spanning all of space-time (see Def. 10.9 in Section 10.7).

We showed above that the relation $\mathrm{CP}_C(\cdot, \cdot)$ on W is an equivalence relation. It follows that the restriction of $\mathrm{CP}_C(\cdot, \cdot)$ to any history h is also an equivalence relation. We may thus use the restricted $\mathrm{CP}_{C|h}(\cdot, \cdot)$ to carve the dynamic present of an event e from a history containing e. Excluding

---

[18] One might perhaps criticize our definition because in the presence of modal correlations, it allows for events to be co-present while their obvious alternatives fail to be co-present. For a pertinent example, consider two ternary (outcomes 1, 2, 3) choice sets $\ddot{c}$ and $\ddot{e}$, with elements c_1, c_2, c_3 and e_1, e_2, e_3, respectively, whose 1-outcomes are strictly correlated, while the 2- and 3-outcomes are uncorrelated, leading to the five (instead of nine) histories $h^{11}, h^{22}, h^{23}, h^{32}, h^{33}$. Here c_1 and e_1 count as co-present (they occur exactly in history h^{11}), but the alternative events c_2 and e_2 do not count as co-present. We are not aware of a thorough discussion of whether the dynamic present should be modally robust, and we do not view the mentioned situation as a failure of our definition, as a fairly straightforward sharpening is available. To define co-presentness such that only modally robust pairs of events are included, we need a formal notion of alternatives, which is easily available for elements of a choice set: we just declare any two elements of a choice set to be alternatives. For such alternatives, a modified definition of co-presentness is not hard to come by; see Exercise 10.3. The hints to that exercise also provide some suggestions for how to define alternatives in the general case.

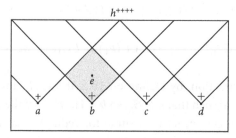

Figure 10.1 The region of events co-present with event e in one history of a BST structure. There are four binary $(+/-)$ choice sets $\ddot{a}$, $\ddot{b}$, $\ddot{c}$, and $\ddot{d}$, and no modal correlations. Thus there are sixteen possible histories, of which h^{++++} is shown. Event e and all events in the shaded region have just a single *causa causans*, $\ddot{b} \rightarrowtail b$. They occur in exactly those eight histories in which the choice set $\ddot{b}$ has outcome $+$.

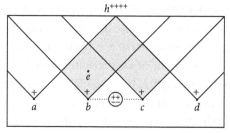

Figure 10.2 The region of events co-present with event e in one history of a BST structure. There are four binary $(+/-)$ choice sets $\ddot{a}$, $\ddot{b}$, $\ddot{c}$, and $\ddot{d}$, and outcomes of $\ddot{b}$ and of $\ddot{c}$ are modally correlated. Thus there are eight possible histories, of which h^{++++} is shown. Event e and all events in the shaded region have a set of two *causae causantes*, $\{\ddot{b} \rightarrowtail b, \ddot{c} \rightarrowtail c\}$. They occur in exactly those four histories in which choice set $\ddot{b}$ (and thus, by modal correlation, also choice set $\ddot{c}$) has outcome $+$.

featureless structures (see Def. 10.4), the restricted equivalence relation $\mathrm{CP}_{C|h}(\cdot,\cdot)$ is neither the identity nor the universal relation on the history in question. We can sum up this result as Theorem 10.1, which shows that we have indeed defined a non-frame-dependent, non-trivial equivalence relation of co-presentness on relativistic space-times, based on spatio-temporal indeterminism.

To be formally precise, here is the definition that singles out those $\mathrm{BST_{NF}}$ structures in which our construction only yields a trivial equivalence relation:

Definition 10.4. A BST_{NF} structure is called *featureless* if either (i) it contains just a single history (i.e., no indeterminism), or (ii) it has at least one history consisting wholly of elements of choice sets and in which there are no modal correlations.

Our Theorem then reads as follows:

Theorem 10.1. *Let $\langle W, < \rangle$ be a BST_{NF} structure that is not featureless. Then for any history $h \in \mathrm{Hist}(W)$, the relation of co-presentness $CP_C(\cdot, \cdot)$ restricted to h, $CP_{C|h}(\cdot, \cdot)$, is a non-trivial equivalence relation on h, i.e., neither the identity nor the universal relation on h.*

Proof. Let h be a history in a BST_{NF} structure $\langle W, < \rangle$. As mentioned, $CP_{C|h}(\cdot, \cdot)$ is a restriction of an equivalence relation and thus is itself an equivalence relation. For non-triviality, assume first that $CP_{C|h}(\cdot, \cdot)$ is the universal relation on h. This means that all $e, e' \in h$ satisfy $H_e = H_{e'}$ (i.e., h contains no element of a choice set). This is only possible in a BST_{NF} structure with just one history, which is featureless according to Def. 10.4. The second type of triviality would be that $CP_{C|h}(\cdot, \cdot)$ is the identity relation on h. This implies that for any $e, e' \in h$, $H_e \neq H_{e'}$. This is only possible in a BST_{NF} structure in which each element of h is a member of an uncorrelated choice set, which again is a featureless structure according to Def. 10.4.   □

Having defined co-presentness, we can define the dynamic present of a given event e as the set of events that are co-present with e; we preserve the subscript "C" to indicate that we are dealing with a notion defined in terms of *causae causantes*:

Definition 10.5 (Dynamic present based on *causae causantes*). Let e be an event in a BST_{NF} structure $\mathscr{W} = \langle W, < \rangle$. The dynamic present of e is defined as

$$\mathrm{Present}_C(e) =_{\mathrm{df}} \{e' \in W \mid CP_C(e, e')\} = \{e' \in W \mid H_{e'} = H_e\}.$$

In order to provide a full explication of dynamic time based on *causae causantes*, so that we can match our approach with the list of desiderata given above (Def. 10.1), we need to spell out definitions of the dynamic past and the dynamic future of an event e as well. Clearly, the dynamic future of e must come after the dynamic present of e and must not overlap with it. Analogously, the dynamic past of e must be before the dynamic present of e and must not overlap with it either. Accordingly, $e' \in$ Future(e) iff $e' \notin$ Present$_C(e)$ and there is $e'' \in$ Present$_C(e)$ such that $e'' < e'$. The last two conditions imply $H_{e'} \subseteq H_{e''} = H_e$, so by $e' \notin$ Present$_C(e)$, we get $H_{e'} \subsetneq H_e$. By an analogous argument we have $e' \in$ Past(e) iff $H_e \subsetneq H_{e'}$. So these are our definitions for Past(e) and Future(e) (we put no subscript on these definitions as they turn out to coincide with the definitions on the semantic approach of Section 10.6, see Fact 10.5(1,3)):

Definition 10.6 (Dynamic past and future based on *causae causantes*). The dynamic future of e is Future(e) $=_{\mathrm{df}} \{e' \in W \mid H_{e'} \subsetneq H_e\}$. The dynamic past of e is Past(e) $=_{\mathrm{df}} \{e' \in W \mid H_e \subsetneq H_{e'}\}$.

Given these definitions, it is easy to see that any event e has a non-empty dynamic present that contains it; $e \in$ Present$_C(e)$. It might transpire, however, that Past(e) is empty, or Future(e) is empty, or both. By Theorem 10.1, the latter case obtains in a deterministic world (i.e., in a BST structure containing just one history): in a deterministic world there is no real change, and any two events are co-present. The cases of empty dynamic past or future can be interpreted as restricted forms of determinism. Quite generally, if indeterminism is scarce, then a large region of a history contains no real change and belongs fully to the present of an appropriate event.

Past(e), Present$_C(e)$, and Future(e) as just defined have many welcome features. They are defined in purely modal terms, based on the inclusion relation among sets of histories, and they satisfy desiderata (D1)–(D4) and (D6) of our list from Section 10.4, as shown by the following Fact:

Fact 10.2. *Let e be an event in a BST$_{NF}$ structure $\mathscr{W} = \langle W, < \rangle$. Then with respect to the notions of Present$_C(e)$, Past(e), and Future(e) from Defs. 10.5 and 10.6, the following desiderata of Def. 10.1 are fulfilled:*

(D1) *Present$_C(e)$, Past(e), and Future(e) are defined in a relativity-friendly way,*

(D2) *any two of Past(e), Present$_C(e)$, and Future(e) have an empty overlap;*

(D3) *$e' \in$ Past(e) iff $e \in$ Future(e');*

(D4) *the dynamic past and present are modally settled:*
$e \notin$ Past(e) and for every $h \in H_e$: Past(e) $\subseteq h$; furthermore,
$e \in$ Present$_C(e)$ and for every $h \in H_e$: Present$_C(e) \subseteq h$; and

(D6) *the co-presentness relation $CP_C(\cdot, \cdot)$ is reflexive, symmetric, and transitive.*

Proof. (D1) This holds because we work in terms of the primitive notions of BST, as argued in Section 10.4.

(D2) Immediate from Defs. 10.5 and 10.6.

(D3) Immediate from Def. 10.6.

(D4) From Def. 10.6 it follows that $e \notin \text{Past}(e)$ for any $e \in W$. The settledness of the past follows from the downward closure of histories: Let $h \in H_e$ and $e' \in \text{Past}(e)$. As $H_e \subsetneq H_{e'}$ by the definition of $\text{Past}(e)$, we have $e' \in h$. Hence $\text{Past}(e) \subseteq h$.

From the reflexivity of $\text{CP}_C(\cdot, \cdot)$, via Def. 10.3, we have $e \in \text{Present}_C(e)$ for every $e \in W$. Finally, to show the modal settledness of the present, pick some $h \in H_e$. Since for every $e' \in \text{Present}_C(e), H_{e'} = H_e$, we have $\text{Present}_C(e) \subseteq h$.

(D6) The reflexivity, symmetry, and transitivity of $\text{CP}_C(\cdot, \cdot)$ all follow from its definition via an identity (Def. 10.3), as remarked above. $\square$

Here are some further welcome features of our definitions: an event's dynamic future, if non-empty, is never fully contained in a single history, it is thus open. Furthermore, the dynamic past is closed downward, whereas the dynamic future is closed upward. Item (4) of the following Fact points out the conditions under which desideratum (D5) of Def. 10.1, the full partitioning of any history into past, present, and future of any of its events, is satisfied.

Fact 10.3. *Let e be an event in a BST_{NF} structure $\mathscr{W} = \langle W, < \rangle$. Then the following holds:*

(1) *If $\text{Future}(e) \neq \emptyset$, then there is no history $h \in \text{Hist}$ such that $\text{Future}(e) \subseteq h$;*

(2) *$\text{Future}(e)$ is closed upward: if $e' \in \text{Future}(e)$ and $e' \leqslant e''$, then $e'' \in \text{Future}(e)$;*

(3) *$\text{Past}(e)$ is closed downward: if $e' \in \text{Past}(e)$ and $e'' \leqslant e'$, then $e'' \in \text{Past}(e)$.*

(4) *Let $h \in H_e$. Then desideratum (D5) of Def. 10.1 holds (i.e., $h \subseteq (\text{Past}(e) \cup \text{Present}_C(e) \cup \text{Future}(e)))$ iff for every $e' \in h$: $H_e \subseteq H_{e'}$ or $H_{e'} \subseteq H_e$.*

Proof. (1) Let $f \in \text{Future}(e)$. By Def. 10.6, this means that $H_f \subsetneq H_e$. Take $h_f \in H_f$ and $h \in H_e \setminus H_f$; by this choice, $h \neq h_f$. So we can pick some $e' \in h \setminus h_f$. By PCP_{NF}, there is thus some choice set $\ddot{c}$ with unique elements $c \in \ddot{c} \cap h$ and $c_f \in \ddot{c} \cap h_f, c \neq c_f$, for which $c < e'$. By directedness of h, there is some $f' \in h$ for which $c \leqslant f'$ and $e \leqslant f'$. The latter implies $H_{f'} \subseteq H_e$ by

Fact 2.2(2), and as $c_f \in h_f$, we have $h_f \notin H_{f'}$ (else also $c \in h_f$, contradicting unique intersection of choice sets with histories). So we have $H_{f'} \subsetneq H_e$, i.e., $f' \in \text{Future}(e)$. Similarly, by directedness of h_f, there is some $f'' \in h_f$ for which $e \leqslant f''$, $c_f \leqslant f''$, and $f \leqslant f''$. As $H_f \subsetneq H_e$ and $H_{f''} \subseteq H_f$ by $f \leqslant f''$, we have $H_{f''} \subsetneq H_e$, i.e., $f'' \in \text{Future}(e)$. Now there can be no history h' that contains both f' and f'', because any such history would have to contain the inconsistent elements c and c_f of the choice set $\ddot{c}$.

(2) and (3) follow from the observation that if $e' \leqslant e''$, then $H_{e''} \subseteq H_{e'}$, invoking the transitivity of $\subseteq$.

(4) Let $e' \in h \in H_e$. The conclusion is immediate from Defs. 10.5 and 10.6: $e' \in (\text{Past}(e) \cup \text{Present}_C(e) \cup \text{Future}(e))$ iff $(H_e \subseteq H_{e'}$ or $H_{e'} \subseteq H_e)$.   □

Fact 10.3(4) shows that the only thing that our definitions of dynamic time in terms of *causae causantes* leave open in general is desideratum (D5), the full partitioning of any history $h \in H_e$ by $\text{Present}_C(e)$, $\text{Past}(e)$, and $\text{Future}(e)$. As Figure 10.3 shows, our definitions imply that in general, the dynamic present of an event need not reach all across space. And as the dynamic past and future only contain events that are properly before or after an event in the dynamic present, that situation implies that a portion of an event's so-called *elsewhere* (i.e., a portion of the events *SLR* to it), will be neither dynamically past, nor present, nor future. In terms of *causae causantes*, the reach of causation in such structures does not extend far enough across space to allow one to classify these events one way or another.

As we are working toward a characterization of structures in which all of our desiderata are fulfilled, we do not enter into a lengthy discussion of the significance of a history not being partitioned into $\text{Past}(e)$, $\text{Present}_C(e)$, and $\text{Future}(e)$. One might argue that such a failure of (D5) makes possible the following odd situation: an object is in the past of e, continues to exist for a very long time, but avoids the present of e completely. A full discussion of such cases would require a theory of enduring objects in BST and probably also an explication of the notion of self-moving agents, two topics that we have to leave to the wayside here.[19]

The exhaustiveness of $\text{Past}(e)$, $\text{Present}_C(e)$, and $\text{Future}(e)$ can be guaranteed, but there is a price to be paid for that, viz., a strong form of modal correlations reaching all across space. We will consider the general situation

---

[19] See Belnap (2003a, 2005a) for some pertinent remarks.

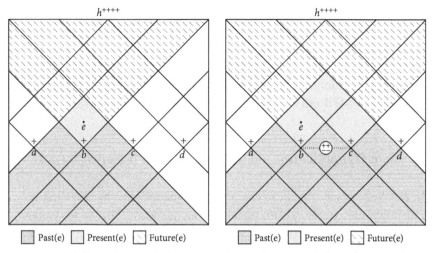

Figure 10.3 Illustration of the *causae causantes*-based notions of dynamic past, present, and future for the structures depicted in Figure 10.1 (left) and Figure 10.2 (right). The events in the white regions are neither in the dynamic past, nor in the present, nor in the future of e.

in Section 10.7. Here we show that desideratum (D5) can be violated in rather simple structures:

Fact 10.4. *Let $\mathscr{W} = \langle W, < \rangle$ be a BST_{NF} structure without MFB that contains two choice sets $\ddot{c}_1$ and $\ddot{c}_2$ such that $c_1 \, SLR \, c_2$. Then there is a history h in $\mathscr{W}$ and $e, e' \in h$ such that neither $H_e \subseteq H_{e'}$ nor $H_{e'} \subseteq H_e$. Thus, h is not partitioned by $Past(e), Present_C(e)$, and $Future(e)$.*

Proof. For concreteness, take a 4-history BST_{NF} structure with no MFB, with two binary choice sets, $\ddot{c}_1 = \{c_1, c_1'\}$ and $\ddot{c}_2 = \{c_2, c_2'\}$, where $c_1 \, SLR \, c_2$. (The case in which the choice sets have more members is exactly parallel.) Pick $e > c_1$ and $e \, SLR \, c_2$ and, symmetrically, $e' > c_2$ and $e' \, SLR \, c_1$. By no MFB, there are histories $h \in H_e \setminus H_{e'}$ and $h' \in H_{e'} \setminus H_e$. Thus, neither $H_e \subsetneq H_{e'}$ nor $H_{e'} \subsetneq H_e$, nor $H_{e'} = H_e$, so e' is neither in $Past(e)$, nor in $Future(e)$, nor in $Present_C(e)$. $\qquad\square$

So, unless a structure is trivial, MFB is required in order to fulfill desideratum (D5) for $CP_C(\cdot, \cdot)$: the existence of pairs of histories h and h' as in the above proof must be prohibited by a specific form of MFB, by requiring that $\Pi_{\ddot{c}_1}(e) \in \Pi_{\ddot{c}_1}$ be correlated (in the sense of having a non-empty intersection) with $\Pi_{\ddot{e}_2}(e') \in \Pi_{\ddot{e}_2}$ only. So our illustration already suggests that not just *any* type of MFB is enough to secure that all histories are fully partitioned into the dynamic past, present, and future of any of their events. The construction of a structure with MFB that still violates (D5) is the subject of Exercise 10.4. To ensure the partitioning of histories, we need a specific form of MFB, which we call Sticky MFB and which is the subject of Definition 10.9 in Section 10.7.

10.6 Dynamic time via the semantics of the open future

In this Section we attempt to characterize dynamic time by reflecting how one might reasonably explain futurity in modal terms. That is, we imagine someone sympathetic to linking the dynamic future to its being open. To find the link, we look for the truth conditions for the sentence "f belongs to the future of e", which we then analyze in the semantic apparatus constructible on BST structures. We will find that this semantic take on dynamic time yields a structure similar to what we unearthed above, thus supporting our concept of dynamic time.

There are two intuitions that seem relevant to explaining the past, present, and future in modal terms: the settledness of the past intuition (**SP**), and the openness of the future intuition (**OF**). To clarify them, we turn our attention to how we speak about future events, and consider what explanations of futurity in modal terms are acceptable. So we will investigate schematic explanations of the form "event f belongs to the future of event e because ...". Once we find an acceptable explanation of this kind, we will turn it into truth-conditions for "f belongs to the future of e", and research the consequences of these truth-conditions. In our discussion we will focus on concrete token events.

Consider, therefore, two concrete events that appear to be good candidates for one being in the future of the other: the Summer Solstice in Prague in 2019 (s) and a rainy sunrise on Nov 20, 2018 in Del Mar (r). The first intuition sees the past as settled. That is, although before that particular sunrise in Del Mar things could have turned out differently (it might have been rainy, and it might have been sunny, or foggy, etc.), from the perspective of a future event, like s, it is settled (fixed, inevitable) that there was this rainy sunrise. The settledness of the past intuition suggests the following schematic explanation:

SP Event s belongs to the future of event r because at event s it is settled that r has happened.

The openness of the future intuition is more elusive; we begin tentatively with this proposal:

OF$_1$ s belongs to the future of r because it is not settled at r that s will occur.

Our schematic explanation **OF$_1$** seems too strict, however. Perhaps this is overly optimistic, but we are inclined to think that no matter how the world evolves from its conditions in November 2018, there is the 2019 solstice in each of its possible evolutions. Answer **OF$_1$** sounds bad because in this case either the explanans is false, or s mysteriously does not lie in the future of r after all. Let us therefore try another one:

OF$_2$ s belongs to the future of r because the way s will occur is not settled at r.

Although answer **OF$_2$** does not look immediately incorrect if applied to the Summer Solstice 2019, it is still counter-intuitive. Think of your grandfather's Swiss watch (mechanical, almost perfect, always wound); suppose it sits in an isolating contraption, and ask yourself if it is already settled how it will signal tomorrow's noon. Our intuition is that this is already settled, no matter what the watch's *surroundings* are—the watch is isolated, after all. Like with answer **OF$_1$**, in this case either the explanans is false, or s does not lie in the future of r. The moral is that we need to accommodate the surroundings of s, which is what proposal **OF$_3$** does.

OF$_3$ s belongs to the future of r because before s there is an event and some aspect of it that is not settled at r.

In other words, for s to belong to the future of r, one needs some, however small, contingency, like the presence of a radioactive particle that may or may not decay, to obtain before s (but not necessarily after r). According to this proposal, a small but properly located contingency makes tomorrow's event involving your grandfather's watch belong to the future of your reading these words, no matter how well the watch is isolated. **OF$_3$** thus commends itself as being sufficiently weak, while still linking the dynamic future to contingency.

Since the schematic explanations **OF$_3$** and **SP** of dynamic futurity seem acceptable, we will use them as truth-conditions for the sentence "f is in the dynamic future of e":

SPOF f is in the dynamic future of e iff at f it is settled that e has happened, and there is some event e' weakly before f and a subject matter A such that at e it is contingent that A obtains at the location of e'.

In what follows, we translate the schema **SPOF** into a regimented language amenable to a BST analysis, as introduced in Chapter 4.5. We can then find out the shape of the particular regions of the dynamic present, past, and future of a given possible event e.

We now formulate our schema **SPOF** in a semantic model based on BST$_{NF}$ with space-time locations (see Def. 2.9); Ψ is the model's interpretation function.

Definition 10.7 (Dynamic future, semantic style). Let $\mathcal{M} = \langle \mathcal{W}, \Psi \rangle$ be a semantic model based on a BST$_{NF}$ structure with space-time locations $\mathcal{W} = \langle W, <, Loc \rangle$. For events $e, f \in W$ we say that f belongs to the future of e, written $f \in \text{Future}(e)$, iff there is an event $e' \in W$ and an atomic formula A such that

1. $e' \leqslant f$ and
2. $e \models Poss : At_{Loc(e')} : A$ and
3. $e \models Poss : At_{Loc(e')} : \neg A$, and
4. for every history h, if $f \in h$, then $e \in h$.

To explain, the first clause requires that a witness e' for f belonging to the future of e occur before f or be f itself. The meaning of clauses (2) and (3) is that from the perspective of e, it is contingent whether A is true at the location of the witness e'. The last clause encapsulates settledness of the past, **SP**. Note that we restrict our definition to atomic formulas, as other formulas might implicitly or explicitly refer to the past or the future, which would jeopardize our definition.

We next define the dynamic past and present of an event e by explicitly invoking desiderata (D3) and (D5) of Def. 10.1. We use the subscript "S" for "semantic-style":

Definition 10.8 (Dynamic past and present, semantic style). An event e' belongs to the past of event e, written $e' \in \text{Past}(e)$, iff $e \in \text{Future}(e')$.
An event e belongs to the present of event e', written $e \in \text{Present}_S(e')$, iff e and e' are compatible and neither $e \in \text{Future}(e')$, nor $e \in \text{Past}(e')$. We write $CP_S(e, e')$ for co-presentness of e and e', defined as $CP_S(e, e') =_{df} e \in \text{Present}_S(e')$.

In order to spell out Defs. 10.7 and 10.8 in our BST framework, we need a semantic model based on a BST_{NF} structure with a set *Loc* of spatio-temporal locations, as explained in Chapter 4.5. In addition, we need to impose two constraints on the interpretation function.

To recall, a model's interpretation function Ψ maps the set *Sent* of sentences of our language $\mathscr{L}$ to the set of sets of indexes of evaluation, i.e., $\Psi : Sent \to \mathscr{P}(E/\mathrm{Hist})$, where $E/\mathrm{Hist} =_{\mathrm{df}} \{e/h \mid e \in W \wedge h \in H_e\}$. This definition thus allows for an atomic formula to be true at some e/h but false at e/h', for different histories h, h' passing through the same event e. If histories h and h' do not split at the point event e, the interpretation function should, however, not discern between the two indices for *atomic* sentence A. Now, in BST_{NF} structures, histories do *not* split at point events, as there are no maximal elements in the overlap of histories. Thus, in a semantic model based on a BST_{NF} structure it is reasonable to require the following:

Postulate 10.1. *For any atomic sentence A and for any $e \in W$:*

$$\{e/h \mid h \in H_e\} \subseteq \Psi(A) \quad or \quad \{e/h \mid h \in H_e\} \cap \Psi(A) = \emptyset.$$

Our next assumption concerns the relation between branching histories of BST_{NF} and qualitative differences between histories, the latter being induced by the interpretation function. Uncontroversially, two branching histories should be qualitatively different somewhere. BST_{NF} structures introduce a good candidate as to where the differences are to be located, as they have a well-defined notion of minimal elements in the difference of any two histories—such minimal elements form a choice set. Our second postulate says that there is some qualitative difference at those minimal elements:

Postulate 10.2. *For histories h_1, h_2 in a BST_{NF} structure, if $h_1 \perp_{\ddot{c}} h_2$, then there is an atomic sentence A such that*

$$(c_1/h_1 \in \Psi(A) \quad but \quad c_2/h_2 \notin \Psi(A)) \quad or \quad (c_1/h_1 \notin \Psi(A) \quad but$$
$$c_2/h_2 \in \Psi(A)),$$

where $\{c_i\} = \ddot{c} \cap h_i$ $(i = 1, 2)$.

We are now in a position to prove that the dynamic future, present, and past are characterized modally, that is, in terms of the inclusion of histories, exactly like in Defs. 10.5 and 10.6:

Fact 10.5. *Let $\mathcal{M} = \langle W, <, Loc, \Psi \rangle$ be a semantic model based on a BST_{NF} structure with spatio-temporal locations $\mathcal{W} = \langle W, <, Loc \rangle$, where the interpretation function Ψ satisfies Postulates 10.1 and 10.2. Then for every $e, f \in W$:*

1. *$f \in Future(e)$ iff $H_f \subsetneq H_e$;*
2. *if $Future(e) \neq \emptyset$, then there are incompatible $f', f'' \in Future(e)$;*
3. *$f \in Past(e)$ iff $H_e \subsetneq H_f$;*
4. *$f \in Present_S(e)$ iff e and f are compatible and $\neg(H_e \subsetneq H_f)$ and $\neg(H_f \subsetneq H_e)$;*
5. *for every history $h \in H_e$: $h \subseteq (Past(e) \cup Present_S(e) \cup Future(e))$;*
6. *$e \in Present_S(e)$;*
7. *if $f \in Present_S(e)$ then $e \in Present_S(f)$;*
8. *any two of $Past(e)$, $Present_S(e)$, and $Future(e)$ have an empty overlap.*

Proof. (1) "$\Rightarrow$": ($\star$) $H_f \subseteq H_e$ is just clause (4) of Def. 10.7. To prove the strict inclusion, by clauses (2) and (3) of this definition, there are $h, h' \in H_e$ and an atomic sentence A such that $\{c\} = Loc(e') \cap h$ and $\{c'\} = Loc(e') \cap h'$ and $c/h \models A$ but $c'/h' \not\models A$. Now, if $e' \in h \cap h'$, we would have $e' = c' = c$, and hence, from the above, $e'/h \models A$ but $e'/h' \not\models A$, contradicting Postulate 10.1. Thus, $e' \notin h \cap h'$, and hence, by clause (1) of Def. 10.7, at least one of h, h' does not belong to H_f. By ($\star$), since $h, h' \in H_e$, we get $H_f \subsetneq H_e$.

(1) "$\Leftarrow$" $H_f \subsetneq H_e$ means that clause (4) of Def. 10.7 is satisfied and furthermore that there is h such that $e, f \in h$ and h' such that $e \in h'$ but $f \notin h'$. Thus, $f \in h \setminus h'$, so by PCP_{NF} there is a choice set $\ddot{c}$ such that $h \perp_{\ddot{c}} h'$ and $c \leqslant f$, where $\{c\} = \ddot{c} \cap h$ and $\{c'\} = \ddot{c} \cap h'$. By Postulate 10.2 there is an atomic sentence A such that $c/h \models A$ but $c'/h' \not\models A$. Since $c, c' \in Loc(c)$ and $h, h' \in H_e$, we have $e \models Poss : At_{Loc(c)} : A$ and $e \models Poss : At_{Loc(c)} : \neg A$, so clauses (2) and (3) of Def. 10.7 are satisfied, with c playing the role of the witness e'. Finally, clause (1) of this definition holds as well (by $c \leqslant f$).

(2) This follows by (1) as in the proof of Fact 10.3(1).

(3)–(8): As every clause is an immediate consequence of Def. 10.8 and item (1) of this Fact, the proofs are left as Exercise 10.1. $\qquad\square$

Note that clause (2) means that unless the dynamic future of an event e is degenerate (i.e., empty), it is modally open in the sense of containing incompatible events: there is no history that fully contains it. In contrast, item (3) amounts to the settledness of the dynamic past, i.e., for every $h \in H_e$:

$\text{Past}(e) \subseteq h$. Clause (5) means that $(\text{Past}(e) \cap h)$, $(\text{Present}_S(e) \cap h)$, and $(\text{Future}(e) \cap h)$ partition each history from H_e. Clauses (6) and (7) mean that the co-presentness relation $\text{CP}_S(\cdot, \cdot)$ is reflexive and symmetric on W.

Let us compare the results of the semantics-inspired approach to defining dynamic time with our list of desiderata from Def. 10.1. The success we can announce is that all items (D1)–(D5) are fulfilled: (D1) holds in virtue of working in BST, (D2) follows from Fact 10.5(8), (D3) follows from clauses (1) and (3) of that Fact, (D4) follows from clause (3), and (D5) is implied by clause (5). The only item on the list of desiderata that we cannot tick off immediately is (D6), the transitivity of co-presentness.

It is not difficult to come up with BST_{NF} structures in which co-presentness as defined in Def. 10.8 *is* transitive, however. For the simplest case, consider a (deterministic) one-history structure, in which for any e, the dynamic past and future are empty and every event belongs to the dynamic present of e. There are also more interesting examples.

Fact 10.6. *There are BST_{NF} structures in which there are multiple histories and SLR choice sets and in which (1) the dynamic present is modally settled and (2) the notion of co-presentness is transitive.*

Proof. As an example, we can take a BST structure that has two histories h, h' that split at the two choice sets $\ddot{a} = \{a, a'\}$ and $\ddot{c} = \{c, c'\}$, a SLR c such that there is maximal MFB in the structure: h contains a and c, h' contains a' and c', and there are no histories containing a and c' or a' and c (compare Figure 5.1).

Note that by Fact 10.5(4), we have $\text{CP}_S(e, e)$ iff either $H_e = H_{e'}$ or $(H_e \cap H_{e'} \neq \emptyset$ and $H_e \setminus H_{e'} \neq \emptyset$ and $H_{e'} \setminus H_e \neq \emptyset)$. The second disjunct is impossible to fulfill in our two-history structure, so that we have, for all events e, e', that $e' \in \text{Present}_S(e)$ iff $H_e = H_{e'}$.

(1) We need to show that for any $e \in W$ and any history h'', if $e \in h''$, then $\text{Present}_S(e) \subseteq h''$. Pick some $e \in W$ and some $e' \in \text{Present}_S(e)$. Consider a history h''. If $e \in h''$, then $h'' \in H_e = H_{e'}$, i.e., $e' \in h''$. So indeed, $\text{Present}_S(e) \subseteq h''$.

(2) As $\text{CP}_S(e, e')$ iff $H_e = H_{e'}$, the transitivity of $\text{CP}_S(\cdot, \cdot)$ follows by the transitivity of identity. $\square$

In general, however, we cannot guarantee that co-presentness as defined in Def. 10.8 is always transitive. There are in fact two different questions we can ask regarding the transitivity of $CP_S(\cdot,\cdot)$. On the *causae causantes* analysis, the co-presentness relation $CP_C(\cdot,\cdot)$ is defined via an identity (Def. 10.3), and so that relation is an equivalence relation on all of W. Its restriction to any history is then of course also an equivalence relation. Desideratum (D6) of Def. 10.1, on the other hand, only requires that the notion of co-presentness *in any given history* be an equivalence relation. It is possible that that is so while the union of all the history-relative relations is *not* transitive. In fact, the simple example of a structure with two *SLR* choice sets and no modal correlations discussed in the proof of Fact 10.4 provides an example (see Exercise 10.5).

The following Fact shows that there are also cases in which the history-relative notion of co-presentness $CP_{S|h}(\cdot,\cdot)$ fails to be transitive.

Fact 10.7. *There are BST_{NF} structures in which the semantics-based relation of being co-present, $CP_{S|h}(\cdot,\cdot)$, is not transitive on some history h.*

Proof. Consider a structure in which there are three compatible binary choice sets $\ddot{c}_1$, $\ddot{c}_2$, and $\ddot{c}_3$, and in which there is no MFB. The relation of the choice sets is such that in history h, the elements c_1, c_2, and c_3 occur (think of these as the '+' outcomes) and $c_1\,SLR\,c_2$ and $c_2\,SLR\,c_3$, while $c_1 < c_3$. Thus, for $\ddot{c}_3$ to occur, $\ddot{c}_1$ has to have outcome c_1. Given no MFB, there are six histories in this structure (mnemonically we can write them as h^{+++}, h^{++-}, h^{+-+}, h^{+--}, h^{-+}, and h^{--}). It is easy to verify that $CP_{S|h}(c_1,c_2)$ and $CP_{S|h}(c_2,c_3)$: the respective sets of histories do not properly nest (e.g., $h^{+-+} \in H_{c_1} \setminus H_{c_2}$). By the ordering relation $c_1 < c_3$ and as $\ddot{c}_3$ is a choice set, however, the sets of histories H_{c_1} and H_{c_3} do properly nest ($H_{c_3} \subsetneq H_{c_1}$), i.e., $c_3 \in \mathrm{Future}(c_1)$, whence $\neg CP_{S|h}(c_1,c_3)$. This shows that $CP_{S|h}(\cdot,\cdot)$ is not transitive. $\square$

This example indicates the price to be paid for the transitivity of co-presentness on the semantic approach, and thus, for fulfilling all the six desiderata of Def. 10.1: the trouble here was connected to the existence of pairs of *SLR* events whose respective sets of histories do not nest by set inclusion either way. This observation is analogous to our diagnosis from the end of Section 10.5; see Fact 10.3(4). We will now show that these observations generalize to provide a useful characterization of those BST structures in which all of the desiderata for the definition of dynamic time can be satisfied.

10.7 The way to guarantee satisfactory dynamic time in BST: Sticky modal funny business

Let us start by remarking that with both of our approaches to defining dynamic time, we ended up with definitions in terms of the interrelation of sets of histories, which is to be expected given that, with a view to desideratum (D1) of Def. 10.1, we are working on the basis of the primitive notions of BST. On both of our approaches,

- $f \in \text{Past}(e)$ iff $H_e \subsetneq H_f$, and
- $f \in \text{Future}(e)$ iff $H_f \subsetneq H_e$.

These conditions are mirror images, as required by desideratum (D3). Now we can note that quite generally, for any $e, f \in W$, there are five different ways in which their sets of histories can be interrelated, which are mutually exclusive and jointly exhaustive. The first four are simple: (1) $H_e \cap H_f = \emptyset$, i.e., e and f are incompatible; (2) $H_e = H_f$, i.e., e and f occur on exactly the same histories (this was the basis for defining co-presentness on the *causae causantes*-based approach of Section 10.5); (3) $H_e \subsetneq H_f$ (analyzed to mean that f is in the dynamic past of e), and (4) $H_f \subsetneq H_e$ (analyzed to mean that f is in the dynamic future of e). It is easy to see that these four cases are mutually exclusive (note that the sets of histories H_e and H_f must be non-empty). The remaining fifth case came up in problematic cases on both of our approaches to defining the dynamic present. We will call the condition "(NN)" for "non-nesting". It is formally simply the negation of (1), (2), (3), and (4), which can also be written as follows:

$$H_e \cap H_f \neq \emptyset \quad \wedge \quad \neg(H_e \subseteq H_f) \quad \wedge \quad \neg(H_f \subseteq H_e). \qquad \text{(NN)}$$

As we showed, instances of (NN) can cause trouble for both of our analyses. We can systematize the respective observations in the form of an equivalence between three conditions on BST_{NF} structures.

Fact 10.8. *Let $\mathscr{W}\langle W, < \rangle$ be a BST_{NF} structure. Then the following three conditions are equivalent:*

1. *There are no $e, f \in W$ that satisfy condition (NN).*

2. On $\mathcal{W}$, the notions of $Present_C(\cdot)$ and $Present_S(\cdot)$ coincide (and accordingly, the relations $CP_C(\cdot,\cdot)$ and $CP_S(\cdot,\cdot)$ coincide).
3. On $\mathcal{W}$, the notions of $Past(\cdot)$, $Present_C(\cdot)$, and $Future(\cdot)$ satisfy all the desiderata (D1)–(D6) of Def. 10.1.

Proof. "(1) $\Leftrightarrow$ (2)": We have $f \in Present_C(e)$ (and thus, $CP_C(e,f)$) iff $H_e = H_f$ by Def. 10.5. On the other hand, by Def. 10.8 and Fact 10.5(1,3), we have $f \in Present_S(e)$ (and thus, $CP_S(e,f)$) iff e and f are compatible and neither $H_e \subsetneq H_f$ nor $H_f \subsetneq H_e$. So in any $\mathrm{BST_{NF}}$ structure, if $f \in Present_C(e)$, then $f \in Present_S(e)$. A case in which $f \in Present_S(e)$ but not $f \in Present_C(e)$ has to be one in which (a) e and f are compatible, (b) $H_e \subsetneq H_f$, (c) $H_f \subsetneq H_e$ (by the definition of $Present_S(\cdot)$), but (d) not $H_e = H_f$. We can pull together (b) and (d) and (c) and (d), so that we can characterize such a case via the three conditions (i) $H_e \cap H_f \neq \emptyset$, (ii) $\neg(H_e \subseteq H_f)$, and (iii) $\neg(H_f \subseteq H_e)$. These are exactly the three conjuncts of (NN). So we have established that if there is no instance of (NN) on W, then $Present_C(\cdot)$ and $Present_S(\cdot)$ coincide on all of W, and if $Present_C(\cdot)$ and $Present_S(\cdot)$ coincide on all of W, then there can be no instance of (NN).

"(1) $\Leftrightarrow$ (3)": By Fact 10.2, $Present_C(\cdot)$ fulfills desiderata (D1)–(D4) and (D6) in any case, and by Fact 10.3(4), desideratum (D5) holds in addition iff for any compatible e and f, we have $H_e \subseteq H_f$ or $H_f \subseteq H_e$, i.e., iff there are no e and f for which (i) $H_e \cap H_f \neq \emptyset$, (ii) $\neg(H_e \subseteq H_f)$, and (iii) $\neg(H_f \subseteq H_e)$. These are again exactly the three conjuncts of (NN). So, desideratum (D5) holds for $Present_C(\cdot)$ iff there is no instance of (NN). $\square$

There is also a relevant condition on $Present_S(\cdot)$ that is implied by (1) but the converse implication does not hold.

Fact 10.9. *Let $\mathcal{W}\langle W, < \rangle$ be a $\mathrm{BST_{NF}}$ structure. Then condition (1) implies condition (4), but not vice versa:*

(1) *There are no $e, f \in W$ that satisfy condition (NN).*
(4) *On $\mathcal{W}$, the notions of $Past(\cdot)$, $Present_S(\cdot)$, and $Future(\cdot)$ satisfy all the desiderata (D1)–(D6) of Def. 10.1.*

Proof. "(1) $\Rightarrow$ (4)": If there are no instances of (NN), then by (3), $Past(\cdot)$, $Present_C(\cdot)$, and $Future(\cdot)$ fulfill all the desiderata (D1)–(D6), and by (2), $Present_C(\cdot)$ and $Present_S(\cdot)$ coincide, so $Past(\cdot)$, $Present_S(\cdot)$, and $Future(\cdot)$ fulfill all the desiderata (D1)–(D6) as well.

"(4) $\nRightarrow$ (1)": To show the failure of the converse implication, consider a 2-dimensional MBS specified by three labels, i.e., $\Sigma = \{\sigma, \gamma, \eta\}$. The sets of splitting points are $S_{\sigma\gamma} = \{y\}$, $S_{\sigma\eta} = \{x\}$, and $S_{\gamma\eta} = \{x, y\}$, where $x = (0, -1)$ and $y = (0, 1)$; see Figure 10.4. (Note that by our convention, the first, temporal coordinate is depicted vertically.) Accordingly $[\sigma x] = [\gamma x] \neq [\eta x]$, whereas $[\sigma y] = [\eta y] \neq [\gamma y]$. The structure contains thus three histories h_σ, h_γ, and h_η and harbors MFB, as $H_{[\eta x]} \cap H_{[\gamma y]} = \emptyset$ (note that $[\gamma x] \in [\eta x]$ and $[\gamma x] \, SLR \, [\gamma y]$). Now, $[\sigma x]$ and $[\sigma y]$ produce an instance of (NN) as (†) $[\sigma x], [\sigma y] \in h_\sigma$, $\neg(H_{[\sigma x]} \subseteq H_{[\sigma y]})$ (witnessed by history h_γ), and $\neg(H_{[\sigma y]} \subseteq H_{[\sigma x]})$ (witnessed by history h_η). It is relatively easy to establish that $CP_S(\cdot, \cdot)$ is transitive for any given instance, but we need to check a number of cases. Here we consider explicitly only one of the harder cases. Let us attempt to falsify the transitivity claim by taking $[\sigma x], [\eta x]$, and $[\sigma y]$. Since the first two events are incompatible, we would falsify transitivity of $CP_S(\cdot, \cdot)$ if $CP_S([\sigma x], [\sigma y])$ and $CP_S([\sigma y], [\eta x])$. Note that each pair of these events is compatible. By (†) and the observation that strict nesting implies nesting, we have $CP_S([\sigma x], [\sigma y])$. However, for $CP_S([\sigma y], [\eta x])$ we need $\neg(H_{[\eta x]} \subsetneq H_{[\sigma y]})$, which is false since, due to MFB, $H_{[\eta x]} = \{h_\eta\}$ and $H_{[\sigma y]} = \{h_\sigma, h_\eta\}$. We ask the reader to check the remaining cases to establish transitivity in Exercise 10.7. So (D6) is satisfied, and the remaining desiderata (D1)–(D5) hold by Fact 10.5 . Thus, we have a case in which there is an instance of (NN) despite the satisfaction of (D1)–(D6), which falsifies the implication from (4) to (1).   □

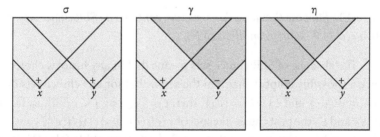

Figure 10.4 Illustration of the Minkowskian Branching Structure invoked in the proof of Fact 10.9. The structure contains three histories h_σ, h_γ, and h_η. Pluses and minuses indicate different outcomes at the splitting points x and y, and differences in shading indicate where the histories fail to overlap. Note that the structure exhibits a case of MFB, as $H_{[\eta x]} \cap H_{[\gamma y]} = \emptyset$.

We have seen in a few places above that the satisfaction of all the desiderata (D1)–(D6) for $CP_S(\cdot,\cdot)$ is facilitated by the presence of certain instances of modal funny business. It is thus interesting to learn what kind of MFB it takes to secure these desiderata in any BST_{NF} structure. A fifth condition that we put forward below is to serve precisely this purpose: to secure the desiderata in any possible BST_{NF} structure. We will call this condition Sticky MFB, see Def. 10.9.

In working toward that definition, we investigate which BST_{NF} structures contain no instances of condition (NN). First, as one might expect, deterministic BST_{NF} structures are too simple to support cases of (NN):

Fact 10.10. *If a BST_{NF} structure $\langle W, < \rangle$ contains one history h only, then no $e, f \in W$ satisfy condition (NN).*

Proof. In a deterministic structure $\langle W, < \rangle$, for any $e, f \in W$, we have $H_e = H_f = \{h\}$, where $h = W$ is the only history in the structure. $\qquad\square$

Somewhat more interestingly, condition (NN) also fails in a BST_{NF} structure without SLR choice sets. To prove this, we first establish an auxiliary fact:

Fact 10.11. *For any $e \in W$,*

$$H_e = \bigcap\{\Pi_{\ddot{e}}\langle H_e \rangle \mid \ddot{e} \in cll(e) \wedge \exists c \in \ddot{e}\,[c \leqslant e]\}. \qquad (\star)$$

Proof. See Exercise 10.2. $\qquad\square$

We now show that structures without *SLR* choice sets cannot contain instances of (NN). This generalizes Fact 10.10:

Fact 10.12. *Suppose that a BST_{NF} structure contains no SLR choice sets. Then no two $e, e' \in W$ satisfy condition (NN).*

Proof. The absence of SLR choice sets means that in any history, choice sets induce a (possibly empty) chain, in the sense that for any choice sets $\ddot{c}_1, \ddot{c}_2$, if $\ddot{c}_1 \cap h = \{c_1\}$ and $\ddot{c}_2 \cap h = \{c_2\}$, then $c_1 \leqslant c_2$ or $c_2 < c_1$. Thus, for any events e and e', the past cause-like loci of e ($cll(e)$) and e' ($cll(e')$) constitute chains l_e and $l_{e'}$. Assume now that e and e' are compatible (else (NN) fails anyway). By the above observation, every element of l_e and every element of $l_{e'}$ are comparable, as otherwise a pair of such incomparable elements would belong to a pair of *SLR* choice sets (remember that each element of l_e or $l_{e'}$ belongs to a choice set). Thus, since by the assumption no choice sets are

SLR, (†) $l_e \subseteq l_{e'}$ or $l_{e'} \subseteq l_e$. Recall now the identity $(\star)$ of Fact 10.11. By (†), the sets defining the intersection for H_e and $H_{e'}$ in this identity are related by the inclusion relation. Hence either $H_{e'} \subseteq H_e$ or $H_e \subseteq H_{e'}$, so condition (NN) is false. $\qquad\square$

Finally, if a BST_{NF} structure contains SLR choice sets, there is a specific form of MFB that implies that no two events satisfy condition (NN). We call this specific form "Sticky MFB" because it binds together SLR outcomes more stringently than required for MFB alone. We define Sticky MFB as follows:

Definition 10.9 (Sticky MFB). Let $\mathscr{W} = \langle W, < \rangle$ be a BST_{NF} structure. We say that two SLR events $c_1 \in \ddot{c}_1$ and $c_2 \in \ddot{c}_2$, $\ddot{c}_1, \ddot{c}_2 \subseteq W$, form an instance of *Sticky MFB* in $\mathscr{W}$ iff they satisfy the following:
if $H_{c_1} \cap H_{c_2} \neq \emptyset$, then
  for every $H_1 \in \Pi_{\ddot{c}_1}$ with $H_1 \neq H_{c_1}$: $H_1 \cap H_{c_2} = \emptyset$ or
  for every $H_2 \in \Pi_{\ddot{c}_2}$ with $H_2 \neq H_{c_2}$: $H_{c_1} \cap H_2 = \emptyset$.

Here is a useful Fact that will help us to see the connection between Sticky MFB and the nesting conditions that occur in condition (NN), as well as in our definition of $\text{CP}_S(\cdot, \cdot)$:

Fact 10.13. *For compatible $c_1 \in \ddot{c}_1$ and $c_2 \in \ddot{c}_2$:*

$$H_{c_1} \subseteq H_{c_2} \quad \text{iff for every } H_2 \in \Pi_{\ddot{c}_2} \text{ with } H_2 \neq H_{c_2} : H_{c_1} \cap H_2 = \emptyset.$$

Proof. "$\Rightarrow$": Let $H_{c_1} \subseteq H_{c_2}$. Any $h \in H_{c_1}$ must be in H_{c_2}, hence h cannot be in any outcome of $\ddot{c}_2$ other than H_{c_2}.
"$\Leftarrow$": In the opposite direction, if $h \in H_{c_1}$ and for any outcome H_2 of $\ddot{c}_2$ other than H_{c_2}, $h \notin H_2$, then $h \in H_{c_2}$ (by compatibility of c_1 and c_2). $\qquad\square$

Now, universal Sticky MFB guarantees that no two events satisfy condition (NN). Via the equivalences stated in Fact 10.8, this guarantees that $\text{Present}_C(\cdot)$ and $\text{Present}_S(\cdot)$ coincide, and the dynamic Past, Present, and Future satisfy all the desiderata (D1)–(D6). Emphatically, this concerns *all* BST_{NF} structures: for an apparently intuitive dynamic time, universal Sticky MFB is sufficient.

Fact 10.14. *Suppose that in a BST_{NF} structure $\mathscr{W} = \langle W, < \rangle$, there is universal Sticky MFB, i.e., for any SLR pair of choice sets $\ddot{c}_1, \ddot{c}_2$, any two elements*

$c_1 \in \ddot{c}_1$, $c_2 \in \ddot{c}_2$ *form an instance of Sticky MFB. Then no two* $e, e' \in W$ *satisfy condition (NN).*

Proof. Observe first that in a structure without choice sets, or without SLR choice sets, the claim holds vacuously. Let us thus assume that $\mathscr{W}$ is as in the premise, and argue for the contraposition. Assume thus that there is an instance of (NN), i.e., there are two compatible $e, e' \in W$ such that $H_e \not\subseteq H_{e'} \wedge H_{e'} \not\subseteq H_e$. Thus, there are histories $h, h_1, h_2 \in \mathrm{Hist}$ such that $e, e' \in h$ and $(\dagger)$, $e' \in h \setminus h_1$, $e \in h_1$ and $e \in h \setminus h_2$, $e' \in h_2$. By PCP$_{\mathrm{NF}}$, there is a choice set $\ddot{c}_1$, with $c_1, c_1' \in \ddot{c}_1$ such that $h \perp_{\ddot{c}_1} h_1$ and $\ddot{c}_1 \cap h = \{c_1\}$, $\ddot{c}_1 \cap h_1 = \{c_1'\}$, and $c_1 \leqslant e'$. Accordingly, $H_{c_1} = \Pi_{\ddot{c}_1}\langle h \rangle = \Pi_{\ddot{c}_1}\langle h_2 \rangle$ and $H_{c_1'} = \Pi_{\ddot{c}_1}\langle h_1 \rangle$. There is also a choice set $\ddot{c}_2$ with $c_2, c_2' \in \ddot{c}_2$ such that $h \perp_{\ddot{c}_2} h_2$ and $\ddot{c}_2 \cap h = \{c_2\}$, $\ddot{c}_2 \cap h_2 = \{c_2'\}$, $c_2 \leqslant e$. Accordingly, $H_{c_2} = \Pi_{\ddot{c}_2}\langle h \rangle = \Pi_{\ddot{c}_2}\langle h_1 \rangle$ and $H_{c_2'} = \Pi_{\ddot{c}_2}\langle h_2 \rangle$. We claim next that $c_1 \, SLR \, c_2$. For, if $c_1 \leqslant c_2$, then $c_1 \leqslant e$, and hence $e \notin h_1$, contradicting $(\dagger)$. For a similar reason it impossible that $c_2 \leqslant c_1$. And as $c_1, c_2 \in h$, they must be SLR. Furthermore, we have $\Pi_{\ddot{c}_1}\langle h \rangle \cap \Pi_{\ddot{c}_2}\langle h \rangle \neq \emptyset$, $\Pi_{\ddot{c}_1}\langle h_1 \rangle \cap \Pi_{\ddot{c}_2}\langle h \rangle = \Pi_{\ddot{c}_1}\langle h_1 \rangle \cap \Pi_{\ddot{c}_2}\langle h_1 \rangle \neq \emptyset$, and $\Pi_{\ddot{c}_1}\langle h \rangle \cap \Pi_{\ddot{c}_2}\langle h_2 \rangle = \Pi_{\ddot{c}_1}\langle h_2 \rangle \cap \Pi_{\ddot{c}_2}\langle h_2 \rangle \neq \emptyset$. This shows that two events $c_1 \in \ddot{c}_2$ and $c_2 \in \ddot{c}_2$, which are SLR, do not form an instance of Sticky MFB in $\mathscr{W}$. This proves the consequence of the contraposition. $\qquad\square$

We would like to obtain an implication in the opposite direction as well: from the fact that no events satisfy (NN), to Sticky MFB. As a preparation we state and prove the following Fact. Observe that this Fact concerns a somewhat richer class of BST$_{\mathrm{NF}}$ structures; see Figure 10.5 for illustration.

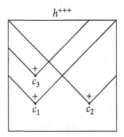

Figure 10.5 Illustration of the type of structure invoked in Facts 10.15 and 10.16: there are three choice sets $\ddot{c}_1, \ddot{c}_2$, and $\ddot{c}_3$, and $c_1 \, SLR \, c_2$, $c_2 \, SLR \, c_3$, but $c_1 < c_3$.

Fact 10.15. *Let* $\mathscr{W} = \langle W, < \rangle$ *be a BST*$_{\mathrm{NF}}$ *structure in which the notions of Past$(\cdot)$, Present$_S(\cdot)$, and Future$(\cdot)$ satisfy all the desiderata (D1)–(D6) of Def. 10.1. Let* $\mathscr{W}$ *contain at least three choice sets* $\ddot{c}_1, \ddot{c}_2$, *and* $\ddot{c}_3$ *such that*

$\{c_1, c_2, c_3\}$ *is consistent,* $c_1 < c_3$, $c_1 \, SLR \, c_2$, *and* $c_2 \, SLR \, c_3$. *Then there is an instance of Sticky MFB in* $\mathscr{W}$.

Proof. In our structure, $c_3 \in \text{Future}(c_1)$, so clearly $\neg CP_S(c_1, c_3)$. We must thus have $\neg CP_S(c_1, c_2)$ or $\neg CP_S(c_2, c_3)$—otherwise transitivity fails, contrary to the assumption. Consider $\neg CP_S(c_1, c_2)$ (the case of the other disjunct is analogous). To recall, $CP_S(c_1, c_2)$ is the conjunction:

$$H_{c_1} \cap H_{c_2} \neq \emptyset \wedge \neg(H_{c_1} \subsetneqq H_{c_2}) \wedge \neg(H_{c_2} \subsetneqq H_{c_1}),$$

so $\neg CP_S(c_1, c_2)$ is equivalent to:

$$(H_{c_1} \cap H_{c_2} = \emptyset) \vee (H_{c_1} \subsetneqq H_{c_2}) \vee (H_{c_2} \subsetneqq H_{c_1}).$$

As c_1 is compatible with c_2 by assumption, the first disjunct is false, and by Fact 10.13, since strict inclusion implies inclusion, $\neg CP_S(c_1, c_2)$ implies

$$\forall H_2 \in \Pi_{\ddot{c}_2} \; [H_2 \neq H_{c_2} \rightarrow H_{c_1} \cap H_2 = \emptyset] \quad \text{or}$$
$$\forall H_1 \in \Pi_{\ddot{c}_1} \; [H_1 \neq H_{c_1} \rightarrow H_1 \cap H_{c_2} = \emptyset].$$

Thus, as $c_1 \in \ddot{c}_1$ and $c_2 \in \ddot{c}_1$ are SLR, they form an instance of Sticky MFB. □

The preceding Fact, interesting on its own, helps us to clarify the relation from (NN) to Sticky MFB:

Fact 10.16. *Let* $\mathscr{W} = \langle W, < \rangle$ *be a* BST_{NF} *structure in which no* $e, f \in W$ *satisfy condition (NN). Let* $\mathscr{W}$ *contain at least three choice sets* $\ddot{c}_1, \ddot{c}_2$, *and* $\ddot{c}_3$ *such that* $\{c_1, c_2, c_2\}$ *is consistent,* $c_1 < c_3$, $c_1 \, SLR \, c_2$, *and* $c_2 \, SLR \, c_3$. *Then there is an instance of Sticky MFB in* $\mathscr{W}$.

Proof. By Fact 10.9 we have that the notions of $\text{Past}(\cdot)$, $\text{Present}_S(\cdot)$, and $\text{Future}(\cdot)$ satisfy all the desiderata (D1)–(D6) of Def. 10.1. By the proof of Fact 10.16, either $c_1 \in \ddot{c}_1$ and $c_2 \in \ddot{c}_2$, or $c_2 \in \ddot{c}_2$ and $c_3 \in \ddot{c}_3$ form an instance of Sticky MFB in $\mathscr{W}$. □

Having these Facts on the table, we can describe the logical landscape they point to as follows:

First, we have the equivalence of three conditions (by Fact 10.8): (1) There are no $e, f \in W$ that satisfy condition (NN). (2) The notions of $\mathrm{Present}_C(\cdot)$ and $\mathrm{Present}_S(\cdot)$ coincide. (3) The notions of $\mathrm{Past}(\cdot)$, $\mathrm{Present}_C(\cdot)$, and $\mathrm{Future}(\cdot)$ satisfy all the desiderata (D1)–(D6) of Def. 10.1. Each of these conditions implies (4): The notions of $\mathrm{Past}(\cdot)$, $\mathrm{Present}_S(\cdot)$, and $\mathrm{Future}(\cdot)$ satisfy all the desiderata. The converse implication, from (4) to each of (1)–(3), does not hold, however; see Fact 10.9.

Second, universal Sticky MFB is a sufficient condition of each of (1)–(3) and of (4), by Fact 10.14. As for necessary conditions, they depend on the complexity of the $\mathrm{BST}_{\mathrm{NF}}$ structure under consideration. We do not offer a maximally fine-grained characterization of $\mathrm{BST}_{\mathrm{NF}}$, but we can observe the following: (a) Structures without SLR choice sets (including deterministic structures, which contain no choice sets at all) do not support cases of (NN), so they satisfy (1)–(3) trivially, and by implication also (4), the transitivity of $\mathrm{CP}_S(\cdot, \cdot)$ as a relation on W. (b) Structures that contain choice sets that are all pairwise *SLR* need a strong form of MFB to exclude cases of (NN). Interestingly, however, in the absence of MFB, for any history h, the relation $\mathrm{CP}_{S|h}(\cdot, \cdot)$ is transitive on such structures. In the presence of MFB, $\mathrm{CP}_{S|h}(\cdot, \cdot)$ can be transitive or fail to be transitive—these observations are the subject of Exercise 10.6. Of course, by Fact 10.14, universal Sticky MFB is sufficient to enforce transitivity of $\mathrm{CP}_S(\cdot, \cdot)$ in all $\mathrm{BST}_{\mathrm{NF}}$ structures, including those with exclusively *SLR* choice sets. (c) If a structure lies outside of cases (a) and (b); that is, if it is neither such that it contains no *SLR* choice sets nor such that it contains choice sets all of which are *SLR*, then it has to contain instances both of *SLR* choice sets and of choice sets that are order related. The structures considered in Fact 10.15 and in Fact 10.16 are a subtype of type (c): they contain a triple of choice sets, with two pairs being *SLR* and one pair being related by $<$. For these structures we showed that for (1), (2), and (3), but also for the weaker condition (4), a necessary condition is the existence of an instance of Sticky MFB.

Having this landscape before our eyes, how should we estimate the price for having an intuitively satisfying notion of dynamic time (i.e., dynamic time that satisfies all the desiderata (D1)–(D6)), either on the $\mathrm{Present}_C(\cdot)$ analysis or on the $\mathrm{Present}_S(\cdot)$ analysis? It seems the response depends on whether one considers the question from the perspective of a creator of all possible $\mathrm{BST}_{\mathrm{NF}}$ universes, or from the perspective of a dweller in one of such universes. For the former perspective, as the $\mathrm{BST}_{\mathrm{NF}}$ postulates alone permit a large variety of structures, it seems prudent to require the constraint of universal Sticky MFB—this will guarantee that (D1)–(D6) are satisfied in all BST$_{NF}$ structures. The dweller's perspective seems less constrained: perhaps her world is such that it tolerates some failures of (D1)–(D6). Perhaps there were such failures long ago in the past, or in remote regions—why should she care? Or, perhaps, there are such failures in her vicinity, but they are small, well-localized, and hardly visible. Our dweller might strike you as overly optimistic. The reasons for her optimism, however, are facts of dynamic time, and these can be debated. After all, such facts fully supervene on what is possible in our world.

In the next section we turn to the representation of dynamic time in Minkowskian Branching Structures (MBSs), which allow for a perspicuous representation of the welcome consequences of the types of MFB that we discussed earlier.

10.8 What does dynamic time look like in MBSs?

In Minkowskian Branching Structures, our notions of the dynamic past, present, and future of a given event are defined in terms of the Minkowski ordering $<_M$, which is clearly invariant with respect to the automorphisms of Minkowski space-time, the space-time of special relativity. Thus, the regions of histories that our definitions single out are invariant with respect to these automorphisms, and our constructions are directly relevant to the discussion of the problem of the present in special relativity (see Section 10.2).

A salient feature of our definitions is that the shape of the dynamic past, present, and future of an event e in a history h depends on the location of elements of choice sets in h. In MBSs, these choice sets are induced by the pattern of qualitative differences between histories.

In Section 10.3 we discussed and defended the general idea of making room for a dynamic present that is extended in the coordinate-temporal dimension. We will now apply our definitions to some selected MBSs to visualize what the dynamic future, present and past of a given event look like. Our focus is on cases in which the *causae causantes*-based approach and the semantics-based approach deliver the same verdict.

Example 1: Simple cases. The simplest case is a deterministic MBS (i.e., a structure with just one history). In such a structure, there is no indeterminism, and the co-presentness relation on both of our accounts turns out to be the universal relation: any event is co-present with any other event. Dynamic time is trivial under determinism. We do not provide an illustration for this case. We also leave out of our considerations another kind of structure that is featureless according to Def. 10.4 (i.e., an MBS in which there is a history that consists wholly of members of uncorrelated choice sets). In such a structure, the relation of co-presentness on that history is simply the identity relation, which is also trivial.

The first simple but non-trivial case is an MBS with two histories σ and η, which split at a single point c, so $\{c\} = S_{\sigma\eta}$ (see Figure 10.6). For $x \ngeq c$ (which excludes $x = c$), $\text{Present}([\sigma x]) = \{[\sigma y] \mid y \ngeq c\} = \{[\eta y] \mid y \ngeq c\}$ and $\text{Future}([\sigma x]) = \{[\sigma y], [\eta y] \mid y \geq c\}$. Note that $\text{Past}([\sigma x]) = \emptyset$. The situation is analogous for $x > c$: $\text{Present}([\sigma x]) = \{[\sigma y] \mid y \geq c\}$, $\text{Past}([\sigma x]) = \{[\sigma y] \mid y \ngeq c\} = \{[\eta y] \mid y \ngeq c\}$, and $\text{Future}([\sigma x]) = \emptyset$.

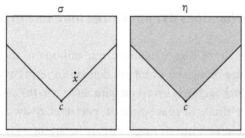

Figure 10.6 Illustration of the simplest non-trivial MBS from Example 1, which includes a single binary splitting point c. Different shading indicates where the histories differ. The present of $e = [\sigma x]$ is the future light cone above c in history h_σ; the future of e is, accordingly, empty.

Example 2: Two time-like splitting points. Consider next an MBS with three histories, $\Sigma = \{\sigma, \gamma, \eta\}$, in which there are two splitting points $c_1, c_2 \in \mathbb{R}^4$ such that $c_1 <_M c_2$ and $S_{\sigma\gamma} = S_{\sigma\eta} = \{c_1\}$ and $S_{\gamma\eta} = \{c_2\}$. That is, γ and η split from σ at c_1, and then η splits from γ at c_2. Figure 10.7 represents these three histories as squares with a common bottom region, taking the history labelled by σ to be our reference history. The shading convention is that a difference in shading indicates that the corresponding regions are not to be identified.

Now pick an event $e = [\gamma x]$ that is above c_1 but not above c_2, $c_1 <_M x$ and $x \nRightarrow_M c_2$, and ask: (1) What is the dynamic future of e? (2) What is its dynamic past? (3) And what is its dynamic present?

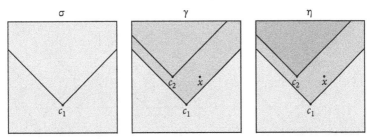

Figure 10.7 Illustration of Example 2: Two time-like splitting points. Different shading indicates where the histories differ, thus showing the past, the present, and the future of $e = [\gamma x]$. See text for details.

Fact 10.5 (as well as Fact 10.3) yields the following verdicts:

1. The dynamic future of $e = [\gamma x]$ is the set of events that are weakly above $[\gamma c_2] = [\eta c_2]$: Future($[\gamma x]$) = $\{[\xi z] \mid c_2 \leqslant_M z \wedge \xi \in \{\gamma, \eta\}\}$. Note that this region is the union of two future light cones of c_2, in histories labeled by γ and η. The light cones include their boundaries. And the future of $e = [\gamma x]$ is above $[\gamma c_2]$ rather than above e.
2. The dynamic past of $e = [\gamma x]$ is the set of events that are in history γ and not weakly above $[\gamma c_1]$: Past($[\gamma x]$) = $\{[\gamma z] \mid c_1 \not\leqslant_M z\}$. In contrast to the future, the past is shared by all of the three histories.
3. The dynamic present of $e = [\gamma x]$ is the set of events in history γ and "between" c_1 and c_2 in the sense: Present($[\gamma x]$) = $\{[\gamma z] \mid c_1 \leqslant_M z \wedge c_2 \not\leqslant_M z\}$. The present of e is shared by the two histories to which e belongs, γ and η.

Note that the present of $e = [\gamma x]$ turns out to be a spatially extended and temporally thick collection of events, whose temporal thickness depends on the distance between c_1 and c_2.

Example 3: Four splitting points, layered in two SLR pairs. Consider an MBS with three histories, $\Sigma = \{\sigma, \gamma, \eta\}$, with $S_{\sigma\gamma} = S_{\sigma\eta} = \{c_1, c_2\}$ and $S_{\gamma\eta} = \{c_3, c_4\}$, where $c_1 <_M c_3$ and $c_2 <_M c_4$—see Figure 10.8. We consider an event $[\gamma x]$ with x sliced between two pairs of splitting points, that is, $(c_1 <_M x$ or $c_2 <_M x)$ and $(c_3 \not\leqslant_M x$ and $c_4 \not\leqslant_M x)$. Applying Fact 10.5 we arrive at the following result:

$$\text{Present}([\gamma x]) = \{[\gamma y] \mid (y \geqslant_M c_1 \vee y \geqslant_M c_2) \wedge (y \not\geqslant_M c_3 \wedge y \not\geqslant_M c_4)\}.$$

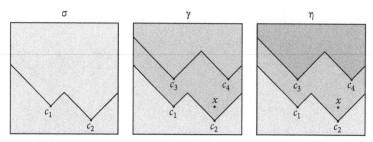

Figure 10.8 Illustration of Example 3: Three histories with four splitting points $c_1, \ldots, c_4$ and with two instances of Sticky MFB, one involving c_1 and c_2 and the other involving c_3 and c_4. The dynamic present of event $e = [\gamma x]$ is the shaded W-shaped region containing x.

Thus, the present of $[\gamma x]$ turns out to have the shape of a thick letter W. Note that by adding more splitting points to $S_{\sigma\gamma}$ and to $S_{\gamma\eta}$ we obtain as Present(e) a "generalized" letter W, with more top corners and more bottom corners. And by making the separation between $S_{\sigma\gamma}$ and $S_{\gamma\eta}$ smaller, we can make the generalized W arbitrarily thin. We can make it to maximally extend through the whole history as well. A suggestive construction can be based on universal Sticky MFB for infinitely many space-like related choice sets. Given densely packed choice sets and universal Sticky MFB, the dynamic present of an event can approximate its static present.

10.9 Conclusions

In this chapter we have investigated the concept of dynamic time that is inherently related to real change. We argued that if our world harbors no real change, there is no real time, or at least real time is thoroughly trivialized. We contrasted real, dynamic time with static coordinate time. Given this contrast between two notions of time, from the very start we acknowledged that the notions of dynamic past, present, and future might be different from their static counterparts. For instance, the dynamic present of some event might be thick along coordinate time.

Real change has a definitely modal ring to it: real change means that from among a family of alternative possible outcomes, one is actualized later on, while the remaining outcomes have ceased to be possible. BST, by being a theory of local indeterminism playing out in space and time, and with its causal and semantic resources, is perfectly suited to analyzing indeterministic real change, and hence to explicate dynamic time.

There is more than one way to link dynamic time to the phenomenon of the passing of possibilities, and one might worry that this could stand in the way of a unified account. We offered two approaches that at first glance are quite different, one based on causal concepts, and the other building on intuitions concerning our talk about future events. The former's focus is the relation of co-presentness, which is defined in terms of the identity of sets of *causae causantes*. Unsurprisingly, co-presentness so defined is an equivalence relation. The dynamic present, past, and future of a given event are then naturally defined. It turns out that the three notions can be characterized in purely modal terms, i.e., by inclusion or identity of sets of histories. Moreover, these notions are relativity-proof as the analysis is rooted in the primitive notions of BST.

The second approach begins by explicating what it means, in modal terms, that one event belongs to the dynamic (open) future of another event. It then defines the dynamic past and present of a given event by assuming certain natural desiderata on how the three notions are related. The resulting notions are relativity-proof and characterized in purely modal terms, as in the first approach. Moreover, the two approaches agree about their concepts of dynamic past and future. They can disagree, however, on the analysis of the dynamic present, and hence, of co-presentness.

When confronted with our list of intuitive desiderata for a notion of dynamic time (Def. 10.1), the two analyses show different strengths and weaknesses. In the first (causal) approach, the co-presentness relation is automatically transitive, but a history might fail to be fully carved into the past, the present, and the future of an event from this history. In the second (semantic) approach, in contrast, any history is automatically partitioned into the dynamic past, present, and future of any of its events, but the relation of co-presentness is not necessarily transitive. As our discussion showed, both issues have similar reasons: In non-trivial structures, a certain strong form of modal funny business is required to fulfill all the desiderata on any of the two analyses. A sufficient condition for a satisfying notion of dynamic time is universal Sticky MFB: If a BST structure is such that there are modal correlations all across space, then dynamic time based on indeterminism is relativity-proof whilst retaining all other intuitive features as well. This proves our main point: A BST analysis of local indeterminism in space-time leads to formal structures that are rich enough to accomplish what the structure of a single space-time cannot: to anchor a satisfying notion of real, dynamic time.

10.10 Exercises to Chapter 10

Exercise 10.1. Prove items (3)–(8) of Fact 10.5.

Exercise 10.2. Prove the strengthened version of identity ($*$) from the proof of Fact 10.1, which restricts cll of e to those lying in the past of e (i.e., prove Fact 10.11):

For any $e \in W$,

$$H_e = \bigcap \{\Pi_{\ddot{c}}\langle H_e \rangle \mid \ddot{c} \in cll(e) \wedge \exists c \in \ddot{c} \, [c \leqslant e]\}. \qquad (*)$$

Hint: A proof is provided in Appendix B.10.

Exercise 10.3. Provide formal details for the strengthened definition of co-presentness discussed in footnote 18.

Hint: As a first approximation, consider this definition:

Definition 10.10 (Co-presentness strengthened). $CP(e_1, e_2)$ iff $H_{e_1} = H_{e_2}$ and (if $\ddot{e}_i \neq \{e_i\}$ for $i = 1, 2$, then $\Pi_{\ddot{e}_1} = \Pi_{\ddot{e}_2}$).

Show how this definition resolves the problem of the non-robust co-presentness of c_1 and c_2 introduced in footnote 18. Discuss why the antecedent, "if $\ddot{e}_i \neq \{e_i\}$ for $i = 1, 2$", is needed. As a suggestion for a full-fledged notion of alternatives, consider the requirement: each alternative is above a different element of some one choice set, but in the same outcome of any other choice set.

Exercise 10.4. Construct a BST_{NF} structure $\mathcal{W}$ with MFB but in which desideratum (D5) of Def. 10.1 is still violated for $CP_C(\cdot, \cdot)$; that is, such that for some history h and $e, e' \in h$, neither $H_e \subseteq H_{e'}$ nor $H_{e'} \subseteq H_e$.

Hint: Consider a structure with only two SLR binary choice sets $\ddot{e} = \{e, e_1\}$ and $\ddot{e}' = \{e', e'_1\}$, and with MFB given by $H_{e_1} \cap H_{e'_1} = \emptyset$. The remaining intersections are non-empty, so $e, e' \in h$ for some history h. And there is MFB in this structure but $H_e \not\subseteq H_{e'}$ and $H_{e'} \not\subseteq H_e$.

Exercise 10.5. Show that in the structure with two *SLR* choice sets and no modal correlations discussed in the proof of Fact 10.4, for any history, the restricted relation of co-presentness $CP_{C|h}(\cdot, \cdot)$ is transitive, while the unrestricted relation $CP_C(\cdot, \cdot)$ is not transitive.

Hint: Show that different elements of a choice set, which are incompatible and thereby not co-present, can be co-present to the same *SLR* event.

Exercise 10.6. Consider a BST_{NF} structure with exactly three pairwise *SLR* binary choice sets. Show that in this structure (a) in the absence of MFB (the case of 8 histories), the relation $CP_S(\cdot, \cdot)$ is transitive. (b) Exhibit a case in which there is MFB but $CP_S(\cdot, \cdot)$ is not transitive.

Hint: For (b), a structure with 6 histories will do.

Exercise 10.7. Check that $CP_S(\cdot, \cdot)$ is satisfied in the cases that were left open in the proof of Fact 10.9.

PART III
APPENDIX

Selected Proofs and Additional Material

A.1 Dedekind continuity

Postulate 2.4 of the existence of infima for lower-bounded chains (p. 36) guarantees that a maximal chain in $\langle W, < \rangle$ is continuous in the sense of Dedekind: a Dedekind cut of a maximal chain has neither gaps nor jumps. A Dedekind cut completely partitions a maximal chain C into two non-empty sub-chains A and B such that every element of A is strictly below every element of B. There is a gap if A has no maximal element and B has no minimal element, whereas there is a jump if A has a maximal element and B has a minimal element. Clearly, a jump is excluded by density and the chain's maximality. Further, a gap is excluded as well (by Postulate 2.4), which is the topic of the following Fact: given Postulate 2.4, maximal chains are continuous.

Fact A.1. *Postulate 2.4 implies that a Dedekind cut $\{A, B\}$ of a maximal chain C has no gaps.*

Proof. Let $\{A, B\}, A < B$ be a Dedekind cut of the maximal chain $C = A \cup B$. Since $A \neq \emptyset$, B is lower bounded, so B has the infimum $\inf B$ by Postulate 2.4. Clearly, $\inf B \leqslant B$ and $A \leqslant \inf B$ since every element of A is a lower bound for B (by the definition of infima). Hence $\inf B \in C$, as otherwise C would not be a *maximal* chain. Since $\{A, B\}$ completely partitions C, either $\inf B \in A$ or $\inf B \in B$. If the former, by the definition of infima $\inf B$ is a maximal element of A; if the latter, $\inf B$ is a minimal element of B. This proves that there is no gap. $\square$

Since a maximal chain is a subset of some history (see Fact 2.1(3)), Postulate 2.5 of the existence of history-relative suprema provides another guarantee that a maximal chain is Dedekind continuous; the argument is analogous to the proof of Fact A.1.

A.2 Formal details of the interrelation of BST_{92} and BST_{NF}

This section complements Chapter 3 by offering details of the interrelation between the two alternative BST frameworks, BST_{92} and BST_{NF}, that were described there. Among other things, we provide proofs to the translatability results for BST_{92} and BST_{NF} that were formulated in Chapter 3.6.2. We establish two translation mappings, the Λ- and the Υ-transform, which will be shown to preserve the basic indeterministic structure: given a BST_{92} structure, its Υ-transform replaces each choice point by the set of transitions from that choice point, leading to a choice set and, accordingly, to a BST_{NF} structure. In the other direction, the Λ transform operates on a BST_{NF} structure, replacing each choice set by a single point, which will be a choice point in the resulting BST_{92} structure. In this way, as announced, BST_{92} and BST_{NF} can be seen as two alternative representations of an underlying indeterministic structure. This means that we can represent indeterministic scenarios without really having to decide between the different prior choice principles of BST_{92} and BST_{NF}, and that we can take a pragmatic attitude toward the topological consequences of BST_{92} vs. BST_{NF} as well.

A.2.1 Characterizing the transition structure of a BST_{92} structure

We are often only interested in indeterministic transitions, as deterministic transitions make no difference to the branching of histories[1]. In the present context, however, it is important to consider *all* transitions, including those that are trivial from the point of view of indeterminism. We repeat the definition of the Υ *transform* of a BST_{92} structure:

Definition 3.17 (The Υ transform as the full transition structure of a BST_{92} structure.). Let $\langle W, < \rangle$ be a BST_{92} structure. Then we define the transformed structure, $\Upsilon(\langle W, < \rangle)$, to be the full transition structure (including trivial transitions) together with the transition ordering $\prec$ from Def. 3.10, as follows:

$$\Upsilon(\langle W, < \rangle) =_{df} \langle W', \prec \rangle, \quad \text{where} \quad W' =_{df} TR_{full}(W) = \{e \mapsto H \mid e \in W, H \in \Pi_e\}.$$

From here on we will mark transformed elements with primes.

Having defined the transition structure, we now characterize its properties. It turns out that the full transition structure $\Upsilon(\langle W, < \rangle)$ looks very much like the original BST_{92} structure $\langle W, < \rangle$, except for what happens at the choice points. In fact, we will be able to show that, apart from the prior choice postulate, all defining properties of BST_{92}; that is, the whole list of properties of a common BST structure from Def. 2.10, continue to hold; see Lemma 3.3, which was stated in Chapter 3 and which we will repeat and prove below. With respect to the choice points, the difference is the following. In BST_{92}, the branching of histories is from a choice point, shared among the histories that branch, to the immediate possibilities for the future at that choice point. There are no first points in these different possible futures, and this fact leads to the failure of local Euclidicity in the global BST_{92} topology (see Section 4.4). In $\Upsilon(\langle W, < \rangle)$, on the other hand, each choice point is replaced by all the transitions that have that choice point as an initial. Therefore, where in BST_{92} there was a last point that was shared between the different possibilities, in the transition structure there are now multiple first points characterizing these different possibilities, and there is no longer any last shared point.[2] In the structures of Figure 3.1 (p. 44), the move from (a) to (b) exactly corresponds to the move from the BST_{92} structure M_a to its transition structure M_b.[3] For the topological consequences, see Section 4.4.

In order to prove that $\Upsilon(\langle W, < \rangle)$ is a common BST structure, we first need to establish the form that histories (i.e., maximal directed sets), have in that ordering. Their form is quite intuitive, even though it turns out that the proof of that fact is somewhat lengthy. We first establish a useful general fact about directed sets of transitions:

---

[1] For a study along those lines, see Müller (2010).

[2] This image of fanning out the transitions from a choice point motivates our notation, Υ.

[3] To be precise, the transition structure of M_a is order-isomorphic to M_b. See Chapter A.2.3 for a formal discussion.

Fact A.2. *Let $T \subseteq \Upsilon(\langle W, < \rangle)$ be a set of transitions, and let there be $e \in W$ and $H_1, H_2 \in \Pi_e$, $H_1 \neq H_2$, such that both $\tau_1 = e \rightarrowtail H_1$ and $\tau_2 = e \rightarrowtail H_2$ are members of T. Then T is not directed.*

Proof. Assume otherwise; that is, assume that there is some $\tau^* = e^* \rightarrowtail H^* \in T$ for which $\tau_1 \preccurlyeq \tau^*$ and $\tau_2 \preccurlyeq \tau^*$. By Fact 3.11(1), this implies $H_{e^*} \subseteq H_1$ and $H_{e^*} \subseteq H_2$. But as H_1 and H_2 are different elements of the partition Π_e, we have $H_1 \cap H_2 = \emptyset$, contradicting $H_{e^*} \subseteq H_1 \cap H_2$. (Note that $H_{e^*} \neq \emptyset$ by Fact 2.1(3).) □

Now we can tackle the form of histories in $\Upsilon(\langle W, < \rangle)$.

Lemma A.1. *Let $\langle W, < \rangle$ be a BST$_{92}$ structure, and let $\langle W', \prec \rangle =_{\text{df}} \Upsilon(\langle W, < \rangle)$. The histories (maximal directed sets) in $\langle W', \prec \rangle$ are exactly the sets*

$$T_h =_{\text{df}} \{ e \rightarrowtail \Pi_e \langle h \rangle \mid e \in h \}$$

for h in $\text{Hist}(W)$.

Proof. First we establish that such sets are indeed histories in $\langle W', \prec \rangle$. Thus, take some $h \in \text{Hist}(W)$, and let $T_h =_{\text{df}} \{ e \rightarrowtail \Pi_e \langle h \rangle \mid e \in h \} \subseteq W'$. The set T_h is directed: take $e_1 \rightarrowtail \Pi_{e_1} \langle h \rangle$ and $e_2 \rightarrowtail \Pi_{e_2} \langle h \rangle$ from T_h, whence $e_1, e_2 \in h$. As h is directed, there is $e_3 \in h$ such that $e_1, e_2 \leqslant e_3$. By construction, $e_3 \rightarrowtail \Pi_{e_3} \langle h \rangle \in T_h$. And as to the ordering, $H_{e_3} \subseteq \Pi_{e_1} \langle h \rangle$ because $e_3 \in h$ and $e_1 \leqslant e_3$. Analogously, $H_{e_3} \subseteq \Pi_{e_2} \langle h \rangle$. So indeed (noting Fact 3.11(1)), $e_i \rightarrowtail \Pi_{e_i} \langle h \rangle \preccurlyeq e_3 \rightarrowtail \Pi_{e_3} \langle h \rangle$ ($i = 1, 2$), establishing the common upper bound. Moreover, T_h is maximal directed. To prove this, take some $\tau^* \in (W' \setminus T_h)$; this transition has the form $\tau^* = e^* \rightarrowtail H^*$ for some $e^* \in W, H^* \in \Pi_{e^*}$. There are two cases.

Case 1: There is some $\tau = e \rightarrowtail \Pi_e \langle h \rangle \in T_h$ for which $e = e^*$, i.e., $e^* \in h$. Then, as $\tau \neq \tau^*$, by Fact A.2, $T_h \cup \{ \tau^* \}$ cannot be directed.

Case 2: There is no $\tau = e \rightarrowtail \Pi_e \langle h \rangle \in T_h$ for which $e = e^*$, i.e., $e^* \notin h$. Then we have $e^* \in h'$ for a different $h' \in \text{Hist}(W)$, and by the BST$_{92}$ prior choice principle, there is some $c \in h \cap h'$ such that $c < e^*$ and $h \perp_c h'$. As $c \in h$, we have $\tau_c =_{\text{df}} c \rightarrowtail \Pi_c \langle h \rangle \in T_h$. We can now show that $T_h \cup \{ \tau^* \}$ is not directed: there can be no common upper bound for τ_c and τ^* in W'. Assume for reductio that there is some $\tau' = e' \rightarrowtail H' \in T_h \cup \{ \tau^* \}$ for which $\tau_c \preccurlyeq \tau'$ and $\tau^* \preccurlyeq \tau'$. We can rule out $\tau' = \tau^*$: we have $\tau_c \neq \tau^*$ by $c < e^*$, and $\tau_c \not\preccurlyeq \tau^*$ as $H_{e^*} \not\subseteq \Pi_c \langle h \rangle$ (as $H_{e^*} \subseteq \Pi_c \langle h' \rangle$). So we must have $\tau' \in T_h$. By the definition of $\prec$, the assumed ordering relations imply $H_{e'} \subseteq \Pi_c \langle h \rangle$ and $H_{e'} \subseteq H^* \subseteq \Pi_c \langle h' \rangle$. But we have $\Pi_c \langle h \rangle \cap \Pi_c \langle h' \rangle = \emptyset$.

So, having shown that the sets T_h are indeed histories in $\langle W', \prec \rangle$, we need to show that *all* histories in $\langle W', \prec \rangle$ are of that form. Thus, let $g \subseteq W'$ be a history in $\langle W', \prec \rangle$, maximal directed with respect to $\prec$. By Fact A.2, there is no $e \in W$ for which g contains two transitions $e \rightarrowtail H_1$ and $e \rightarrowtail H_2$, $H_1 \neq H_2$, so that we can write

$$g = \{ e \rightarrowtail H(e) \mid e \in E \}$$

for some set $E \subseteq W$, where $H(e) \in \Pi_e$. We first show that E is directed: Take $e_1, e_2 \in E$, so that $\tau_i =_{\text{df}} e_i \rightarrowtail H(e_i) \in g$. By directedness of g, there is some $\tau_3 = e_3 \rightarrowtail H(e_3) \in g$ for which $\tau_i \preccurlyeq \tau_3$ ($i = 1, 2$), which implies $e_3 \in E$, $e_1 \leqslant e_3$, and $e_2 \leqslant e_3$. This proves that E is directed, and therefore there is some history $h \in \text{Hist}(W)$ for which $E \subseteq h$. We now show that for all $e \in E$, we have $h \in H(e)$. Thus, take some $e \in E$, which is the initial of some $\tau = e \rightarrowtail H(e) \in g$. There are two cases.

Case 1: τ is maximal in g. By Fact 2.1(9), τ therefore is also maximal in W', which implies that e is maximal in W. By Fact 2.1(10), there is a unique history containing e, and as $e \in h$, that unique history must be our h. We therefore have $\Pi_e = \{\{h\}\}$. As $H(e) \in \Pi_e$, this implies $H(e) = \{h\}$, whence $h \in H(e)$.

Case 2: τ is not maximal in g, i.e., there is some $\tau' = e' \rightarrowtail H(e') \in g$ for which $\tau \prec \tau'$. By Fact 3.11(1), this implies that $H_{e'} \subseteq H(e)$, and as $e' \in E \subseteq h$, we have $h \in H_{e'}$ and therefore also $h \in H(e)$.

As $h \in H(e)$ for all $e \in E$, we have $H(e) = \Pi_e \langle h \rangle$ for all $e \in E$. This implies $g \subseteq T_h$, and by Fact 2.1(7), we have established $g = T_h$. $\square$

Given these facts, we can now prove that switching from a BST$_{92}$ structure to its full transition structure preserves the common BST structure axioms. We will need to assume, however, that the BST$_{92}$ structure contains no minima otherwise the historical connection might fail: If a minimal element $e \in W$ is a choice point, its image under the Υ transformation consists of two or more minimal elements in the resulting order, which accordingly have no common lower bound.

Lemma 3.3. *Let $\langle W, < \rangle$ be a BST$_{92}$ structure without minima. Then its full transition structure $\Upsilon(\langle W, < \rangle)$ is still a common BST structure according to Definition 2.10.*

Proof. We need to check that $\langle W', \prec \rangle =_{\mathrm{df}} \Upsilon(\langle W, < \rangle)$ satisfies all the properties (1)–(7) listed in Definition 2.10.

1. W' is non-empty. See Fact 3.10(1).

2. $\langle W' \prec \rangle$ is a strict partial ordering. See Fact 3.10(2).

3. $\prec$ is dense.
   Let $(e_1 \rightarrowtail H_1) \prec (e_3 \rightarrowtail H_3)$, which means that $e_1 < e_3$ and $H_3 \subseteq H_1$. By density of $<$, there is $e_2 \in W$ for which $e_1 < e_2 < e_3$. Take some $h \in H_3$, so that $\{e_1, e_2, e_3\} \subseteq h$. Let $H_2 =_{\mathrm{df}} \Pi_{e_2} \langle h \rangle$. We claim that the transition $e_2 \rightarrowtail H_2$ is $\prec$-sliced between the two transitions above. We have to show that $H_2 \subseteq H_1$ and $H_3 \subseteq H_2$. For the former, take some $h_2 \in H_2$; we have $h_2 \equiv_{e_1} h$ as witnessed by e_2. As $H_1 \in \Pi_{e_1}$, therefore, $h_2 \in H_1$ iff $h \in H_1$. Now as $h \in H_3$ and $H_3 \subseteq H_1$, we have $h \in H_1$, so that indeed, $H_2 \subseteq H_1$. The latter claim is established analogously.

4. Any lower bounded chain in $\langle W', \prec \rangle$ has an infimum in $\prec$.
   Let $l' = \{e_i \rightarrowtail H_i \mid i \in \Gamma\}$ (Γ some index set) be a chain that is lower bounded by $e^* \rightarrowtail H^*$. Then the set $l =_{\mathrm{df}} \{e_i \mid i \in \Gamma\}$ of initials of l' is a chain lower bounded by e^*, and there is a history $h \subseteq W$ for which $l \subseteq h$. By the BST$_{92}$ postulate of infima, l has an infimum v in $<$. The infimum v gives rise to the transition $v' =_{\mathrm{df}} v \rightarrowtail \Pi_v \langle h \rangle \in W'$. Let $e_i \rightarrowtail H_i \in l'$. We have $v \leqslant e_i$ (as $e_i \in l$), and $H_i \subseteq \Pi_v \langle h \rangle$ because $e_i \in h$ and $v \leqslant e_i$. Thus, $v' \preccurlyeq (e_i \rightarrowtail H_i)$, so v' is a lower bound of l'. Let now $e \rightarrowtail H$ be any lower bound of l', whence e is a lower bound of l. As v is the infimum of l, we have $e \leqslant v$, and as $l \subseteq h$, we have $H = \Pi_e \langle h \rangle$, which implies $H_v \subseteq H$. Thus $e \rightarrowtail H \preccurlyeq v'$, i.e., v' is indeed the greatest lower bound of l'.

5. Any upper-bounded chain in $\langle W', \prec \rangle$ has a history-relative supremum in each history to which it belongs.
   Let the chain l' be upper bounded by u' in $\langle W', \prec \rangle$ and $l' \cup \{u'\} \subseteq h'$ for $h' \in \mathrm{Hist}(W', \prec)$. Given the form of histories in $\mathrm{Hist}(W', \prec)$ (see Lemma A.1), $h' = \{e \rightarrowtail \Pi_e \langle h \rangle \mid e \in h\}$ for some $h \in \mathrm{Hist}(W)$. It follows that for the set l of initials of l' and for u the initial of u', $l \cup \{u\} \subseteq h$; additionally, $l \leqslant u$. By the BST$_{92}$

axiom of history-relative suprema, there is a history-relative supremum $s = \sup_h l$ of l in h. Consider now the transition $s' = s \rightarrowtail \Pi_s\langle h \rangle \in h'$. That transition is an upper bound of l': for any $e \rightarrowtail H \in l'$, we have $e \leqslant s$ and $H_s \subseteq H = \Pi_e\langle h \rangle$. Furthermore, s' is the least upper bound (i.e., the supremum) of l' in h': let $s^{*\prime} \in h'$ be an upper bound of l' in h'; by the form of histories, $s^{*\prime} = s^* \rightarrowtail \Pi_{s^*}\langle h \rangle$ with $s^* \in h$. Thus, $s \leqslant s^*$ (as s is the h-relative supremum of l), and as $s', s^{*\prime} \in h'$, by Fact 3.11(1), we have $s' \preccurlyeq s^{*\prime}$.

Summing up, we have established that for $l' \subseteq h'$ an upper-bounded chain in a W'-history $h' = \{e \rightarrowtail \Pi_e\langle h \rangle \mid e \in h\}$ (where h is the corresponding W-history), the initials satisfy $l \subseteq h$, and there exists the h'-relative supremum

$$\sup_{h'} l' = s \rightarrowtail \Pi_s\langle h \rangle,$$

where $s = \sup_h l$.

6. $\langle W', \prec \rangle$ satisfies Weiner's postulate.

   We will employ the claim established at the end of the previous item (5), and the fact that $\langle W, < \rangle$ satisfies Weiner's postulate.

   Let $h'_1, h'_2 \in \text{Hist}(W')$, and let $h'_i = \{e \rightarrowtail \Pi_e\langle h_i \rangle \mid e \in h_i\}$ for $h_i \in \text{Hist}(W)$ $(i = 1, 2)$. We consider two chains $l', k' \subseteq h'_1 \cap h'_2$, their respective chains of initials $l, k \subseteq h_1 \cap h_2$, and their history-relative suprema $s'_i = s_i \rightarrowtail \Pi_{s_i}\langle h_i \rangle = \sup_{h'_i}(l')$ and $c'_i = c_i \rightarrowtail \Pi_{c_i}\langle h_i \rangle = \sup_{h'_i}(k')$, where $s_i = \sup_{h_i} l$ and $c_i = \sup_{h_i} k$, for $i = 1, 2$. Suppose that $s'_1 \preccurlyeq c'_1$, i.e., $s_1 \leqslant c_1$ and $\Pi_{c_1}\langle h_1 \rangle \subseteq \Pi_{s_1}\langle h_1 \rangle$ (see Fact 3.11(3)). By Weiner's postulate of BST_{92} applied to the chains l and k, from $s_1 \leqslant c_1$ we may infer $s_2 \leqslant c_2$. Note that $s_2 \in h_2$. Hence $\Pi_{c_2}\langle h_2 \rangle \subseteq \Pi_{s_2}\langle h_2 \rangle$. In terms of the transition ordering, this means that $s'_2 \preccurlyeq c'_2$.

7. Historical connection. Note that by assumption, W has no minima. Let $h'_1, h'_2 \in \text{Hist}(W')$ be histories, which by Lemma A.1 correspond to $h_1, h_2 \in \text{Hist}(W)$. By historical connection for W, there is some $e \in h_1 \cap h_2$, and by no minima, there is some $e^* \in W$ for which $e^* < e$. It follows that $e^* \in h_1 \cap h_2$. Let $\tau =_{\text{df}} e^* \rightarrowtail \Pi_{e^*}\langle h_1 \rangle$. By Lemma A.1, we have $\tau \in h'_1$. Now $e > e^*$ is a witness for $h_1 \equiv_{e^*} h_2$, so that $\Pi_{e^*}\langle h_1 \rangle = \Pi_{e^*}\langle h_2 \rangle$, i.e., $\tau \in h'_2$ as well. $\qquad\square$

A.2.2 BST_{92} transition structures are BST_{NF} structures

We can now show that the full transition structure of a BST_{92} structure without minima is indeed a BST_{NF} structure. The only thing that is still missing is to show that the new prior choice principle PCP_{NF} is satisfied. To this end we need an auxiliary fact that shows how BST_{92} choice points give rise to BST_{NF} choice sets.

Fact A.3. *Let $\langle W, < \rangle$ be a BST_{92} structure without minima, and let $h_1 \perp_c h_2$ for $h_1, h_2 \in \text{Hist}(W)$. Then $c'_1 =_{\text{df}} c \rightarrowtail \Pi_c\langle h_1 \rangle$ and $c'_2 =_{\text{df}} c \rightarrowtail \Pi_c\langle h_2 \rangle$ belong to $\Upsilon(\langle W, < \rangle)$ and are elements of a choice set $\ddot{c}$ for which $h'_1 \perp_{\ddot{c}} h'_2$, where $h'_i = \{e \rightarrowtail \Pi_e\langle h_i \rangle \mid e \in h_i\}$ $(i = 1, 2)$ are the histories in $\Upsilon(\langle W, < \rangle)$ corresponding to h_1 and h_2.*

Proof. Let $\langle W', <' \rangle =_{\text{df}} \Upsilon(\langle W, < \rangle)$, let $h_1, h_2 \in \text{Hist}(W)$, and let $c \in h_1 \cap h_2$ be such that $h_1 \perp_c h_2$. By Lemma A.1, $h'_i = \{e \rightarrowtail \Pi_e\langle h_i \rangle \mid e \in h_i\} \in \text{Hist}(W')$ $(i = 1, 2)$. Let $c'_i =_{\text{df}} c \rightarrowtail \Pi_c\langle h_i \rangle$, so that $c'_i \in h'_i$ $(i = 1, 2)$. In order to show that c'_1, c'_2 are elements of a choice set $\ddot{c}$

that fulfills $h'_1 \perp_{\breve{e}} h'_2$, we need to show that every chain $l' \in \mathscr{C}_{c'_1}$, for which $\sup_{h'_1} l' = c'_1$, has c'_2 as another history-relative supremum, and vice versa. Since c is not a minimal element of W, $\mathscr{C}_{c'_1} \neq \emptyset$. Pick an arbitrary chain $l' \in \mathscr{C}_{c'_1}$, and note that it has the form $l' = \{e \rightarrowtail \Pi_e \langle h_1 \rangle \mid e \in l\}$ for some chain $l \subseteq h_1$, with $c = \sup_{h_1} l$. Since $h_1 \perp_c h_2$, $l \subseteq h_2$ as well, and as $l < c$, we have that for every $e \in l$, $\Pi_e \langle h_1 \rangle = \Pi_e \langle h_2 \rangle$. Hence $l' \subseteq h'_1 \cap h'_2$. It follows that $\sup_{h'_1} l' = c'_1$ and $\sup_{h'_2} l' = c'_2$ (note that the $c'_i \in h'_i$ are upper bounds of l' and that their initial, c, is the h_1- as well as the h_2-relative supremum of l). Since l' is an arbitrary chain in $\mathscr{C}_{c'_1}$, we showed that every chain in $\mathscr{C}_{c'_1}$ has at least two history-relative suprema, c'_1 and c'_2, i.e., there is a choice set $\breve{e}'$ such that $\{c'_1, c'_2\} \subseteq \breve{e}'$. Since $h'_1 \cap \breve{e}' = c'_1 \neq c'_2 = h'_2 \cap \breve{e}'$, we have $h'_1 \perp_{\breve{e}'} h'_2$. □

Given this auxiliary fact, we can establish our lemma:

Lemma 3.4. *Let $\langle W, < \rangle$ be a BST$_{92}$ structure without minima. Then that structure's full transition structure $\langle W', <' \rangle =_{\mathrm{df}} \Upsilon(\langle W, < \rangle)$ satisfies the PCP$_{\mathrm{NF}}$ as in Definition 3.14.*

Proof. Let l' be a chain in $\langle W', <' \rangle$ that is lower bounded by u', and let $h'_1, h'_2 \in \mathrm{Hist}(W')$ be such that (†) $l' \subseteq h'_1$ but (‡) $l' \cap h'_2 = \emptyset$. By Lemma A.1, $h'_i = \{e \rightarrowtail \Pi_e \langle h_i \rangle \mid e \in h_i\}$ for some $h_i \in \mathrm{Hist}(W)$, $i = 1, 2$. By (†) we have that $l' = \{e \rightarrowtail \Pi_e \langle h_1 \rangle \mid e \in l\}$ for a chain $l \subseteq h_1$ that is lower bounded by u, where u is the initial of u'. By (‡), for every $e \rightarrowtail \Pi_e \langle h_1 \rangle \in l'$, either $e \notin h_2$, or ($e \in h_2$ but $\Pi_e \langle h_1 \rangle \neq \Pi_e \langle h_2 \rangle$). There are now four cases, depending on the form of l': (i) l' has a minimal element $v' = v \rightarrowtail \Pi_v \langle h_1 \rangle$, $v \in h_1 \cap h_2$, and $\Pi_v \langle h_1 \rangle = \Pi_v \langle h_2 \rangle$, or (ii) l' has a minimal element $v' = v \rightarrowtail \Pi_v \langle h_1 \rangle$, $v \in h_1 \cap h_2$, but $\Pi_v \langle h_1 \rangle \neq \Pi_v \langle h_2 \rangle$, or (iii) l' has a minimal element $v' = v \rightarrowtail \Pi_v \langle h_1 \rangle$, and $v \in h_1 \setminus h_2$, or (iv) l' has no minimal element at all. Case (i) is impossible as it contradicts (‡). Consider then case (ii): since $h_1 \perp_v h_2$, by Fact A.3 and no minimal elements in $\langle W, < \rangle$, the two transitions, $v'_1 = v \rightarrowtail \Pi_v \langle h_1 \rangle$ and $v'_2 = v \rightarrowtail \Pi_v \langle h_2 \rangle$ are distinct elements of a choice set $\breve{v}'$ at which histories h'_1 and h'_2 split, $h'_1 \perp_{\breve{v}'} h'_2$. Furthermore, as v'_1 is the minimal element of l', $v'_1 \preccurlyeq l'$, as required by PCP$_{\mathrm{NF}}$. In cases (iii) and (iv), no element $e \in l$ can belong to h_2: (iii) if l' has a minimum $v' = v \rightarrowtail \Pi_v \langle h_1 \rangle$ with $v \in h_1 \setminus h_2$, no point above v can belong to h_2 (Fact 2.1(6)), and (iv) in case l' has no minimum at all, the assumption that $e \in l \cap h_2$ implies that there is also some other $e_1 \in l \cap h$ with $e_1 < e$, and hence $e_1 \rightarrowtail \Pi_{e_1} \langle h_2 \rangle \in l' \cap h'_2$, contradicting (‡). Thus, $l \cap h_2 = \emptyset$ in cases (iii) and (iv). Applying the PCP of BST$_{92}$ to the chain $l \subseteq h_1 \setminus h_2$ that is lower bounded by u, we get $v \in W$ such that $v < l$ and $h_1 \perp_v h_2$. Exactly as in case (ii) we thus invoke the assumption of no minimal elements in $\langle W, < \rangle$ and Fact A.3 to produce the sought-after choice set $\breve{v}'$ containing $v \rightarrowtail \Pi_v \langle h_1 \rangle$ and $v \rightarrowtail \Pi_v \langle h_2 \rangle$, for which $h'_1 \perp_{\breve{v}'} h'_2$. Since $v < l$ and given the form of l', we have $v \rightarrowtail \Pi_v \langle h_1 \rangle \preccurlyeq l'$ as well. □

Given this result, we have shown that the full transition structure of a BST$_{92}$ structure is a BST$_{\mathrm{NF}}$ structure:

Theorem 3.2. *Let $\langle W, < \rangle$ be a BST$_{92}$ structure without minima. Then that structure's full transition structure $\Upsilon(\langle W, < \rangle)$ is a BST$_{\mathrm{NF}}$ structure.*

Proof. From Lemma 3.3 and Lemma 3.4. □

For ease of reference, we repeat as a separate numbered Fact the result of Exercise 3.2, the proof of which is given in Appendix B.3:

Fact A.4. *Let $\langle W, < \rangle$ be a BST$_{92}$ structure without maxima or minima. Then its Υ-transform, $\langle W', <' \rangle$, has no maxima or minima either.*

A.2.3 From new foundations BST$_{NF}$ to BST$_{92}$

We have seen how the move from a BST$_{92}$ structure to its full transition structure brings us from BST$_{92}$ to BST$_{NF}$. In the other direction, there is also a fairly simple translation, viz., combining all the elements of a choice set to form a single point.

The Λ transformation from BST$_{NF}$ to BST$_{92}$ is defined in Chapter 3, Def. 3.18, which we repeat here for convenience's sake before we establish some pertinent facts:

Definition 3.18 (The Λ transformation from BST$_{NF}$ to BST$_{92}$.). Let $\langle W, < \rangle$ be a BST$_{NF}$ structure. Then we define the companion Λ-transformed ("collapsed") structure as follows:

$$\Lambda(\langle W, < \rangle) =_{df} \langle W', <' \rangle, \quad \text{where}$$
$$W' =_{df} \{ \ddot{e} \mid e \in W \};$$
$$\ddot{e}_1 <' \ddot{e}_2 \quad \text{iff} \quad e_1' < e_2' \quad \text{for some} \quad e_1' \in \ddot{e}_1, e_2' \in \ddot{e}_2.$$

We extend the Λ notation to points and subsets of W setting $\Lambda(e) =_{df} \ddot{e}$ for $e \in W$, and $\Lambda(A) =_{df} \{ \ddot{e} \mid e \in A \}$ for $A \subseteq W$.

Fact A.5 (Facts about Definition 3.18). *The following holds:*

1. *Let $e_1, e_2 \in W$ and let $e_1 < e_2$. Then $\Lambda(e_1) <' \Lambda(e_2)$.*
2. *Let $t \subseteq W$ be a chain (with respect to $<$). Then $\Lambda(t)$ is a chain (with respect to $<'$).*
3. *Let $E \subseteq W$ be directed. Then $\Lambda(E)$ is also directed.*

Proof. (1): holds by the definition of $<'$. (2) and (3) follow immediately. □

Fact A.6 (Justification of the notation in Definition 3.18). *In BST$_{NF}$: (1) If $e_1 < e_2$ and $e_2' \in \ddot{e}_2$, then $e_1 < e_2'$. So we can write $e_1 <' \ddot{e}_2$. (2) If $e_1 <' \ddot{e}_2$ and $e_1^* \in \ddot{e}_1$, $e_1 \neq e_1^*$, then $e_1^* \not<' \ddot{e}_2$. So given $\ddot{e}_1 <' \ddot{e}_2$, there is a unique $e_1 \in \ddot{e}_1$ for which $e_1 <' \ddot{e}_2$. (3) If $\ddot{e}_1 <' \ddot{e}_2$, then there are no $e_i^* \in \ddot{e}_i$ ($i = 1, 2$) for which $e_2^* < e_1^*$.*

Proof. (1): By Fact 3.23.

(2): Let $e_1 <' \ddot{e}_2$ as witnessed by e_2 (i.e., $e_1 < e_2$), and $e_1^* \in \ddot{e}_1$, $e_1 \neq e_1^*$. Assume for reductio $e_1^* <' \ddot{e}_2$, then there has to be some witness $e_2^* \in \ddot{e}_2$ for which $e_1^* < e_2^*$. Then by (1), we also have $e_1^* < e_2$, so that (by downward closure of histories) there is a history containing e_1 and e_1^*, contradicting Fact 3.13(1).

(3): Let $\ddot{e}_1 <' \ddot{e}_2$ as witnessed by e_1 and e_2 (i.e., $e_1 \in \ddot{e}_1, e_2 \in \ddot{e}_2$, and $e_1 < e_2$). Assume for reductio that there are $e_1^* \in \ddot{e}_1$ and $e_2^* \in \ddot{e}_2$ for which $e_2^* < e_1^*$. Then by (1), we have $e_1 < e_2^*$, and by transitivity, $e_1 < e_1^*$. Thereby e_1 and e_1^*, being different elements of $\ddot{e}_1$ (by irreflexivity of $<$), would have to belong to one history, contradicting Fact 3.13(1). □

Similarly to what we established about the properties of the transition structure of a BST$_{92}$ structure, we can characterize the Λ-transform of a BST$_{NF}$ structure. It transpires that, as announced, the Λ-transform leads us back to BST$_{92}$. As above, we split the proof into a number of steps.

Fact A.7. *Let $\langle W, < \rangle$ be a BST$_{NF}$ structure. Then its Λ-transform, $\langle W', <' \rangle =_{df} \Lambda(\langle W, < \rangle)$, is (1) non-empty and (2) a strict partial ordering.*

Proof. (1) By construction, W' is non-empty (given that W was non-empty).

(2) Asymmetry follows from Fact A.6(3). For transitivity, let $\ddot{e}_1 <' \ddot{e}_2$ and $\ddot{e}_2 <' \ddot{e}_3$. Then by Fact A.6(2), there is a unique $e_2 \in \ddot{e}_2$ for which $e_2 <' \ddot{e}_3$, and a unique $e_1 \in \ddot{e}_1$ for which $e_1 <' \ddot{e}_2$. So by $e_2 \in \ddot{e}_2$ and by transitivity of $<$ we have $e_1 <' \ddot{e}_3$, which proves $\ddot{e}_1 <' \ddot{e}_3$.   □

Before we can establish the history-relative suprema of upper bounded chains, we need to prove a lemma about the form of histories in W'. From here on, we need to work under the assumption that the BST_{NF} structures under consideration have no maxima, as otherwise the transformed structure may contain fewer histories.

Lemma A.2. *Let $\langle W, < \rangle$ be a BST_{NF} model without maxima, and let $\langle W', <' \rangle =_{\text{df}}$ $\Lambda(\langle W, < \rangle)$. The histories (maximal directed sets) in $\langle W', <' \rangle$ are exactly the sets $\Lambda(h)$, for $h \in \text{Hist}(W)$. That is, (1) for $h \in \text{Hist}(W)$, the set $\Lambda(h)$ is maximal directed and (2) for any maximal directed set $h' \in \text{Hist}(W')$ there is a unique history $h \in \text{Hist}(W)$ such that $h' = \Lambda(h)$.*

Proof. (1) The set $\Lambda(h) \subseteq W'$ is directed by Fact A.5(3), so there is some maximal directed $h' \in \text{Hist}(W')$ for which $\Lambda(h) \subseteq h'$. By Fact A.7, $\langle W', <' \rangle$ is a non-empty strict partial ordering. Thus, by Fact 2.1(9), h' cannot have a maximum. This allows us to define a function $f : h' \mapsto W$ that establishes the converse of Λ on h', in the following sense: (i) for any $\ddot{e} \in h'$, $\Lambda(f(\ddot{e})) = \ddot{e}$, (ii) for any $\ddot{e}_1, \ddot{e}_2 \in h'$, we have $\ddot{e}_1 <' \ddot{e}_2$ iff $f(\ddot{e}_1) < f(\ddot{e}_2)$, and (iii) for any $e \in h$, $f(\Lambda(e)) = e$. (Note that the primed histories and the primed ordering refer to $\Lambda(\langle W, < \rangle)$, not to the BST_{NF} structure.)

To define f, let $\ddot{e}_1 \in h'$. Let $\ddot{e}_2 \in h'$ such that $\ddot{e}_1 <' \ddot{e}_2$; such an element exists as h' has no maxima. By Fact A.6(2), there is a unique $v \in W$ for which $\ddot{e}_1 = \Lambda(v)$ and $v <' \ddot{e}_2$. That v is, moreover, independent of the chosen upper bound $\ddot{e}_2 \in h'$: let $v^* \in \ddot{e}_1$ be such that $v^* <' \ddot{e}_3$ for some $\ddot{e}_3 \in h'$ for which $\ddot{e}_1 <' \ddot{e}_3$. Then by the directedness of h', there is some common upper bound $\ddot{e}_4$ of $\ddot{e}_2$ and $\ddot{e}_3$, and again invoking Fact A.6(2), we have $v^* = v$. So we can set $f(\ddot{e}_1) = v$ as specified. Note that thereby, $f(\ddot{e}_1) \in \ddot{e}_1$. Constraint (i) holds by construction, as $\Lambda(f(\ddot{e})) = \Lambda(v) = \ddot{e}$. For (ii, "$\Rightarrow$"), let $\ddot{e}_1, \ddot{e}_2 \in h'$ satisfy $\ddot{e}_1 <' \ddot{e}_2$. Then $f(\ddot{e}_1) <' \ddot{e}_2$ by construction (as $\ddot{e}_2$ is an upper bound of $\ddot{e}_1$ in h'), and the claim follows by Fact A.6(1), noting that $f(\ddot{e}_2) \in \ddot{e}_2$. For (ii, "$\Leftarrow$"), let $\ddot{e}_1, \ddot{e}_2 \in h'$ be such that $f(\ddot{e}_1) < f(\ddot{e}_2)$. By (i) and by the definition of the ordering $<'$, this implies $\ddot{e}_1 <' \ddot{e}_2$. For (iii), let $e_1 \in h$. As h contains no maxima, there is some $e_2 \in h$ for which $e_1 < e_2$. Let $\ddot{e}_i = \Lambda(e_i)$ $(i = 1, 2)$, so that (by the definition of $<'$) we have $\ddot{e}_1 <' \ddot{e}_2$. By the definition of f, $f(\ddot{e}_1) = e^*$ for the unique member $e^* \in \ddot{e}_1$ for which $e^* <' \ddot{e}_2$. Given $e_1 < e_2$, we have $e^* = e_1$, i.e., $f(\Lambda(e_1)) = f(\ddot{e}_1) = e_1$.

Now for maximality of the directed set $\Lambda(h) \subseteq h'$, assume for reductio that $h' = \Lambda(h) \cup A'$ with $\Lambda(h) \cap A' = \emptyset$ and $A' \neq \emptyset$ (i.e., assume that $\Lambda(h)$ is not maximal directed). Let $A =_{\text{df}} \{f(\ddot{e}) \mid \ddot{e} \in A'\}$, so that $A \neq \emptyset$ and $A \cap h = \emptyset$. By property (ii) of f, $h \cup A$ is directed, violating the maximality of h.

(2) Let $h' \in \text{Hist}(W')$, and define f as above. Let $E =_{\text{df}} \{f(\ddot{e}) \mid \ddot{e} \in h'\}$, so that by (i), $\Lambda(E) = h'$. By (ii), E is directed, so that there is some $h \in \text{Hist}(W)$ with $E \subseteq h$. It follows that $h' = \Lambda(E) \subseteq \Lambda(h)$. By (1), we have that $\Lambda(h) = h''$ for some $h'' \in \text{Hist}(W')$. So we have two histories $h', h'' \in \text{Hist}(W')$ for which $h' \subseteq h''$, whence, by Fact 2.1(7), $h' = h''$. This means that we have found some $h \in \text{Hist}(W)$ for which $\Lambda(h) = h'$. For uniqueness of h, let $h_1, h_2 \in \text{Hist}(W)$ be such that $\Lambda(h_1) = \Lambda(h_2) = h'$. Then $h_1 = f(h')$ and $h_2 = f(h')$, establishing $h_1 = h_2$.   □

We can now prove the two main Lemmas and the resulting Theorem for this translatability direction, as claimed in Chapter 3.

Lemma 3.5. *Let $\langle W, < \rangle$ be a BST_{NF} structure without maxima. Then its Λ-transform, $\langle W', <' \rangle =_{df} \Lambda(\langle W, < \rangle)$, is a common BST structure.*

Proof. Our task is to show that $\langle W', <' \rangle$ satisfies the postulates of Definition 2.10 of a common BST structure.

1. Non-emptiness: By Fact A.7(1).

2. Partial ordering: By Fact A.7(2)

3. $<'$ is dense.
   Let $\ddot{e}_1 <' \ddot{e}_2$, so by Fact A.6(2), there is a unique $e_1 \in \ddot{e}_1$ for which $e_1 <' \ddot{e}_2$. Let $e_2 \in \ddot{e}_2$; in particular, $e_1 < e_2$. By density of $<$, we have $e^* \in W$ such that $e_1 < e^* < e_2$. By the definition of the $<'$ ordering, this establishes $\ddot{e}_1 = \Lambda(e_1) <' \Lambda(e^*) <' \Lambda(e_2) = \ddot{e}_2$, which proves density of $<'$.

4. Lower bounded chains have infima in $<'$.
   Let $t' \subseteq W'$ be a lower bounded chain, and let $\ddot{b} \in W'$ be a lower bound for t'. The elements of t' are of the form $\ddot{e} = \Lambda(e)$ with $e \in W$. We distinguish two cases. (a) If t' has a least element (which covers the case that t' has only one element), then that least element is the infimum of t' with respect to $<'$, by definition. (b) If t' has no least element, pick some $\ddot{e} \in t'$, and let $t^{*'} =_{df} \{x \in t' \mid x <' \ddot{e}\}$. We have $\inf t^{*'} = \inf t'$ by the definition of the infimum. And by Fact A.6(2), for all $\ddot{e}_1 \in t^{*'}$ there are unique $e_1 \in W$ for which $e_1 \in \ddot{e}_1$ and $e_1 < \ddot{e}$, and there is a unique $b^* \in \ddot{b}$ for which $b^* < \ddot{e}$. So there is a unique set $t^* \subseteq W$ given by

   $$t^* = \{e_1 \in W \mid e_1 < \ddot{e} \wedge \ddot{e}_1 \in t^{*'}\},$$

   which is a chain since $t^{*'}$ is a chain; furthermore, t^* is lower bounded by $b^* \in W$. By the properties of BST_{NF}, t^* therefore has an infimum $a =_{df} \inf t^*$, $a \in W$. We claim that $\ddot{a}$ is the infimum of t' with respect to $<'$. As $a < t^*$, we have $\ddot{a} <' t'$ by the definition of $<'$. Now let $\ddot{c} \leqslant' t'$. Again by Fact A.6(2), there is a unique $c \in \ddot{c}$ for which $c < \ddot{e}$. By the fact that a is the infimum of t^*, we have $c \leqslant a$, which implies $\ddot{c} \leqslant' \ddot{a}$. So $\ddot{a}$ is indeed the greatest lower bound (i.e., the infimum), of t'.

5. Upper bounded chains have history-relative suprema in $<'$.
   Let t' be an upper bounded chain with $\ddot{b}$ an upper bound, and let $\ddot{b} \in h'$ for h' some history in $\langle W', <' \rangle$, so that $t' \subseteq h'$ as well. As h' has a unique pre-image h under Λ (by Lemma A.2), also $\ddot{b}$ and t' have unique pre-images $b \in h$ and $t \subseteq h$. So by the BST_{NF} axioms, t has an h-relative supremum $s \in h$. By Fact A.5(1), we have $t' \leqslant' \ddot{s}$, and for any $\ddot{a} \in h'$ for which $t' \leqslant' \ddot{a}$, we can consider the unique pre-image $a \in h \cap \ddot{a}$, for which $t \leqslant a$. By the fact that s is the h-relative supremum of t, we have $s \leqslant a$, which translates into $\ddot{s} \leqslant' \ddot{a}$; that is, $\ddot{s}$ is the least upper bound in h', and therefore the h'-relative supremum, of t'.

6. Weiner's postulate.
   Consider two histories $h'_1, h'_2 \in \text{Hist}(W')$, two chains $l', k' \subseteq h'_1 \cap h'_2$, and their history-relative suprema $\ddot{s}_i = \sup_{h'_i} l'$ and $\ddot{c}_i = \sup_{h'_i} k'$ $(i = 1, 2)$. Assume that $\ddot{s}_1 \leqslant' \ddot{c}_1$. We denote the unique pre-images of $h'_1, h'_2, l', k', \ddot{s}_1, \ddot{s}_2, \ddot{c}_1$, and $\ddot{c}_2$ under Λ by $h_1, h_2, l, k, s_1, s_2, c_1$, and c_2, respectively. By the uniqueness of pre-images

and properties of $<'$, we have $l, k \subseteq h_1 \cap h_2$, $s_i = sup_{h_i}(l)$, $c_i = sup_{h_i}(k)$ $(i = 1, 2)$, and $s_1 \leqslant c_1$. Then by Weiner's postulate of BST_{NF}, $s_2 \leqslant c_2$. Since $\breve{s}_2 = \Lambda(s_2)$ and $\breve{c}_2 = \Lambda(c_2)$, we have $\breve{s}_2 \leqslant' \breve{c}_2$.

7. Historical connection. Pick h'_1, h'_2 histories in $\langle W', <' \rangle$, and consider their unique pre-images h_1, h_2 by Λ, which by Lemma A.2 are histories in the BST_{NF} structure $\langle W, < \rangle$. By historical connection for BST_{NF} structures, we have $h_1 \cap h_2 \neq \emptyset$. Hence $\Lambda(h_1) \cap \Lambda(h_2) \neq \emptyset$, i.e., $h'_1 \cap h'_2 \neq \emptyset$. $\square$

Lemma 3.6. *The Λ-transform $\Lambda(\langle W, < \rangle)$ of a BST_{NF} structure without maxima $\langle W, < \rangle$ satisfies the BST_{92} prior choice principle.*

Proof. Let h'_1, h'_2 be histories in $\langle W', <' \rangle$, and let $t' \subseteq h'_1 \setminus h'_2$ be a lower bounded chain in h'_1 that contains no element of h'_2. We have to find a maximal element $c \in h'_1 \cap h'_2$ that lies below t', $c <' t'$, and for which $h'_1 \perp_c h'_2$. The histories h'_1, h'_2 have as unique pre-images the $\langle W, < \rangle$-histories h_1, h_2. As $t' \subseteq h'_1$, the unique pre-image $t \subseteq h_1$. Furthermore, $t \cap h_2 = \emptyset$, for an element $e \in t \cap h_2$ would give rise to $\breve{e} \in t' \cap h'_2$, violating our assumption about t'. So $t \subseteq h_1 \setminus h_2$. From the BST_{NF} prior choice principle, we have a choice set $\breve{s}$ and $s_1 \in h_1 \cap \breve{s}$ for which $s_1 \leqslant t$, while there is some $s_2 \in \breve{s} \cap h_2$. Let $c' =_{df} \Lambda(s_1) = \breve{s}$; we claim that c' is the sought-for choice point. (a) By Lemma 3.2 we have $\breve{s}_1 = \breve{s}_2$, and as $s_i \in h_i$, we have $\breve{s}_i \in h'_i$ $(i = 1, 2)$, so that $c' = \breve{s}_1 = \breve{s}_2 \in h'_1 \cap h'_2$. (b) As c' lies at the intersection of h'_1 and h'_2, it cannot be that $\breve{s}_1 \in t'$. This excludes $s_1 \in t$, so that in fact $s_1 < t$. This in turn implies $c' = \breve{s}_1 <' t'$. (c) For the maximality of c' in $h'_1 \cap h'_2$, assume that there is $\breve{a} \in h'_1 \cap h'_2$ for which $c' < \breve{a}$. Then we have a unique pre-image $a_1 \in h_1 \cap h_2$ for which both $s_1 < a_1$ and $s_2 < a_1$, so that both s_1 and s_2 belong to history h_1. This contradicts Fact 3.13(1). So $c' = \breve{s}$ is in fact maximal in $h'_1 \cap h'_2$. (d) By the definition of $\perp_{c'}$, we therefore have $h'_1 \perp_{c'} h'_2$. $\square$

Theorem 3.3. *The Λ-transform $\Lambda(\langle W, < \rangle)$ of a BST_{NF} structure without maxima $\langle W, < \rangle$ is a BST_{92} structure.*

Proof. By Lemma 3.5 and Lemma 3.6. $\square$

For the Λ transform we can prove, just like for the Υ transform, that a structure without maxima or minima is transformed into one that has no maxima or minima either.

Fact A.8. *Let $\langle W, < \rangle$ be a BST_{NF} structure without maxima and minima. Then its Λ-transform, $\langle W', <' \rangle$, has no maxima and minima either.*

Proof. Let $\breve{e} \in W'$. There is some $e_1 \in W$ for which $\breve{e} = \Lambda(e_1)$. As W has no maximal nor minimal elements, there are $e_2, e_3 \in W$ for which $e_2 < e_1 < e_3$. Then by the definition of the ordering, $\breve{e}_2 <' \Lambda(e_1) = \breve{e}$, establishing that $\breve{e}$ cannot be a minimum, and $\Lambda(e_1) = \breve{e} <' \breve{e}_3$, establishing that $\breve{e}$ cannot be a maximum. $\square$

A.2.4 Going full circle

We have now established that there is a way to get from BST_{92} structures without minima to BST_{NF} structures and from BST_{NF} structures without maxima to BST_{92} structures. This leads to the question of where we land when we concatenate these transformations (restricted to structures without minima and maxima). We can show that, as one might hope, we end up where we started: the resulting structures are order-isomorphic to the ones we started with.

A.2.4.1 From BST_{92} to BST_{NF} to BST_{92}

We can prove the following Theorem, already presented in Chapter 3:

Theorem A.4. *The function $\Lambda \circ \Upsilon$ is an order isomorphism of BST_{92} structures without maximal or minimal elements: Let $\langle W_1, <_1 \rangle$ be a BST_{92} structure without maximal or minimal elements, let $\langle W_2, <_2 \rangle =_{df} \Upsilon(\langle W_1, <_1 \rangle)$, and let $\langle W_3, <_3 \rangle =_{df} \Lambda(\langle W_2, <_2 \rangle)$. Then there is an order isomorphism φ between $\langle W_1, <_1 \rangle$ and $\langle W_3, <_3 \rangle$, i.e., a bijection between W_1 and W_3 that preserves the ordering. Accordingly, $\langle W_3, <_3 \rangle$ has no minima and no maxima.*

Proof. We claim that we can use the mapping φ, defined for $e \in W_1$ to be

$$\varphi(e) =_{df} \{ e \rightarrowtail H \mid H \in \Pi_e \}.$$

We have to show (1) that φ is indeed a mapping from W_1 to W_3, (2) that φ is injective, (3) that φ is surjective, and (4) that φ preserves the ordering.

(1) Mapping: We have to show that for any $e \in W_1$, $\varphi(e) = \{ e \rightarrowtail H \mid H \in \Pi_e \} \in W_3$. Since $\langle W_1, <_1 \rangle$ has no minimal elements, the set W_2 is the full transition structure of $\langle W_1, <_1 \rangle$, so that for $e \in W_1$ and for any $H \in \Pi_e$, the transition $e \rightarrowtail H \in W_2$. Thus, for $e \in W_1$, the set $\varphi(e) \subseteq W_2$. The set W_3 contains, for any $\tau \in W_2$, the set $\Lambda(\tau) = \ddot{\tau} \in W_3$, and $\ddot{\tau} \subseteq W_2$ as well. Let now $e \in W_1$, and pick some $H^* \in \Pi_e$, which fixes some $\tau = e \rightarrowtail H^* \in W_2$. We claim that

$$\ddot{\tau} = \{ e \rightarrowtail H \mid H \in \Pi_e \},$$

which establishes $\ddot{\tau} = \varphi(e)$, so that indeed, $\varphi(e) \in W_3$. The claim is an equality between subsets of W_2, so that we show inclusion both ways.

"$\subseteq$": Let $\tau' = e' \rightarrowtail H' \in \ddot{\tau}$; we have to show that $\tau' \in \{ e \rightarrowtail H \mid H \in \Pi_e \}$. We have $\tau \in h$ and $\tau' \in h'$ for $h, h' \in \text{Hist}(W_2)$. By Lemma A.1 we know that these histories are of the form

$$h = \{ e_1 \rightarrowtail \Pi_{e_1} \langle h_1 \rangle \mid e_1 \in h_1 \}; \quad h' = \{ e'_1 \rightarrowtail \Pi_{e'_1} \langle h'_1 \rangle \mid e'_1 \in h'_1 \}$$

for some $h_1, h'_1 \in \text{Hist}(W_1)$. The set $\ddot{\tau}$ is defined as the intersection of all sets of history-relative suprema of any chain $l \subseteq W_2$ ending in, but not containing τ ($l \in \mathscr{C}_\tau$), so that for any $l \in \mathscr{C}_\tau$, we have $\sup_h l = \tau$ and $\sup_{h'} l = \tau'$, since $\tau' \in \ddot{\tau}$. As $\tau = e \rightarrowtail H^* \in h$, we have $e \in h_1$. We now claim that $e \in h'_1$ as well. Assume otherwise, so that $e \in h_1 \setminus h'_1$. By PCP_{92}, there is then some $c <_1 e$ for which $h_1 \perp_c h'_1$. Let $\tau_c =_{df} c \rightarrowtail \Pi_c \langle h_1 \rangle$, so that $\tau_c \in h$ and $\tau_c <_2 \tau$. There is thus some chain $l \in \mathscr{C}_\tau$ for which $\tau_c \in l$. As $\sup_{h'} l = \tau'$ (by $\tau' \in \ddot{\tau}$, see above), we have $l \subseteq h'$, which implies $\tau_c \in h'$ and $c \in h'_1$. Elements of h' are of the form $e'_1 \rightarrowtail \Pi_{e'_1} \langle h'_1 \rangle$; we therefore must have $\Pi_c \langle h_1 \rangle = \Pi_c \langle h'_1 \rangle$, contradicting $h_1 \perp_c h'_1$. So indeed $e \in h_1 \cap h'_1$.

Now take some $l \in \mathscr{C}_\tau$; we have $l \subseteq W_2$ and indeed $l \subseteq h \cap h'$. Let l_1 be the set of initials of the elements of l, i.e., $l_1 \subseteq W_1$ and $l = \{ e_1 \rightarrowtail \Pi_{e_1} \langle h_1 \rangle \mid e_1 \in l_1 \}$. Note that $\sup_{h_1} l_1 = e = \sup_{h'_1} l_1$ from $\sup_h l = \tau$ and $e \in h_1 \cap h'_1$, and as $l \subseteq h \cap h'$, we have $\Pi_{e_1} \langle h_1 \rangle = \Pi_{e_1} \langle h'_1 \rangle$ for all $e_1 \in l_1$. We now claim that $\sup_{h'} l = \tau'' =_{df} e \rightarrowtail \Pi_e \langle h'_1 \rangle$. We have $\tau'' \in h'$ because

$e \in h'_1$, and $l <_2 \tau''$ because $l_1 <_1 e$, so τ'' is an upper bound of l in h'. Let now $\tau^* = e^* \rightarrowtail \Pi_{e^*}\langle h'_1 \rangle \in h'$ be some upper bound of l in h'. Then e^* is an upper bound of l_1 in h'_1, and thus $e \leqslant_1 e^*$ as $\sup_{h'_1} l_1 = e$, so that $\tau'' \leqslant_2 \tau^*$, proving that τ'' is the h'-relative supremum of l. So we have shown that $\tau'' = e \rightarrowtail \Pi_e \langle h'_1 \rangle = \sup_{h'} l = \tau'$. So indeed, $\tau' \in \{e \rightarrowtail H \mid H \in \Pi_e\}$.

"$\supseteq$": Given $\tau = e \rightarrowtail H^*$, consider an arbitrary $\tau' \in \{e \rightarrowtail H \mid H \in \Pi_e\}$, i.e., $\tau' = e \rightarrowtail H$ for the e in question and for some $H \in \Pi_e$. We have to show that $\tau' \in \ddot{\tau}$. We have $\tau \in h$ and $\tau' \in h'$ for some $h, h' \in \mathrm{Hist}(W_2)$, which are again of the form

$$ h = \{e_1 \rightarrowtail \Pi_{e_1}\langle h_1 \rangle \mid e_1 \in h_1\}; \quad h' = \{e'_1 \rightarrowtail \Pi_{e'_1}\langle h'_1 \rangle \mid e'_1 \in h'_1\} $$

for some $h_1, h'_1 \in \mathrm{Hist}(W_1)$, so that $\tau' = e \rightarrowtail \Pi_e \langle h'_1 \rangle$.

Let $l \in \mathscr{C}_\tau$; we have $\sup_h l = \tau = e \rightarrowtail H^*$ by assumption. We now claim that $\sup_{h'} l = \tau'$, which establishes $\tau' \in \ddot{\tau}$. To prove that τ' is the h'-relative supremum of l, as above, let l_1 be the set of initials of the elements l, so that $l_1 \subseteq W_1$ and $l = \{e_1 \rightarrowtail \Pi_{e_1}\langle h_1 \rangle \mid e_1 \in l_1\}$. Again as above, $l_1 <_1 e$, and thus τ' is an upper bound of l in h'. Let now $\tau^* = e^* \rightarrowtail \Pi_{e^*}\langle h'_1 \rangle \in h'$ be some upper bound of l in h'. Then $e^* \in h'_1$ is an upper bound of l_1 in h'_1, and thus $e \leqslant_1 e^*$ as $\sup_{h'_1} l_1 = e$ (note that $e \in h'_1$ as $\tau' \in h'$). Therefore $\tau' \leqslant_2 \tau^*$, proving that τ' is the h'-relative supremum of l. As l was an arbitrary chain from $\mathscr{C}_\tau$, we have indeed $\tau' \in \ddot{\tau}$.

(2) Injectivity: Let $e, e' \in W_1$ with $e \neq e'$. Then $\varphi(e) \neq \varphi(e')$. This is clear as the sets $\varphi(e)$ and $\varphi(e')$ have different members.

(3) Surjectivity: Let $a \in W_3$. We have to find some $e \in W_1$ for which $\varphi(e) = a$. As $a \in W_3$, we have $a = \ddot{\tau}$ for some $\tau = e \rightarrowtail H \in W_2$, where $e \in W_1$ and $H \in \Pi_e$. Above under (1) we have established that for $\tau = e \rightarrowtail H \in W_2$, we have $\ddot{\tau} = \{e \rightarrowtail H \mid H \in \Pi_e\}$, i.e., $a = \ddot{\tau} = \varphi(e)$.

(4) Order preservation: We have to show that for $e_1, e_2 \in W_1$, $e_1 <_1 e_2$ iff $\varphi(e_1) <_3 \varphi(e_2)$. (The claim about equality follows from the fact that φ is a bijection.) We know from the definition of φ that $\varphi(e_i) = \{e_i \rightarrowtail H \mid H \in \Pi_{e_i}\}$ ($i = 1, 2$).

"$\Rightarrow$": Let $e_1, e_2 \in W_1$ with $e_1 <_1 e_2$, and let $h_2 \in H_{e_2}$. Let $\tau_1 =_{\mathrm{df}} e_1 \rightarrowtail \Pi_{e_1}\langle h_2 \rangle$ and $\tau_2 =_{\mathrm{df}} e_2 \rightarrowtail H$ for some $H \in \Pi_{e_2}$; we have $\tau_1, \tau_2 \in W_2$ and $\varphi(e_i) = \ddot{\tau}_i$ ($i = 1, 2$). By the definition of the transition ordering $<_2$, we have $\tau_1 <_2 \tau_2$, and by the definition of $<_3$ in terms of instances, we thus have $\ddot{\tau}_1 <_3 \ddot{\tau}_2$, i.e., $\varphi(e_1) <_3 \varphi(e_2)$.

"$\Leftarrow$": Let $\varphi(e_1) <_3 \varphi(e_2)$, i.e., there are some $\tau_1 \in \varphi(e_1), \tau_2 \in \varphi(e_2)$ for which $\tau_1 <_2 \tau_2$. These transitions have the form $\tau_i = e_i \rightarrowtail H_i$ for some $e_i \in W_1$ and $H_i \in \Pi_{e_i}$ ($i = 1, 2$). Thus, in particular, from $\tau_1 <_2 \tau_2$ we have that $e_1 <_1 e_2$. $\qquad \square$

A.2.4.2 From BST$_{\mathrm{NF}}$ to BST$_{92}$ to BST$_{\mathrm{NF}}$

Before we can tackle the main Theorem below (which has also been presented in Chapter 3), we need to establish an additional fact.

Fact A.9. *Let $\langle W_1, <_1 \rangle$ be a BST$_{\mathrm{NF}}$ structure without maxima and $\langle W_2, <_2 \rangle =_{\mathrm{df}} \Lambda(\langle W_1, <_1 \rangle)$ the corresponding BST$_{92}$ structure. Then for any $h_1, h_2 \in \mathrm{Hist}(W_1)$, we have $h_1 \perp^1_{\ddot{e}} h_2$ iff $\Lambda(h_1) \perp^2_{\ddot{e}} \Lambda(h_2)$, where $\perp^i_{\ddot{e}}$ is the relation of splitting for histories in W_i.*

Proof. "$\Rightarrow$" Let $h_1, h_2 \in \mathrm{Hist}(W_1)$, and let $h_1 \perp^1_{\ddot{e}} h_2$. Then there are $e_1, e_2 \in \ddot{e}$ such that $e_1 \neq e_2$ and $h_i \cap \ddot{e} = \{e_i\}$ ($i = 1, 2$). Then $\Lambda(e_1) = \Lambda(e_2) = \ddot{e}$, so $\ddot{e} \in \Lambda(h_1) \cap \Lambda(h_2)$. Moreover, $\ddot{e}$ is maximal in $\Lambda(h_1) \cap \Lambda(h_2)$, which establishes $\Lambda(h_1) \perp^2_{\ddot{e}} \Lambda(h_2)$. To prove this, assume for reductio that there is some $\ddot{e}' >_2 \ddot{e}$ in the intersection of $\Lambda(h_1)$ and $\Lambda(h_2)$. This means that there are some $e'_1, e'_2 \in \ddot{e}'$ with $e'_1 \in h_1$, $e'_2 \in h_2$, and $\Lambda(e'_1) = \Lambda(e'_2) = \ddot{e}'$. The ordering $\ddot{e} <_2 \ddot{e}'$ implies that for some $e^* \in \ddot{e}$, $e^* <_2 \ddot{e}'$, which further implies (by

Fact A.6(1)) that $e^* <_1 e'_1$ and $e^* <_1 e'_2$. But then $e^* \in h_1 \cap h_2$, and by Fact A.6(2), it must be that $e^* = e_1 = e_2$, which contradicts $h_1 \perp^1_{\ddot{e}} h_2$.

"$\Leftarrow$" Let $\Lambda(h_1) \perp^2_{\ddot{e}} \Lambda(h_2)$, which implies that $\ddot{e} \in \Lambda(h_1) \cap \Lambda(h_2)$. Note that there are e_i such that $e_i \in h_i$ and $\Lambda(e_i) = \ddot{e}$ $(i = 1, 2)$. Therefore, $h_i \cap \ddot{e} \neq \emptyset$, so that h_1 and h_2 fulfill the precondition for either $h_1 \equiv^1_{\ddot{e}} h_2$ or $h_1 \perp^1_{\ddot{e}} h_2$ (see Def. 3.13). For reductio, assume the former, which means that $h_1 \cap \ddot{e} = h_2 \cap \ddot{e}$, i.e., $e_1 = e_2$. As there are no maxima in the intersection of histories in BST$_{\text{NF}}$ (Fact 3.17), there is some $e^* \in h_1 \cap h_2$ for which $e_1 <_1 e^*$. Now for $\ddot{e}^* =_{\text{df}} \Lambda(e^*)$ we have $\ddot{e}^* \in \Lambda(h_1) \cap \Lambda(h_2)$, and $\ddot{e} <_2 \ddot{e}^*$. This, however, contradicts the maximality of $\ddot{e}$ implied by $\Lambda(h_1) \perp^2_{\ddot{e}} \Lambda(h_2)$. So in fact, we have $h_1 \perp^1_{\ddot{e}} h_2$. Note that by contraposing the above Fact (and making a simple observation) we have that for any $h_1, h_2 \in \text{Hist}(W_1)$, $h_1 \equiv^1_{\ddot{e}} h_2$ iff $\Lambda(h_1) \equiv^2_{\ddot{e}} \Lambda(h_2)$. $\square$

Theorem 3.5. *The function $\Upsilon \circ \Lambda$ is an order isomorphism of BST$_{\text{NF}}$ structures without maximal or minimal elements: Let $\langle W_1, <_1 \rangle$ be a BST$_{\text{NF}}$ structure without maximal or minimal elements, let $\langle W_2, <_2 \rangle =_{\text{df}} \Lambda(\langle W_1, <_1 \rangle)$, and let $\langle W_3, <_3 \rangle =_{\text{df}} \Upsilon(\langle W_2, <_2 \rangle)$. Then there is an order isomorphism φ between $\langle W_1, <_1 \rangle$ and $\langle W_3, <_3 \rangle$, i.e., a bijection between W_1 and W_3 that preserves the ordering. Accordingly $\langle W_3, <_3 \rangle$ has no minima and no maxima.*

Proof. We claim that we can use the mapping φ, defined for $e \in W_1$ to be

$$\varphi(e) =_{\text{df}} \ddot{e} \longmapsto \Pi_{\ddot{e}} \langle \Lambda(h) \rangle \text{ for arbitrary } h \in H_e \subseteq \text{Hist}(W_1).$$

First we show that $\varphi(e)$ is well-defined. Thus, let $h, h' \in H_e$; we need to show that $\Pi_{\ddot{e}} \langle \Lambda(h) \rangle = \Pi_{\ddot{e}} \langle \Lambda(h') \rangle$. By Lemma A.2 (1), $\Lambda(h), \Lambda(h') \in \text{Hist}(W_2)$. Also, by Fact 3.13, $h \equiv_{\ddot{e}} h'$, and so by Fact A.9, $\Pi_{\ddot{e}} \langle \Lambda(h) \rangle = \Pi_{\ddot{e}} \langle \Lambda(h') \rangle$.

We now have to show (1) that φ is indeed a mapping from W_1 to W_3, (2) that φ is injective, (3) that φ is surjective, and (4) that φ preserves the ordering.

(1) Mapping: We have to show that for any $e \in W_1$, $\varphi(e) = \ddot{e} \longmapsto \Pi_{\ddot{e}} \langle \Lambda(h) \rangle \in W_3$, where $h \in H_e \subseteq \text{Hist}(W_1)$. The set W_3 is defined via W_2, and the set $W_2 = \Lambda[W_1]$, which means that for every $e \in W_1$, $\ddot{e} \in W_2$. By Lemma A.2, $\Lambda(h)$ is a history in $\langle W_2, <_2 \rangle$ for any $h \in \text{Hist}(W_1)$. Since for any $h \in H_e$, $\ddot{e} = \Lambda(e) \in \Lambda(h)$, we get that $\Pi_{\ddot{e}} \langle \Lambda(h) \rangle$ is an basic outcome of $\ddot{e}$, so indeed $\ddot{e} \longmapsto \Pi_{\ddot{e}} \langle \Lambda(h) \rangle \in W_3$.

(2) Injectivity: Let $e, e' \in W_1$ and $e \neq e'$. If $\ddot{e} \neq \ddot{e}'$, then obviously $\varphi(e) \neq \varphi(e')$, as these two transitions then have different initials. If $\ddot{e} = \ddot{e}'$ but $e \neq e'$, then e and e' are incompatible elements of the choice set $\ddot{e}$, and moreover, for any $h, h' \in \text{Hist}(W_1)$, if $e \in h, e' \in h'$, then $h \perp_{\ddot{e}} h'$, and hence by Fact A.9, $\Lambda(h) \perp_{\ddot{e}} \Lambda(h')$. Accordingly, $\Pi_{\ddot{e}} \langle \Lambda(h) \rangle \neq \Pi_{\ddot{e}} \langle \Lambda(h') \rangle$, and hence $\varphi(e) \neq \varphi(e')$.

(3) Surjectivity: Let $a \in W_3$. We have to find some $e \in W_1$ for which $\varphi(e) = a$. As $a \in W_3$, we have $a = \ddot{e}' \longmapsto H$, where $\ddot{e}' \in W_2$ and $H \in \Pi_{\ddot{e}'}$. Since $\langle W_2, <_2 \rangle$ is the result of Λ-transform applied to $\langle W_1, <_1 \rangle$, there is (possibly more than one) $e^* \in W_1$ for which $\Lambda(e^*) = \ddot{e}'$. We need to find which of these is the sought-after e. Clearly, there is some $h^* \in \text{Hist}(W_2)$ for which $H = \Pi_{\ddot{e}'} \langle h^* \rangle$. By Lemma A.2(2), there is a unique $h \in \text{Hist}(W_1)$ such that $h^* = \Lambda(h)$, and hence $H = \Pi_{\ddot{e}'} \langle \Lambda(h) \rangle$. For the sought-after e we thus take the unique $e \in \ddot{e}' \cap h$; clearly $\ddot{e}' = \ddot{e}$. It follows that $\varphi(e) = \ddot{e} \longmapsto H$, where $H = \Pi_{\ddot{e}} \langle \Lambda(h) \rangle$.

(4) Order preservation: We have to show that for $e_1, e_2 \in W_1$, $e_1 <_1 e_2$ iff $\varphi(e_1) <_3 \varphi(e_2)$. (The claim about equality follows from the fact that φ is a bijection.)

402 BRANCHING SPACE-TIMES

"⇒": Let $e_1, e_2 \in W_1$ with $e_1 <_1 e_2$. We show that $\varphi(e_1) <_3 \varphi(e_2)$. Since for $\varphi(e_1)$ we may pick an arbitrary member of H_{e_1}, we pick $h_2 \in H_{e_2} \subseteq H_{e_1}$, so that $e_1, e_2 \in h_2$. We get, as required, $\ddot{e}_1 <_2 \ddot{e}_2$ and hence, as the basic outcomes $\varphi(e_i)$ are defined by the same history $\Lambda(h_2)$, we get $H_{\ddot{e}_2} \subseteq \Pi_{\ddot{e}_1}\langle \Lambda(h_2)\rangle$. Hence, $\ddot{e}_1 \rightarrowtail \Pi_{\ddot{e}_1}\langle\Lambda(h_2)\rangle <_3 \ddot{e}_2 \rightarrowtail \Pi_{\ddot{e}_2}\langle\Lambda(h_2)\rangle$.

"⇐": Let $\varphi(e_1) <_3 \varphi(e_2)$, i.e., $(\ddot{e}_1 \rightarrowtail \Pi_{\ddot{e}_1}\langle\Lambda(h_1)\rangle) <_3 (\ddot{e}_2 \rightarrowtail \Pi_{\ddot{e}_2}\langle\Lambda(h_2)\rangle)$, for $h_i \in H_{e_i}, e_i \in \ddot{e}_i$. Hence for some $e_1' \in \ddot{e}_1$: (i) $e_1' <_2 \ddot{e}_2$, and hence $H_{\ddot{e}_2} \subseteq H_{e_1'}$, so $h_2 \in H_{e_1'}$ (because $H_{e_2} \subseteq H_{\ddot{e}_2}$). Since $h_2 \in H_{e_2} \subseteq H_{e_1}$, and it is impossible that $\{e_1, e_1'\} \subseteq h_2$ (by Fact 3.13(1)), it must be that $e_1 = e_1'$ and hence $e_1 < e_2$ (by (i)). □

A.2.5 The translatability of some notions pertaining to MFB

We assume in this section that there is a BST_{NF} structure $\mathscr{W} = \langle W, <\rangle$ with no maximal elements, and we consider its Λ-transform, $\mathscr{W}' = \langle W', <'\rangle =_{df} \Lambda(\mathscr{W})$, which is a BST_{92} structure. We use relational symbols with primes for relations on $\mathscr{W}'$, and relational symbols without primes for relations on $\mathscr{W}$.

We define what we will claim to be the transform of basic transitions in $\mathscr{W}$ to basic transitions in $\mathscr{W}'$. Note that for set-theoretical reasons, since $\Lambda(\ddot{c}) = \{\ddot{c}\}$ rather than $\ddot{c}$, we cannot identify the transform of basic transitions with our standard transform Λ. For this new transform we use $\tilde{\Lambda}$. We write transitions $X \rightarrowtail Y$ as pairs $\langle X, Y\rangle$ for clarity here.

Definition A.1. Let $\tau = \langle\ddot{c}, H_{c'}\rangle$, with $c' \in \ddot{c}$, be a basic transition in $\mathscr{W}$. Then $\tilde{\Lambda}(\tau) =_{df} \langle\ddot{c}, \Lambda(H_{c'})\rangle = \langle\ddot{c}, \{\Lambda(h) \mid c' \in h\}\rangle$. We extend this notation to sets of basic transitions, so we write $\tilde{\Lambda}(T) = \{\tilde{\Lambda}(\tau) \mid \tau \in T\}$.

Observe that in the deterministic case $\ddot{c}$ is a singleton, $\ddot{c} = \{c\}$, so $\tau = \langle\{c\}, H_c\rangle$, and its transform is $\tilde{\Lambda}(\tau) = \langle\{c\}, \Lambda(H_c)\rangle$. We next prove the claim announced above:

Fact A.10. *Let* $\tau = \langle\ddot{c}, H_{c'}\rangle$ *with* $c' \in \ddot{c}$ *be a basic transition in* $\mathscr{W}$. *Then* $\tilde{\Lambda}(\tau)$ *is a basic transition in* $\mathscr{W}'$.

Proof. We note that for every $h \in H_{c'}$ we have $\ddot{c} \in \Lambda(h)$. Next, we observe that no histories in $\Lambda(H_{c'})$ split at $\ddot{c}$ in the sense of $\perp'_{\ddot{c}}$. This follows from Fact A.9, since no histories in $H_{c'}$ split at $\ddot{c}$ in the sense of $\perp_{\ddot{c}}$. □

Our next lemma says that MFB-related notions translate between $\mathscr{W}$ and $\Lambda(\mathscr{W})$. To recall, while each: consistency, downward closure of a set of transitions, explanatory funny business, and combinatorial funny business is defined exactly the same in BST_{92} and BST_{NF}, combinatorial consistency is defined somewhat differently, by Def. 5.5 for BST_{92} and Def. 5.11 for BST_{NF}.

Lemma A.3. *Let* $\mathscr{W} = \langle W, <\rangle$ *be a BST_{NF} structure with no maximal elements and let* $\mathscr{W}' = \langle W', <'\rangle =_{df} \Lambda(\mathscr{W})$ *be the corresponding BST_{92} structure. Let* $T \subseteq TR(W)$ *be a set of basic transitions, and let* $T' =_{df} \{\tilde{\Lambda}(\tau) \mid \tau \in T\} \subseteq TR(W')$. *Then:*

1. *T is consistent iff T' is consistent.*

2. *For $\tau_1, \tau_2 \in TR(W)$, $\tau_1 \prec \tau_2$ iff $\tilde{\Lambda}(\tau_1) \prec' \tilde{\Lambda}(\tau_2)$. Hence τ belongs to the downward extension of T iff $\tilde{\Lambda}(\tau)$ belongs to the downward extension of T'.*

3. *T is combinatorially consistent in the sense of Def. 5.11 iff T' is combinatorially consistent in the sense of Def. 5.5.*

4. T is a case of combinatorial funny business (CFB) iff T' is a case of CFB (see Def. 5.6).

5. T is a case of explanatory funny business (EFB) iff T' is a case of EFB (see Def. 5.8).

Proof. (1) By the form of histories in $\mathcal{W}'$, h witnesses the consistency of $T \subseteq \mathrm{TR}(W)$ iff $\Lambda(h)$ witnesses the consistency of $T' \subseteq \mathrm{TR}(W')$, from which the claim follows.

(2) Let $\tau_i = \langle \ddot{e}_i, H_{e'_i} \rangle$ for $e'_i \in \ddot{e}_i$ and $i = 1, 2$. Since the initial of τ_i is the same as the initial of $\tilde{\Lambda}(\tau_i)$, we need only look at their outcomes. However, $H_{e'_1} \subseteq H_{e'_2}$ is equivalent to $\Lambda(H_{e'_1}) \subseteq \Lambda(H_{e'_2})$, so the claim follows.

(3) We need to check whether the following equivalences hold: a pair $\tau_1, \tau_2 \in \mathrm{TR}(W)$ satisfies a clause of Def. 5.11 iff the pair $\tilde{\Lambda}(\tau_1), \tilde{\Lambda}(\tau_2) \in \mathrm{TR}(W')$ satisfies the corresponding clause of Def. 5.5. To begin with clause (1), since τ_i and $\tilde{\Lambda}(\tau_i)$ share the same initial, the absence of blatant inconsistency in the former pair means the absence of blatant inconsistency in the latter pair. Turning to clause (2), its antecedent in Def. 5.11 is equivalent to its antecedent in Def. 5.5, since the $\tilde{\Lambda}$ transform leaves the initials intact. The consequent of clause (2) in Def. 5.11 is $H_{\ddot{e}_2} \subseteq H_{e'_1}$, which holds iff e'_1 is the unique member of $\ddot{e}_1$ for which all members e'_2 of $\ddot{e}_2$ satisfy $e'_1 < e'_2$ (see Fact 4.9). Furthermore, by Theorem 3.1 we have, for all $e'_2, e''_2 \in \ddot{e}_2$, that $e'_1 < e'_2$ iff $e'_1 < e''_2$. So $H_{\ddot{e}_2} \subseteq H_{e'_1}$ iff $H_{e'_2} \subseteq H_{e'_1}$, where $H_{e'_2}$ is the given outcome of τ_2. Clause (2) of Def. 5.5 follows as $H_{e'_2} \subseteq H_{e'_1}$ iff $\Lambda(H_{e'_2}) \subseteq \Lambda(H_{e'_1})$. Thus, a pair $\tau_1, \tau_2 \in \mathrm{TR}(W)$ satisfies clause (2) (or (3), which is proved in exactly the same way) of Def. 5.11 iff the pair $\tilde{\Lambda}(\tau_1), \tilde{\Lambda}(\tau_2) \in \mathrm{TR}(W')$ satisfies the corresponding clause of Def. 5.5. Turning to condition (4), it is satisfied because of the following equivalence: for any $e'_1 \in \ddot{e}_1$ and $e'_2 \in \ddot{e}_2$, e'_1 and e'_2 are incomparable in the sense of $<$ and $e'_1, e'_2 \in h$ for some history h iff $\ddot{e}_1$ and $\ddot{e}_2$ are incomparable in the sense of $<'$ and $\ddot{e}_1, \ddot{e}_2 \in \Lambda(h)$.

(4) Let T be a case of CFB (i.e., it is combinatorially consistent but inconsistent). Then by (1) and (3) above, T' is combinatorially consistent as well, but inconsistent (i.e., a case of CFB). The opposite direction follows analogously.

(5) To recall, EFB means inconsistency plus no downward extension being blatantly inconsistent. From (2) we have that T^* is a downward extension of T iff $\tilde{\Lambda}(T^*)$ is a downward extension of T'. By item (3) of this Lemma we have that there is a blatantly inconsistent pair in T iff there is a blatantly inconsistent pair in T'. Together with (1) this implies that T is a case of EFB iff T' is a case of EFB. $\square$

A.3 Proof of Theorem 5.1

In Chapter 5, we defined two different notions of modal funny business, namely combinatorial funny business (Def. 5.6) and explanatory funny business (Def. 5.8). We discussed their interrelation in Chapter 5.2.4, announcing the main Theorem 5.1 that states that the two notions are equivalent at the level of BST_{92} structures. Here we present a proof of that theorem. First, we repeat the definitions for convenience's sake.

Definition 5.6 (Combinatorial funny business). A set of basic transitions T constitutes a case of *combinatorial funny business* (CFB) iff T is combinatorially consistent (Def. 5.5), but inconsistent ($H(T) = \emptyset$).

Definition 5.8 (Explanatory funny business). A set T of transitions is a case of *explanatory funny business* (EFB) iff (1) T is inconsistent ($H(T) = \emptyset$) and (2) there is no downward extension T^* of T that is blatantly inconsistent.

Theorem 5.1 (There is combinatorial funny business iff there is explanatory funny business). *Let $\langle W, < \rangle$ be a BST$_{92}$ structure. For its set of basic indeterministic transitions, TR(W), the following holds: There is a subset $T_1 \subseteq$ TR(W) exhibiting combinatorial funny business iff there is a subset $T_2 \subseteq$ TR(W) exhibiting explanatory funny business.*

Proof. "$\Rightarrow$": This has been established via Lemma 5.3.

"$\Leftarrow$": Assume that there is no CFB in the given BST$_{92}$ structure. We will show that the assumption that there is EFB in that structure leads to a contradiction. Thus, assume for reductio that a set of transitions T in the given structure witnesses EFB (i.e., it is inconsistent), but no downward extension of T is blatantly inconsistent. Since there is no CFB, T must be combinatorially inconsistent. Let T^* be the maximal downward extension of T, which contains all transitions τ^* for which for some $\tau \in T$, we have $\tau^* \preccurlyeq \tau$. By Lemma 5.2, T^* is also inconsistent and combinatorially inconsistent. Combinatorial inconsistency means that there are transitions $\tau_1 = e_1 \rightarrowtail H_1$ and $\tau_2 = e_2 \rightarrowtail H_2$ in T^* that fail at least one of the four clauses of Definition 5.5. By our assumption that T witnesses EFB, T^* is not blatantly inconsistent. The fact that T^* is not blatantly inconsistent shows that the first three clauses of Definition 5.5 cannot be what accounts for the combinatorial inconsistency: If $e_1 = e_2$, then by no blatant inconsistency, $H_1 = H_2$ (i.e., $\tau_1 = \tau_2$). And if $e_1 < e_2$, then $\tau =_{df} e_1 \rightarrowtail \Pi_{e_1}(e_2) \in T^*$ by the fact that T^* is maximally downward extended (note that $\tau \prec \tau_2$), and as τ_1 also has the initial e_1 and T^* is not blatantly inconsistent, we have $\tau = \tau_1$, so that $\tau_1 \prec \tau_2$. For $e_2 < e_1$ we argue in the same way. Thus, the combinatorial inconsistency of T^* must be due to the existence of some $\tau_1 = e_1 \rightarrowtail H_1$ and $\tau_2 = e_2 \rightarrowtail H_2$ in T^* for which e_1 and e_2 are not order related and do not share any history (are not *SLR*)—all other ways for a set of transitions to witness combinatorial inconsistency are excluded.

We let E^* be the set of initials of transitions from T^*. By no blatant inconsistency, for any $e \in E^*$ there is exactly one transition $\tau = e \rightarrowtail H \in T^*$. We will denote that transition by τ_e. Let now $e_1, e_2 \in E$ be two initials of transitions from T^* that witness its combinatorial inconsistency; that is, e_1, e_2 are incomparable and not *SLR*, so there is no history $h \supseteq \{e_1, e_2\}$. We will now find a set of transitions $T_C \subseteq T^*$ that is combinatorially consistent but inconsistent, violating our initial assumption of no CFB (by Lemma 5.2). By this we will have established the right to left direction of our theorem. Note that, to establish the combinatorial consistency of a set of transitions $T_C \subseteq T^*$, it suffices to show that all initials of transitions from T_C are pairwise consistent (they share a history), as having inconsistent initials of transitions was the only way for T^*, and therefore any of its subsets to be combinatorially inconsistent.

Let now C_i be the set of choice points in the past of e_i splitting off some e_i-history from some non-e_i-history ($i = 1, 2$).[4] That is, we define

$$C_i = \{c < e_i \mid \exists h_i \in H_{e_i} \exists h \notin H_{e_i} [h_i \perp_c h]\}.$$

We have $C_1 \cup C_2 \subseteq E^*$ as T^* is downward maximal, and we know that C_1 and C_2 are each consistent, and thus in particular, if $e, e' \in C_i$, then there exists h such that $e, e' \in h$: since $C_i < e_i$, for any h_i containing e_i we have $C_i \subseteq h_i$ ($i = 1, 2$). We set (again, for $i = 1, 2$)

---

[4] In previous papers on MFB, the term "past causal loci" was used for the members of C_i, with the notation $C_i = pcl(e_i)$.

$$T_i =_{df} \{\tau_c \mid c \in C_i\}, \quad \tau_c = c \longmapsto \Pi_c(e_i),$$

where the form of τ_c (which, to recall, was unique in T^* for any $e \in E^*$) follows from the fact that $e_i \in E^*$ and T^* is downward maximal.

We will now define a maximal pairwise consistent subset C of $C_1 \cup C_2$ (which might be equal to $C_1 \cup C_2$), as follows. Let

$$A =_{df} \{c \in C_2 \mid \forall c_1 \in C_1 \exists h \in \text{Hist} [c \in h \wedge c_1 \in h]\}.$$

We claim that $C =_{df} C_1 \cup A$ is pairwise consistent, and maximally so as a subset of $C_1 \cup C_2$. For pairwise consistency, let $e, e' \in C$. There are three cases: if $e, e' \in C_1$, the claim follows by the consistency of C_1; similarly for $e, e' \in A \subseteq C_2$ by the consistency of C_2, and for $e \in C_1, e' \in A$, the claim follows from the definition of A. For maximality, let $e \in (C_1 \cup C_2) \setminus C$, i.e., $e \in C_2 \setminus A$. By the definition of A, there is then some $c_1 \in C_1$ for which there is no $h \in \text{Hist}$ containing both e and c_1, i.e., $C \cup \{e\}$ is not pairwise consistent.

We now let $T_C =_{df} \{\tau_c \mid c \in C\}$, which implies $T_C \subseteq T_1 \cup T_2$. As C is pairwise consistent and $T_C \subseteq T^*$, the set T_C is combinatorially consistent. We claim that T_C is inconsistent. Once we have established this, then our work here is done. So assume for reductio that T_C is consistent, so that there is some $h \in H(T_C)$. We can write $C = C_1' \cup C_2'$ with $C_i' \subseteq C_i$, via $C_i' = C_i \cap C$ ($i = 1, 2$). The fact that $h \in H(T_C)$ then implies that

$$h \in \bigcap_{c \in C_1'} \Pi_c(e_1) \cap \bigcap_{c \in C_2'} \Pi_c(e_2),$$

by the form of T_1 and T_2 noted above. Note that $C \subseteq h$, as C constitutes the set of initials for the transitions from T_C. We now show that $e_1 \in h$ and $e_2 \in h$, which is the sought-for contradiction, as no history can contain both e_1 and e_2. To establish $e_i \in h$ ($i = 1, 2$), assume that $e_i \notin h$, so that $e_i \in h_i \setminus h$ for some $h_i \in H_{e_i}$. Now by PCP92, there is some $c^* < e_i$ for which $h_i \perp_{c^*} h$. By the definition of C_i, we have $c^* \in C_i$. As $C \subseteq h$ and $c^* \in h$, we have that $C \cup \{c^*\}$ is consistent (as witnessed by h), and thereby also pairwise consistent. As C was maximally pairwise consistent, we thus must have $c^* \in C$, implying $c^* \in C_i'$. The fact that $h_i \perp_{c^*} h$ implies that $h \notin \Pi_{c^*}(h_i) = \Pi_{c^*}(e_i)$. But then $h \notin H(T_C)$ after all, contradicting our assumption.

So, bringing all of these disparate strands together, we have shown that T_C, a set of transitions in our BST92 structure, exhibits CFB, contrary to our initial assumption. So the set of transitions T cannot witness EFB after all, and we have shown that if there is no CFB in a BST92 structure, there is also no EFB.   □

A.4 Additional material for Chapter 8

A.4.1 Extensions by one point or by multiple points?

The Bell-Aspect setup discussed in Chapter 8.4.4 contains four cases of PFB. In this section we justify our decision to analyze this setup using only a single new choice point $\langle e^*, 0 \rangle$. The alternative option would be to consider a surface structure with four candidates for new choice points, e_{ij}^*, one for each case of PFB. Each e_{ij}^* would then need to be placed below a_i and below b_j. This choice of "one vs. many" is the BST version of the distinction between multiple separate screener-off systems and a single common screener-off system in the purely probabilistic framework of Hofer-Szabó, Rédei, and Szabó (see Hofer-Szabó, 2008).[5]

Let us investigate the option with many e_{ij}^*'s, that is, let us suppose that in a surface structure for the Bell-Aspect experiment there are four point events e_{ij}^*, each intended to take care of a single case of PFB. Observe that e_{ij}^* cannot be above L or R; otherwise it would prohibit the occurrence of one of the settings a_i or b_j. By the same observation, every point e_{ij}^* must belong to every history to which $a_i \cap b_j$ belongs, for every $i = 1, 2$, $j = 3, 4$. That means that all these points have to be SLR to or below both selection events L and R.

In addition, if one of the e_{ij}^* is below some other $e_{i'j'}^*$, this makes the bottom one irrelevant for PFB, so they all need to be pairwise SLR. And as they are introduced in order to explain instances of PFB, they had better not lead to additional cases of PFB in the multiplied structure. The latter structure is produced by making four subsequent multiplications, each with respect to a different e_{ij}^*. Given that e_{ij}^* is associated with an N_{ij}-multiplication, we produce the $N_{13} \cdot N_{14} \cdot N_{23} \cdot N_{24}$-multiplication. The SLR relation between any two e_{ij}^* and $e_{i'j'}^*$ ensures that a subsequent multiplication leaves $\langle e_{ij}^*, \emptyset \rangle$ intact, so eventually we obtain an extended event $E_0^* =_{df} \{ \langle e_{ij}^*, \emptyset \rangle \mid i = 1, 2, j = 3, 4 \}$ with $N_{13} \cdot N_{14} \cdot N_{23} \cdot N_{24}$ elementary outcomes. Each elementary outcome of E_0^* is identified with the intersection $H_{13}^n \cap H_{23}^{n'} \cap H_{14}^{n''} \cap H_{24}^{n'''}$ of some four outcomes of events $\langle e_{13}^*, \emptyset \rangle, \langle e_{14}^*, \emptyset \rangle, \langle e_{23}^*, \emptyset \rangle$, and $\langle e_{24}^*, \emptyset \rangle$, respectively. (Here the superscripts n, n', n'', n''' point to elementary outcomes of events $\langle e_{ij}^* \emptyset \rangle$, i.e., n points to the n-th outcome of $\langle e_{13}^*, \emptyset \rangle$, n' points to the n'-th outcome of $\langle e_{14}^*, \emptyset \rangle$, etc.). These intersections, being elementary outcomes, provide the most fine-grained partition of outcomes involved in the cases of PFB, as offered by a $N_{13}N_{14}N_{23}N_{24}$-multiplied BST_{92} structure.

Now, if Outcome Independence is satisfied with respect to the elementary outcomes of E_0^*, we just have an $N_{13}N_{14}N_{23}N_{24}$-multiplied probabilistic structure with one extended event E_0^* that serves the role of a single hidden variable. Its values are given by transitions $E_0^* \rightarrowtail H_{13}^n \cap H_{23}^{n'} \cap H_{14}^{n''} \cap H_{24}^{n'''}$. The difference between this structure and an N-multiplied structure with a single *point* event $\langle e^*, 0 \rangle$ is inessential.

On the remaining option, Outcome Independence is not satisfied by the elementary outcomes of E_0^*, but is satisfied by the elementary outcomes of each $\langle e_{ij}^*, 0 \rangle$. However, an elementary outcome of $\langle e_{ij}, 0 \rangle$ is identifiable with a *non-elementary* outcome of E_0^*. Accordingly, Outcome Independence is not satisfied by elementary outcomes of E_0^*, but is satisfied by some non-elementary outcomes of it. We may re-phrase this fact in terms of partitions: the condition is satisfied on a less than maximally fine-grained level, while failing at the most fine-grained level. This looks like a fluke, and in any case, does not explain the four cases of PFB we started with.

To sum up, in the context of the Bell-Aspect experiment, the construction of a structure with many hidden variables for PFB either reduces to the construction with a single hidden variable, or abandons any explanation of PFB.[6]

---

[5] The distinction was, however, first introduced and argued for in the BST framework (see Belnap and Szabó, 1996), where it was phrased it in terms of common causes and *common* common causes. The framework of Hofer-Szabó, Rédei, and Szabó was introduced in Hofer-Szabó et al. (1999). For a recent presentation of their results in this framework, see Hofer-Szabó et al. (2013).

[6] For a similar diagnosis in a purely probabilistic framework, see Wroński et al. (2017, p. 95).

A.4.2 Proofs for Chapter 8

Lemma 8.4. *Let $\mathscr{W} = \langle W, <, \mu, e^*, E, C \rangle$ be a probabilistic BST$_{92}$ surface structure and $\mathscr{W}'$—the N-multiplied structure corresponding to $\mathscr{W}$. Then*

(1) *For every history $h \in \mathrm{Hist}(W)$ the set $\varphi^n(h)$ is a maximal directed subset of W', i.e., a history in W'.*

(2) *For every maximal directed subset $A' \subseteq W'$ there is a history $h \in \mathrm{Hist}(W)$ and $n \in \{1, \ldots, N\}$ for which $A' = \varphi^n(h)$.*

Proof. (1) It is easy to see that φ^n is an order-preserving bijection between h and $\varphi^n(h)$, which implies that $\varphi^n(h)$ is directed. To establish maximal directedness, take a directed set $A' \subseteq W'$ for which $\varphi^n(h) \subseteq A'$. Note that by the definition of the ordering, there are no upper bounds for elements $\langle x_1, n_1 \rangle$ and $\langle x_2, n_2 \rangle$ if $n_1, n_2 \in \{1, \ldots, N\}$ and $n_1 \neq n_2$. It follows that A' can be written as the union of $A' = \{\langle x, 0 \rangle \mid e^* \not< x \wedge x \in W\} \cup \{\langle x, n \rangle \mid e^* < x \wedge x \in W\}$ for some $n \in \{1, \ldots, N\}$. We claim now that the set $A =_{\mathrm{df}} \{x \in W \mid \langle x, 0 \rangle \in A'\} \cup \{x \in W \mid \langle x, n \rangle \in A'\}$ is directed: Let $e_1, e_2 \in A$, so that there are unique $\langle e_1, n_1' \rangle, \langle e_2, n_2' \rangle \in A'$, with $n_1', n_2' \in \{0, 1, \ldots, N\}$. By directedness of A', these elements have a common upper bound $\langle e_3, n_3' \rangle \in A'$, so that $e_3 \in A$, and by the definition of the ordering, $e_1 \leqslant e_3$ and $e_2 \leqslant e_3$, so A is directed, indeed. Finally, by $\varphi^n(h) \subseteq A'$ we have $h \subseteq A$. Now as h is maximal directed, it must be that $A = h$, whence $A' = \varphi^n(h)$.

(2) Let A' be a maximal directed subset of W'. In (1) above we established that $A' = \{\langle x, 0 \rangle \mid e^* \not< x \wedge x \in W\} \cup \{\langle x, n \rangle \mid e^* < x \wedge x \in W\}$ for some $n \in \{1, \ldots, N\}$ and that the set $A =_{\mathrm{df}} \{x \in W \mid \langle x, 0 \rangle \in A'\} \cup \{x \in W \mid \langle x, n \rangle \in A'\}$ is directed. We now claim that there is $h \in \mathrm{Hist}$ st $A = h$. Since A is a directed subset of W and histories are maximal directed subsets of W, there is an $h \in \mathrm{Hist}(W)$ st $A \subseteq h$. Suppose that $A \subsetneq h$. But then

$$A' \subsetneq \{\langle x, 0 \rangle \mid e^* \not< x \wedge x \in h\} \cup \{\langle x, n \rangle \mid e^* < x \wedge x \in h\},$$

As the set on the RHS is directed, A' is not maximally directed, which contradicts the premise. Hence $A = h$, which implies, given the form of A', that $A' = \varphi^n(h)$. $\square$

Fact 8.15. *Let $\mathscr{W} = \langle W, <, \mu, e^*, E, C \rangle$ be a probabilistic BST$_{92}$ surface structure and let $\mathscr{W}'$ be the N-multiplied structure corresponding to $\mathscr{W}$. Then*

(1) *for every $n \in \{1, \ldots, N\}$ and every $h_1, h_2 \in H_{e^*}$: $\varphi^n(h_1) \equiv_{\langle e^*, 0 \rangle} \varphi^n(h_2)$.*

(2) *for every $n, m \in \{1, \ldots, N\}$ such that $n \neq m$ and every $h \in H_{e^*}$: $\varphi^n(h) \perp_{\langle e^*, 0 \rangle} \varphi^m(h)$.*

(3) *$\langle e^*, 0 \rangle$ is a choice point with N outcomes $\Pi_{\langle e^*, 0 \rangle} \langle \varphi^n(h) \rangle$, where h is an arbitrary history from H_{e^*};*

(4) *for every $n \in \{1, \ldots, N\}$, every $e \in W$, and every $h_1, h_2 \in \mathrm{Hist}(W)$: $h_1 \perp_e h_2$ iff $\varphi^n(h_1) \perp_{\langle e, l \rangle} \varphi^n(h_2)$, where $l = n$ iff $e^* < e$, and $l = 0$ otherwise;*

(5) *for every $m, n, l \in \{1, \ldots, N\}$ with $m \neq n$, every $e \in W$ such that $e^* < e$, and every $h_1, h_2 \in \mathrm{Hist}(W)$: neither $\varphi^m(h_1) \equiv_{\langle e, l \rangle} \varphi^n(h_2)$, nor $\varphi^m(h_1) \perp_{\langle e, l \rangle} \varphi^n(h_2)$;*

(6) *for every $m, n \in \{1, \ldots, N\}$ with $m \neq n$, every $e \not> e^*$ and every $h \in H_e$: $\varphi^m(h) \equiv_{\langle e, 0 \rangle} \varphi^n(h)$.*

Proof. (1) Since e^* is deterministic, $h_1 \equiv_{e^*} h_2$, so there is $e > e^*$ such that $e \in h_1 \cap h_2$. Accordingly $\langle e, n \rangle >' \langle e^*, 0 \rangle$ and $\langle e, n \rangle \in \varphi^n(h_1) \cap \varphi^n(h_2)$, which proves $\varphi^n(h_1) \equiv_{\langle e, 0 \rangle} \varphi^n(h_2)$.

(2) Take any $h \in H_{e^*}$. Then $\langle e^*, 0 \rangle \in \varphi^n(h) \cap \varphi^{n'}(h)$ for any $n, n' \in \{1, \ldots, N\}$. For any $e > e^*$ any m and any different n, n', event $\langle e, m \rangle$ cannot be shared by $\varphi^n(h)$ and $\varphi^{n'}(h)$. This proves that $\varphi^n(h) \perp_{\langle e^*, 0 \rangle} \varphi^{n'}(h)$, and hence that N histories split at $\langle e^*, 0 \rangle$.

(3) By (1) and (2) above.

(4) Take $h_1, h_2 \in \text{Hist}(W)$ such that $h_1 \perp_e h_2$. It follows that $\langle e, l \rangle \in \varphi^n(h_1) \cap \varphi^n(h_2)$. Also, there is no $e' > e$ such that $e' \in h_1 \cap h_2$. Hence there is no $\langle e', l' \rangle$ with $\langle e', l' \rangle >'$ $\langle e, l \rangle$ such that $\langle e', l' \rangle \in \varphi^n(h_1) \cap \varphi^n(h_2)$, which proves the $\Rightarrow$ direction. In the opposite direction, assume for reductio that $h_1 \not\perp_e h_2$. If $e \notin h_1 \cap h_2$, then $\langle e, l \rangle \notin \varphi^n(h_1) \cap \varphi^n(h_2)$, which contradicts $\varphi^n(h_1) \perp_{\langle e, l \rangle} \varphi^n(h_2)$. So let $h_1 \equiv_e h_2$. There is then $\langle e', l' \rangle >' \langle e, l \rangle$ such that $\langle e', l' \rangle \in \varphi^n(h_1) \cap \varphi^n(h_2)$, which contradicts $\varphi^n(h_1) \perp_{\langle e, l \rangle} \varphi^n(h_2)$.

(5) Since $e > e^*$, it can only be associated with some $l \in \{1, \ldots, N\}$. Furthermore, for different l, l', $\langle e, l \rangle$ and $\langle e, l' \rangle$ have no upper bound with respect to $<'$, and hence there is no history to which they belong. Thus, neither $\varphi^m(h_1) \equiv_{\langle e, l \rangle} \varphi^n(h_2)$, nor $\varphi^m(h_1) \perp_{\langle e, l \rangle} \varphi^n(h_2)$.

(6) Let $e \not> e^*$. Then $\langle e, 0 \rangle \in \varphi^n(h)$ for every $n \in \{1, \ldots, N\}$ and every $h \in H_e$. If e is maximal in W, $\langle e, 0 \rangle$ is maximal in W', and hence $\varphi^m(h) \equiv_{\langle e, 0 \rangle} \varphi^n(h)$ for every $m, n \in \{1, \ldots, N\}$ and every $h \in H_e$. If e is not maximal in W, pick an arbitrary $h \in H_e$; there is then $e' \in h$ such that $e' > e$ and $e' \not> e^*$ as well (by density). Each e and e' is associable with 0 only, $\langle e, 0 \rangle <' \langle e', 0 \rangle$ and $\langle e', 0 \rangle \in \varphi^n(h)$ for any $n \in \{1, \ldots, N\}$. Hence $\varphi^m(h) \equiv_{\langle e, 0 \rangle} \varphi^n(h)$ for any $m, n \in \{1, \ldots, N\}$. $\square$

Lemma 8.6. *Let $\mathscr{W} = \langle W, <, \mu, e^*, E, C \rangle$ be a probabilistic BST$_{92}$ surface structure in which transitions $\{I_1 \rightarrowtail 1_1, \ldots, I_K \rightarrowtail 1_K\}$ with random variables $X_1, \ldots, X_K$ exhibit PFB, where $1_k = \{\hat{O}_{k, \gamma(k)} \mid \gamma(k) \in \Gamma(k)\}$, $\Gamma(k)$ are index sets and $1 \leqslant k \leqslant K$. Let $cll(I_k \rightarrowtail \hat{O}_{k, \gamma(k)}) \subseteq E$ for every $k \leqslant K$ and every $\gamma(k) \in \Gamma(k)$ and $C = \emptyset$. Then there exists a structure with a probabilistic hidden variable for this case of PFB. Moreover, the structure satisfies C/E propensity independence.*

Proof. Let us suppose that there is $e^* \in W$ that is below every I_k. We will explicitly exhibit a structure $\mathscr{W} = \langle W', <', \mu' \rangle$ with a probabilistic hidden variable, in which event $\langle e^*, 0 \rangle$ has N outcomes $H^1, H^2, \ldots, H^N$ and there are N sets of uncorrelated random variables $\{X_1^n, \ldots, X_K^n\}$ $(1 \leqslant n \leqslant N)$ corresponding to $\{X_1, \ldots, X_K\}$, resp. Since the cardinality of S is N, we may number elements of S as $T_1, T_2, \ldots T_N$. Note that each $T_n \in S$ is determined as well by a sequence of values of random variables, $X_1(T_n) = \gamma(1), X_2(T_n) = \gamma(2), \ldots, X_K(T_n) = \gamma(K)$ with $\gamma(k) \in \Gamma(k)$, which is the observation we use below. By means of Outcome Independence we get

$$p^n(X_1^n = \gamma(1))p^n(X_2^n = \gamma(2)) \ldots p^n(X_K^n = \gamma(K)) =$$
$$p^n(X_1^n = \gamma(1) \wedge X_2^n = \gamma(2) \wedge \ldots \wedge X_K^n = \gamma(K)) = \mu'(T_l^n),$$

where T_l^n is a unique element of S^n st $X_k^n(T_l^n) = \gamma(k)$ for $k = 1, 2, \ldots K$ and $\gamma(k) \in \Gamma(k)$. By adequate probabilistic assignment we get

$$\mu(T_l) = \sum_{n=1}^N \mu'(\langle e^* 0 \rangle \rightarrowtail H^n)\mu'(T_l^n), \text{ where } T_l \in S \text{ corresponds to } T_l^n \in S^n.$$

Given this correspondence, T_l is given by the same values of the corresponding random variables as T_l^n, so we have $\mu(T_l) = p(X_1 = \gamma(1) \wedge X_2 = \gamma(2) \wedge \ldots \wedge X_K = \gamma(K))$; then the above equations yield

$$p(X_1 = \gamma(1) \wedge X_2 = \gamma(2) \wedge \ldots \wedge X_K = \gamma(K)) =$$

$$\sum_{n=1}^{N} \mu'(\langle e^*0 \rangle \rightarrowtail H^n) p^n(X_1^n = \gamma(1)) p^n(X_2^n = \gamma(2)) \ldots p^n(X_K^n = \gamma(K)).$$

Let us abbreviate the above formula as:

$$Z_{\gamma(1)\gamma(2)\ldots\gamma(K)} = \sum_{n=1}^{N} \alpha^n q_{1,\gamma(1)}^n q_{2,\gamma(2)}^n \cdots q_{K,\gamma(K)}^n, \qquad (\text{A.1})$$

where $Z_{\gamma(1)\gamma(2)\ldots\gamma(K)} = p(X_1 = \gamma(1) \wedge X_2 = \gamma(2) \wedge \ldots \wedge X_K = \gamma(K))$, $q_{k,\gamma(k)}^n = p^n(X_k^n = \gamma(k))$, and $\alpha^n = \mu'(\langle e^*0 \rangle \rightarrowtail H^n)$. Formula A.1 encapsulates $|\Gamma(1)| \times \cdots \times |\Gamma(K)|$ equations. To construct a sought-after structure means to solve these equations for N unknown variables α^n and $N(|\Gamma(1)| + |\Gamma(2)| + \ldots + |\Gamma(K)|)$ unknown variables $q_{k,\gamma(k)}^n$. Here is a simple set of solutions (there are other sets of solutions):

$$\alpha^n = p(T_n)$$

$$q_{k,\gamma(k)}^n = \begin{cases} 1 & \text{iff } X_k(T_n) = \gamma(k) \\ 0 & \text{otherwise.} \end{cases}$$

It is easy to calculate that this ascription of values satisfies the above equations. The LHS is equal to: $p(T_l)$, where $T_l \in S$ is such that for every $k \in \{1, \ldots, K\}$: $X_k(T_l) = \gamma(k)$. The RHS is $\alpha^{l'} q_{1,\gamma(1)}^{l'} q_{2,\gamma(2)}^{l'} \cdots q_{K,\gamma(K)}^{l'} = \alpha^{l'} = p(T_{l'})$, where $T_{l'}$ is such that for every $k \in \{1, \ldots, K\}$: $X_k(T_{l'}) = \gamma(k)$. Thus, $T_l = T_{l'}$ and so the equation holds. $\qquad\square$

Lemma 8.7. *Let $\mathcal{W} = \langle W, <, \mu, e^*, E, C \rangle$ be a probabilistic BST_{92} surface structure harboring multiple cases of PFB for which $C = \emptyset$. Let μ be defined on every subset of $\text{TR}(W)$. Then there is an N-multiplied probabilistic BST_{92} structure corresponding to $\mathcal{W}$ that provides a hidden variable for every case of PFB in $\mathcal{W}$. Moreover, that extended structure satisfies C/E propensity independence.*

Proof. We first decide how large a multiplication we will introduce. We are after a quasi-deterministic hidden variable. By the finitistic assumptions, each $\tilde{T}_E$ and S_E is finite, and they contain all basic transitions in $\mathcal{W}$, and all maximal consistent subsets of basic transitions in $\mathcal{W}$, respectively. For N, the size of the multiplication, we take the cardinality of S_E. By the above there are finitely many cases of PFB as well. Note also that e^* is below the initial of every transition in $\tilde{T}_E$. In what follows we will need some bijection $f : S_E \mapsto \{1, 2, \ldots, N\}$.

We consider next an N-multiplied BST_{92} structure $\mathcal{W}' = \langle W', <' \rangle$ corresponding to $\mathcal{W}$. We construct μ', which is intended to be adequate for $\mathcal{W}$ and N. In accord with Def. 8.20, it is enough to specify what μ' yields for every element of S_E and new transitions. We define μ' as follows:[7]

1. For every basic transition $\langle e^*, 0 \rangle \rightarrowtail H^n$: $\mu'(\{\langle e^*, 0 \rangle \rightarrowtail H^n\}) = \mu(f^{-1}(n))$, where $H^n \in \Pi_{\langle e^*, 0 \rangle}$;

2. For every $T \in S_E$ and $n \in \{1, \ldots, N\}$: $\mu'(T^n) = \delta_n^{f(T)}$ (where δ is Kronecker's delta).

---

[7] Note the simplification below due to the fact that every $e \in E_T$ is above e^*.

It is immediately discernible that μ' is adequate for $\mathcal{W}$ and N. To calculate a single non-trivial clause of Def. 8.20: $\sum_{n \leqslant N} \mu'(\langle e^*, 0 \rangle \rightarrowtail H^n) \cdot \mu'(T^n) = \sum_{n \leqslant N} \mu(f^{-1}(n)) \delta_n^{f(T)} = \mu(T)$. Note that the rules above induce a zero-one assignment to sets corresponding to subsets of $\tilde{T}_E$: for any $Y \subseteq \tilde{T}_E$ and any $n \in \{1, \ldots, N\}$,

$$\mu'(Y^n) = \begin{cases} 1 & \text{if there is } T \in S_E \text{ such that } Y \subseteq T \text{ and } \mu'(T^n) = 1 \\ 0 & \text{otherwise.} \end{cases}$$

To check if Outcome Independence is satisfied, pick an arbitrary case of PFB in $\mathcal{W}$. This is given by a set of transitions $\{I_1 \rightarrowtail \mathbf{1}_1, \ldots, I_K \rightarrowtail \mathbf{1}_K\}$ with SLR initials, each transition being associated with a random variable, $X_1, \ldots, X_K$, respectively, and where $\mathbf{1}_k = \{\hat{O}_{k,\gamma(k)} \mid \gamma(k) \in \Gamma(k)\}$, with finite index sets $\Gamma(k)$, and $1 \leqslant k \leqslant K$. Then these random variables $X_1, \ldots, X_K$ exhibit PFB, We abbreviate: $Y_{\langle \gamma(1), \ldots, \gamma(K) \rangle} =_{df} CC(G \rightarrowtail \hat{O}_{1,\gamma(1)} \cup \ldots \cup \hat{O}_{K,\gamma(K)})$ and $Y_{\langle k,\gamma(k) \rangle} =_{df} CC(I_k \rightarrowtail \hat{O}_{k,\gamma(k)})$, where $G = \bigcup_{k \leqslant K} I_k$. Our random variables are defined on CPS, for which the base set is: $S = \{Y_{\langle \gamma(1), \ldots, \gamma(K) \rangle} \mid \gamma(1) \in \Gamma(1), \ldots, \gamma(K) \in \Gamma(K)\}$. We write $Y^n_{\langle \gamma(1), \ldots, \gamma(K) \rangle}$ and $Y^n_{\langle k,\gamma(k) \rangle}$ for the sets corresponding to $Y_{\langle \gamma(1), \ldots, \gamma(K) \rangle}$ and $Y_{\langle k,\gamma(k) \rangle}$, respectively.

Consider now a corresponding CPS, $\langle S^n, \mathscr{A}^n, p^n \rangle$ with an arbitrary $n \leqslant N$. We need to check if the following identity is satisfied:

$$p^n(X_1^n = \gamma(1) \wedge \ldots \wedge X_K^n = \gamma(K)) = p^n(X_1^n = \gamma(1)) \cdot \ldots \cdot p^n(X_K^n = \gamma(K)), \qquad (\dagger)$$

which is equivalent to

$$\mu'(Y^n_{\langle \gamma(1) \ldots \gamma(K) \rangle}) = \mu'(Y^n_{\langle 1,\gamma(1) \rangle}) \cdot \ldots \cdot \mu'(Y^n_{\langle K,\gamma(K) \rangle}).$$

By the definition of μ', each value of μ' above must be either 0 or 1. Then the argument that this identity ($\dagger$) holds is exactly the same as in the proof of Lemma 8.6. This means that Outcome Independence is satisfied. As the propensity assignment μ' is adequate as well, $\langle W', <', \mu' \rangle$ is an N-multiplied BST$_{92}$ probabilistic structure corresponding to $\mathcal{W}$ that provides a hidden variable for every case of PFB in $\mathcal{W}$. Furthermore, the N-multiplied structure satisfies independence with respect to images of C and E, since C is assumed to be empty. $\square$

APPENDIX B

Answers to Selected Exercises

Here we provide answers to selected exercises from Chapters 1–10.

B.1 Answers to selected exercises from Chapter 1

Exercise 1.1. Lewis (1986a, p. 208) assumes that all elements of a possible world are to "stand in suitable external relations, preferably spatiotemporal". Somewhat similarly, in Branching Space-Times any two point events from *Our World* are linked by appropriately combined instances of the pre-causal relation $<$ (see the M property, Fact 2.4). Discuss whether the pre-causal relation (which is formally explained in Chapter 2.1) is a "suitable external relation" from Lewis's perspective.

Answer: It is useful to focus on the notion of "spatiotemporal relations", Lewis's paradigm for a relation to be used to draw a distinction between objects that inhabit one world, and objects that do not inhabit one world. Lewis acknowledges that there is an ambiguity in this notion. The exercise asks for an assessment from Lewis's perspective. This calls for some familiarity with his metaphysics, as we need to grasp the notion as he understood it. The central tenet of Lewis's metaphysics is the reduction of modal claims to mereological relations obtaining between certain maximal objects called "possible worlds". Thus, within a given possible world there are no (non-trivial) modal relations. On the hypothesis of so-called Humean Supervenience, a world can be identified with a "Humean mosaic", which is an ascription of properties ("perfectly natural properties") to point-like bearers. A natural option to understand such bearers is to identify them with spatio-temporal points (this appears to be Lewis's preferred option). Such points stand in various spatio-temporal relations. Paradigmatically, a spatio-temporal relation is identified with some distance between points in space-time. As to the ascription of properties, each point has exactly one set of jointly instantiable properties assigned. No property is modal, so that, for example, "possibly going up" is not an admissible property. Now, the M-property in the form of a real M (rather than just a part of it) relates objects (idealized to be point-like) that are incompatible; such objects cannot occur together. By Lewis's central tenet, these objects cannot inhabit one and the same mosaic. If they could, they would be related by the spatio-temporal relations that underlie a given mosaic, and hence they would be compatible. And if they do not belong to one and the same mosaic, they do not belong to one and the same possible world. Thus, from Lewis's perspective, the M-property does not relate inhabitants of one world.

A reflection on this argument shows that it assumes that each mosaic has its own space-time, its own space-time points, and its own spatio-temporal relations. This make the argument against the M-property as a demarcation principle fall short. In contrast, our everyday claims concerning what could happen at a given location at a given time in an alternative course of events suggests that alternative scenarios share the same space-time. That idea of a shared space-time also underlies the mathematical description of indeterminism in classical physics. On this view, then, spatio-temporal relations relate incompatible events. Of course, we do not want to claim that any two incompatible events, which are spatio-temporally related in the wide sense, belong to our world. But some such events, like tomorrow's possible outcomes *tails up* and *heads up* of a particular coin toss do seem to belong to our world. (At least this is the intuition underlying BST.) The M-property, with the particular shape of the letter "M", should thus relate objects like the two incompatible results of a certain toss: these objects are incompatible, and yet they belong to our single, modally thick world.

B.2 Answers to selected exercises from Chapter 2

Exercise 2.2. Prove the M property (Fact 2.4):
For every pair e_1, e_5 of point events in $\mathcal{W}$, there are e_2, e_3, e_4 in $\mathcal{W}$ such that $e_1 \leqslant e_2, e_5 \leqslant e_4$ and $e_3 \leqslant e_2, e_3 \leqslant e_4$.

Proof. Let $e_1 \in h_1$ and $e_5 \in h_5$. By Historical Connection (Postulate 2.2), there is $e_3 \in h_1 \cap h_5$. Since histories are directed, there is e_2 such that $e_1 \leqslant e_2$ and $e_3 \leqslant e_2$ (via directedness of h_1) and $e_5 \leqslant e_4$ and $e_3 \leqslant e_4$ (via directedness of h_5). ☐

Exercise 2.3. Let $\langle W, < \rangle$ be a partially ordered set satisfying Postulates 2.1 and 2.2. Prove that if every history of W is downward directed, then so is W as a whole. (Note that the assumption is true, for example, if each history is isomorphic to Minkowski space-time.)

Proof. Pick any $e_1, e_5 \in W$. If these events share a history, we are done. If not, let us invoke the M property. There are thus e_2, e_3, e_4 such that $e_1 \leqslant e_2, e_5 \leqslant e_4$ and $e_3 \leqslant e_2, e_3 \leqslant e_4$. Let $e_2 \in h_2$ and $e_4 \in h_4$, so $e_3 \in h_2 \cap h_4$. By the assumption, histories are downward directed, so there are $e_6, e_7 \in h_2 \cap h_4$ that are lower bounds of, respectively, e_1, e_3 and e_3, e_5. Since e_6, e_7 share a history, they have a lower bound e_9. By transitivity of $\leqslant$, $e_9 \leqslant e_1$ and $e_9 \leqslant e_5$. ☐

B.3 Answers to selected exercises from Chapter 3

Exercise 3.2. Prove the following extension of Fact 3.10: For a BST_{92} structure $\langle W, < \rangle$ that has neither maximal nor minimal elements, its full transition structure $\langle W', <' \rangle =_{\text{df}} \langle \text{TR}_{\text{full}}(W), \prec \rangle$ has no maxima nor minima either.

Proof. For no maxima, let $\tau = e \rightarrowtail H \in W'$, and let $h \in H \subseteq H_e$. As W contains no maxima, h contains no maxima either (Fact 2.1(9)), so there is $e_1 \in h$ for which $e < e_1$. Accordingly we have $\tau' =_{\text{df}} e_1 \rightarrowtail \Pi_{e_1}\langle h \rangle \in W'$. It is easy to check that $\tau \prec \tau'$, which establishes that τ is not maximal in W'.

For no minima, similarly, let $\tau = e \rightarrowtail H \in W'$. As W contains no minima, there is $e_1 \in W$ for which $e_1 < e$. Let $h \in H_e$. By downward closure, $e_1 \in h$, i.e., $h \in H_{e_1}$. So there is $\tau' =_{\text{df}} e_1 \rightarrowtail \Pi_{e_1}\langle h \rangle \in W'$, and $\tau' \prec \tau$. Thus, τ is not minimal in W'. ☐

Exercise 3.7. Let $\langle W, < \rangle$ satisfy Postulates 2.1–2.5. Let l be an upper-bounded chain, and let $e =_{\text{df}} \sup_{h'}(l)$. Then for every history h of W containing the chain l, if e lies in h, then $e = \sup_h(l)$.

Proof. Suppose that $e \in h$. Since $e = \sup_{h'}(l)$, by definition e upper-bounds l. Now, suppose toward a contradiction that e is not the least upper bound of l in h, that is, suppose that there is some e' in h such that $l \leqslant e' < e$. By Fact 2.1 (5) histories are downwards closed, which means that the element e' also lies in h', contradicting $e = \sup_{h'}(l)$. Therefore there is no such e' in h, and consequently, $e = \sup_h(l)$. $\square$

B.4 Answers to selected exercises from Chapter 4

Exercise 4.4. Prove that the diamond topology of Def. 4.13 and the history-relative diamond topologies of Def. 4.14 are indeed topologies for both BST_{92} and BST_{NF}; that is, prove that both (1) the base set (W or h, respectively) and (2) the empty set are open, (3) arbitrary unions of open sets are open, and (4) finite intersections of open sets are open.

Proof. In both definitions, (1) is explicitly required to hold, and the form of the definitions, which is universal, guarantees that (2) and (3) hold as well (note that the condition is vacuous for the empty set). The only condition that needs a proof is the finite intersection property, (4). It suffices to prove that the intersection of any two open sets is open, as the finite case then follows by simple induction.

The proof of (4) is the same for BST_{92} and for BST_{NF}, as no instance of a prior choice principle is needed. We give the proof for $\mathcal{T}$; the proof for $\mathcal{T}_h$ is exactly analogous, replacing the set of chains $MC(e)$ with $MC_h(e)$. Let thus $Z_1, Z_2 \in \mathcal{T}$, let $Z =_{df} Z_1 \cap Z_2$, and take some $e \in Z$ and $t \in MC(e)$. To show that Z is open, we have to find $e_1, e_2 \in t$ with $e_1 < e < e_2$ and such that the diamond $D_{e_1,e_2} \subseteq Z$. Now as Z_i is open ($i = 1.2$), there are e_1^i, e_2^i with $e_1^i < e < e_2^i$ and such that $D_{e_1^i,e_2^i} \subseteq Z_i$. Let $e_1 = \max(e_1^1, e_1^2)$ and $e_2 = \min(e_1^1, e_1^2)$ (note that these elements are comparable as they belong to the same chain). Then $D_{e_1,e_2} \subseteq Z_1$ and $D_{e_1,e_2} \subseteq Z_2$, which implies that $D_{e_1,e_2} \subseteq Z$. This shows that Z is indeed open. $\square$

B.5 Answers to selected exercises from Chapter 5

Exercise 5.1. Let O be an outcome chain. Prove that if for all $e \in cll(O)$, we have $e < O$, then $H_{\langle O \rangle} = \bigcap_{e \in cll(O)} \Pi_e \langle O \rangle$.

Proof. For any $e \in cll(O)$, $e < O$ implies $H_{\langle O \rangle} \subseteq \Pi_e \langle O \rangle$ (see the proof of Fact 5.4), so the inclusion "$\subseteq$" is straightforward. For the opposite inclusion, we argue indirectly; that is, we assume for reductio that there is $h \in \text{Hist}$ such that $h \notin H_{\langle O \rangle}$ but that for every $e \in cll(O), h \in \Pi_e \langle O \rangle$. Take some $h_O \supseteq O$. As $h \notin H_{\langle O \rangle}$, we have $O \subseteq h_O \setminus h$, so by PCP_{92} there is $c \in W$ such that $c < O$ and $h \perp_c h_O$. By Fact 3.8, $h \perp_c H_{\langle O \rangle}$. Thus, $c \in cll(O)$, so by our assumption: $h \in \Pi_c \langle O \rangle$. Since $H_{\langle O \rangle} \subseteq \Pi_c \langle O \rangle$, we get $h \equiv_c H_{\langle O \rangle}$, which contradicts $h \perp_c H_{\langle O \rangle}$. $\square$

Exercise 5.3. Prove a version of Facts 5.4 and 4.7(2) for $\hat{O}$ a scattered outcome. That is, prove the following facts:
Let $\hat{O}$ be a scattered outcome. (1) If there is no-MFB, then for all $e \in cll(\hat{O})$, we have $e < \hat{O}$. (2) If $e < \hat{O}$, then there is a unique basic outcome of e that is consistent with $H_{\langle \hat{O} \rangle}$, which we denote $\Pi_e \langle \hat{O} \rangle$.

Proof. (1) By no-MFB we know from Fact 5.4 that for each $O \in \hat{O}$, any $e \in cll(O)$ is in the past of O. By the same Fact and by Exercise 5.1, for any $O \in \hat{O}$ we have $H_{\langle O \rangle} = \bigcap_{e \in cll(O)} \Pi_e \langle O \rangle$. Thus, by the definition of $H_{\langle \hat{O} \rangle}$,

$$H_{\langle \hat{O} \rangle} = \left(\bigcap_{O \in \hat{O}} \bigcap_{e \in cll(O)} \Pi_e \langle O \rangle \right). \tag{B.1}$$

Let us now suppose for reductio that there is $h \in$ Hist and $c \in W$ such that $h \perp_c H_{\langle \hat{O} \rangle}$ (i.e., $c \in cll(\hat{O})$) but $c \not< \hat{O}$, which means that for every $O \in \hat{O}$: $c \not< O$. Accordingly, $c \notin \bigcup_{O \in \hat{O}} cll(O)$ but $c \in cll(\hat{O})$. Consider now the union of outcomes of c that are consistent with $H_{\langle \hat{O} \rangle}$, which is $\tilde{H} =_{df} \bigcup \{ H \in \Pi_c \mid H \cap H_{\langle \hat{O} \rangle} \neq \emptyset \}$. Observe next that since $h \perp_c H_{\langle \hat{O} \rangle}$, we have $H_{\langle \hat{O} \rangle} \subseteq H_c$, and hence $H_{\langle \hat{O} \rangle} \subseteq \tilde{H}$. Note also that for every $O \in \hat{O}$ and for every $e \in cll(O)$, $e < O$, so that $H_{\langle O \rangle} \subseteq \Pi_e \langle O \rangle$. Accordingly, for every $O \in \hat{O}$ and for every $e \in cll(O)$, we have $H_{\langle \hat{O} \rangle} \subseteq H_{\langle O \rangle} \subseteq \Pi_e \langle O \rangle$. Now consider the following intersection of sets of histories, H':

$$H' =_{df} \left(\bigcap_{O \in \hat{O}} \bigcap_{e \in cll(O)} \Pi_e \langle O \rangle \right) \cap \tilde{H}. \tag{B.2}$$

By $H_{\langle \hat{O} \rangle} \subseteq \tilde{H}$ and by Eq. (B.1) we have $H_{\langle \hat{O} \rangle} = H'$, i.e.,

$$\left(\bigcap_{O \in \hat{O}} \bigcap_{e \in cll(O)} \Pi_e \langle O \rangle \right) \cap \tilde{H} = H_{\langle \hat{O} \rangle}. \tag{B.3}$$

Since $\Pi_c \langle h \rangle \subseteq H_c \setminus \tilde{H}$, Eqs. (B.3) and (B.1) imply

$$\left(\bigcap_{O \in \hat{O}} \bigcap_{e \in cll(O)} \Pi_e \langle O \rangle \right) \cap \Pi_c \langle h \rangle = \emptyset. \tag{B.4}$$

We now claim that Eq. (B.4) implies that our structure contains an instance of modal funny business, which contradicts our premise of no-MFB.

Let $\tau_c =_{df} (c \rightarrowtail \Pi_c \langle h \rangle)$, and consider the following set of transitions, which we claim is combinatorially consistent (Def. 5.5):

$$T =_{df} \{ e \rightarrowtail \Pi_e \langle O \rangle \mid O \in \hat{O}, e \in cll(O) \} \cup \{ \tau_c \}.$$

The subset to the left is combinatorially consistent since it is consistent by Eq. (B.1). So to check combinatorial consistency, we only need to consider pairs of τ_c and some $\tau_e = (e \rightarrowtail \Pi_e \langle O \rangle)$, for some $O \in \hat{O}$ and some $e \in cll(O)$. Note that for any $h' \in H_{\langle \hat{O} \rangle}$, $\{e, c\} \subseteq h'$. It follows that if $e < c$, then $\Pi_c \langle h \rangle \subseteq \Pi_e \langle O \rangle$, and if e, c are incomparable, then they are *SLR*. Finally, it cannot happen that $c \leqslant e$, since this implies $c < \hat{O}$, contrary to our assumption that $c \not< \hat{O}$. Thus, T is combinatorially consistent, but inconsistent (by Eq. (B.4)). This shows that there is CFB in the structure, contrary to the premise of no-MFB. We thus have established that any element of $cll(\hat{O})$ is in the past of $\hat{O}$.

(2) Let $e \in cll(\hat{O})$ and $e < \hat{O}$, i.e., there is some $O \in \hat{O}$ for which $e < O$. Then by Fact 4.7(2), $\Pi_e\langle O \rangle$ is uniquely defined, and we can set $\Pi_e\langle \hat{O} \rangle =_{\mathrm{df}} \Pi_e\langle O \rangle$, as $H_{\langle \hat{O} \rangle} \subseteq H_{\langle O \rangle} \subseteq \Pi_e\langle O \rangle$. □

B.6 Answers to selected exercises from Chapter 6

Exercise 6.1. Prove clause (2) of Fact 6.1.

That is, given a transition $I \rightarrowtail \hat{O}$ to a scattered outcome $\hat{O}$, we have to prove $cll(I \rightarrowtail \hat{O}) = \bigcup_{O \in \hat{O}} cll(I \rightarrowtail O)$.

Proof. "⊇": Let $O \in \hat{O}$, and let $e \in cll(I \rightarrowtail O)$. Then for $h \in H_{[I]}$, $h \perp_e H_{\langle O \rangle}$. Since $H_{\langle \hat{O} \rangle} \subseteq H_{\langle O \rangle}$, we have $h \perp_e H_{\langle \hat{O} \rangle}$, and hence $e \in cll(I \rightarrowtail \hat{O})$.

"⊆": As there is no MFB in $\mathscr{W}$, for every $e \in cll(I \rightarrowtail \hat{O})$ we have $e < \hat{O}$ (see Exercise 5.3). Thus, $e \in cll(I \rightarrowtail \hat{O})$ implies that there is $O \in \hat{O}$ for which $e < O$. Let $h' \in H_{\langle \hat{O} \rangle} \subseteq H_{\langle O \rangle}$, and let $h'' \in H_{\langle O \rangle}$. We have $h' \equiv_e h''$. As $h \perp_e h'$ for some $h \in H_{[I]}$, it follows that $h \perp_e H_{\langle O \rangle}$, and hence $e \in cll(I \rightarrowtail O)$. □

Exercise 6.2. Prove clauses (4) and (5) of Fact 6.4.

That is, making no assumptions about the presence or absence of MFB and given $e \in cll(I \rightarrowtail \mathscr{O}^*)$, (4) for $\mathscr{O}^* = \hat{O}$ a scattered outcome, we have to show that there is some initial segment O' of some $O \in \hat{O}$ such that for every $e' \in O'$, we have $e \leqslant e'$ or $e\,SLR\,e'$, and (5) for $\mathscr{O}^* = \breve{O}$ a disjunctive outcome, we have to show that there is some $\hat{O} \in \breve{O}$ and some initial segment O' of some $O \in \hat{O}$ such that for every $e' \in O'$, we have $e \leqslant e'$ or $e\,SLR\,e'$,

Proof. (4) Let $e \in cll(I \rightarrowtail \hat{O})$. If $e \in cll(I \rightarrowtail O)$ for some $O \in \hat{O}$, then by clause (3) of Fact 6.4, for every $e' \in O$ either $e < e'$ or $e\,SLR\,e'$, so we are done. The other case is that $e \in cll(I \rightarrowtail \hat{O})$ but $e \notin cll(I \rightarrowtail O)$ for all $O \in \hat{O}$. Note that $cll(I \rightarrowtail \hat{O})$ is consistent, and pick some $O \in \hat{O}$ such that for some $h \in H_{[I]}$: $h \cap O = \emptyset$; thus by PCP$_{92}$ there is $c < O$ such that $h \perp_c H_{\langle O \rangle}$, and hence $c \in cll(I \rightarrowtail O) \subseteq cll(I \rightarrowtail \hat{O})$. Since elements of $cll(I \rightarrowtail \hat{O})$ are SLR, we have that (†) $e\,SLR\,c$.

As $e \in cll(I \rightarrowtail \hat{O})$, there is some h in $H_{[I]}$ for which $h \perp_e H_{\langle \hat{O} \rangle}$. Pick some $h_O \in H_{\langle \hat{O} \rangle}$. Thus, in particular, $h_O \in H_{\langle O \rangle}$. We have $e \in h_O$ by $h \perp_e H_{\langle \hat{O} \rangle}$, and as $h_O \in H_{\langle O \rangle}$, there is an initial segment $O' \subseteq O$ such that $O \subseteq h_O$. As $c < O$, we have $c < O$, and $c \in h_O$ holds by downward closure of histories. The claim then follows: for $e \in O$, we have $e, e \in h_O$, so e and e' are either comparable or SLR. It cannot be that $e \leqslant e$, as from $c < e'$ we could conclude $c < e$, contradicting (†), so either $e < e$ or $e\,SLR\,e'$.

(5) follows immediately from the above by noting that if $e \in cll(I \rightarrowtail \breve{O})$, then $e \in cllr(I \rightarrowtail \hat{O}) \subseteq cll(I \rightarrowtail \hat{O})$ for some $\hat{O} \in \breve{O}$. □

Exercise 6.6. Prove Theorem 6.3.

Theorem 6.3. (nns for transitions to outcome chains or scattered outcomes in BST$_{92}$ with MFB) *Let $\mathscr{O}^*$ be an outcome chain or a scattered outcome. Then the* causae causantes *of $I \rightarrowtail \mathscr{O}^*$ satisfy the following inus-related conditions:*

1. *joint sufficiency – nns:* $\bigcap_{e\in cll(I\rightarrowtail\mathscr{O}^*)} H_{e\rightarrowtail\check{\mathbf{H}}_e\langle H_{\langle\mathscr{O}^*\rangle}\rangle} \subseteq H_{I\rightarrowtail\mathscr{O}^*}$;

2. *joint necessity – nns:* $H_{\langle\mathscr{O}^*\rangle} = H_{[I]}\cap H_{I\rightarrowtail\mathscr{O}^*} \subseteq \bigcap_{e\in cll(I\rightarrowtail\mathscr{O}^*)} H_{e\rightarrowtail\check{\mathbf{H}}_e\langle H_{\langle\mathscr{O}^*\rangle}\rangle}$;

3. *non-redundancy – nns:* for every $(e_0 \rightarrowtail \check{\mathbf{H}}) \in CC(I \rightarrowtail \mathscr{O}^*)$ and every $\check{\mathbf{H}}'$ such that $\check{\mathbf{H}}\cap\check{\mathbf{H}}' = \emptyset$, where $\check{\mathbf{H}}, \check{\mathbf{H}}' \subseteq \Pi_{e_0}$,

$$\text{either} \quad \bigcup\check{\mathbf{H}}'\cap \bigcap_{e\in cll(I\rightarrowtail\mathscr{O}^*)\setminus\{e_0\}} (H_e\cap H_{e\rightarrowtail\check{\mathbf{H}}_e\langle H_{\langle\mathscr{O}^*\rangle}\rangle}) = \emptyset, \quad \text{or} \qquad (6.5)$$

$$\bigcup\check{\mathbf{H}}'\cap \bigcap_{e\in cll(I\rightarrowtail\mathscr{O}^*)\setminus\{e_0\}} (H_e\cap H_{e\rightarrowtail\check{\mathbf{H}}_e\langle H_{\langle\mathscr{O}^*\rangle}\rangle}) \not\subseteq H_{[I]}\cap H_{I\rightarrowtail\mathscr{O}^*}. \qquad (6.6)$$

Proof. We give a proof for transitions to scattered outcomes, which can easily be simplified to cover transitions to outcome chains as well.

(1) If $I \rightarrowtail \mathscr{O}^*$ is deterministic, then the theorem holds as then $H_{I\rightarrowtail\mathscr{O}^*} = \text{Hist}$. We thus assume that the transition is indeterministic (i.e., $H_{[I]} \setminus H_{\langle\mathscr{O}^*\rangle} \neq \emptyset$) and argue for the contraposition, so we take $h \in H_{[I]} \setminus H_{\langle\mathscr{O}^*\rangle}$, and hence we get that for some $c: h \perp_c H_{\langle\mathscr{O}^*\rangle}$. Accordingly, $c \in cll(I \rightarrowtail \mathscr{O}^*)$. Further, for every $H \in \check{\mathbf{H}}_c\langle H_{\langle\mathscr{O}^*\rangle}\rangle: h \perp_c H$. Thus, for every H of that sort, $h \notin H$, and hence $h \notin \bigcup\check{\mathbf{H}}_c\langle H_{\langle\mathscr{O}^*\rangle}\rangle$. Since $h \in H_c$, it follows that $h \notin H_{c\rightarrowtail\check{\mathbf{H}}_c\langle H_{\langle\mathscr{O}^*\rangle}\rangle}$.

(2) Pick an arbitrary $h \in H_{\langle\mathscr{O}^*\rangle}$ and an arbitrary $e \in cll(I \rightarrowtail \mathscr{O}^*)$. Clearly, as $h \in H_e$, h belongs to some $H \in \Pi_e$ such that $H \cap H_{\langle\mathscr{O}^*\rangle} \neq \emptyset$, Thus $h \in \bigcup\check{\mathbf{H}}_e\langle H_{\langle\mathscr{O}^*\rangle}\rangle$, and hence $h \in H_{e\rightarrowtail\check{\mathbf{H}}_e\langle H_{\langle\mathscr{O}^*\rangle}\rangle}$.

(3) Since $(e_0 \rightarrowtail \check{\mathbf{H}}) \in CC(I \rightarrowtail \mathscr{O}^*)$, it must be that $\check{\mathbf{H}} = \check{\mathbf{H}}_{e_0}\langle H_{\langle\mathscr{O}^*\rangle}\rangle$. Pick then an arbitrary $\check{\mathbf{H}}' \subseteq \Pi_{e_0}$ such that $\check{\mathbf{H}}'\cap\check{\mathbf{H}}_{e_0}\langle H_{\langle\mathscr{O}^*\rangle}\rangle = \emptyset$. Hence $(\bigcup\check{\mathbf{H}}')\cap(\bigcup\check{\mathbf{H}}_{e_0}\langle H_{\langle\mathscr{O}^*\rangle}\rangle) = \emptyset$. By Fact 6.3 we have $H_{\langle\mathscr{O}^*\rangle} \subseteq \bigcup\check{\mathbf{H}}_e\langle H_{\langle\mathscr{O}^*\rangle}\rangle$, which implies (*) $\bigcup\check{\mathbf{H}}' \cap H_{\langle\mathscr{O}^*\rangle} = \emptyset$. Let us abbreviate $H^- = \bigcap_{e\in cll(I\rightarrowtail\mathscr{O}^*)\setminus\{e_0\}} H_e \cap H_{e\rightarrowtail\check{\mathbf{H}}_e\langle H_{\langle\mathscr{O}^*\rangle}\rangle}$ and consider two cases: (i) $\bigcup\check{\mathbf{H}}' \cap H^- = \emptyset$ and (ii) $\bigcup\check{\mathbf{H}}' \cap H^- \neq \emptyset$. If (i), we are done. If (ii), by (*) we have $H^-\cap(\bigcup\check{\mathbf{H}}')\cap H_{\langle\mathscr{O}^*\rangle} = \emptyset$. Since $H_{\langle\mathscr{O}^*\rangle} \neq \emptyset$, it follows that $H^-\cap(\bigcup\check{\mathbf{H}}') \not\subseteq H_{\langle\mathscr{O}^*\rangle}$, which is Eq. 6.6. $\qquad\square$

Exercise 6.7. Prove Theorem 6.4.

Theorem 6.4. (nus for transitions to disjunctive outcomes in BST$_{92}$ with MFB) *Let* $\check{\mathbf{O}} = \{\hat{O}_\gamma \mid \gamma \in \Gamma\}$ *be a disjunctive outcome consisting of more than one scattered outcome. The set of causae causantes of* $I \rightarrowtail \check{\mathbf{O}}$, *i.e.,* $\{CCr(I \rightarrowtail \hat{O}_\gamma)\}_{\gamma\in\Gamma}$ *as well as each* $CC(I \rightarrowtail \hat{O}_\gamma)$, *satisfy the following inus-related conditions:*

1. *each* $CCr(I \rightarrowtail \hat{O}_\gamma)$ *is sufficient – nus: for every* $\gamma \in \Gamma$:
$\bigcap_{e\in cllr(I\rightarrowtail\hat{O}_\gamma)} H_{e\rightarrowtail\check{\mathbf{H}}_e\langle H_{\langle\hat{O}_\gamma\rangle}\rangle} \subseteq H_{I\rightarrowtail\check{\mathbf{O}}}$;

2. *each* $CC(I \rightarrowtail \hat{O}_\gamma)$ *is unnecessary – nus: for every* $\gamma \in \Gamma$:
$H_{\langle\check{\mathbf{O}}\rangle} = H_{[I]}\cap H_{I\rightarrowtail\check{\mathbf{O}}} \not\subseteq \bigcap_{e\in cll(I\rightarrowtail\hat{O}_\gamma)} H_{e\rightarrowtail\check{\mathbf{H}}_e\langle H_{\langle\hat{O}_\gamma\rangle}\rangle}$.

3. *for each* $\gamma \in \Gamma$, *each* $\tau_0 = (e_0 \rightarrowtail \check{\mathbf{H}}) \in CCr(I \rightarrowtail \hat{O}_\gamma)$ *is non-redundant – nus. That is, for every* $\check{\mathbf{H}}' \subseteq \Pi_{e_0}$ *such that* $\check{\mathbf{H}}\cap\check{\mathbf{H}}' = \emptyset$:

$$either \quad \bigcup \check{\mathbf{H}}' \cap \bigcap_{e \in cllr(I \rightarrowtail \hat{O}_\gamma) \setminus \{e_0\}} (H_e \cap H_{e \rightarrowtail \check{\mathbf{H}}_e \langle H_{\langle \hat{O}_\gamma \rangle}\rangle) = \emptyset, \tag{6.7}$$

$$or \quad \bigcup \check{\mathbf{H}}' \cap \bigcap_{e \in cllr(I \rightarrowtail \hat{O}_\gamma) \setminus \{e_0\}} (H_e \cap H_{e \rightarrowtail \check{\mathbf{H}}_e \langle H_{\langle \hat{O}_\gamma \rangle}\rangle) \not\subseteq H_{[I]} \cap H_{I \rightarrowtail \hat{O}_\gamma}. \tag{6.8}$$

Proof. (1) This is almost exactly the same proof as for Theorem 6.2(1). Just like in this proof, from $h \notin H_{\hat{O}_\gamma}$ we arrive at $h \perp_e \Pi_e \langle \hat{O}_\gamma \rangle$ and get that $e \notin DET_{I \rightarrowtail \check{O}}$ since $h \in (H_e \cap H_{[I]}) \setminus H_{\langle \check{O} \rangle}$, which entails $e \in cllr(I \rightarrowtail \hat{O}_\gamma)$. As we allow for MFB, $h \notin H_{\langle \hat{O}_\gamma \rangle}$ implies $h \notin \{H \in \Pi_e \mid H \cap H_{\langle \hat{O}_\gamma \rangle}\} = \check{\mathbf{H}}_e \langle H_{\langle \hat{O}_\gamma \rangle}\rangle$. Accordingly, $h \notin H_{e \rightarrowtail \check{\mathbf{H}}_e \langle H_{\langle \hat{O}_\gamma \rangle}\rangle}$. (2) Since $\check{O}$ has at least two scattered outcomes, there is $\hat{O}_\gamma \in \check{O}$ and $h \in H_{[I]}$ such that $h \notin H_{\langle \hat{O}_\gamma \rangle}$. By Theorem 6.3(1) $h \notin \bigcap_{e \in cllr(I \rightarrowtail \hat{O}_\gamma)} H_{e \rightarrowtail \check{\mathbf{H}}_e \langle H_{\langle \hat{O}_\gamma \rangle}\rangle}$.

(3) Pick an arbitrary $\gamma \in \Gamma$ and an arbitrary $(e_0 \rightarrowtail \check{\mathbf{H}}) \in CCr(I \rightarrowtail \hat{O}_\gamma)$, so $cllr(I \rightarrowtail \hat{O}_\gamma) \neq \emptyset$. It must be that $\check{\mathbf{H}} = \check{\mathbf{H}}_{e_0} \langle H_{\langle \hat{O}_\gamma \rangle}\rangle$. Pick then a $\check{\mathbf{H}}' \subseteq \Pi_{e_0}$ such that $\check{\mathbf{H}}' \cap \check{\mathbf{H}}_{e_0} \langle H_{\langle \hat{O}_\gamma \rangle}\rangle = \emptyset$. Hence $(\bigcup \check{\mathbf{H}}') \cap (\bigcup \check{\mathbf{H}}_{e_0} \langle H_{\hat{O}_\gamma} \rangle) = \emptyset$. By Fact 6.3 we have $H_{\langle \hat{O}_\gamma \rangle} \subseteq \bigcup \check{\mathbf{H}}_e \langle H_{\langle \hat{O}_\gamma \rangle}\rangle$, which implies (*) $\bigcup \check{\mathbf{H}}' \cap H_{\langle \hat{O}_\gamma \rangle} = \emptyset$. Let us abbreviate $H^- =_{df} \bigcap_{e \in cllr(I \rightarrowtail \hat{O}^*) \setminus \{e_0\}} H_e \cap H_{e \rightarrowtail \check{\mathbf{H}}_e \langle H_{\langle \hat{O}_\gamma \rangle}\rangle}$ and consider two cases: (i) $\bigcup \check{\mathbf{H}}' \cap H^- = \emptyset$ and (ii) $\bigcup \check{\mathbf{H}}' \cap H^- \neq \emptyset$. If (i), we are done. If (ii), by (*) we have $H^- \cap (\bigcup \check{\mathbf{H}}') \cap H_{\langle \hat{O}_\gamma \rangle} = \emptyset$. Since $H_{\langle \hat{O}_\gamma \rangle} \neq \emptyset$, it follows that $H^- \cap (\bigcup \check{\mathbf{H}}') \not\subseteq H_{\langle \hat{O}_\gamma \rangle}$, which is Eq. 6.8. $\square$

Exercise 6.9. Prove Theorem 6.5

Theorem 6.5. *(nns for transitions to outcome chains or scattered outcomes in BST$_{NF}$ with no MFB) Let $\mathcal{O}^*$ be an outcome chain or a scattered outcome. The* causae causantes *of $I \rightarrowtail \mathcal{O}^*$ satisfy the following inus-related conditions:*

1. *joint sufficiency – n̲n̲s̲:* $\bigcap_{\ddot{e} \in cll(I \rightarrowtail \mathcal{O}^*)} H_{\ddot{e} \rightarrowtail \Pi_{\ddot{e}} \langle \mathcal{O}^* \rangle} \subseteq H_{I \rightarrowtail \mathcal{O}^*}$;

2. *joint necessity – n̲n̲s̲:* $H_{\langle \mathcal{O}^* \rangle} = H_{[I]} \cap H_{I \rightarrowtail \mathcal{O}^*} \subseteq \bigcap_{\ddot{e} \in cll(I \rightarrowtail \mathcal{O}^*)} H_{\ddot{e} \rightarrowtail \Pi_{\ddot{e}} \langle \mathcal{O}^* \rangle}$;

3. *non-redundancy – n̲n̲s̲: for every $(\ddot{e}_0 \rightarrowtail H) \in CC(I \rightarrowtail \mathcal{O}^*)$ and every $H' \in \Pi_{\ddot{e}_0}$ such that $H' \cap H = \emptyset$:*

$$either \; H' \cap \bigcap_{\ddot{e} \in cll(I \rightarrowtail \mathcal{O}^*) \setminus \{\ddot{e}_0\}} \Pi_{\ddot{e}} \langle \mathcal{O}^* \rangle = \emptyset, \tag{6.9}$$

$$or \; H' \cap \bigcap_{\ddot{e} \in cll(I \rightarrowtail \mathcal{O}^*) \setminus \{\ddot{e}_0\}} \Pi_{\ddot{e}} \langle \mathcal{O}^* \rangle \not\subseteq H_{[I]} \cap H_{I \rightarrowtail \mathcal{O}^*}. \tag{6.10}$$

Proof. (1) If $H_{[I]} = H_{\langle \hat{O} \rangle}$, then $H_{I \rightarrowtail \hat{O}} = $ Hist, and we are done. Let us thus suppose that there is $h \in$ Hist such that $h \in H_{[I]}$ but $h \notin H_{\langle \hat{O} \rangle}$. Thus, there is $O \in \hat{O}$ s.t $h \cap O = \emptyset$. There is also $h' \in H_{\langle \hat{O} \rangle}$, which implies $h' \in H_{[I]}$ and $h' \in H_{\langle O \rangle}$. Accordingly, there is $O' \subseteq O$ such that $O' \subseteq h' \setminus h$. By PCP$_{NF}$ there is $\ddot{c}$, with $c \in \ddot{c}$ and $c \leqslant O'$ such that $h \perp_{\ddot{c}} H_{\langle O' \rangle}$. Since $H_{\langle \hat{O} \rangle} \subseteq H_{\langle O \rangle} \subseteq H_{\langle O' \rangle}$, we get $h \perp_{\ddot{c}} H_{\langle \hat{O} \rangle}$; thus $\ddot{c} \in cll(I \rightarrowtail \hat{O})$. On the other hand, $h \in H_{\ddot{c}}$, and $h \perp_{\ddot{c}} H_{\langle \hat{O} \rangle}$ implies $h \notin \Pi_{\ddot{c}} \langle \hat{O} \rangle$. For if $h \in \Pi_{\ddot{c}} \langle \hat{O} \rangle$, then (since $c \leqslant \hat{O}$) $h \equiv_{\ddot{c}} h_1$ for every $h_1 \in H_{\langle \hat{O} \rangle}$, contradicting $h \perp_{\ddot{c}} H_{\langle \hat{O} \rangle}$. Hence $h \notin H_{\ddot{c} \rightarrowtail \Pi_{\ddot{c}} \langle \hat{O} \rangle}$. By simplifying this proof appropriately, one obtains the argument for transitions to outcome chains.

(2) Note that $H_{[I]} \cap H_{I \rightarrow \mathcal{O}^*} = H_{\langle \mathcal{O}^* \rangle}$ and for every $c \leqslant \mathcal{O}^*$: $H_{\langle \mathcal{O}^* \rangle} \subseteq H_c \subseteq \Pi_{\ddot{e}} \langle \mathcal{O}^* \rangle$. Thus, $H_{[I]} \cap H_{I \rightarrow \mathcal{O}^*} \subseteq H_{\ddot{e} \rightarrow \Pi_{\ddot{e}} \langle \mathcal{O}^* \rangle}$.

(3) Pick an arbitrary H' such that $H' \cap H = \emptyset$, where $H, H' \in \Pi_{\ddot{e}_0}$ and $\ddot{e}_0 \rightarrowtail H \in CC(I \rightarrowtail \hat{O})$. Since $H_{\langle \hat{O} \rangle} \subseteq H$, (*) $H' \cap H_{\langle \hat{O} \rangle} = \emptyset$. Let us next abbreviate: $H^- = \bigcap_{\ddot{e} \in cll(I \rightarrowtail \hat{O}) \setminus \{\ddot{e}_0\}} \Pi_{\ddot{e}} \langle \hat{O} \rangle$ and consider then two cases: (i) $H' \cap H^- = \emptyset$ and (ii) $H' \cap H^- \neq \emptyset$. If (i), since it is identical to Eq. 6.9, we are done. In (ii), by (*) we have $H' \cap H^- \cap H_{\langle \hat{O} \rangle} = \emptyset$. Since $H_{\langle \hat{O} \rangle} \neq \emptyset$, it follows that $H' \cap H^- \not\subseteq H_{\langle \hat{O} \rangle}$, which is Eq. 6.10. $\quad\square$

B.7   Answers to selected exercises from Chapter 7

Exercise 7.1. Prove Lemma 7.2.

Lemma 7.2. *Let the conditions of Def. 7.5 hold for a transition $I \rightarrowtail \check{O}$ to a disjunctive outcome $\check{O}$, and consider $CPS(I \rightarrowtail \check{O}) = \langle S, \mathscr{A}, p \rangle$. That triple is in fact a probability space satisfying Def. 7.1. That is, $CPS(I \rightarrowtail \check{O})$ is well defined and p is a normalized measure on $\mathscr{A}$. Furthermore, we have that*

$$p(CC(I \rightarrowtail \hat{O}_\gamma)) = \mu(\{T \in S \mid CC(I \rightarrowtail \hat{O}_\gamma) \subseteq T\}) = \sum_{T \in S, CC(I \rightarrowtail \hat{O}_\gamma) \subseteq T} \mu(T);$$

$$p(CC(I \rightarrowtail \check{O})) = \sum_{\gamma \in \Gamma} p(CC(I \rightarrowtail \hat{O}_\gamma)).$$

Proof. The observation underlying this proof is that $T \in S$ iff $T \in S_\gamma$ for some γ such that $\hat{O}_\gamma \in \check{O}$, and where S_γ is the set of causal alternatives to $CC(I \rightarrowtail \hat{O}_\gamma)$.

We first prove that for every $T \in S$, $\mu(T)$ is defined. If $T \in S$ then $T \in S_\gamma$ for some γ such that $\hat{O}_\gamma \in \check{O}$, so by Def 7.5 and Postulate 7.2, $\mu(T)$ is defined. Thus $p(T)$ is defined via Def. 7.4, which induces measure p on the whole $\mathscr{A}$.

We need to show that p is a normalized probability measure. It suffices to show that the probabilities assigned to the different elements of S sum to one, as then

$$p(\mathbf{1}_{\mathscr{A}}) = \sum_{T \in S} p(T) = 1.$$

Our proof uses the law of total probability in the form of Postulate 7.3. We thus have to show that the elements of S partition the set of histories in which the initial I occurs. First, we have to show that for two different $T_1, T_2 \in S$, $T_1 \neq T_2$, we have $H(T_1) \cap H(T_2) = \emptyset$. If $T_1, T_2 \in S_\gamma$, the claim follows immediately since T_1, T_2 are maximal consistent subsets. If there is no $\hat{O}_\gamma \in \check{O}$ such that $T_1, T_2 \in\in \hat{O}_\gamma$, the claim follows from the fact that elements of $\check{O}$ are inconsistent as $\check{O}$ is a disjunctive outcome.

Second, we show $H_{[I]} \subseteq \bigcup_{T \in S} H(T)$. In the proof of Lemma 7.1, we already showed that $H_{[I]} \subseteq \bigcup_{T \in S_\gamma} H(T)$. Since $S_\gamma \subseteq S$, the claim follows. With the premises of the law of total causal probability (Postulate 7.3) satisfied, we therefore have

$$\sum_{T \in S} p(T) = \sum_{T \in S} \mu(T) = 1,$$

showing that the measure p is indeed normalized. $\quad\square$

B.8 Answers to selected exercises from Chapter 8

Exercise 8.3. Prove Fact 8.11.

Proof. Since each $T \in S$ is maximal consistent, for any two $T_1, T_2 \in S$, T_1-histories and T_2-histories have to split. Pick $T_1 \in S$. If for any other $T_2 \in S$, T_1-histories split with T_2-histories at members of E only, then $\{T_1\} = \lambda \in \mathfrak{I}_c$, and hence $T_1 \in \bigcup \lambda$. Suppose there is $T_2 \in S$ such that T_1-histories and T_2-histories split at a member of C. If for any other $T_3 \in S$, T_3-histories split with T_1-histories at E, or T_3-histories split with T_2-histories at E, we put $\{T_1, T_2\} = \lambda$. If not, we consider a triple. Given the finiteness of S, we will find a maximal subset of S with respect to the defining condition for contextual instruction sets. $\qquad\square$

B.9 Answers to selected exercises from Chapter 9

Exercise 9.3. Furnish the detail of the chain construction in the proof of Fact 9.2(2).

Proof. We define a function 'up': for $a = \langle a^0, a^1, a^2, a^3 \rangle \in \mathbb{R}^4$, $b = \langle b^0, b^1, b^2, b^3 \rangle \in \mathbb{R}^4$, we let $up(a,b) =_{\mathrm{df}} \langle a^0 + (\sum_1^3 (a^i - b^i)^2)^{1/2}, a^1, a^2, a^3 \rangle \in \mathbb{R}^4$.

The chain E is constructed in the following way:

Step 1. $z_0 = up(y, x_0)$, $z_1 = up(z_0, x_1)$, and generally, $z_{k+1} = up(z_k, x_{k+1})$. Note that $x_k = T \circ \Theta(\sigma_k)$.

Step 2. Suppose ρ is a limit ordinal. Define $A_\rho := \{z_m \mid m < \rho\}$. A_ρ is the part of our chain we have managed to construct so far. We need to distinguish two cases:

Case 1: A_ρ is upper bounded with respect to $\leqslant_M$. Then it has to have 'vertical' upper bounds $t_0, t_1, \dots$ with spatial coordinates $t_n^i = z_0^i$ ($i = 1, 2, 3$). In this case, we use the function T to choose one of the upper bounds of A_ρ:

$$t_\rho := T(\{t \in \mathbb{R}^4 \mid \forall m < \rho \ z_m \leqslant_M t \wedge t^i = z_0^i (i = 1,2,3)\}). \tag{B.5}$$

Then we put $z_\rho =_{\mathrm{df}} up(t_\rho, x_\rho)$, arriving at the next element of our chain E.

Case 2: If A_ρ is not upper bounded with respect to $\leqslant_M$, then no matter which point in $\mathbb{R}^4$ we choose, it is possible to find a point from A_ρ above it (since A_ρ is time-like). Therefore, the set

$$B_\rho = \{t \in A_\rho \mid x_\rho \leqslant_M t\} \tag{B.6}$$

is not empty. We put $z_\rho =_{\mathrm{df}} T(B_\rho)$, arriving at the next element of our chain E. $\qquad\square$

B.10 Answers to selected exercises from Chapter 10

Exercise 10.2. Prove the strengthened version of identity $(*)$ from the proof of Fact 10.1, which restricts *cll* of e to those lying in the past of e (i.e., prove Fact 10.11):
For any $e \in W$,

$$H_e = \bigcap \{\Pi_{\ddot{c}} \langle H_e \rangle \mid \ddot{c} \in cll(e) \wedge \exists c \in \ddot{c} \, [c \leqslant e]\}. \tag{$*$}$$

Proof. For the "$\subseteq$" direction, since $c \leqslant e$, $H_e \subseteq H_c$, and H_c is identical to $\Pi_{\ddot{c}}\langle H_e \rangle$, hence $H_e \subseteq \Pi_{\ddot{c}}\langle H_e \rangle$.

For the "$\supseteq$" direction, let us assume for reductio that there is (†) $h \in \bigcap\{\Pi_{\ddot{c}}\langle H_e \rangle \mid \ddot{c} \in cll(e) \wedge \exists c \in \ddot{c}\ c \leqslant e\}$, but $h \notin H_e$. Take some $h' \in H_e$. As $e \in h' \setminus h$, by PCP$_{NF}$ there is a choice set $\ddot{c}$ at which $h \perp_{\ddot{c}} h'$, and $c \in \ddot{c}$ such that $c \leqslant e$. From the last relations $h \perp_{\ddot{c}} H_e$ follows, so that $\ddot{c} \in cll(e)$. Since $h \perp_{\ddot{c}} H_e$, we get $h \notin \Pi_{\ddot{c}}\langle H_e \rangle$, and hence $h \notin \bigcap_{\ddot{c} \in cll(e)} \Pi_{\ddot{c}}\langle H_e \rangle$, which contradicts (†). $\qquad\square$

Bibliography

Abellán, C., Acín, A., Alarcón, A., and The BIG Bell Test Collaboration (2018). Challenging local realism with human choices. *Nature*, 557:212–216. doi:10.1038/s41586-018-0085-3.

Adlam, E. C. (2018). Quantum mechanics and global determinism. *Quanta*, 7(1):40–53.

Anscombe, G. E. M. (1971). *Causality and Determination*. Cambridge University Press, Cambridge.

Aspect, A., Dalibard, J., and Roger, G. (1982a). Experimental test of Bell's inequalities using time-varying analyzers. *Physical Review Letters*, 49:1804–1807.

Aspect, A., Grangier, P., and Roger, G. (1982b). Experimental realization of Einstein-Podolsky-Rosen-Bohm *Gedankenexperiment*: A new violation of Bell's inequalities. *Physical Review Letters*, 49(2):91–94.

Balashov, Y. (2010). *Persistence and Spacetime*. Oxford University Press, Oxford.

Barnes, E., and Cameron, R. P. (2011). Back to the open future. *Philosophical Perspectives*, 25(1):1–26.

Baron, S. (2012). Presentism and causation revisited. *Philosophical Papers*, 41(1):1–21.

Beer, M. (1994). Temporal indexicals and the passage of time. In Oaklander, L. N., and Smith, Q., editors, *The New Theory of Time*, pages 87–93. Yale University Press, New Haven, CT.

Bell, J. S. (1964). On the Einstein Podolsky Rosen paradox. *Physics*, 1:195–200.

Bell, J. S. (1987a). *Speakable and Unspeakable in Quantum Mechanics*. Cambridge University Press, Cambridge.

Bell, J. S. (1987b). The theory of local beables. In Bell (1987a), pages 52–62.

Bell, J. S. (1997). Indeterminism and nonlocality. In Driessen, A., and Suarez, A., editors, *Mathematical Undecidability, Quantum Nonlocality and the Question of the Existence of God*, pages 83–100. Springer, Berlin.

Belnap, N. (1992). Branching space-time. *Synthese*, 92(3):385–434. See also the postprint, Belnap (2003).

Belnap, N. (1999). Concrete transitions. In Meggle, G., editor, *Actions, Norms, Values: Discussions with Georg Henrik von Wright*, pages 227–236. de Gruyter, Berlin.

Belnap, N. (2002). EPR-like "funny business" in the theory of branching space-times. In Placek, T., and Butterfield, J., editors, *Non-locality and Modality*, pages 293–315. Dordrecht: Kluwer.

Belnap, N. (2003a). Agents in branching space-times. *Journal of Sun-Yatsen University*, 43:147–166.

Belnap, N. (2003b). Branching space-time. *Pittsburgh PhilSci-Archive*. http://philsci-archive.pitt.edu/1003/.

Belnap, N. (2003c). No-common-cause EPR-like funny business in branching space-times. *Philosophical Studies*, 114:199–221.

Belnap, N. (2005a). Agents and agency in branching space-times. In Vanderveken, D., editor, *Logic, Thought and Action*, pages 291–313. Springer, Berlin.

Belnap, N. (2005b). A theory of causation: *Causae causantes* (originating causes) as inus conditions in branching space-times. *British Journal for the Philosophy of Science*, 56:221–253.

Belnap, N. (2007). Propensities and probabilities. *Studies in History and Philosophy of Modern Physics*, 38:593–625.

Belnap, N. (2011). Prolegomenon to norms in branching space-times. *Journal of Applied Logic*, 9(2):83–94.

Belnap, N. (2012). Newtonian determinism to branching space-times indeterminism in two moves. *Synthese*, 188:5–21. doi:10.1007/s11229-012-0063-5.

Belnap, N., Müller, T., and Placek, T. (2021). New foundations for branching space-times. *Studia Logica*, 109:239–284.

Belnap, N., Perloff, M., and Xu, M. (2001). *Facing the Future: Agents and Choices in Our Indeterminist World*. Oxford University Press, Oxford.

Belnap, N., and Szabó, L. E. (1996). Branching space-time analysis of the GHZ theorem. *Foundations of Physics*, 26(8):982–1002.

Belot, G. (1995). New work for counterpart theorists: determinism. *British Journal for the Philosophy of Science*, 46:185–195.

Belot, G. (2011). Background-independence. *General Relativity and Gravitation*, 43(10):2865–2884.

Brierley, S., Kosowski, A., Markiewicz, M., Paterek, T., and Przysiężna, A. (2015). Non-classicality of temporal correlations. *Physical Review Letters*, 115:120404.

Briggs, R. A., and Forbes, G. A. (2019). The future, and what might have been. *Philosophical Studies*, 176(2):505–532.

Butterfield, J. (1984). Seeing the present. *Mind*, 93:161–176.

Butterfield, J. (1989). The hole truth. *British Journal for the Philosophy of Science*, 40(1): 1–28.

Butterfield, J. (1992). Bell's theorem: What it takes. *British Journal for the Philosophy of Science*, 43(1):41–83.

Butterfield, J. (2005). Determinism and indeterminism. In *Routledge Encyclopedia of Philosophy*, volume 3. Routledge, London.

Butterfield, J. (2018). Peaceful coexistence: Examining Kent's relativistic solution to the quantum measurement problem. In Ozawa, M., Butterfield, J., Halvorson, H., Rédei, M., Kitajima, Y., and Buscemi, F., editors, *Reality and Measurement in Algebraic Quantum Theory*, pages 277–314. Springer, Singapore.

Callender, C. (2017). *What Makes Time Special?* Oxford University Press, Oxford.

Choquet-Bruhat, Y., and Geroch, R. (1969). Global aspects of the Cauchy problem in general relativity. *Communications in Mathematical Physics*, 14:329–335.

Chruściel, P. T. (1991). On uniqueness in the large of solutions of Einstein's equations ("Strong Cosmic Censorship"). In *Proceedings of the Centre for Mathematics and its Applications*, volume 27. Australian National University, Canberra.

Chruściel, P. T. (2011). Elements of causality theory. Technical report, University of Vienna. arXiv:1110.6706v1 [gr-qc] 31 Oct 2011.

Chruściel, P. T., and Isenberg, J. (1993). Nonisometric vacuum extensions of vacuum maximal globally hyperbolic spacetimes. *Physical Review D*, 48(4):1616–1628.

Clarke, C. (1976). Space-time singularities. *Communications in Mathematical Physics*, 49:17–23.

Clauser, J. F., and Horne, M. A. (1974). Experimental consequences of objective local theories. *Physical Review D*, 10:526–535.

Clifton, R., and Hogarth, M. (1995). The definability of objective becoming in Minkowski spacetime. *Synthese*, 103(3):355–387.

Colbeck, R., and Renner, R. (2011). No extension of quantum theory can have improved predictive power. *Nature Communications*, 2:1–5.

Colbeck, R., and Renner, R. (2012). Free randomness can be amplified. *Nature Physics*, 8(6):450–453.

Costa, J., Girão, P. M., Natário, J., and Silva, J. (2015). On the global uniqueness for the Einstein–Maxwell-scalar field system with a cosmological constant. Part 3: Mass inflation and extendability of solutions. arXiv:1406.7261v1 [gr-qc].

Diaconis, P., Holmes, S., and Montgomery, R. (2007). Dynamical bias in the coin toss. *SIAM Review*, 49(2):211–235. doi:10.1137/S0036144504446436.

Earman, J. (2006). Aspects of determinism in modern physics. In Butterfield, J., and Earman, J., editors, *Handbook of the Philosophy of Physics*, pages 1369–1434. Elsevier, Amsterdam.

Earman, J. (2008). Pruning some branches from branching space-times. In Dieks, D., editor, *The Ontology of Spacetime II*, pages 187–206. Elsevier, Amsterdam.

Eddington, A. S. (1949). *The Nature of Physical World*. Cambridge University Press, Cambridge.

Eddington, A. S. (1953). *Space, Time, Gravitation*. Cambridge University Press, Cambridge.

Ehlers, J., and Geroch, R. (2004). Equation of motion of small bodies in relativity. *Annals of Physics*, 309(1):232–236.

Einstein, A., Born, M., and Born, H. (1971). *The Born-Einstein Letters: Correspondence Between Albert Einstein and Max and Hedwig Born from 1916–1955, with Commentaries by Max Born*. Macmillan, London.

Einstein, A., and Grossmann, M. (1913). Entwurf einer verallgemeinerten Relativitätstheorie und eine Theorie der Gravitation. *Zeitschrift für Mathematik und Physik*, 62: 225–261.

Einstein, A., Podolsky, B., and Rosen, N. (1935). Can quantum-mechanical description of physical reality be considered complete? *Physical Review*, 47(10):777–780.

Ellis, G. F. R. (2006). Physics in the real universe: time and space-time. *General Relativity and Gravitation*, 38:1797–1824.

Esfeld, M. (2015). Bell's theorem and the issue of determinism and indeterminism. *Foundations of Physics*, 45(5):471–482.

Fine, A. (1982). Hidden variables, joint probability, and the Bell inequalities. *Physical Review Letters*, 48:291–295.

Fine, K. (2005). *Modality and Tense*, chapter 7, "The Varieties of Necessity," pages 235–260. Oxford University Press, Oxford.

Gale, R. M. (1963). Some metaphysical statements about time. *Journal of Philosophy*, 60(9):225–237.

Gigerenzer, G., Swijtink, Z., Porter, T., Daston, L., Beatty, J., and Krüger, L. (1989). *The Empire of Chance. How Probability Changed Science and Everyday Life*. Cambridge University Press, Cambridge.

Giustina, M., Mech, A., Ramelow, S., Wittmann, B., Kofler, J., Beyer, J., Lita, A., Calkins, B., Gerrits, T., Nam, S. W., Ursin, R., and Zeilinger, A. (2013). Bell violation using entangled photons without the fair-sampling assumption. *Nature*, 497:227.

Gödel, K. (1949). Some observations about the relationship between relativity theory and Kantian philosophy. In Feferman, S., Dawson, J. W., Goldfarb, W., Parsons, C., and

Solovay, R. M., editors, *Collected Works*, volume 3, pages 230–260. Oxford University Press, Oxford. Edition published 1995.

Goldstein, S., Norsen, T., Tausk, D. V., and Zanghi, N. (2011). Bell's theorem. *Scholarpedia*, 6(10):8378. revision #91049.

Greenberger, D. M., Horne, M. A., and Zeilinger, A. (1989). Going beyond Bell's theorem. In Kafatos, M., editor, *Bell's Theorem, Quantum Theory, and Conceptions of the Universe*, pages 69–72. Kluwer, Dordrecht. Reprint available as arXiv:0712.0921.

Hacking, I. (2006). *The Emergence of Probability*. Cambridge University Press, Cambridge, 2nd edition.

Hájíček, P. (1971a). Bifurcate space-time. *Journal of Mathematical Physics*, 12(1):157–160.

Hájíček, P. (1971b). Causality in non-Hausdorff space-times. *Communications in Mathematical Physics*, 21:75–84.

Hawking, S. W., and Ellis, G. F. R. (1973). *The Large Scale Structure of Space-Time*. Cambridge University Press, Cambridge.

Hensen, B., Bernien, H., Dréau, A. E., Reiserer, A., Kalb, N., Blok, M. S., Ruitenberg, J., Vermeulen, R. F. L., Schouten, R. N., Abellán, C., Amaya, W., Pruneri, V., Mitchell, M. W., Markham, M., Twitchen, D. J., Elkouss, D., Wehner, S., Taminiau, T. H., and Hanson, R. (2015). Loophole-free Bell inequality violation using electron spins separated by 1.3 kilometres. *Nature*, 526:682.

Hestevold, H. S. (2008). Presentism: Through thick and thin. *Pacific Philosophical Quarterly*, 89:325–347.

Hirsch, M. W. (1976). *Differential Topology*. Springer Verlag, Berlin.

Hofer-Szabó, G. (2008). Separate- versus common-common-cause-type derivations of the Bell inequalities. *Synthese*, 163(2):199–215.

Hofer-Szabó, G., Rédei, M., and Szabó, L. E. (1999). On Reichenbach's common cause principle and Reichenbach's notion of common cause. *British Journal for the Philosophy of Science*, 50:377–399.

Hofer-Szabó, G., Rédei, M., and Szabó, L. E. (2013). *The Principle of the Common Cause*. Cambridge University Press, Cambridge.

Humphreys, P. (1985). Why propensities cannot be probabilities. *Philosophical Review*, 94:557–570.

Humphreys, P. (2004). Some thoughts on Wesley Salmon's contributions to the philosophy of probability. *Philosophy of Science*, 71:942–949.

James, W. (1884). The dilemma of determinism. *Unitarian Review*, 22(3):193–224. Reprinted as Chapter 5 of his *The Will to Believe*. New York 1896 and later reprints.

Jarrett, J. (1984). On the physical significance of the locality conditions in the Bell arguments. *Noûs*, 18:569–589.

Kervaire, M. A. (1960). A manifold which does not admit any differentiable structure. *Commentarii Mathematici Helvetici*, 34(1):257–270.

Kripke, S. (1980). *Naming and Necessity*. Basil Blackwell, London. Based on lectures presented in 1970 and first published in 1972.

Lee, J. (2012). *Introduction to Smooth Manifolds*. Springer-Verlag, New York.

Lewis, D. K. (1973). *Counterfactuals*. Harvard University Press, Cambridge MA.

Lewis, D. K. (1983). New Work for a Theory of Universals. *Australasian Journal of Philosophy*, 61:343–377.

Lewis, D. K. (1986a). *On the Plurality of Worlds*. Blackwell, Oxford.

Lewis, D. K. (1986b). *Philosophical Papers: Volume II*. Oxford University Press, Oxford.

Luc, J. (2020). Generalised manifolds as basic objects of General Relativity. *Foundations of Physics*, 50(6):621–643.

Luc, J., and Placek, T. (2020). Interpreting non-Hausdorff (generalized) manifolds in General Relativity. *Philosophy of Science*, 87(1): 21–42.

Mackie, J. L. (1974). *The Cement of the Universe*. Oxford University Press, Oxford.

Malament, D. (2012). *Topics in the Foundation of General Relativity and Newtonian Gravitation Theory*. University of Chicago Press, Oxford.

Malpass, A., and Wawer, J. (2012). A future for the thin red line. *Synthese*, 188(1):117–142.

Margalef-Bentabol, J., and Villaseñor, E. J. S. (2014). Topology of the Misner space and its *g*-boundary. *General Relativity and Gravitation*, 46(7):1755.

Maudlin, T. (2019). *Philosophy of Physics: Quantum Theory*. Princeton University Press, Princeton.

McCall, S. (1994). *A Model of the Universe*. Oxford University Press, Oxford.

Melia, J. (1999). Holes, haecceitism and two conceptions of determinism. *British Journal for the Philosophy of Science*, 50(4):639–664.

Mermin, D. (1981). Quantum mysteries for anyone. *Journal of Philosophy*, 78(7):397–408.

Mermin, D. (1990). Quantum mysteries revisited. *American Journal of Physics*, 58(8): 731–734.

Miller, D. (1994). *Critical Rationalism. A Restatement and Defense*. Open Court Publishing Company, Chicago and La Salle.

Misner, C. W., and Taub, A. H. (1969). A singularity-free empty universe. *Soviet Physics JEPS*, 28(1):122–133. Originally in ZhETF 55(1), 233–255 (July 1968).

Müller, T. (2002). Branching space-time, modal logic and the counterfactual conditional. In Placek, T., and Butterfield, J., editors, *Nonlocality and Modality*, NATO Science Series, pages 273–291. Kluwer Academic Publisher, Dordrecht.

Müller, T. (2005). Probability theory and causation: a branching space-times analysis. *British Journal for the Philosophy of Science*, 56(3):487–520.

Müller, T. (2006). On the problem of defining the present in special relativity: a challenge for tense logic. In Stadler, F., and Stöltzner, M., editors, *Time and History. Proceedings of the 28. International Ludwig Wittgenstein Symposium, Kirchberg am Wechsel, Austria 2005*, pages 441–458. Ontos Verlag, Frankfurt a.M.

Müller, T. (2010). Towards a theory of limited indeterminism in branching space-times. *Journal of Philosophical Logic*, 39:395–423.

Müller, T. (2013). A generalized manifold topology for branching space-times. *Philosophy of science*, 80:1089–1100.

Müller, T. (2014). Alternatives to histories? Employing a local notion of modal consistency in branching theories. *Erkenntnis*, 79:343–364. doi:10.1007/s10670-013-9453-4.

Müller, T. (2020). Defining a relativity-proof notion of the present via spatio-temporal indeterminism. *Foundations of Physics*, 50:644–664.

Müller, T., Belnap, N., and Kishida, K. (2008). Funny business in branching space-times: infinite modal correlations. *Synthese*, 164:141–159.

Müller, T., and Briegel, H. J. (2018). A stochastic process model for free agency under indeterminism. *dialectica*, 72(2):219–252.

Müller, T., and Placek, T. (2001). Against a minimalist reading of Bell's theorem: lessons from Fine. *Synthese*, 128(3):343–379.

Müller, T., and Placek, T. (2018). Defining determinism. *British Journal for the Philosophy of Science*, 69(1):215–252.

Mundy, B. (1986). Optical axiomatization of Minkowski space-time geometry. *Philosophy of Science*, 53(1):1–30.

Munkres, J. R. (2000). *Topology*. Prentice Hall, New York.

Myrvold, W., Genovese, M., and Shimony, A. (2019). Bell's theorem. In Zalta, E. N., editor, *The Stanford Encyclopedia of Philosophy*. Metaphysics Research Lab, Stanford University, spring 2019 edition.

Newman, E., Tamburino, L., and Unti, T. (1963). Empty-space generalization of the Schwarzschild metric. *Journal of Mathematical Physics*, 4:915–925.

Øhrstrøm, P. (2009). In defence of the Thin Red Line: a case for Ockhamism. *Humana.mente*, 8:17–32.

Ometto, D. (2016). *Freedom and Self-Knowledge*. PhD thesis, Utrecht University. Department of Philosophy, Utrecht University, *Quaestiones Infinitae*, vol. 95.

Penrose, R. (1969). Gravitational collapse: the role of general relativity. *La Rivista del Nuovo Cimento*, 1:252–276.

Perry, J. (1979). The problem of the essential indexical. *Noûs*, 13(1):3–21.

Pitowsky, I. (1989). *Quantum Probability — Quantum Logic*. Springer, Berlin.

Placek, T. (2000). *Is Nature Deterministic? A Branching Perspective on EPR Phenomena*. Jagiellonian University Press, Cracow.

Placek, T. (2004). Quantum state holism: a case for holistic causation. *Studies in History and Philosophy of Modern Physics*, 35(4):671–692.

Placek, T. (2010). On propensity-frequentist models for stochastic phenomena with applications to Bell's theorem. In Czarnecki, T., Kijania-Placek, K., Poller, O., and Woleński, J., editors, *The Analytic Way. Proceedings of the European Congres on Analytic Philosophy, Kraków, August 2008*, pages 105–144. London: College Publications.

Placek, T. (2011). Possibilities without possible worlds/histories. *Journal of Philosophical Logic*, 40:737–765.

Placek, T. (2014). Branching for general relativists. In Müller, T., editor, *Nuel Belnap on Indeterminism and Free Action*, pages 191–221, Cham. Outstanding Contributions to Logic 2.

Placek, T. (2019). Laplace's demon tries on Aristotle's cloak: on two approaches to determinism. *Synthese*, 196(1):11–30.

Placek, T. (2021). Past, present and future modally introduced. *Synthese*, 198(4): 3603–3624.

Placek, T., and Belnap, N. (2012). Indeterminism is a modal notion: branching spacetimes and Earman's pruning. *Synthese*, 187(2):441–469.

Placek, T., Belnap, N., and Kishida, K. (2014). On topological issues of indeterminism. *Erkenntnis*, 79:403–436.

Ploug, T., and Øhrstrøm, P. (2012). Branching time, indeterminism and tense logic. Unveiling the Prior-Kripke letters. *Synthese*, 188:367–379. doi:10.1007/s11229-011-9944-2.

Popper, K. (1959). The propensity interpretation of probability. *British Journal for the Philosophy of Science*, 10:25–42.

Popper, K. (1982). *Quantum Theory and the Schism in Physics*. Hutchinson, London.

Prior, A. N. (1967). *Past, Present, and Future*. Oxford University Press, Oxford.

Prior, A. N. (1970). The notion of the present. *Studium Generale*, 23:245–48.

Prior, A. N. (1996). Some free thinking about time. In Copeland, B. J., editor, *Logic and Reality: Essays on the Legacy of Arthur Prior*, pages 47–52. Oxford University Press, Oxford. Undated text, edited posthumously by Copeland.

Rakić, N. (1997a). *Common Sense Time and Special Relativity*. PhD thesis, ILLC, Universiteit van Amsterdam.

Rakić, N. (1997b). Past, present, future, and special relativity. *British Journal for the Philosophy of Science*, 48:257–280.

Reichenbach, H. (1952). Les fondements logiques de la mécanique des quanta. *Annales de l'institut Henri Poincaré*, 13:109–158. Quoted from the English translation, "The logical foundations of quantum mechanics", in *Hans Reichenbach. Collected writings 1909–1953, Vol. 2*, ed. by M. Reichenbach and R. S. Cohen, Dordrecht: Reidel 1978, pages 237–278.

Reichenbach, H. (1956). *The Direction of Time*. University of California Press, Berkeley, CA.

Ringström, H. (2009). *The Cauchy Problem in General Relativity*. European Mathematical Society Publishing House, Zürich.

Robb, A. A. (1914). *A Theory of Space and Time*. Cambridge University Press, Cambridge.

Robb, A. A. (1936). *Geometry of Time and Space*. Cambridge University Press, Cambridge.

Rumberg, A. (2016a). Transition semantics for branching time. *Journal of Logic, Language and Information*, 25:77–108.

Rumberg, A. (2016b). *Transitions: Toward a Semantics for Real Possibility*. PhD thesis, University of Utrecht.

Salmon, W. C. (1984). *Scientific Explanation and the Causal Structure of the World*. Princeton University Press, Princeton, NJ.

Salmon, W. C. (1989). Four decades of scientific explanation. *Minnesota Studies in the Philosophy of Science* 13:3–219.

Sattig, T. (2015). Pluralism and determinism. *Journal of Philosophy*, 111:135–150.

Scheidl, T., Ursin, R., Kofler, J., Ramelow, S., Ma, X. S., Herbst, T., and Zeilinger, A. (2010). Violation of local realism with freedom of choice. *Proceedings of the National Academy of Sciences*, 107:19708–19713.

Shalm, L. K., Meyer-Scott, E., Christensen, B. G., Bierhorst, P., Wayne, M. A., Stevens, M. J., Gerrits, T., Glancy, S., Hamel, D. R., Allman, M. S., Coakley, K. J., Dyer, S. D., Hodge, C., Lita, A. E., Verma, V. B., Lambrocco, C., Tortorici, E., Migdall, A. L., Zhang, Y., Kumor, D. R., Farr, W. H., Marsili, F., Shaw, M. D., Stern, J. A., Abellán, C., Amaya, W., Pruneri, V., Jennewein, T., Mitchell, M. W., Kwiat, P. G., Bienfang, J. C., Mirin, R. P., Knill, E., and Nam, S. W. (2015). Strong loophole-free test of local realism. *Physics Review Letters*, 115:250402.

Shimony, A. (1978). Metaphysical problems in the foundations of quantum mechanics. *International Philosophical Quarterly*, 8:2–17.

Shimony, A. (1984). Controllable and uncontrollable non-locality. In Kamefuchi, S., et al., editors, *Foundations of Quantum Mechanics in Light of the New Technology*, pages 225–230. Physical Society of Japan, Tokyo.

Smeenk, C. (2013). Time in cosmology. In Bardon, A., and Dyke, H., editors, *The Blackwell Companion to the Philosophy of Time*, pages 201–219. Blackwell, Oxford.

Spinoza, B. (1677). *Ethica Ordine Geometrico Demonstrata*. Jan Rieuwertsz, Amsterdam. Quoted from the translation by M. Silverthorne, ed. by M. J. Kisner, *Ethics Proved in Geometrical Order*. Cambridge: Cambridge University Press 2018.

Spohn, W. (1983). Deterministic and probabilistic reasons and causes. In Hempel, C. G., Putnam, H., and Essler, W., editors, *Methodology, Epistemology, and Philosophy of Science*, pages 371–396. Kluwer, Dordrecht. doi:10.1007/978-94-015-7676-5_20.

Spohn, W. (2012). *The Laws of Belief: Ranking Theory and Its Philosophical Applications.* Oxford University Press, Oxford.

Spohn, W. (2017). The epistemology and auto-epistemology of temporal self-location and forgetfulness. *Ergo,* 4(13):359–418.

Stalnaker, R. (1976). Possible worlds. *Noûs,* 10(1):65–75.

Stalnaker, R. (2012). *Mere Possibilities: Metaphysical Foundations.* Princeton University Press, Princeton, NJ.

Stalnaker, R. (2015). Counterfactuals and Humean reduction. In Loewer, B., and Schaffer, J., editors, *A Companion to David Lewis,* pages 411–424. Wiley-Blackwell, Oxford.

Stein, H. (1991). On relativity theory and openness of the future. *Philosophy of Science,* 58(2):147–167.

Strobach, N. (2007). *Alternativen in der Raumzeit. Eine Studie zur philosophischen Anwendung multidimensionaler Aussagenlogiken,* volume 16 of *Logische Philosophie.* Logos, Berlin.

Tarski, A. (1933). *Pojęcie prawdy w językach nauk dedukcyjnych.* Prace Towarzystwa Naukowego Warszawskiego, Warsaw.

Tarski, A. (1935). Der Wahrheitsbegriff in den formalisierten Sprachen. *Studia Philosophica,* 1:261–405.

Tarski, A. (1956). *Logic, Semantics, Metamathematics: Papers from 1923 to 1938,* chapter "The Concept of Truth in Formalized Languages", pages 152–278. Hackett, Indianapolis, 2nd, 1983 edition.

Thomason, R. H. (1970). Indeterminist time and truth-value gaps. *Theoria,* 36:264–281.

Van Benthem, J. F. A. K. (1983). *The logic of time. A model-theoretic investigation into the varieties of temporal ontology and temporal discourse,* volume 156 of *Synthese library.* D. Reidel, Dordrecht.

Van Fraassen, B. (1980). *The Scientific Image.* Oxford University Press, Oxford.

Vetter, B. (2015). *Potentiality: From Dispositions to Modality.* Oxford University Press, Oxford.

von Kutschera, F. (1986). Bewirken. *Erkenntnis,* 24(3):253–281.

von Wright, G. H. (1963). *Norm and Action. A Logical Inquiry.* Routledge, London.

von Wright, G. H. (1974). *Causality and Determinism.* Columbia University Press, New York.

Wald, R. M. (1984). *General Relativity.* Chicago University Press, Chicago, IL.

Weiner, M., and Belnap, N. (2006). How causal probabilities might fit into our objectively indeterministic world. *Synthese,* 149:1–36.

Whitehead, A. N. (1925). *Science and the Modern World.* Macmillan, New York.

Whitrow, G. J. (1961). *The Natural Philosophy of Time.* Nelson, London.

Wroński, L. (2014). *Reichenbach's Paradise. Constructing the Realm of Probabilistic Common "Causes".* De Gruyter Open, Warsaw.

Wroński, L., and Placek, T. (2009). On Minkowskian branching structures. *Studies in History and Philosophy of Modern Physics,* 40:251–258.

Wroński, L., Placek, T., and Godziszewski, M. T. (2017). Separate common causes and EPR correlations: An "almost no-go" result. In Hofer-Szabó, G., and Wroński, L., editors, *Making it Formally Explicit,* pages 85–107. Springer, Charm.

Wüthrich, C. (2020). When the actual world is not even possible. In Glick, D., Darby, G., and Marmodoro, A., editors, *The Foundation of Reality,* pages 233–254. Oxford University Press, Oxford.

Xu, M. (1997). Causation in branching time (I): Transitions, events and causes. *Synthese*, 112(2):137–192.

Yin, J., Cao, Y., Li, Y.-H., Liao, S.-K., Zhang, L., Ren, J.-G., Cai, W.-Q., Liu, W.-Y., Li, B., Dai, H., Li, G.-B., Lu, Q.-M., Gong, Y.-H., Xu, Y., Li, S.-L., Li, F.-Z., Yin, Y.-Y., Jiang, Z.-Q., Li, M., Jia, J.-J., Ren, G., He, D., Zhou, Y.-L., Zhang, X.-X., Wang, N., Chang, X., Zhu, Z.-C., Liu, N.-L., Chen, Y.-A., Lu, C.-Y., Shu, R., Peng, C.-Z., Wang, J.-Y., and Pan, J.-W. (2017). Satellite-based entanglement distribution over 1200 kilometers. *Science*, 356(6343):1140–1144.

Name Index

Subject Index